BASIC CONCEPTS *of* INDUSTRIAL HYGIENE

Ronald Scott

Acquiring Editor: Ken McCombs
Project Editor: Albert W. Starkweather, Jr.
Cover design: Dawn Boyd

Library of Congress Cataloging-in-Publication Data

Scott, Ronald McLean, 1933–
Basic concepts of industrial hygiene / Ronald M. Scott.
p. cm.
Includes bibliographical references and index.
ISBN 1-56670-292-5 (alk. paper)
1. Industrial hygiene. I. Title.
RC967.S356 1997
616.9′803—dc21

97-28597
CIP

International Standard Book Number 1-56670-292-5
Library of Congress Card Number 97-28597
Printed in the United States of America 1 2 3 4 5 6 7 8 9 0
Printed on acid-free paper

The Author

Ronald M. Scott received a B.S. in chemistry from Wayne State University and a Ph.D. in biochemistry from the University of Illinois. Since 1959, he has lectured in biochemistry and, more recently, in toxicology and industrial hygiene at Eastern Michigan University in Ypsilanti.

He has been a visiting professor at the University of Warwick, England. He taught for OSHA at the OSHA Training Institute, and directs the industrial hygiene program for the regional OSHA Training Institute near Detroit.

Dr. Scott has consulted on problems of metal ion toxicity, and has published research papers in professional journals on chromatographic methods of trace analysis, techniques for clinical analysis, and work relating to the use of freshwater clams as environmental monitors of industrial effluents. onmental monitors of industrial effluents. onmental monitors of industrial effluents. Dr. Scott has written three chromatography reference books — a biochemistry textbook, three industrial hygiene references or textbooks, and was technical editor of a reference book on trace Dr. Scott has written three chromatography reference books — a biochemistry textbook, three industrial hygiene references or textbooks, and was technical editor of a reference book on trace Dr. Scott has written three chromatography reference books — a biochemistry textbook, three industrial hygiene references or textbooks, and was technical editor of a reference book on trace analysis.

Preface

This book is a nonencyclopedic textbook of industrial hygiene, the field dedicated to the protection of workers. The book is based on years of teaching an industrial hygiene course. It is a broad survey of the field and addresses the typical student. Extra discussion is provided where experience has shown that some students will not be strongly prepared, given the variety of backgrounds of students enrolled in such a class. At the end of each chapter, material covered is summarized in a Key Points section. References are provided both to material that helps the student who has little background in the topic of that chapter and to sources that expand beyond the scope of the chapter. The problem sets used often have a practical basis and lead students into the CFR (*Code of Federal Regulations*) to familiarize them with the contents and the manner of locating information in this source. Extensive appendices provide practical information to support the instructional message and to allow the text to be a reference of value to the student later.

The U.S. needs well-trained and qualified people to serve as industrial hygienists. The maintenance of high standards for health and safety in the workplace involves controlling a variety of working conditions, including noise level, radiation level, temperature, and potential for physical injury. There is increasing concern about bacterial and viral infections resulting from job contacts. Industrial toxicology focuses on the threat of injury due to contact with chemicals used in the workplace. This threat is not restricted to workers in the chemical industry because the use of chemicals is a ubiquitous part of modern industry. Many commercial processes involve painting or packaging a product, degreasing or otherwise cleaning machinery, assembling with adhesives, dyeing or printing with dissolved pigments, and a host of other processes that expose workers to chemicals. Furthermore, it can be argued that the need for regulation is greater outside chemical manufacturing, because management there is less likely to include chemistry specialists.

The book is divided into five sections. The first section introduces the field from a historical standpoint and describes the legal basis of health and safety in the U.S. The second section focuses on the chemical hazards. The basics of toxicology are presented, problems arising from skin contact or inhalation of chemicals are described, the detection and control of airborne contaminants is outlined, and the threat of fire or explosion is discussed. The third section is concerned with injury from other causes, including sound, radiation, heat, biological agents, and accidents. Ergonomics, the study of problems due to the manner of performing the job, is introduced. Finally, the fourth section describes a range of industries, ones that are major sources both of jobs and of potential injury, and illustrates and expands on the principles presented in earlier sections.

Ronald Scott
Ypsilanti, Michigan

Acknowledgments

The author wishes to recognize and express appreciation to Ralph Pedersen, an engineer specializing in noise problems, Dr. Krish Rengan, a radiochemist, and Dr. Edith Hurst, an anatomist, for their help with the manuscript.

Contents

Chapter 5

SECTION III
CHEMICAL HAZARDS IN THE WORKPLACE

Chapter 6

Section I
The Scope of Industrial Hygiene

Chapter 1 introduces the field of industrial hygiene from a historical standpoint and describes the range of concerns of the professional industrial hygienist. Chapter 2 continues with a description of the role of the federal government in occupational health and safety concerns. In the U.S. since the passage of the OSH Act, OSHA and industrial health and safety have become totally entwined. Therefore, a foundation for understanding the responsibilities, approaches, powers, and procedures of OSHA is presented. As a general pattern throughout the book, discussions of hazards are followed by descriptions of OSHA regulations.

CHAPTER 1

History and Basis of Industrial Hygiene: The Historical View

The first occupational injury victim may well have been a cave man with a chip of stone in the eye from trying to form a spear point or an ax head. The development of civilized society led to a division of labor with individuals providing specialized services. In these ancient civilizations, the variety of ways to injure oneself increased with the sophistication of jobs. Citizens had a disdain for manual labor in classical societies, so high-risk work such as mining and construction was done by slaves, provided in plentiful supply by wars. There was thus a minimum of concern with job injuries or illnesses. The embryonic medical professions targeted cure of the citizen, not the slave. Plinius Secundus described the use of a bladder across the face to filter out dusts during mercury or lead processing. Galen, a physician of the second century AD, observed hazards of mining copper, but did not suggest ways to protect or treat workers.

During the middle ages medical knowledge waned in Europe, and earlier gains would have been lost had it not been for the Arab societies. Other than farming, work was primarily cottage industries, and workers essentially protected themselves.

In 1473 Ellenbog published a pamphlet about occupational diseases, and in 1556 Agricola, a mining town physician, wrote a book that included descriptions of injuries and sicknesses of miners, describing both treatments and preventions. Bernardo Ramazzini published the first real comprehensive work concerned with occupational diseases in 1700. A few other advances were made in the 18th century, particularly the observation by Percival Pott that the scrotal warts of chimney sweeps were caused by the soot they rarely washed away. This was the first linkage of cancer to chemical exposure.

The industrial revolution altered the nature of work and sharply increased the numbers of industrial workers. Large industrial sites exposed many more people to an increased variety of both physical and chemical work hazards. Laws at that time protected the employer by placing blame for work injury on the employee. Government finally became concerned with the welfare of workers with the English Factory Acts of 1833. These acts focused on worker's compensation rather than safe practices, but led to improved safety because they gave employers a financial incentive to prevent accidents.

During the first decade of the 20th century Alice Hamilton, an American physician dedicated to the welfare of workers, identified hazards and raised public consciousness about the safety of workers. Universities began studies of and courses in occupational health. At about that time the U.S. Public Health Service and the Bureau of Mines initiated studies of workplace hazards. In 1911 New York passed the first worker's compensation law, and within a decade all but a few states passed similar laws. In 1918 the first college degree in industrial hygiene was earned.

The 1930s saw significant advances in promotion of worker safety. The industrial hygiene profession formed and gained recognition. A few university programs trained specialists in the field. The American Public Health Association created an industrial hygiene section, the American Conference of Governmental Industrial Hygienists organized in 1938, and the American Industrial Hygiene Association originated in 1939.

Although the U.S. moved rapidly to improve the health and safety of workers through regulation and research, public expectations grew faster. As the workforce became increasingly unionized, issues of job safety became the subject of labor–management negotiations. The social activism of the 1960s led to three significant pieces of legislation. The Metal and Nonmetal Mine Safety Act of 1966 created advisory committees with representatives of mine owners, miners, and mine inspectors, required reporting of accidents and occupational diseases, and regularized mine inspection. The Federal Coal Mine Health and Safety Act of 1969 empowered the federal government to close unsafe mines. It set standards for respirable dust, roof supports, and electrical distribution systems, and required medical examinations of miners. Finally the Occupational Safety and Health Act of 1970 (OSH Act) created the Occupational Safety and Health Administration (OSHA) to establish and enforce uniform national work safety standards, and generated a research agency, the National Institute for Occupational Safety and Health (NIOSH) to develop and recommend new standards. The OSH Act has the goal to "assure so far as possible every working man and woman in the nation safe and healthful working conditions and to preserve our human resources." Since its passage, a succession of new regulations have defined aspects of safe working conditions in quantitative terms.

OSHA inspects a great many workplaces each year. The right-to-know law mandates that workers are trained and informed about workplace hazards and have direct access to OSHA to voice concerns. Research results improve our understanding of hazardous situations, create new protective equipment, and lead to safer work practices. Initial concerns regarding hazards of equipment and chemicals are now expanded to include the spread of disease, injury from awkward task design or repetitive operations, air quality in offices, and many other issues. Today's workplace is safer in many ways than when the OSH Act was passed. Deaths in the workplace have been cut in half since the OSH Act was passed (Pompei, 1995).

OSHA has vastly expanded the need for occupational safety professionals. For example, membership in the American Industrial Hygiene Association has grown from 160 in 1940 to more than 10,000 today (Cralley et al., 1995).

INDUSTRIAL HYGIENISTS

The American Industrial Hygiene Association defines industrial hygiene as "that science or art devoted to the anticipation, recognition, evaluation, and control of those environmental factors (stresses) arising in or from the workplace which may cause sickness, impaired health and well being, or significant discomfort among workers or among citizens of the community." An industrial hygienist is one of a group of health professionals trained to address these concerns. More than 30 university programs provide formal professional training in industrial hygiene. In 1960 the American Board of Industrial Hygiene was formed to test and certify individuals as meeting high professional standards. This testing is not done immediately on completion of university training, as is the general case in specialties in the medical sciences, but comes after a mandatory period of practice in the field and only after recommendation of the candidate by other certified professionals. Successful completion of this examination entitles individuals to place the letters CIH (Certified Industrial Hygienist) after their name. Once certified, the individual may join the American Academy of Industrial Hygiene. This group has established a Code of Ethics which describes worker protection as the primary responsibility of industrial hygiene professionals.

Academic requirements of university industrial hygiene programs are interdisciplinary in nature, reflecting the broadly based requirements of the profession. The focus of the industrial hygienist is not a narrow specialized field, but is the workplace itself. The industrial hygienist must therefore be able to recognize the hazards inherent in a wide range of workplace circumstances. These include harm from contact with chemicals, whether due to their short- or long-term toxic characteristics, corrosive or irritant impact on the skin or eyes, or behavior-altering properties such as narcotic effects. Physical hazards involve exposure to harmful radiation (ionizing or nonionizing), excessive sound levels, vibration, or extremes of temperature. Biological hazards arise from the possibility of infectious, allergic, or other damaging effects from organisms or their products found in the workplace, such as bacteria, viruses, fungi, or plant products. Many jobs are performed outdoors, and others bring the indoor worker into contact with living or once living materials. Finally, the worker may be harmed by the design of the workplace or the nature of the work itself. Good workplace design includes guards around moving machinery, handrails on stairs, nonslip surfaces on floors, and a host of other accident-preventing installations. Procedures should exist to prevent injury from start-up of machinery during repair or from entering confined spaces. Work itself can cause bodily damage, for example, because of the need for repetitive hand or arm movements, lifting objects that are too heavy, or lifting in a fashion that stresses the back, leading to back injury.

There is a wide variation in the level of hazard of each type of job (Table 1.1). Once a potential hazard is identified, it is often necessary for the industrial hygienist to measure the level of hazard to determine whether or not it is within tolerable limits. Some sound is unavoidable around machinery, but is the sound above or below levels that might injure hearing? Workers are likely to have some contact with a chemical in use in a factory, but are levels encountered below those which cause harmful effects? Has exposure to X-rays reached a level such that an airport security guard should take breaks from luggage checking? An industrial hygienist must be trained in a variety of analytical methods and techniques.

Table 1.1. Incidence Rates in Selected Industries.

Industry	Loss of Work Time or Death (incidents per 100 workers)
Chemical	0.64
Textile	0.69
Communication	0.79
Agri-chemicals	1.01
Electrical equipment	1.12
Primary metals (nonferrous)	1.18
Steel	1.32
Auto manufacturing	1.36
Furniture building	1.75
Cement	1.88
Average — all industries	2.11
Polymers	2.32
Metal fabrication	2.33
Paper	2.65
Foundries (iron and steel)	2.66
Printing	2.75
Construction	2.92
Mining (metals)	3.02
Mining (coal)	3.40
Food processing	3.65
Lumber	4.66
Meat processing	5.18
Trucking	6.54

Data from National Safety Council, Elk Grove Village, IL, 1982.

Because worker protection is a legal matter, the industrial hygienist must understand the extent and limits of legal protection provided workers and the manner of enforcement of these laws. Correct and thorough recordkeeping is essential. Records of worker health are required by OSHA, and these are useful in identifying unsafe conditions that might not have been recognized otherwise. Measuring or otherwise estimating levels of hazard is therefore the collecting of evidence, which must be protected, interpreted, and presented properly, perhaps even in court.

Once the degree of hazard is determined, the industrial hygienist must understand what corrective measures can be taken. There is a hierarchy of approaches to reducing job hazard. Begin with the most desirable, and move through the list in priority order:

1. Reduce hazards or stresses by application of engineering controls. Redesign the job or workplace to eliminate, reduce, or modify the stress. For example, close an open vat to reduce solvent evaporation, remove airborne contaminants with properly designed ventilation, repair or replace a noisy machine, change the job process to isolate the worker from a threat, install proper shielding to block harmful radiation, and lower sound levels with sound barriers or by isolating the worker from the source. Engineering solutions are the preferred means of providing a safe and healthful work environment.
2. Institute administrative controls to reduce stress. Without changing the design of the workplace, reduce the time a worker experiences damaging stress. For example, move a worker in a noisy location to a quieter work station after a period of time, or have employees trade assembly jobs at intervals to change the repetitive motions performed, reducing the strain on a single part of the body.
3. Provide personal protective equipment. Thus, in a noisy environment, require workers to wear ear protection. In a place where vapors or airborne particulate are found, provide correctly designed respirators to purify breathing air. Prevent dermal exposure to chemicals by using protective clothing. Recognize that protective equipment might be required for a worker who is not presently experiencing excessive exposure, but needs protection in the case of an accident. Eye protection, an excellent example, should be worn by any worker in the vicinity of chemicals just in case a chemical exposure occurs. If such devices are necessary, the industrial hygienist ensures that employers provide them, instructs the worker in their importance and proper use, and enforces their use on the job.

Increasingly the law defines conditions necessary for worker safety. Employers must meet these standards, and for many industrial hygienists the daily task is to determine if a specific workplace is "in compliance".

OTHER OCCUPATIONAL HEALTH PROFESSIONALS

Other safety professionals work to ensure safe working conditions. Most important is the occupational physician. Whereas the focus of the industrial hygienist is on the workplace, that of the occupational physician is on the worker. The two cooperate, with the industrial hygienist providing a description of the worker's environment, which can be used by the occupational physician to determine if the worker is avoiding excessive workplace exposures, if observed problems have an occupational origin, or if a particular worker is or is not well suited to a specific job. Organizations of occupational physicians may contract with industries to deal not only with the treatment of injuries, but with their prevention (Kornecki, 1995).

Nurses are trained with occupational problems as their speciality. Other individuals train in narrow aspects of health and safety. Health physicists specialize in radiation problems. Workers may take special training courses to qualify for special responsibilities in case of accidental chemical spills or other emergency situations.

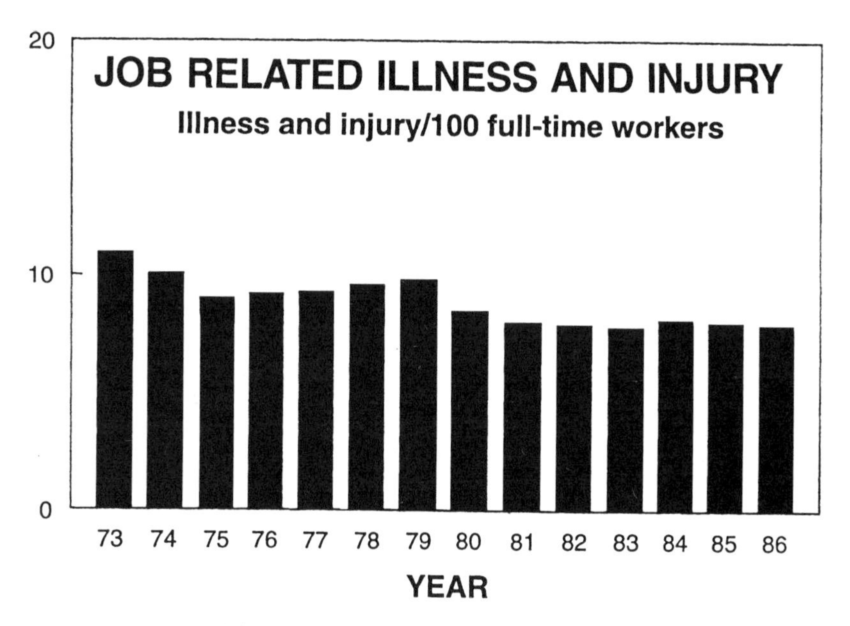

Figure 1.1. Job-related illness and injury.

Absolute safety for the worker is as unattainable as absolute safety for motorists. The objective is to approach that goal as closely as possible, which is the purpose of laws and their enforcement. Today we have less than one-sixth the occupational death rate we had in 1912 (U.S. Office of Technology Assessment, 1985). Problems exist, but data show that progress has been made (Figure 1.1).

BIBLIOGRAPHY

L. J. Cralley, L. V. Cralley, and J. S. Bus, "Rationale," in *Patty's Industrial Hygiene and Toxicology,* Vol. III, Part B, John Wiley & Sons, New York, 1995.

L. Kornecki, "Targeted Medical Care Reduces Lost-Time Injuries," *Occupational Health and Safety,* Sept. 1995, p. 56.

J. Pompei, "A 21st Century Approach to Health and Safety," *Occupational Health and Safety,* Oct. 1995, p. 52.

U.S. Office of Technology Assessment, *Preventing Illness and Injury in the Workplace,* Washington, D.C., 1985.

CHAPTER 2

Government Regulation

In this chapter, we deal with the legal apparatus that has been established by the federal government to protect the health and safety of workers across the entire U.S. These laws and the apparatus to enforce them is of paramount importance to industrial hygienists. We begin by describing the laws, standards, and provisions established by the federal government, and explain how to gain access to the details of these standards.

THE OCCUPATIONAL SAFETY AND HEALTH ACT

Federal involvement in worker health and safety begins with the Williams-Steiger Occupational Safety and Health Act of December 1970 (OSH Act; Public Law 91-596). In the House of Representatives, the Health and Safety Subcommittee of the Education and Labor Committee, and in the Senate, the Labor Subcommittee of the Labor and Human Resources Committee have jurisdiction.

BEFORE THE OSH ACT

Before passage of the OSH Act, regulation of health and safety matters in the workplace was a state responsibility, with a few exceptions, such as contractors to the federal government. Standards were not uniform across the country, nor was the level of enforcement. Private groups such as the American Conference of Government Industrial Hygienists (ACGIH) provided some common ground in setting standards. For example, their threshold limit values (TLVs) were widely, though not uniformly, employed as a guide for establishing maximum safe levels of exposure to chemicals. In discussing the history of this legislation, Mary Worobec (1986) described the situation as it existed before passage of the OSH Act, quoting these statistics:

> Each year 14,000 workers died and 2.2 million were disabled by accidents in the workplace, according to former Labor Secretary George P. Schultz.
>
> Work-related deaths and injuries were causing annual losses of $1.5 billion in wages and $8 billion in gross national product.
>
> About 65 percent of U.S. workers were being exposed to harmful physical agents, yet only about 25 percent of them were adequately protected.

A total of 390,000 new cases of occupational disease occurred each year.

Chemical agents, including lead, mercury, asbestos, and cotton dust, posed known health threats to workers, but were not controlled.

A new, potentially toxic chemical was being introduced in industry every 20 minutes, according to a Public Health Service estimate.

PROVISIONS OF THE OSH ACT

When Congress passes new laws such as the OSH Act, they are drafted in broad, sweeping language that includes goals and purposes of the legislation. It is then the function of the administrative branch of the government to establish the machinery needed to meet these goals and purposes. Such a major effort as demanded by the OSH Act required the establishment of an administrative unit to meet the goals of the new law. This agency first had to set official standards to be met by employers. Once it was clear what behavior was expected of employers, a mechanism for enforcement of these standards was needed.

Responsibility for regulation of worker safety and health is assigned to the Occupational Safety and Health Administration (OSHA), created for this purpose within the Department of Labor of the federal government (Figure 2.1). The OSH Act intends to prevent injury or illness among all workers. This is a huge responsibility, more than can be reasonably achieved with the manpower and funds available. Employers already regulated by another federal agency (farms, mines[1]), the activities of governmental agencies,[2] and the self-employed were therefore not regulated by OSHA. The immediate goal was to generate as great an improvement in working conditions as the resources available could provide. As a realistic starting point, full application was limited to workplaces with more than 10 employees. This freed the agency to concentrate efforts on larger employers where agency efforts could yield greater dividends.

Figure 2.1. OSHA is a division of the U.S. Department of Labor, housed in this building in Washington, D.C. (Photo courtesy of Billy Rose.)

[1] Responsibility for mine health and safety has now been transferred to OSHA.

[2] Although federal, state, and local government employees are not protected by OSHA, as a result of Executive Order 12196 signed by President Carter, governmental units must maintain a program consistent with that of OSHA.

In brief, the powers and responsibilities of OSHA include the following:

1. To establish safety and health standards.
2. To conduct workplace inspections and issue citations for health and safety violations.
3. To require records of safety and health to be kept by employers, and, in conjunction with the Department of Health and Human Services, keep occupational health and safety statistics.
4. To train employers, employees, and personnel employed to enforce the act.

The act also establishes a research agency, the National Institute for Occupational Safety and Health (NIOSH). Originally NIOSH was organized under the Department of Health, Education, and Welfare, and, when a separate Department of Education was established, this became the Department of Health and Human Services. The National Institutes of Health (NIH) is the umbrella organization through which the U.S. government performs health-related research, and is subdivided into groups focused on cancer research, mental health research, and other such specialized research. NIOSH is the NIH subdivision specializing in occupational health and safety problems. NIOSH studies may be laboratory centered or they may focus on review of data from the workplace. From their results, NIOSH recommends new standards or changes in existing standards, but their recommendations do not have the weight of law behind them. Only OSHA can set the legal standards. However, in the absence of a legal standard in some area, OSHA may choose to use the NIOSH recommendation as the reasonable standard employers should meet to provide a safe workplace.

The OSH Act did not intend to create in OSHA an agency with absolute powers. Employers can challenge an OSHA ruling. For this purpose an independent legal group, the Occupational Safety and Health Review Commission, was established. The commission has three members who are appointed by the president for 6-year terms, and are confirmed by the Senate. Their function is to investigate and rule on the appropriateness of actions of OSHA, such as standards, penalties, and citations, when complaints are lodged. Decisions of the commission can clarify or modify OSHA standards, establishing a body of case law to guide later actions, but the committee cannot initiate a review of OSHA laws unless it is dealing with an appeal. The commission is the highest authority within OSHA in these matters, but its rulings can be reviewed by federal courts.

The OSH Act is intended to protect workers from harmful agents or circumstances in the workplace.[3] Examples include physical agents such as noise, extreme temperature, and types of radiation. Rules are established for contact with and disposal of harmful biological agents. Chemical agents are regulated by setting limits on exposure and by an OSHA cancer policy. Hazards due to chemicals must be indicated by labeling, and workers must be trained regarding safe use of chemicals. Safety standards protect the worker from such accidents as fires, cuts, falls, electrocution, and construction collapse.

THE ACCUMULATED REGULATIONS — THE *FEDERAL REGISTER* AND THE *CODE OF FEDERAL REGULATIONS* (CFR)

You might assume if you want information about some aspect of the law governing workplace health and safety, that the ultimate source is the original law as passed by Congress. In fact, the original law is a statement of the intent of the legislation with broad outlines as to purposes and limitations. The real impact of the law arises when the administrative branch of government translates the generalizations into specific regulations.

Every weekday information about all the new rules, meetings, and other congressional proceedings are printed by the National Archives and Records Service in the *Federal Register*.

[3] The agency is challenged to protect 90,000,000 workers in 6,000,000 places of employment with a budget of $294,000,000 (1994).

When new laws are entered in the *Federal Register*, they are often accompanied by extensive information explaining reasons for the new law and aspects of its intent. This is a huge collection of volumes that is indexed so that original submissions can be located and read, but it is unwieldy to use at best.

Once a year the regulations are transferred from the *Federal Register* to the *Code of Federal Regulations* (CFR), adding them to the accumulated existing regulations. Statements of the regulations now are presented in compact form. The organization is not changed from year to year, so new rules are simply fit into the established framework at the appropriate location. This is a great improvement in organization over the entries in the *Federal Register* because it compiles all related rules in one place. However, the organization is not as straightforward as these comments make it seem, and some practice is needed to learn to locate information in the CFR.

When in the library facing the CFR, one may notice first that there are a series of 50 titles, each dealing with a specific concern of the federal government (Table 2.1). Everything pertaining to OSHA is found in Title 29, Department of Labor. The numbers 1900 to 1999 are reserved for OSHA. OSHA regulations are found primarily in the two volumes that include parts 1900 to the end of 1910 (Figure 2.2). Section 1910 covers general industry. Additional material is found in section 1926, construction, and section 1915, maritime employment.

The standard reference to locate a specific rule might start out as 29 CFR 1910, which means Title 29 of the CFR, which gets you to the Department of Labor section. If you take the volumes including parts 1900 to 1910 off the shelf and turn the pages starting at the beginning, you come to a page that indicates what each of the major divisions (1900, 1901, 1902, and so forth) includes. Most of the document is found in section 1910. Turn to the first page of 1910 (Occupational Safety and Health Standards) and you find a listing of the contents of each major subdivision of 1910. This list would lead you to the appropriate section to find the area in which you wish to see the standards.

Suppose what you have instead is a reference to a standard such as 29 CFR 1910.133(a)(2)(i–vii). In section 1910 you locate 1910.133,[4] which turns out to be standards for eye and face protection. Heading (a) is called "General", and its inclusion does not mean that there has to be a (b), only that the possibility of a (b) is held open. Subsection (2) lists the minimum requirements eye protection must meet, and seven points are included, i–vii.

THE SOURCE OF STANDARDS

Consider the challenge facing OSHA as a newly established agency. The agency was expected to begin improving the safety of workers as quickly as possible, but there was no existing machinery to accomplish this. Two requirements had to be met to begin functioning: (1) health and safety standards must be established, and (2) a means of enforcing them must be created. Obviously there first had to be standards before enforcement became a concern. To create a completely new set of standards covering all aspects of workplace health and safety was an enormous task.

Standards for safety and health therefore were initially established utilizing either existing federal standards or consensus standards. The latter are standards established by private groups whose expertise is recognized by authorities. Consensus standards were drawn from a variety of sources including the National Fire Protection Association (NFPA), the Compressed Gas Association (CGA), the National Electrical Codes (NEC), the American Conference of Governmental Industrial Hygienists (ACGIH), the American National Standards Institute (ANSI), and the American Society for Testing and Materials (ASTM).

[4] Notice that we are not using the decimal system as a mathematician would. 1910.133 does not belong between 1910.13 and 1910.14, and 1910.99 is followed by 1910.100.

Table 2.1. Code of Federal Regulations Titles.

Title 1	General Provisions
Title 2	Reserved
Title 3	The President
Title 4	Accounts
Title 5	Administrative Personnel
Title 6	Reserved
Title 7	Agriculture
Title 8	Aliens and Nationals
Title 9	Animals and Animal Products
Title 10	Energy
Title 11	Federal Elections
Title 12	Banks and Banking
Title 13	Business Credit and Assistance
Title 14	Aeronautics and Space
Title 15	Commerce and Foreign Trade
Title 16	Commercial Practices
Title 17	Commodity and Securities Exchanges
Title 18	Conservation of Power and Water Resources
Title 19	Customs Duties
Title 20	Employees' Benefits
Title 21	Food and Drugs
Title 22	Foreign Relations
Title 23	Highways
Title 24	Housing and Urban Development
Title 25	Indians
Title 26	Internal Revenue
Title 27	Alcohol, Tobacco Products, and Firearms
Title 28	Judicial Administration
Title 29	Labor
Title 30	Mineral Resources
Title 31	Money and Finance: Treasury
Title 32	National Defense
Title 33	Navigation and Navigable Waters
Title 34	Education
Title 35	Panama Canal
Title 36	Parks, Forests, and Public Property
Title 37	Patents, Trade Markets, and Copyrights
Title 38	Pensions, Bonuses, and Veterans' Relief
Title 39	Postal Service
Title 40	Protection of Environment
Title 41	Public Contracts and Property Management
Title 42	Public Health
Title 43	Public Lands: Interior
Title 44	Emergency Management and Assistance
Title 45	Public Welfare
Title 46	Shipping
Title 47	Telecommunication
Title 48	Federal Acquisitions Regulations System
Title 49	Transportation
Title 50	Wildlife and Fisheries

The OSH Act provides some guidelines for these standards. The act specifies that "The Secretary of Labor will promulgate standards based upon research, demonstrations, experiments, and other such information as may be appropriate. In addition to the attainment of the highest degree of health and safety protection for the employee, other considerations shall be the latest scientific data in the field, the feasibility of the standards, and experience gained in this and other health and safety laws. Wherever practicable, the standard promulgated shall be expressed in terms of objective criteria and performance desired." The latter is very

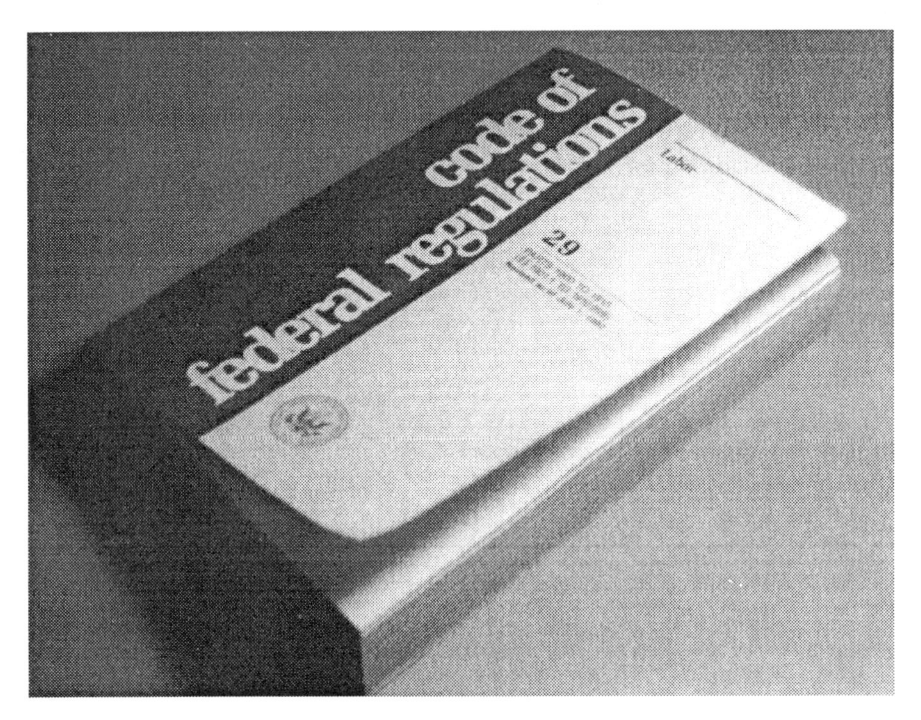

Figure 2.2. Shown here is the *Code of Federal Regulations* containing parts 1900 to 1910.999. This volume, from July 1, 1992, held the complete OSHA regulations. More recent editions require two volumes to contain the total set of regulations. References are organized in the same fashion in all editions, so it is not necessary to change all your reference numbers each time a new edition is printed.

important. It is not productive to pass a vaguely worded standard that leads to misunderstanding and endless enforcement headaches. Needed is a clearly worded, quantitative, measurable mark that is the level of performance the employer must reach.

As an example, the 1968 ACGIH TLVs for chemical exposure were adopted as official exposure standards,[5] as were ANSI acceptable concentration values.[6] These standards state maximum concentrations of chemicals that are permitted in the air breathed by employees. Such standards for exposure to chemicals were then referred to as permissible exposure levels (PEL), and became not recommendations, but legal limits. PEL values and other standards are not carved in stone, but may be changed as new information becomes available, for example, through advice from NIOSH. Some individual standards have been updated. However, in 1989 OSHA revised the original standards in a single sweeping readjustment, which proved to be controversial, and in 1992 a federal court threw out this revision (see Chapter 4).

OSHA adopted the consensus standards in use at the time and continues to refer to the regulation of that time, even though that agency or association may have updated standards. To handle this problem OSHA stated[7] that a de minimus violation is defined as when "An employer's workplace is at the 'state of the art' which is technically advanced beyond the requirements of the applicable standard, and provides equivalent or better safety and health protection. An example would be when OSHA standards, derived from such consensus groups as NEC, NFPA, etc. have been updated in later consensus publications in accord with new technology or equipment and the updated standard provides equal or greater safety and health protection." Such a violation is of less significance than when the existing legal standard is not met.

[5] See 29 CFR 1910.1000 Table Z-1.
[6] See 29 CFR 1910.1000 Table Z-2.
[7] Program Directive #200-67 (Dec. 1, 1977).

CHANGING STANDARDS

The decision to consider a new substance or working condition for regulation may be made by OSHA, or it may originate outside OSHA. For example, labor or employer groups may request the establishment of a standard. NIOSH, since it is authorized to conduct research on problems related to health and safety, is the most likely external source of such a request. Recommendations by NIOSH, since they represent the results of NIOSH research and may foretell new regulatory directions, are often included in references listing standards, but they are legally enforced only when they are adopted by OSHA.

When OSHA plans to create a new standard, it may first publish an early warning notice in the *Federal Register* to allow time for public reaction. Often a proposed new rule is accompanied by extensive information and documentation, which can be useful. If requested, there then may be a public hearing. This early warning approach allows OSHA to gather information and assess public reaction, without itself being under pressure of a deadline.

Whether or not an early notice is published, if OSHA wishes to set a new standard, it is required at some point to publish the potential standard as a proposed rule. In its notice, OSHA indicates when the time to make comments ends and where the comments should be sent. Then, in no less than 30 days or no more than 60 days, after the completion of a public hearing, the final standard is published. Disagreement with the standard from this time on is a matter that must be handled by the courts.

Occasionally, where evidence has been presented of a serious threat not adequately dealt with by the existing standards and delay caused by the usual mechanisms to introduce new standards is risky, the Labor Department has instituted an emergency standard. Such standards go into effect at the time of publication in the *Federal Register.* Emergency standards were set for vinyl chloride, asbestos, and benzene. These were replaced later by standards passed in the conventional fashion. Such replacement must occur within 6 months.

We find in the OSH Act the general duty clause, which states that "Each employer shall furnish to each of his employees employment and a place of employment which are as free from recognized hazards that are causing or are likely to cause death or serious physical harm to his employees."[8] In other words, OSHA compliance officers may issue citations for hazards in the workplace even when the hazard is not specified in legal published standards. The compliance officer needs a justification for application of the general duty clause, and this can come from one of several reliable sources. NIOSH publishes numerous *Criteria Documents*, which list recommendations for maximum exposure levels accompanied by a rationale for the chosen level. These are not legal limits, but are a good example of a source that could be the basis for issuance of a citation under the general duty clause.

ENFORCEMENT OF STANDARDS

INSPECTIONS

Enforcement of standards begins with an inspection of the workplace (see Figure 2.3). In 1994 OSHA performed 42,337 inspections.[9] Inspectors normally visit an establishment without advance notice, and the inspectors have the right to inspect not only the workplace itself, but also the employer's records of injury and illness. The decision to inspect a particular facility may be based on a worker's complaint, occurrence of a serious incident involving the health of the workers, a fatality, or a catastrophe wherein five or more workers are hospitalized. Otherwise inspections are scheduled in a pattern that favors more frequent visitation of higher risk facilities.

[8] OSH Act, Section 5(a)(1).
[9] Source: OSHA.

Figure 2.3. OSHA frequently inspects construction sites. Construction hazards are somewhat different from those of general industry, and are covered in 29 CFR 1926.

OSHA must depend on employers choosing to meet standards because OSHA might visit. OSHA has too few inspectors to permit visiting the majority of workplaces. At a given time many inspectors are inspecting sites where a worker has complained, which OSHA must do by law. This is important, but it reduces OSHA's flexibility to focus on high-risk facilities. The Associated Press analyzed records of worksites that had experienced serious accidents and found that about three-fourths of them had never been visited by OSHA.

Personnel involved in inspections are called industrial hygienist compliance officers. Of OSHA's approximately 2000 employees, more than half are inspectors. These individuals are trained to recognize and evaluate health hazards. They interview workers, collect dust and vapor samples, measure noise levels, and watch for such evidence as eye irritation, odor, or visible emissions.

An OSHA industrial hygiene inspection is complex and formal. OSHA has published a book, the *Field Operations Manual*, that specifically describes the inspection protocol for the benefit of inspectors. This is clearly of interest to the employer also, providing insight into the details of an investigation. Inspections begin with a conference with the plant management, then a walk-through inspection, scrutiny of records, the collecting of samples and measurements according to a standard format modified to be appropriate to the particular plant process, and a closing conference with the employer. In addition to the results of the inspection at this conference, the employer's occupational health program is discussed.

In standards such as noise or chemical exposure, OSHA defines an "action level". For chemical exposure, for example, this level is half the PEL. A number of provisions of the safety standards are not required if levels in the plant are below the action level, including aspects of monitoring, training of employees, and medical surveillance.

If the inspection uncovers violations, citations may be issued and penalties may be imposed. Violations can reflect several levels of seriousness. "Imminent danger" reflects a condition expected immediately to cause death or serious injury. "Serious violation" describes conditions that probably would cause death or serious injury under certain conditions. Less than the above are classed as "other violations".

In 1994, 145,859 violations were discovered, leading to $119,858,261 in penalties.[10] In 1994 OSHA raised the minimum fine for a serious willful violation to $25,000. This caused complaints, especially from state-run programs, and was thought to be too high for very small businesses. In March 1995, the minimum was lowered to $5000 for businesses with 50 or fewer employees.

The Department of Labor, if it finds a very dangerous situation can post imminent danger and request that the facility be closed. However, beyond informing the employees of the hazard, the department has no further authority and must seek the assistance of the courts if the employer refuses to close down the operation.

CONTESTING OSHA CITATIONS

The employer may contest citations, penalties, time allotted to correct a violation, or notices of failure to correct a violation. However, this must be submitted to the area director in writing within 15 federal working days of notification. Such a contest must be reported to the employees. The appeal may be handled by the area director, or in extreme cases is heard by the Occupational Safety and Health Review Commission, an adjudicatory commission independent of the Department of Labor. At such a hearing, the Department of Labor must prove its case against the employer. The decision of the judge in such a hearing may be sent for review by the U.S. Court of Appeals.

THE REVIEW COMMISSION AT WORK

The following case illustrates the function of the Occupational Safety and Health Review Commission. In 1992 there was an explosion at the Arcadian fertilizer plant in Lake Charles, LA, that injured seven people, three of them employees. The company was cited under the general duty clause and fined $50,000 for each employee presumed exposed to the hazard, a total of 87 employees. Arcadian protested the fine, and it was reviewed by the commission. In a 2-to-1 decision, the commission ruled that the Secretary of Labor does not have the authority to issue fines on a per-employee basis when enforcement is based on the general duty clause.

STATE-RUN PROGRAMS

The OSH Act allows individual states to take over enforcement of rules,[11] and at the time of writing 23 states and two territories have elected to do so. State programs must be submitted to the Secretary of Labor and be approved before implementation. The standards enforced must be " . . . at least as effective in providing safe and healthful employment and places of employment as the standards promulgated under section 6 of the Act which relate to the same issues."

RECORDKEEPING

An important aspect of the task of reducing workplace illness, injury, or deaths is the keeping of accurate records regarding the frequency and seriousness of such incidents. This requirement is spelled out in detail in 29 CFR 1904. Such records are used by the Bureau

[10] Source: OSHA. OSHA is not immune to political pressures. In 1995, with Congress more sympathetic to business and determined to cut budgets, the number of inspections dropped 31% to 29,113, violations cited were down 38%, and penalties down 27%. Part of the drop in numbers of inspections may be due to a different method of counting inspections. See D. Hanson, *Chemical and Engineering News*, February 5, 1996, p. 22. See also M. Carmel, "OSHA Needs Remedy for Employers Delinquent with Penalty Payments," *Occupational Health and Safety*, Sept. 1994, p. 30.
[11] 29 CFR 1901–1902. This is based on section 18(e) of the OSH Act.

CASE HISTORY — OSHA OVERSIGHT OF A STATE PLAN

North Carolina ran its own OSHA operation. On September 3, 1991, a fire in the Imperial Food Products poultry plant killed 25 workers. Major safety violations were involved. OSHA evaluated the state plan after this fire and indicated that deficiencies in the program required correction. These included

- insufficient inspections of sites
- inadequate sanctions and penalties
- minimal employer oversight

The AFL-CIO requested withdrawal of OSHA approval of the state plan. OSHA action was to assume joint enforcement in the state while reorganization took place.

North Carolina increased its budgeted enforcement staff from 55 to 115 and modified its health and safety laws. OSHA evaluated state progress at intervals. In 1995, satisfied with the operation of the state agency, OSHA withdrew from joint enforcement.

of Labor Statistics to track trends in workplace health and safety incidents. OSHA is not sent the records routinely. However, inspectors visiting a plant will request to see these records. Deficiencies in recordkeeping led to 3944 citations in 1994 and was the second most common violation.

Employers are required to maintain a Log and Summary of Occupational Injuries and Illnesses, OSHA form 200. Incidents are listed vertically on this form, with room for several on one page. The left side of the 200 log (Figure 2.4A) provides space for a file number, date, and name of the employee involved. The employee's occupation (which is not necessarily what occupied the employee at the time of the incident) and department are listed. Finally, there is a brief description of the injury or illness.

The top of the right hand side (Figure 2.4B) records the identity of the business and its location. Below that are two sections: the left side is for injuries and the right is for illnesses. Information here is very simple, including dates, check marks to indicate the category of the incident, and numbers of days off work or of restricted activity. The purpose of the 200 log is to provide for OSHA inspectors a brief and easy-to-read overview of the frequency and nature of problems in that facility.

For each incident there should be an OSHA form 101 on file providing details of the incident (Figure 2.5). This can be replaced by workman's compensation, insurance, or other forms that include all information called for in form 101.

WHEN AND WHERE MUST INCIDENTS BE REPORTED?

Incidents must be recorded within six working days of the employer learning of their occurrence. Any incident that involves a death or hospitalization of at least five employees must be reported within 48 hr at least orally to the local OSHA area office or to the appropriate state agency in states that run their own program. An annual summary for each calendar year must be posted for the workers in an appropriate place by February 1 of the next year and remain posted until March 1. Workers who move to different locations as they perform their duties should be mailed copies of the summary.

These records must be kept on file for an additional 5 years.[12]

[12] 29 CFR 1904.

Bureau of Labor Statistics
Log and Summary of Occupational
Injuries and Illnesses

NOTE: **This form is required by Public Law 91-596 and must be kept in the establishment for *5 years*. Failure to maintain and post can result in the issuance of citations and assessment of penalties.** ***(See posting requirements on the other side of form.)***

RECORDABLE CASES: You are required to record information about every occupational **death**; every nonfatal occupational **illness**; and those nonfatal occupational **injuries** which involve one or more of the following: loss of consciousness, restriction of work or motion, transfer to another job, or medical treatment (other than first aid). ***(See definitions on the other side of form.)***

Case or File Number	Date of Injury or Onset of Illness	Employee's Name	Occupation	Department	Description of Injury or Illness
Enter a nonduplicating number which will facilitate comparisons with supplementary records.	Enter Mo./day.	Enter first name or initial, middle initial, last name.	Enter regular job title, not activity employee was performing when injured or at onset of illness. In the absence of a formal title, enter a brief description of the employee's duties.	Enter department in which the employee is regularly employed or a description of normal workplace to which employee is assigned, even though temporarily working in another department at the time of injury or illness.	Enter a brief description of the injury or illness and indicate the part or parts of body affected. Typical entries for this column might be Amputation of 1st joint right forefinger, Strain of lower back, Contact dermatitis on both hands, Electrocution--body
(A)	(B)	(C)	(D)	(E)	(F)
					PREVIOUS PAGE TOTALS →
					TOTALS (Instructions on other side of form.) →

OSHA No. 200 *U.S.GPO 1990-262-256/15419

Figure 2.4A. Shown here is the left side of the OSHA 200 form, identifying the employee, the employee's occupation, and a brief description of the injury or illness. Notice that several incidents can be recorded on one page.

WHAT IS "RECORDABLE"?

Not surprising, many questions arise about what is or is not recordable, what category is correct for a specific incident, and the like. Incidents recorded on the 200 log do not correspond exactly to those reported to insurance companies or those for which worker's compensation is provided. The guidelines must not be confused. Further, the Department of Labor states

> Recording an injury or illness under the OSHA system does not necessarily imply that management was at fault, that the worker was at fault, that a violation of an OSHA standard has occurred, or that the injury or illness is compensatable under worker's compensation or other systems.

Desiring to have reporting be done in as uniform a manner as possible, the Bureau of Labor Statistics in 1986 published helpful guidelines in the form of typical questions and their answers.[13] This document does not constitute additional regulations, but serves as guidelines. A "decision tree" from this document is a useful first step, and is shown in Figure 2.6.

U.S. Department of Labor

For Calendar Year 19 ______ Page ___ of ___

Company Name

Establishment Name

Establishment Address

Form Approved
O.M.B. No. 1220-0029
See OMB Disclosure Statement on reverse.

Extent of and Outcome of INJURY						Type, Extent of, and Outcome of ILLNESS												
Fatalities	Nonfatal Injuries					Type of Illness							Fatalities	Nonfatal Illnesses				
Injury Related	Injuries With Lost Workdays				Injuries Without Lost Workdays	CHECK Only One Column for Each Illness *(See other side of form for terminations or permanent transfers.)*							Illness Related	Illnesses With Lost Workdays				Illnesses Without Lost Workdays
Enter **DATE** of death. Mo./day/yr.	Enter a **CHECK** if injury involves days away from work, or days of restricted work activity, or both.	Enter a **CHECK** if injury involves days away from work.	Enter number of **DAYS** *away from work.*	Enter number of **DAYS** of *restricted work activity.*	Enter a **CHECK** if no entry was made in columns 1 or 2 but the injury is recordable as defined above.	Occupational skin diseases or disorders	Dust diseases of the lungs	Respiratory conditions due to toxic agents	Poisoning (systemic effects of toxic materials)	Disorders due to physical agents	Disorders associated with repeated trauma	All other occupational illnesses	Enter **DATE** of death. Mo./day/yr.	Enter a **CHECK** if illness involves days away from work, or days of restricted work activity, or both.	Enter a **CHECK** if illness involves days away from work.	Enter number of **DAYS** *away from work.*	Enter number of **DAYS** of *restricted work activity.*	Enter a **CHECK** if no entry was made in columns 8 or 9.
(1)	(2)	(3)	(4)	(5)	(6)	(7) (a)	(b)	(c)	(d)	(e)	(f)	(g)	(8)	(9)	(10)	(11)	(12)	(13)

Certification of Annual Summary Totals By ______ Title ______ Date ______

FOLD

OSHA No. 200 **POST ONLY THIS PORTION OF THE LAST PAGE NO LATER THAN FEBRUARY 1.**

Figure 2.4B. Here the company is identified. Illnesses and injuries are reported separately. Such details as the number of days off work or on restricted activity are included. This half is the part that must be posted once a year in the plant for a month.

Recordable incidents include "occupational death; every nonfatal occupational illness; and those nonfatal occupational injuries which involve one or more of the following: loss of consciousness, restriction of work or motion, transfer to another job, or medical treatment (other than first aid)."

The following treatments characterize recordable injuries: worker loses consciousness, trauma requires stitches or butterfly adhesive dressings, foreign objects are removed from the eye that won't simply wash out, X-rays reveal injury such as broken bones, or injury leads to admission to hospital. What is "first aid"? Doing any of the following once is considered first aid: whirlpool bath therapy, application of antiseptics, application of hot or cold compresses, or use of prescription medicine. However, if such treatment is repeated, it is recordable.

[13] U.S. Department of Labor, Bureau of Labor Statistics, O.M.B. No. 1220-0029, Chapter 5.

Bureau of Labor Statistics
Supplementary Record of
Occupational Injuries and Illnesses

U.S. Department of Labor

This form is required by Public Law 91-596 and must be kept in the establishment for 5 years. Failure to maintain can result in the issuance of citations and assessment of penalties.	Case or File No.	Form Approved O.M.B. No. 1220 0029

Employer

1. Name
2. Mail address *(No. and street, city or town, State, and zip code)*
3. Location, if different from mail address

Injured or Ill Employee

4. Name *(First, middle, and last)* — Social Security No.
5. Home address *(No. and street, city or town, State, and zip code)*
6. Age — 7. Sex: *(Check one)* Male ☐ Female ☐
8. Occupation *(Enter regular job title, not the specific activity he was performing at time of injury.)*
9. Department *(Enter name of department or division in which the injured person is regularly employed, even though he may have been temporarily working in another department at the time of injury.)*

The Accident or Exposure to Occupational Illness

If accident or exposure occurred on employer's premises, give address of plant or establishment in which it occurred. Do not indicate department or division within the plant or establishment. If accident occurred outside employer's premises at an identifiable address, give that address. If it occurred on a public highway or at any other place which cannot be identified by number and street, please provide place references locating the place of injury as accurately as possible.

10. Place of accident or exposure *(No. and street, city or town, State, and zip code)*
11. Was place of accident or exposure on employer's premises? Yes ☐ No ☐
12. What was the employee doing when injured? *(Be specific. If he was using tools or equipment or handling material, name them and tell what he was doing with them.)*
13. How did the accident occur? *(Describe fully the events which resulted in the injury or occupational illness. Tell what happened and how it happened. Name any objects or substances involved and tell how they were involved. Give full details on all factors which led or contributed to the accident. Use separate sheet for additional space.)*

Occupational Injury or Occupational Illness

14. Describe the injury or illness in detail and indicate the part of body affected. *(E.g., amputation of right index finger at second joint; fracture of ribs; lead poisoning; dermatitis of left hand, etc.)*
15. Name the object or substance which directly injured the employee. *(For example, the machine or thing he struck against or which struck him; the vapor or poison he inhaled or swallowed; the chemical or radiation which irritated his skin; or in cases of strains, hernias, etc., the thing he was lifting, pulling, etc.)*
16. Date of injury or initial diagnosis of occupational illness — 17. Did employee die? *(Check one)* Yes ☐ No ☐

Other

18. Name and address of physician
19. If hospitalized, name and address of hospital

Date of report	Prepared by	Official position

OSHA No. 101 (Feb. 1981)

Figure 2.5. OSHA form 101. For every entry on an OSHA 200 log there should be a more detailed report. An OSHA 101 form is one way to do this.

THE FEDERAL RIGHT-TO-KNOW STANDARD

A new standard was established in 1983 by OSHA requiring that all employees likely to come into contact with hazardous materials be informed of the nature of the hazard and the steps necessary to protect the employee from harm. Called the hazard communication standard or HAZCOM (29 CFR 1910.1200) and referred to as the "right-to-know law", it applies to chemical manufacturers, distributors, importers, and manufacturing industry employers of standard industrial classification (SIC) codes 20–39 (Table 2.2). All materials that represent a hazard in at least one of the classifications are covered by the standard, a list of perhaps 70,000–80,000 commercial materials.

MATERIAL SAFETY DATA SHEETS

An important provision of the standard is the requirement that Material Safety Data Sheets (MSDSs) for chemicals used in the plant be available at all times to the employees (an

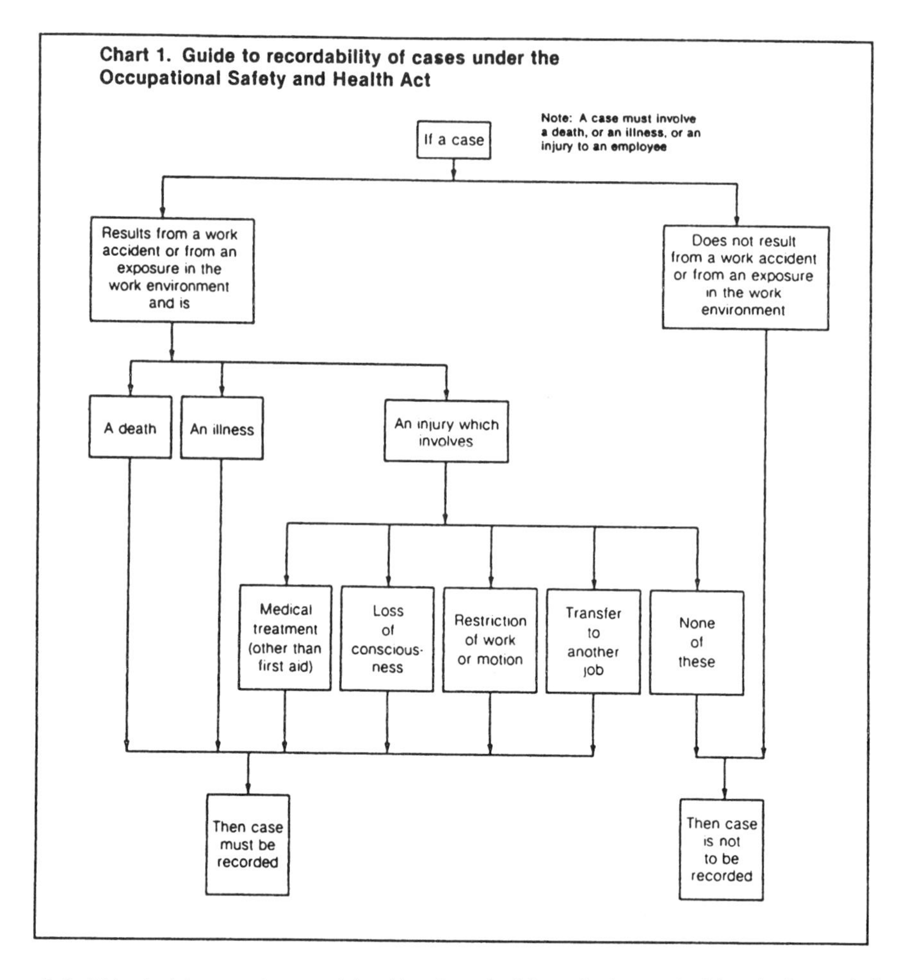

Figure 2.6. This decision tree is a useful guide when deciding whether an incident in the workplace is recordable. (From U.S. Department of Labor, Bureau of Labor Statistics, O.M.B. No. 1220-0029, Chapter 5, 1986.)

example is Appendix B). "Available at all times" means posted or on file in a place that is not locked or otherwise inaccessible any part of the working day. Amendments to the HAZCOM standard (*Federal Register*, Feb. 9, 1994) allow electronic access to centralized MSDS information as an acceptable alternative, as long as there is no barrier to employee access to the information. Thus, if a computer is the source of the MSDS, the employees must be trained to use the computer.

MSDS may be either obtained from the supplier of the chemical or prepared by the manufacturer. The information included on this form includes items in these categories:

1. Identity of the material.
2. Classification of hazard including types of hazard by name, PEL, TLV or other measures of the maximum permitted levels of exposure if available, and other such information.
3. Recommendations for safe usage, including the need for ventilation, protective clothing, and other precautions.
4. Information to guide emergency procedures.
5. Physical hazard data.
6. Chemical hazard data.
7. Physical and chemical properties.
8. Information about the manufacturer or supplier.

Table 2.2. Manufacturing Industries Impacted by the Hazard Communication Standard.

SIC Code	Industry Group
20	Food and kindred products
21	Tobacco manufacturers
22	Textile mill products
23	Apparel and other textile products
24	Lumber and wood products
25	Furniture and fixtures
26	Paper and allied products
27	Printing and publishing
28	Chemicals and allied products
29	Petroleum and coal products
30	Rubber and plastic products
31	Leather and leather products
32	Stone, clay, and glass products
33	Primary metal industries
34	Fabricated metal products
35	Machinery, except electrical
36	Electrical equipment and supplies
37	Transportation equipment
38	Instruments and related products
39	Miscellaneous manufacturing products

From G. G. Lowry and R. C. Lowry, *Handbook of Hazard Communication and OSHA Requirements,* Lewis Publishers, Chelsea, MI, 1985. With permission.

PERMITTING ELECTRONIC ACCESS TO MSDS GENERATES SERVICE BUSINESSES

Companies may now turn to suppliers to handle the HAZCOM requirement for access to MSDS information. An example is the 3E company, which services more than 60,000 customers with a database of MSDSs from 25,000 manufacturers. 3E keeps records of the MSDS needs of each customer. About 7% of these become obsolete each month. On call, 3E mails or faxes information, depending on the urgency of the situation. A 24-hr hotline with 200 incoming lines is staffed with hazardous materials specialists who can supply MSDSs, advise clients about spills, and assist with other emergency situations.

From R. B. Smith, "Outsourcing the Hazcom Paperwork," *Occupational Health and Safety,* June 1995, p. 31.

If the composition of the mixture is a trade secret, it is not required that the manufacturer list the composition on the MSDS. However, all of the hazard information must be provided. OSHA does not supply a standard form, but leaves it to the individual company to design its own form.[14]

[14] Discussions frequently are directed at changing the MSDS format to be more user friendly. Obtaining information from such a sheet requires the education level of a college student, but the intended beneficiary of this information is a factory worker who may not have completed high school or have English as a first language.

LABELING

Hazardous substances must be labeled. The label must include the identity of the substance, information regarding hazards associated with the substance, and the name and address of a person from whom additional information may be obtained. OSHA has an established list of substances for which specific wording is required on the label. The American National Standards Institute has suggested a nine-item label:

1. Identity.
2. CAUTION, WARNING, or DANGER.
3. A statement of the hazards.
4. Safety measures such as protective clothing.
5. Instructions as to procedures to use if exposure occurs.
6. Antidotes for poisoning.
7. Emergency treatment information for physicians.
8. Instructions as to procedures in the case of fire, spill, or leakage.
9. Instructions for handling and storage.

If the name and address of a person to contact were added to this, the list would exceed the OSHA requirements.

Employers are permitted to add their own system of warning labels to the container, which may allow simpler and clearer communication with workers than the manufacturer's label. Systems using color codes for type of hazard and numbers for degree of each type of hazard, as used by the NFPA, are common (Appendix E). Symbols for types of personal protective equipment needed are a useful addition.

TRAINING PROGRAMS

Each manufacturer is required to set up a training program for the employees who work in the area of the plant containing the hazardous materials. They must be informed about the rules established by the standard and how to use the labeling, MSDSs, and other hazards information. The hazard communication standard was revised in February 1994, and a very significant one-word change was made: "Employers shall provide employees with effective information and training on hazardous chemicals in their work area at the time of their initial assignment, ..."[15] This means a visiting compliance officer may assess effectiveness by asking employees questions about hazardous chemicals in their work area, interpretation of warning labels, locations of MSDSs, and protective measures necessary when using these chemicals.

The importance of training programs cannot be overemphasized. They put the worker — the individual with the greatest interest in the success of the safety program — into the worker-protection system, rather than having the responsibility for worker safety reside only in the hands of the employer and the enforcement agency. A worker who understands why a safety rule has been established is less likely to bypass that rule because it is troublesome to follow the guidelines.

The value of safety training can be seen in an example from the mining industry from a period when responsibility for miner safety resided with a separate agency, the Mine Safety and Health Administration (MSHA). In 1977 a mining act required that miners be given extensive safety training before they could work as miners. Congress exempted specified nonmetal mining operations from this requirement, an exemption including about 10,000 mine sites. In the period following the introduction of the training requirement, a dramatic

[15] 29 CFR 1910.1200(h)(1).

reduction in accidents, injuries, and deaths took place at the mines with training programs, but this impressive level of progress was not shared by the exempted operations.[16]

A properly designed training program increases workers' knowledge, sharpens worker skills, or modifies workers' attitudes. Although commercially prepared videos are available and are useful adjuncts to training, workers are more motivated to learn when an informed instructor is involved, one who is knowledgeable and answers questions. At the time of this writing, there is no general requirement focusing on the quality of the instruction or instructors, and no assurance that workers understand the objectives of the training program. However, in the Asbestos Hazard Emergency Response Act (AHERA, 1986) and the Superfund Amendments and Reauthorization Act (SARA, 1986) there is a requirement for accredited instructors. In AHERA the Environmental Protection Agency (EPA) provided a model training program with minimum standards for both the curriculum and the trainer. Trainees were then required to pass a written examination.

OSHA was required by SARA to generate a similar system for training program accreditation. Section 29 CFR 1910.120 is Hazardous Waste Operations and Emergency Response (HAZWOPER), and training requirements are listed in section (e)(1) through (e)(9). Personnel must be trained before they work on a site where they are exposed to hazardous substances or health or safety hazards. Initial training includes 40 hr off site and 3 days of field experience under a trained and experience supervisor, and covers elements specified by OSHA. Managers and supervisors are also to be trained this intensively, plus eight additional hours at the time of assignment. Instructors are "qualified" by having completed a training program as teachers or equivalent experience and credentials. All these people are required to have eight additional hours of refresher training annually. Although guidelines have been proposed, a standard dealing with training program certification or accreditation is not in place at the time of this writing. It is clearly the next frontier in improving worker training.

Worker training has historically not been given high priority in the U.S., certainly not when compared to countries such as Japan, which devote much more time to training than do U.S. companies. Perhaps this is based on the concept that a worker in training is not a worker producing. This is a short-sighted view, since a more knowledgeable or skillful worker is more productive and an injured or ill worker is hardly an economic asset. A survey by the Bureau of Labor Statistics of about 12,000 firms of all sizes indicated that 93% of companies with more than 250 employees, 75% of those with 50 to 250 employees, and 29% of small companies provide health and safety training. In 1994 OSHA issued 3833 citations for violations of the general industry training standard, the third most frequent citation, and 2277 citations for violations of the construction industry training standard (29 CFR 1926.059(h)), the sixth most frequent citation.

WRITTEN HAZARD COMMUNICATION PROGRAM

Finally, the manufacturers must have a written hazard communication program. This program must include

1. A list of hazardous materials in the plant, by plant area in a large operation.
2. MSDSs.
3. A description of the warning labels and how other postings are designed and used.
4. A description of the training program to be used with the employees.

The list of hazardous substances in the plant must be arranged alphabetically and must take account of slang names used by plant workers. OSHA begins inspections by reviewing

[16] J. B. Moran, "Workplace Safety Training Progress Noted, But Additional Work Needed," *Occupational Health and Safety*, March 1994, p. 26.

this document, and issued 4728 citations in 1994 for violation of this requirement, the largest number of citations for any single violation.

THE TOXIC SUBSTANCES CONTROL ACT AND THE ENVIRONMENTAL PROTECTION AGENCY

The Toxic Substances Control Act, known as TSCA, was enacted in 1976 in order to prevent unreasonable chemical risks. TSCA gives the EPA authority to act before harmful chemical substances threaten human health or the environment. The act directs the EPA to make judgments on the safety of new chemicals or chemical mixtures produced by manufacturers.The first major action of the EPA under TSCA was to assemble the most comprehensive inventory in existence of chemical substances in U.S. commerce. The law requires the submission of test data by the manufacturer of a chemical relating both to the health effects and to the effects on the environment, should the agency choose to question its safety. The compound or mixture is new if it is not on the agency's list, but testing may be required if a new use is to be made of an existing compound or mixture that involves a significant change in the exposure of humans or the environment. Notification of intent to manufacture must be given to the agency 90 days or more prior to actual production. Action by the agency can range from a requirement for specific labeling to a ban on manufacture. Where such restrictions are suggested, there is a hearing to allow the various interested groups to have input.

To avoid overlapping of enforcement authority, tobacco, pesticides, ammunition, nuclear material, food and food additives, drugs, and cosmetics, all of which are regulated by other agencies, are exempted. For each compound, setting rules for the workplace is left in the hands of OSHA.

In the CFR, regulations of the EPA are found in Title 40. Referencing is done in the same fashion as in Title 29.

Effective February 6, 1986, the EPA and OSHA established new mechanisms for the sharing of information. Twice yearly the EPA sends OSHA a list of chemical substances under consideration for referral under TSCA. Each agency appoints a staff contact person, and meetings are set to exchange information. Such an arrangement should help greatly in reducing the confusion and overlapping of efforts that are a constant problem in the multi-agency approach to the regulation of toxic substances.

TRENDS IN ENFORCEMENT

In 1995 President Clinton presented a plan for "reinventing OSHA". The objective of the various aspects of the plan is to increase the effectiveness of OSHA at reducing workplace illness and injury, involve OSHA more in a consultive interaction with industry, and rewrite confusing standards.

Maine adopted a revision of its state-run program called Maine 200, which serves as a prototype of some changes proposed. Maine used worker's compensation data to pinpoint employers with bad health and safety records, then used incentives to encourage them to improve their performance.

OSHA resources were stretched thin, including the stress of budget cuts at that time. Companies with poor health and safety records will be increasingly targeted, providing the greatest overall health and safety improvement for the time spent inspecting. During inspections OSHA identifies whether the company has a good program in place. If it does the inspection focuses on more important health and safety issues, spending less time identifying "nitpicking" minor regulations. If violations noted by OSHA are corrected during the inspection — the "quick fix" concept — fines for those violations are reduced. This provides immediate correction of a problem and saves OSHA time reinspecting.

OSHA deals with two types of complaints: formal written and signed requests for inspections by an employee and informal phone messages. To speed response and reduce the number of inspections based on complaints, informal complaints trigger a "phone and fax" response in OSHA offices. The office notifies the employer of the complaint, involve the union and workers to correct the hazard, and verify that it has been resolved without the necessity of a formal inspection.

OSHA directs attention toward classes of business and types of health and safety problems that have not been major concerns in the past. Problems exist in those areas that are as serious as many more traditional enforcement targets.

VOLUNTARY PROTECTION PROGRAMS

The enforcement program had been "all stick and no carrot". A means of encouraging and rewarding employers with good programs and records was desired. OSHA therefore instituted voluntary protection programs as a means of inserting some carrot into the process. OSHA recognizes programs with effective ongoing health and safety programs. Employees must participate in these programs and there must be an annual self-evaluation. The company must have demonstrated good faith in previous dealings with OSHA. The Bureau of Labor Statistics' injury incidence and lost workday injury rates are used as part of the overall evaluation of the program. Comparisons of health and safety records are made within the companies.

OSHA distinguishes three categories of program:

- STAR. The Bureau of Labor Statistics show incident rates for this company to be below the national average for their industry. These rates are reviewed annually, and the participants are reevaluated every 3 years.
- MERIT. These companies do not meet all the requirements for the STAR designation, but have set goals to achieve STAR status.
- DEMONSTRATION. These companies demonstrate promising, innovative, alternative health and safety programs.

OSHA reviews each applicant for one of these awards, including interviews with employees and management. Collective bargaining agents for the workers must agree to the award.

Participants in the program are not subject to random, surprise visits by OSHA inspectors. They may still be visited as a result of employee complaint or a serious incident, however, and are not exempt from OSHA regulations.

HOW SUCCESSFUL HAS OSHA BEEN?

At the time of this writing there is a movement to downsize government, and OSHA is a significant target of conservatives. There is pressure to simplify the regulations, reduce employers' paperwork, and, more significantly, to roll back some of the regulations. Many would like to see OSHA changed from a "policing" to an "advisory" operation. A catchphrase is "inventing a new OSHA". Entering new areas of regulation is resisted strongly. At such a time it is appropriate to consider how effective the agency has been to avoid "throwing away the baby with the bathwater".

Since the passage of the OSH Act, the consciousness of employers about safety and health in the workplace has been increased manifold. Importantly, the typical 14,400 annual work-

EXPLORING NEW OSHA/EMPLOYER RELATIONSHIPS — A CASE STUDY

The Atlanta East office of OSHA approached Argonaut Insurance, a company underwriting worker's compensation insurance, with an experimental proposal to reduce injury and illness incidents at a specific client employer, Horizon Steel Erectors Co. The combination was chosen because both the insurance company and the construction company showed evidence of commitment to reducing accidents and injuries to the roughly 200 Horizon employees.

OSHA did a risk assessment of the company, suspending citations and fines at that time. Testing indicated there was a benefit to using fall protection equipment that exceeded OSHA's standards. A supervisor accountability program was set up at Horizon with Argonaut's help. Argonaut put supervisors through an accident reduction training program. Supervisors were appraised for safety, with potentially substantial bonuses tied to the appraisal. The company safety officer was able to consult with OSHA on safety matters.

The program had a goal of reducing claim costs per work hour by half and to reduce potential OSHA penalty costs by half in 90 days. In fact, claim costs per work hour dropped from $4.26 to $0.18 (to 4%).

Adapted from E. S. Carnevale, "Reinventing an OSHA Relationship," *Occupational Health and Safety*, Oct. 1995, p. 57.

related fatalities of the 1970s was reduced to 6,500 in 1995. According to a report by the AFL-CIO, the greatest successes have been in areas of heavy OSHA concern. OSHA conducted almost half its inspections in construction, and the result has been a 78% decrease in fatalities and a 40% decrease in injuries. In mining, where inspections are mandatory, the fatality rate is down 73% and the injury rate 50%. However, a growing area of concern that has seen an 800% increase in reported injuries in the last decade, repeated trauma disorders or "ergonomics", is not regulated by a formal OSHA rule, and attempts to pass such a rule have been resisted.

Perhaps by the time this book is in print and in use, OSHA's future will be more clear.

KEY POINTS

1. OSHA, founded in 1970 as part of the Department of Labor, brought uniformity to health and safety standards.
2. OSHA sets standards, conducts inspections, requires health and safety recordkeeping, and trains personnel.
3. NIOSH is the health and safety research arm of OSHA.
4. OSHA decisions can be grieved before the Occupational Safety and Health Review Commission.
5. Standards for workplace safety were initially based on existing consensus standards.
6. Rules are first reported in the *Federal Register*, then are condensed into the *Code of Federal Regulations*.
7. The mechanism for changing standards or introducing new regulations allows public input.
8. Rules are enforced by plant inspections. Violations lead to citations and fines.
9. OSHA is empowered to require a safe workplace, even if there is not a specific OSHA regulation covering the safety issue, by the general duty clause.

10. Employers are required to keep logs of fatalities, injuries, and lost time. These are reported to the Bureau of Labor Statistics and may be viewed by OSHA inspectors on demand.
11. The right-to-know standard requires that employees be informed of and trained to avoid workplace hazards. MSDSs, worker training, and hazard labeling are required.
12. TSCA is an act administered by the EPA that judges on the hazards of chemicals.
13. OSHA and the EPA share information.
14. The voluntary protection program encourages companies to work with OSHA in exchange for OSHA's promise not to perform surprise inspections.

BIBLIOGRAPHY

E. J. Baier, "Legislation and Legislative Trends," in G. D. Clayton and F. E. Clayton, Eds., *Patty's Industrial Hygiene and Toxicology,* 4th ed., John Wiley & Sons, New York, 1991.

G. S. Dominquez, *Guidebook: Toxic Substances Control Act,* CRC Press, Boca Raton, FL, 1977.

Guidebook to Occupational Safety and Health, Commerce Clearing House, Chicago, 1978.

J. J. Keller and Associates, Haard Communication Guide, 1988.

G. G. Lowry and R. C. Lowry, *Handbook of Hazard Communication and OSHA Requirements,* Lewis Publishers, Chelsea, MI, 1985.

J. C. Silk, "Hazard Communication: Where Do We Go from Here?" *Applied Industrial Hygiene,* January 1988, p. F27.

Subcommittee on Labor, Committee on Labor and Public Welfare, U.S. Senate, *Legislative History of the Occupational Health and Safety Act of 1970* (S.2193, P.L. 91-596), 92nd Congress, 1st Session, June 1971.

Toxic Substances Control Act, Occupational Safety and Health Act, 95th Congress, 1st Session, Committee Print 95.7, U.S. Government Printing Office, Washington, D.C., 1977.

M. D. Worobec, Toxic Substances Control Primer, 2nd ed., Bureau of National Affairs, Washington, D.C., 1986.

PROBLEMS

1. Contrast the situation in the U.S. before and after passage of the OSH Act. Did the act cause the same degree of change to take place in every state? It is clear that a state that has lax worker protection laws will have more frequent accidents and work-related illness. What is the advantage to a state of having lax worker protection laws?
2. Locate the contents section of 29 CFR 1900–1910. Where would you look for regulations regarding: (a) recording and reporting occupational illnesses and injuries, and (b) OSHA inspections and citations? What is found in section 1900?
3. You are the safety officer in a middle-sized tool company. Your plant has just set up new abrasive cutting wheels, and you want to check if the guards are in compliance. Use the index at the end of 29 CFR 1910 to locate the regulations. Which section contains this information? Do these standards lay out broad general guidelines only, leaving you to decide on the details of compliance, or are they quite specific?
4. The politically liberal individual tends to be an advocate of the worker, whereas the politically conservative person takes the side of the employer. Since the OSH Act is designed to protect workers, it is basically liberal legislation. In a democracy, for legislation or anything else to succeed the needs and interests of both sides must be considered, or an imbalance results that ultimately forces change. Let us consider ways industry has a say in the provision of worker protections.
   A. It is possible that a proposed, more rigorous standard, compliance with which could require manufacturers to spend huge sums of money, would not actually increase workplace safety in a significant fashion or would be no better than a somewhat less rigorous standard that could be more easily met. At what point do manufacturers have an opportunity to present this case, and to whom do they present it?

B. Plant inspections are done by people who can make mistakes or are responding to an agenda that is not strictly enforcement of regulations. Can a manufacturer who is charged with noncompliance protest the charge, and if so, how and to whom (see 29 CFR 1903.17)?

C. What are the limits of OSHA's powers: (1) Can OSHA inspect any worksite it chooses, and can OSHA inspect without warning? (see 29 CFR 1903.3 and 1903.6); (2) If an employer refuses OSHA admission to a worksite, what can OSHA do? (see 29 CFR 1903.4); (3) Can OSHA fine a company for noncompliance, and if so, who decides the size of the fine? (see 29 CFR 1903.15 (a) and (b)); (4) Can an employer send a representative to accompany an inspector during inspection? (see 29 CFR 1903.8); (5) What can OSHA do about continued noncompliance after an inspection and citation: nothing, levy additional fines, or close a plant? (see 29 CFR 1903.18); (6) If the time allowed to correct a situation not in compliance seems unreasonable to the employer, does the employer have any recourse? (see 29 CFR 1903.14a).

5. Learn about the role of the Review Commission by reading 29 CFR 1903.17. How does an employer protest a decision by OSHA?
6. How may employees keep informed of the progress and results of an inspection? Read 29 CFR 1903.8, 1903.16, and 1903.10.
7. Some environmentalists propose banning all chemicals that have chlorine in their structure on the basis that some chlorine compounds are hazardous, and in the case of a few the hazard was not recognized until some damage had been done. In terms of economic cost, contrast that approach to the setting of standards for individual compounds based on laboratory toxicology studies, as is done by OSHA.
8. In 29 CFR 1905 find out what is meant by variance, exemptions, variations, limitations, and tolerances.
9. Turn to the MSDS for benzene (Appendix B). Find the eight classes of information to be included in the MSDS, then locate each of these in the sample sheet.
10. You are the head of your state Department of Labor, and you wish to submit a plan for running your own health and safety program to the federal government. Locate in the CFR how this is done. What is the process for approval or rejection?
11. What is the legal basis in 29 CFR by which OSHA could have terminated North Carolina's state-run program?

Section II
Basics of Toxicology

The general public has little understanding of chemistry and is generally fearful of potential harm from contact with chemicals, a problem chemists sometimes call "chemophobia". One reads statements touting "chemical-free food" and describing ways to keep your body "free of chemicals", obviously written by individuals who do not understand that both the food and the body are totally comprised of chemicals. Industrial hygienists cannot indulge in such simplistic, naive approaches toward chemicals, but must proceed well grounded in the true level of hazard of industrial chemicals, how their properties relate to this hazard, and what measures must be taken to minimize such hazards as exist.

Chapters 3–5 introduce the principles of toxicology. The reader will discover that chemicals present a variety of health problems, become acquainted with the terminology used to describe relative toxicity and assess risk, and learn how chemicals enter and leave the body.

CHAPTER 3

Toxic Effects

WHAT IS TOXIC?

This section directs attention toward health damage to workers who come in contact with toxic chemicals. First we must decide: What is toxic? Chemicals are toxic. Does this mean every chemical is toxic? At the risk of raising the levels of "chemophobia" already loose in the land, there is a sense in which the answer to that is, "yes". When addressing nonscientific groups about concepts of toxicology, this author often asks the group to name a completely safe chemical. Almost always water is named, at which time a newspaper clipping is projected reporting the death of a British man who, believing quantities of water were good for his health, drank so much that he died from dilution of salt concentrations in his body fluids.

If you stop drinking water as a result of reading this, you miss the point. The key words in the above statement are drank "so much". Water is normally a very safe chemical because the quantities we take into the body are well below levels that cause us harm. Although some of the pleasure derived from sitting down to a favorite dish in a good restaurant evaporates when thinking in these terms, all food is merely a mixture of chemicals. Any one of these individual chemicals taken in excess could be harmful, but it is unlikely that one could cause damage with the quantities consumed at dinner. A major responsibility of toxicologists, specialists in toxicity, is to determine what level of exposure to a given chemical can cause harm and what level is safe.

CLASSES OF TOXICITY

In order to assess the toxic hazard of a chemical, one must know how the chemical causes damage. One must put aside the Snow White picture of toxicity in which a poison is an oily green liquid which, when placed on an apple, causes the apple eater to crumple to the floor with the first bite. A specific chemical can cause one or more of a variety of toxic effects. Toxic chemicals fall into such classes as systemic poisons, mutagens, carcinogens, teratogens, or behavioral toxins according to the nature of damage caused. In some cases the damage is apparent immediately, whereas in others the effect is seen only later, perhaps even years later. The symptoms may be obvious, but they also may be subtle and easily missed. One must understand these classes of toxicity and become familiar with the terminology associated with them.

First, some ground rules. Some chemicals are so highly reactive they cause damage just by contacting the skin, eye, or mouth. This sort of damage would be caused by a corrosive agent such as an acid, alkali, or powerful oxidant. These problems, classed as "local effects", are not the object of this discussion. Here we concentrate on chemicals that do harm after entering the body and being distributed by the blood circulation.

SYSTEMIC POISONS

In our bodies, various organs (liver, kidney, digestive tract, heart, and so on) perform specific aspects of necessary body function. These organs contain a variety of cell types, each making a specialized contribution to overall organ function. Such contributions include constituting protective linings or blood vessel structure, serving as contractile tissue (muscle), secreting needed chemicals, and capturing and altering chemicals transported to the organ by the blood. Each cell type has a characteristic array of functional chemicals — biomolecules — suited to the cell purposes.

The most important of these biomolecules are the polymeric structures called proteins. For example, cells contain a vast array of proteins called enzymes. Enzymes are chemical catalysts. As such they create and regulate complex patterns of chemical changes (pathways) that convert dietary material to energy, repair and generate new molecules to replace cell components, stimulate cell growth and reproduction, and perform specialized processes by which a particular cell type contributes to the overall coordinated functioning of the multicellular organism (for example, a gland cell producing a hormone). Simply stated, enzymes are responsible for all the chemical reactions of life processes. Some enzyme types are found in every cell, while others are present only, or are present in higher concentration, in specific cells.

A foreign chemical may reach a concentration in the body at which its presence begins to interfere with one or more of these normal enzyme-catalyzed cell processes. The cell then alters its functions, reduces its contribution to the organism, or dies. This chemical is now acting in a toxic fashion and is termed a systemic poison or systemic toxin.

Interference with other classes of functional proteins also damages or destroys cell function.

Poisoning is not normally the result of a general damaging of all body cells in a uniform fashion. Most toxins damage one specific organ, the target organ, more than others. Symptoms result from malfunction of that organ, even though cells in other organs are also affected. Failure of that organ causes death. The identity of the target organ generates a "class name" for the substance. For example, a liver poison is called a hepatotoxin, nervous system damage is produced by a neurotoxin, and a compound affecting kidney function is a nephrotoxin or renotoxin. Once a target organ is recognized, medical testing to assess the health of that organ monitors the degree of damage by the toxin.

As an example, carbon tetrachloride (CCl_4, a chlorinated hydrocarbon) was once widely used for such functions as degreasing metal parts, dry-cleaning clothes, and extinguishing fires. It was found to produce symptoms of liver damage, such as increased levels of certain marker enzymes released into the blood as liver cells die. Individuals who were heavily exposed showed extensive areas of necrosis (dead cells) in their livers on autopsy. Carbon tetrachloride has since been replaced by less toxic chlorinated hydrocarbons.

Particular tissues or organs become targets for a variety of reasons. Tissues may be most heavily impacted simply because the toxin contacts those tissues first. An airborne toxin may primarily affect the lungs, or a liquid splashed on a worker may primarily affect skin tissues. Heavy metals (e.g., lead, mercury, cadmium) bind to proteins with thiol (–SH) groups. Tissues particularly rich in or unusually dependent on those proteins are selectively damaged. Some organs employ chemistry directed toward a service provided to the rest of the body, and this unique chemistry potentiates the toxin. Liver metabolizes (chemically changes) nonpolar substances into more polar compounds to facilitate their removal from the body. Liver cells are selectively damaged when, in this process, they convert a relatively innocuous chemical into one that is highly toxic. Some chemicals are accumulated or concentrated in a particular organ. Kidneys filter blood into a tubular system, recapture desired substances, then reduce the amount of water in this filtrate. Removing water concentrates the dissolved chemicals. Thus, the kidneys may be damaged as a toxin concentration rises to a dangerous level during the removal of water.

The concentration of a toxicant is an essential component of degree of damage done. Although greater concentrations produce greater damage, the effect is normally not linear (double the concentration leads to double the damage). Often small amounts of toxin are bound, eliminated, or chemically altered rapidly, resulting in no observed toxic effect. Only when these protective systems are overcome at higher toxin concentrations do we see toxic effects.

Organ damage includes cell death. In some organs, notably liver and kidney, dead cells are replaced fairly rapidly, whereas in others, replacement occurs more slowly. In the extreme case, as in neurons of the nervous system, cell loss is nearly permanent. When cell replacement is rapid, there is minimal risk from a single exposure below the level at which the degree of damage seriously interferes with organ function. The organ continues to function, and after clearance of the toxin the organ repairs itself. A person's state of health may alter this resiliency. Health problems can place an increased burden or reliance on the organ in question or reduce its capacity to sustain damage. Previous lung disease or lung damage may leave a worker with lowered lung capacity and a reduced ability to cope with high levels of airborne particulate matter or a lower than normal concentration of oxygen in the air. Habits or practices of the subject such as the use of prescription drugs, alcohol, or tobacco create special risk situations. For example, exposure to asbestos is much more likely to produce lung cancer in a smoker than a nonsmoker, and tolerance of a hepatotoxin is much lower in an alcoholic, whose liver is already under stress.

LETHAL DOSES

When the dose of toxicant is sufficiently high that the target organ is immediately damaged beyond its ability to function adequately, the person dies. This level of exposure, termed an acute lethal dose, is a fundamental measure of the toxicity of the substance. Because of individual differences, the minimum single dose of a particular chemical that causes death is not the same for every individual. Published values that communicate acute lethal dose levels are therefore based on mathematical manipulation of data from several observations.

Although the risk resulting from a single exposure to a chemical at a given level may be low for an individual, that does not mean that repeated exposure to that same level of the compound carries a similarly low risk. A portion of the daily dose of a chemical may still be in the body at the start of the next workday. Thus, the daily intake level may remain constant, but the concentration in the target organ is increasing by some amount each workday. Daily exposure to a chemical in regular use in a workplace is part of many job experiences. An electrician may work for a day repairing wiring at a dry cleaner and inhale solvent vapors, then go home, clear the solvent from his or her system, and feel no ill effects. However, the people operating those cleaning machines every day may be less fortunate.

In a similar fashion, if exposure produces the need to repair tissue and the repair is not complete by the time of the next dose, damage to the body accumulates. Sometimes removal of the chemical from the environment of the worker allows the worker to "catch up" with organ repairs, and symptoms that had appeared then disappear.

However, under conditions where tissue repair is constantly required, another process involving enlargement or irreversible scarring of the organ may occur. Cirrhosis of the liver, common in long-term alcoholics, is an example of the latter. This could also be brought on by regular occupational exposure to one of several chlorinated hydrocarbons used as solvents, or some combination of occupational exposure to the solvents and personal alcohol use. Daily cell replacement does not continue at optimum rates, and collagen formation (scar tissue) occurs instead. Lung tissue is replaced with scar tissue in a number of occupational lung diseases resulting from inhalation of damaging particulate such as silica or asbestos. Nonfunctional tissue gradually replaces healthy tissue, resulting in an organ inadequate to the needs of the victim.

TOXINS AFFECTING DNA

Other toxic effects of chemicals involve other aspects of biochemistry. With few exceptions (notably red blood cells), each cell contains blueprints for the complete organism in the form of giant polymeric molecules called DNA. These blueprints, the genetic message, tell how to build the proteins, the functional molecules of the organism.[1] A functional DNA segment, a gene, codes for a single specific protein. Humans have more than 10,000 different proteins. The "code" is the order of assembly of the four kinds of subunits or monomers comprising the DNA polymer, which the cell translates into correct assembly of the protein molecule.

Simple life forms such as bacteria may have the entire genetic message on a single DNA molecule. In more complex life forms the message requires a number of separate DNA molecules, each termed a chromosome. In humans and other higher life forms each cell contains two complete sets of chromosomes, one set from the male and one from the female parent. Such cells are termed diploid. Because humans are very complex (many different proteins), and because each cell carries two complete messages, the amount of DNA per human cell is relatively large. In fact, if the DNA molecules from a single cell were unfolded and laid end to end, they would stretch approximately the height of the person.

Alteration of even one protein message can damage or kill the cell, so the genetic message must remain undamaged. A first level of protection is provided by the attachment to every DNA molecule of a second DNA molecule called the complement. Complement DNA is a carbon copy of the actual message molecule. The four kinds of monomers in DNA are abbreviated A, T, G, and C. Everywhere on the sense (message-carrying) strand of DNA that an A unit appears, the complement strand has a T, and everywhere the sense DNA has a T, the complement has an A. G and C have the same relationship (Figure 3.1). The A units uniquely hydrogen bond to the T units, and the G units uniquely hydrogen bond to the C units. The two strands of DNA cling to one another by these hydrogen bonds and spontaneously form a spiral assembly called the DNA double helix. Just before a cell divides, these two strands separate. New monomer units assemble next to the message strand, pairing A with T and G with C in response to the hydrogen bonding forces, and zip together into a new complementary molecule of DNA. The DNA now attached to the sense strand is a new complement strand. In the same fashion the original complement strand serves as the template for the formation of a new sense strand. We now have two double-helix DNA molecules each identical to the original, one for each of the cells resulting from cell division. This process, called replication, ensures that each of the cells has a complete correct genetic message (Figure 3.2).

Mutagens

A mutagen is a substance or force that can cause damage to the DNA message. The most important mutagens are either specific chemicals or such radiation as X-rays, radioactive emissions, and ultraviolet light. Spontaneous changes also occur slowly in DNA, reminding us that the components of DNA are not superstable and do react chemically with their

[1] It also codes for RNA molecules, needed to synthesize proteins.

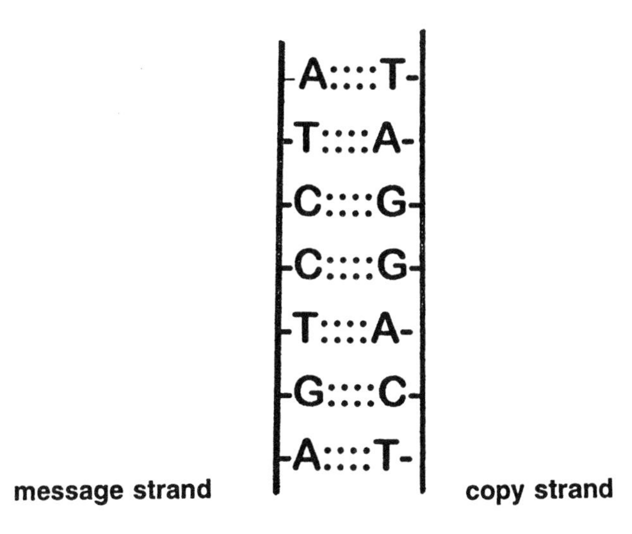

Figure 3.1. The complementary relationship between two strands of DNA. Everywhere one strand has A the other has T, and where one has C, the other has G. A and T join by hydrogen bonds, as do C and G.

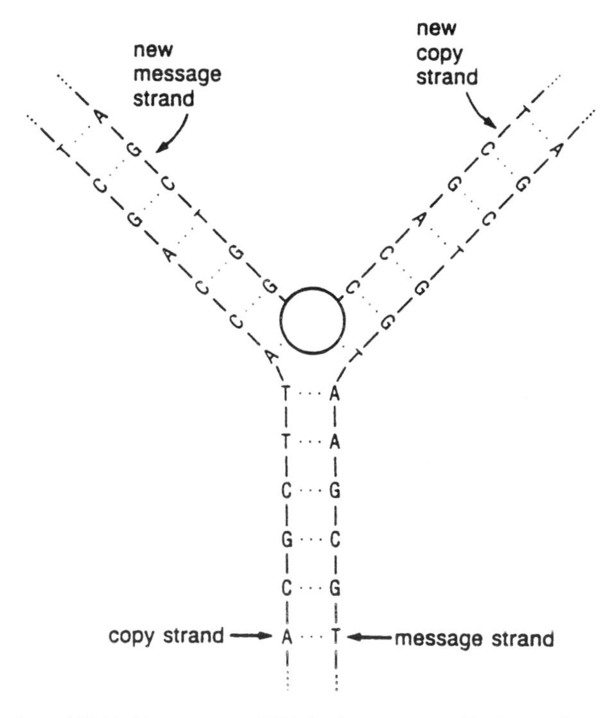

Figure 3.2. Replication of DNA. Here we see DNA in the process of being replicated. The lower arm is DNA that has not yet been acted on by the enzyme, represented by the circle at the fork. Each of the upper arms is one of a pair of identical DNA strands. When replication is complete, there will be two identical sets of DNA, one for each of the cells that will result from cell division. Each of those sets will have one strand of old DNA and one of newly replicated DNA.

environment. Living systems have always had to deal with spontaneous DNA degradation, the effects of radiation, and mutagenic chemicals, so the cell includes systems to check for DNA damage and correct it. This same machinery also locates and corrects mistakes in replication. Because humans are continuously exposed to both chemical mutagens and radiation, a cell experiences such damage frequently, and the repair mechanisms stay very busy.[2] A concern is that some damage escapes notice. If the level of exposure to a mutagen is raised, that likelihood is increased. At high levels of damage, the capacity of the repair system may be exceeded. Damage by mutagens thus shows dependence on dose.

If genetic damage escapes repair before the cell divides, the altered message becomes a permanent feature of the DNA of every descendant of that cell. This results in one of a variety of possible responses. In the most fortunate case the new blueprint produces functional cell components, and the cell suffers no ill effects. However, a whole spectrum of damage can be observed, ranging from production of less efficient cell components to complete failure to produce a functional version of a particular component. The cell so crippled may or may not survive. If it survives, all its descendants will be similarly crippled.

Damage to certain cell types is of particular concern. A crippled liver cell may die, but it is quickly replaced by division of a nearby healthy liver cell. By contrast, alterations to the DNA of either a sperm or egg cell (a germ cell) followed by the damaged germ cell becoming part of a fertilized egg results in the damage being replicated in every cell of the developing embryo, with potentially crippling or lethal effects. Lead aprons are worn during routine medical X-rays to avoid such damage. Mutations in germ cells in earlier times created the hereditary problems now termed genetic diseases, including such difficulties as phenylketonuria (PKU) and sickle cell anemia.

Exposure to either radiation or chemical mutagens is more serious in women than in men, because a woman carries the lifetime supply of egg cells in her ovaries. By contrast, men generate a new set of sperm cells in a period of perhaps 2 weeks. Hazard from exposure is therefore short term in men, but lasts for the entire period of fertility in women.

Exposure to mutagens is also particularly serious during pregnancy. The developing child displays very rapid cell growth and frequent cell division, each cell in early pregnancy giving rise to vast numbers of cells by the time the child is born.

Because women were at greater risk than men in some environments, some companies adopted a policy restricting work in those areas to men. Such policies have been challenged in court, however. In 1990 California courts concluded that such policies violated the Fair Employment and Housing Act of that state. Then in 1991 the U.S. Supreme Court decided that such policies applied to women of child-bearing age were discriminatory, violating Title VII of the Civil Rights Act of 1964.

Carcinogens

The cells of tissues and organs in the bodies of adults are a stable, constant population under normal circumstances. These cells, however, are not immortal, and as a cell dies it is replaced by the division of a nearby cell. Similarly, in the event of an injury, cell division begins to repair the damage and stops when the repair is complete. Clearly, cell division is regulated and occurs only when needed. Occasionally this system of regulation breaks down and cells begin to divide when it is not necessary. This leads to the growth of a cell mass, a tumor or neoplasm, that is not part of the normal pattern of the tissue.

Not all tumors are cancer. Warts, for example, are common examples of tumors. People can develop warts and wear them for most of their lives with only cosmetic damage. Such tumors are termed benign. A benign tumor in the wrong location can be a serious medical problem. For example, the brain, being housed in a rigid box, cannot tolerate a mass of tissue that continues to grow.

Benign tumors threaten us most when they change character and become malignant (cancerous). Unrestrained growth, accompanied by the tendency to metastasize (to spread to other organs and tissues), characterizes a malignant tumor.

[2] Estimates indicate hundreds of thousands of occurrences of damage per cell per day.

Cancer is initiated by a number of different stimuli, including radiation, certain chemicals, diet, or certain viral infections. Individuals may inherit a predisposition to a specific type of cancer; for example, breast cancer often runs in families. The relationship between cancer and specific chemicals (carcinogens) was not always recognized. The original observations were made in the 18th century by a British physician, Sir Percival Potts, who noted that young chimney sweeps were prone to scrotal warts, some of which later transformed into malignant tumors. Suspecting that something in the soot that these young men wore almost as a badge of office caused the warts, he proposed that bathing after work should be a regular practice of the sweeps. The subsequent success of his proposal established the beginning of our present understanding of the relationship between cancer and chemicals. General recognition of the hazard carcinogens present and the development of usable methods to detect carcinogenic potential in a chemical followed slowly, so only more recently has such testing become a routine procedure in toxicology.

Even relatively heavy exposure to a carcinogen does not ensure tumor development. Consider octogenarian lifetime cigarette smokers. Furthermore, of two people exposed to a carcinogen, the one with lighter exposure may be the one to develop cancer. There are differences in susceptibility to the disease based on age, sex, and genetic makeup. However, when surveying a large population, greater exposure to a carcinogen increases the likelihood of the occurrence of cancer.

NAMING TUMORS

Neoplasms are named in a somewhat orderly fashion. In benign tumors we add the "-oma" ending to a derivative of the name of the cell type. Tissues derived from epithelial cells are termed either adenomas (glands) or papillomas (skin), whereas those derived from mesenchymal tissues include osteomas (bone), chondroma (cartilage), lymphangioma (lymph glands), glioma (nervous system glial cells), and so on. Were these benign neoplasms to become malignant, the name would be changed by modification based on the fetal origin of the cells that have transformed (become tumorous). Epithelial cell-based tumors convert the "-oma" to "-carcinoma"; thus an adenoma becomes an adenocarcinoma. Those derived from mesenchymal cells convert the "-oma" to "-sarcoma". Hence osteoma becomes osteosarcoma.

Recently researchers uncovered numerous clues to understanding the causes of cancer. However, the understanding is not complete, and tomorrow's discoveries may change today's concepts. Presently we view cancer as being generated in a multistage process. Initiation involves mutagenic damage to DNA. Certain genes termed oncogenes[3] are key targets for this damage. Oncogenes code for proteins that regulate cell division. If the damage is not repaired before cell division occurs, that damage is passed on to the daughter cells, and we have a population of cells that has passed the first stage of becoming neoplastic.

Additional stages of progression toward malignancy involve additional genetic damage. Once a population of cells takes the first step, they do not spontaneously revert to normal. The likelihood of the additional damage occurring is proportional to the size of this cell population. There are chemicals called promoters that encourage proliferation of the initiated cells. Such chemicals do not alter normal cells, so by themselves they do not initiate new tumors. However, once initiation has happened, by increasing the number of initiated cells they increase the likelihood of the next stages in the progression toward cancer. Animals dosed just with a promoter do not develop tumors as a result of that, and if dosing with a

[3] "Onco-" is a prefix applied to aspects of cancer.

promoter precedes dosing with an initiator, there is no increase in tumor production over that which the initiator produces by itself. If exposure to the promoter follows exposure to initiator, however, the number of tumors produced in the test population increases.

Primary carcinogens are chemicals that cause mutagenic damage leading to cancer upon contact with the tissues. Procarcinogens, or secondary carcinogens, do not cause cancer in the chemical form entering the body. They are metabolically converted in the body into active causative agents for cancer.

Carcinogens display a degree of target organ specificity. One compound induces bladder tumors, another causes growths in the colon. Benzene, an important industrial solvent, generates leukemia, marked by excessive white blood cell proliferation. This specificity may be the result of the manner of exposure, as in the generation of lung tumors by smoking. Target organ specificity could also relate to the tissue most likely to activate a procarcinogen. In other cases a compound always targets a particular organ. The herbicide paraquat targets the lungs regardless of the path of entry to the body.

REPRODUCTIVE TOXINS

Teratogens

Teratogens, like mutagens, cause offspring to be born with abnormalities. We typically see skeletal faults such as improperly formed limbs or vertebral column, soft tissue flaws such as cleft palate, heart malformation, closed segments of intestinal tract, or nervous system imperfections. Unlike mutagens, exposure of the male or female parent to teratogens before conception is of no significance. In early pregnancy the various fetal organs are first formed, the period of organogenesis. The teratogen interferes with proper organ development if a pregnant woman is exposed to a teratogen at that time.

Concern about teratogens is a recent phenomenon. Rubella (measles) infections during pregnancy had long been associated with birth defects, but strictly chemical agents were not generally of concern until 1960 when the drug thalidomide, a potent teratogen in humans, was widely prescribed to pregnant women in Europe and Japan. The defect frequently caused by thalidomide was malformation, even to the point of complete absence, of arms and legs. The U.S. had not yet approved use of thalidomide, sparing that population from this catastrophe.

A CHEMICAL CAUSING MALE INFERTILITY

When male shipping room workers at an agricultural chemicals plant in California became aware of their common fertility problems, investigation uncovered the potential of dibromochloropropane, a chemical used to kill worms in soil, to inhibit sperm production. This case alerted toxicologists to the potential for infertility from workplace exposures, and led to research directed at detection of other problem chemicals.

Toxins Affecting Fertility

Damage to reproductive potential is a special type of systemic toxicity. Here the target organ is the male or female reproductive system. Processes essential to reproduction such as sperm production, normal ovulation, and implantation of the fertilized egg can be affected by exposure to chemicals. Interference with these functions does not produce illness or death either in test animals or humans, and is therefore less readily detected. At this time, however, most chemicals have not been tested for ability to cause infertility.

NEUROTOXINS

The nervous system divides into the central nervous system and the peripheral nervous system. The central nervous system includes the brain and the spinal cord, which connects to the brain at the lower, back end and runs down the hollow center of the vertebral column. The nerves enter and leave the spinal cord through openings at the sides of the vertebral column, and their connections to all parts of the body comprise the peripheral nervous system.

To understand the components and functioning of this system, visualize the response of a gardener bitten by an insect. The insect bite stimulates sensory cells in the skin. These sensory cells convert the stimulus into a pattern of nerve impulses, moving potential changes at the membrane surface of the connecting nerve cell, which is the manner by which the nervous system transfers information. Such a message moves the length of the nerve cell. At the end of the cell the signal transfers to the next nerve cell in the pathway through a synapse. Synaptic transmission involves the release of a chemical called a neurotransmitter from the end of the first cell, diffusion of that chemical across a very small gap between the cells, and binding of the chemical to a receptor on the second cell. The receptor then initiates a nerve impulse in the second cell. In this fashion the message moves from the sensory cells through the nerves of the peripheral nervous system, to a synaptic connection with the spinal cord. The message now moves up the spinal cord to a location in the brain intended to receive this sort of signal. In the brain the signal is passed around a series of nerve pathways that interpret its meaning and determine the correct action. This involves the use of many synapses, and results in the generation of nerve impulses in appropriate motor (muscle-activating) systems. These impulses travel down the spinal cord, exiting at the correct opening in the vertebral column. The message is delivered to the correct muscles, again through a synapse, and the muscles contract to cause the hand to swat the insect.

Functions of the brain go far beyond the simple sting and swat response described above; they generate the whole picture of thought, memory, personality, decision making, emotional response, and more that produce the highly evolved human organism. Scientists do not yet understand all the complexities of this system, but they do understand that it is based on patterns of nerve impulses traveling through cells and across synapses.

A system this complex can be affected by chemicals at a number of points. Most often chemical interference is at the synapses, and can involve any part of the process of release or reception of the neurotransmitter, or the generation of response to its reception. A broad variety of neurotransmitters is found, each with its own chemical peculiarities, and an extended pathway such as the insect bite to swat described above involves several of them. Obviously many different chemicals could alter the functioning of this system.

As the most sensitive and least understood human organ, the brain presents yet another type of difficulty to toxicologists. In the sense that one judges systemic poisons by the presence or absence of visible damage to organs such as liver or kidney, one could look for cell death or tumors appearing in the nervous system following exposure to a chemical. However, this would take into account only one aspect of the possible damage chemicals could cause to the human nervous system. The very sensitive balance maintained between the various signals in the brain that represents normal human behavior can be disrupted by the entry of foreign chemicals without a single cell dying or being transformed. Such an alteration could cause mood and personality changes, loss of intellect, or memory impairment. These same changes might also occur without associated chemical exposure. As a result, changes resulting from a toxic episode might be attributed to another cause, and not stimulate the search for a neurotoxin.

The situation in the central nervous system is complicated by the blood–brain barrier, a term describing the fact that many substances cannot be transferred from the blood into the brain as easily as into most other organs. The effectiveness of this barrier varies among individuals and depends greatly on age. Fetuses and neonates (newborn) in general do not have an effective barrier. This means that potentially harmful substances circulating in the blood of a pregnant woman are blocked from entering her central nervous system by the blood–brain barrier. However, if they transfer from the mother's to the child's blood in the placenta, they might freely enter the unborn child's brain. Variability of the blood–brain barrier among individuals adds further randomness to the already difficult task of evaluating potential neurotoxins.

Testing a chemical with animal surrogates for its potential to cause harm in the central nervous system presents special difficulties. If the toxin produces cell death or stunted development, the animal displays these physical changes. However, how does animal testing assess damage to intelligence, effects on sanity, or alteration of behavior? Some progress has been made in these areas,[4] but the obstacles to full understanding are imposing.

OTHER TOXIC EFFECTS

In addition to substances that cause serious damage, the worker must be guarded against exposure to substances that are simply irritating to the skin, eyes, or respiratory tract. Characterization by animal tests is generally easily performed, and, because we are not concerned with lethal consequences, a degree of experimental human exposure is sometimes an acceptable option.

Clearly the establishment of a profile of the hazards of a chemical in the workplace requires a variety of types of testing and a complex array of decisions. The results of this decision making is a set of standards for the upper limits of acceptable worker exposure.

KEY POINTS

1. All chemicals are toxic at sufficiently high concentrations. Toxicologists determine safe levels.
2. Chemicals cause a variety of toxic effects and are classified by terms such as systemic toxins, mutagens, carcinogens, teratogens, and behavioral toxins.
3. Toxic damage may be immediate or may require an incubation time.
4. Systemic toxins interfere with normal cell function, usually most severely in a specific organ or tissue.
5. DNA carries the genetic information of the organism as a sequence of monomers on a very large polymeric molecule. Higher organisms such as humans have more than one molecular type of DNA, and have two copies of each type, one from each parent.
6. DNA is a double-stranded helical structure, and the sequence on one strand correlates with that on the other in a fashion termed complementary. One strand holds the genetic information and the other is useful when duplicating DNA at the time of cell division or when repairing damage.
7. Mutagens damage DNA monomer sequences. When the damage is to a sperm or egg cell, which then contributes to forming a fertilized egg, the damage is copied in every cell of the developing fetus.
8. Carcinogens increase the likelihood of cancer occurring. Most carcinogens are also mutagens.
9. Teratogens damage a fetus during development, producing birth defects.

[4] J. P. J. Maurissen and J. L. Mattson, "Neurotoxicolgy: An Orientation," in L. J. Cralley, L. V. Cralley, and J. S. Bus, Eds., *Patty's Industrial Hygiene and Toxicology*, 3rd ed., Vol. 3, Part B, John Wiley & Sons, New York, 1995.

10. Some toxicants interfere with sperm production in men or in the processes involved in ovulation or implanting a fertilized egg in women, reducing levels of fertility.
11. Compounds that affect the nervous system, neurotoxins, produce complex effects that are sometimes difficult to assess.

BIBLIOGRAPHY

N. A. Ashford and C. S. Miller, *Chemical Exposures,* Van Nostrand-Reinhold, New York, 1991.
D. Brusich, *Principles of Genetic Toxicology,* Plenum Press, New York, 1980.
I. R. Danse, *Common Sense Toxics in the Workplace,* Van Nostrand-Reinhold, New York, 1991.
B. D. Dinman, *The Nature of Occupational Cancer,* Thomas, Springfield, IL, 1974.
L. Fishbein, *Potential Industrial Carcinogens and Mutagens,* Elsevier, New York, 1979.
W. G. Flamm, *Mutagenesis,* Hemisphere Publishing, New York, 1978.
G. Flamm and M. Mehlman, Eds., "Mutagenesis," in *Advances in Modern Toxicology,* Vol. 5, Halsted Press, New York, 1978.
Health and Welfare-Canada, *The Testing of Chemicals for Carcinogenicity, Mutagenicity, and Teratogenicity,* Ottawa, 1975.
M. Kirsch-Volders, Ed., *Mutagenicity, Carcinogenicity, and Teratogenicity of Industrial Pollutants,* Plenum Press, New York, 1984.
P. Lehmann, *Cancer and the Worker,* New York Academy of Science, New York, 1979.
R. J. Lewis, *Hazardous Chemicals Desk Reference,* 3rd ed., Van Nostrand-Reinhold, New York, 1993.
R. J. Lewis, *Rapid Guide to Hazardous Chemicals in the Workplace,* 3rd ed., Van Nostrand-Reinhold, New York, 1994.
R. J. Lewis, *Sax's Dangerous Properties of Industrial Materials,* 8th ed., Van Nostrand-Reinhold, New York, 1992.
E. Meyer, *Chemistry of Hazardous Materials,* Prentice-Hall, Englewood Cliffs, NJ, 1990.
J. L. O'Donoghue, *Neurotoxicity of Industrial and Commercial Chemicals,* CRC Press, Boca Raton, FL, 1985.
P. Patnaik, *A Comprehensive Guide to the Hazardous Properties of Chemical Substances,* Van Nostrand-Reinhold, New York, 1992.
N. I. Sax, *Cancer Causing Chemicals,* Van Nostrand-Reinhold, New York, 1981.
T. H. Shepard, *Catalog of Teratogenic Agents,* 3rd ed., Johns Hopkins University Press, Baltimore, MD, 1980.

PROBLEMS

1. A wise early investment for a person in training for a career in industrial hygiene is a medical dictionary. There is a large overlap between industrial hygiene, toxicology, and occupational medicine, so many industrial hygiene references use medical terminology with which lay persons and even many science majors are not familiar. The following are direct quotes from an excellent but sophisticated toxicology reference. Locate a suitable dictionary and translate these statements into less technical language, then describe in terms of the major topics of this chapter what sort of toxicant (systemic, carcinogen, etc.) each compound is.
   L. S. Andrews and R. Snyder, "Toxic Effects of Solvents and Vapors," in C. D. Klaassen, M. O. Amdur, and J. Doull, Eds., *Toxicology: The Basic Science of Poisons,* 3rd ed., Macmillan, New York, 1986.
   A. "The renal histopathology, after 6 to 13 days, varied depending on the amount of ethylene glycol consumed. At high dose levels, deposition of calcium oxalate crystals occurred in the proximal renal tubules, and necrotic areas of tubule epithelium occurred adjacent to the crystals"
   B. "Thus, bromobenzene . . . yields massive hepatic necrosis in animals pretreated with phenobarbital."

C. "In humans the adverse effects of benzene are variants of either aplastic anemia or leukemia. It is likely in each case that metabolites of benzene initiate the disease process and also appear to be implicated in mutational events. . . . "

D. "Reports of hematologic effects or peripheral neuropathies (from glue sniffing) are more likely due to the use of glues containing benzene or hexane."

E. "Although there were no indications of maternal toxicity, 100 percent and 50 percent of the fetuses died in the 200- and 100-ppm exposure (to ethylene glycol monomethyl ether) groups, respectively. Cardiovascular and skeletal malformations were increased above unexposed control values in both 50- and 100-ppm exposure groups."

F. ". . . in both rats and rabbits exposed to 300 ppm; every male exhibited degeneration of the testicular germinal epithelium. At the end of the exposure period, male rats were mated to unexposed females and were found to be infertile."

G. "Guinea pigs demonstrated ... hyperplasia of lung cells, as well as squamous cell carcinoma in the nasal cavity."

2. Let's understand DNA function and mutation a little better. Below is a representation of a segment of DNA, the sense strand. Below that is a translation of the genetic message, indicating which amino acid (as a three-letter abbreviation) each three-base sequence designates. (People who have had a biochemistry class will recognize that this is not the genetic code as presented in texts, which translates codons on RNA into amino acids and reads in the opposite direction.

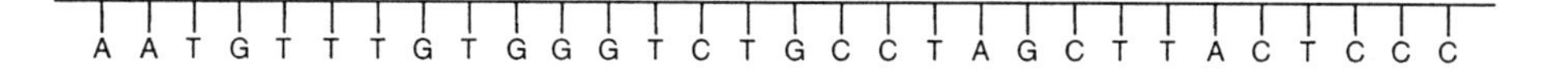

	A		G		T		C	
A	AAA	Phe	AGA	Ser	ATA	Tyr	ACA	Cys
	AAG	Phe	AGG	Ser	ATG	Tyr	ACG	Cys
	AAT	Leu	AGT	Ser	ATT	STOP	ACT	STOP
	AAC	Leu	AGC	Ser	ATC	STOP	ACC	Trp
G	GAA	Leu	GGA	Pro	GTA	His	GCA	Arg
	GAG	Leu	GGG	Pro	GTG	His	GCG	Arg
	GAT	Leu	GGT	Pro	GTT	Gln	GCT	Arg
	GAC	Leu	GGC	Pro	GTC	Gln	GCC	Arg
T	TAA	Ile	TGA	Thr	TTA	Asn	TCA	Ser
	TAG	Ile	TGG	Thr	TTG	Asn	TCG	Ser
	TAT	Ile	TGT	Thr	TTT	Lys	TCT	Arg
	TAC	Ile	TGC	Thr	TTC	Lys	TCC	Arg
C	CAA	Val	CGA	Ala	CTA	Asp	CCA	Gly
	CAG	Val	CGG	Ala	CTG	Asp	CCG	Gly
	CAT	Val	CGT	Ala	CTT	Glu	CCT	Gly
	CAC	Val	CGC	Ala	CTC	Glu	CCC	Gly

Proteins are long polymer chains in which all the monomer units (amino acids) are in a specific order. The characteristics of the protein are determined by the amino acid sequence. In an individual every protein molecule serving a particular function (for example, hemoglobin in red blood cells carrying oxygen) may well have exactly the same sequence. At most there will be two sequences, one from the person's father and one from the mother.

A. First, look over the table. STOP indicates the end of the particular protein chain being assembled. The rest of the DNA sequences are associated with a specific amino acid. Thus the appearance of CGG in the DNA sequence says the amino acid Ala (alanine) belongs next in the protein molecule. How many different amino acids may be incorporated into protein?

B. Write the protein sequence this DNA sense strand describes. Recognize that real proteins have typically hundreds of amino acids, so this is an unrealistically short message.
C. Suppose the 11th base from the left is altered by ionizing radiation so it is now read as A. What happens to the protein sequence?
D. Suppose the 18th base from the left is altered chemically to read as T instead of C. What happens to the protein sequence? Why is this termed a silent mutation?
E. Suppose the 4th base from the left is altered chemically to read as A instead of G. What happens to the protein sequence?
F. Suppose the 27th base from the left (4th from the right) is altered chemically to read as C instead of T. What happens to the protein sequence?
G. All of the above are called point mutations because a single base is changed to another. A base could also be deleted by mutagenic action. Suppose the 4th base from the left were deleted. What happens to the protein sequence?
H. The example in (G) is termed a frame shift mutation. Why? Sometimes when a point mutation changes a single amino acid in the protein sequence, the protein can still perform its designated function because the particular amino acid is not critical to protein function. Will this happen with a frame shift mutation?
I. Write the DNA sequence of the complementary strand to this sense strand using A with T and G with C as the complementary pairings described in the chapter.
J. If the 5th base from the left were damaged by a mutagen, how would repair enzymes know how to fix it?

3. A workplace discrimination issue has arisen because women have been banned from certain jobs because the chemicals involved pose special risks to women. Chemicals displaying what toxic characteristics might generate such an issue? Can you provide examples of jobs from which men might be banned?
4. Explain why a man's exposure to a teratogen before conception would not cause the child to develop malformations. If it did what would that say about the teratogen?

CHAPTER 4

Measuring Relative Toxicity and Assessing Risk

In 1983 the National Research Council published an analysis of the process of controlling risk due to exposure to adverse agents.[1] Four steps were described for risk assessment: (1) recognition of hazard, (2) toxicity testing, (3) estimation of exposure, and (4) decision for management of the risk. This chapter examines aspects of these questions.

RECOGNITION OF HAZARD — THE DECISION TO TEST

When is toxicity testing undertaken? At one time the safety of a chemical was questioned only after exposed workers became ill or died. An alternative that anticipates problems was desired. This is especially important at this time when new chemicals are entering the workplace at an unprecedented rate. One criterion for deciding to test is recognition that exposures, especially on a large scale, are taking place or will soon take place.

TOXICITY TESTING

What are the techniques used to gather information about the health effects of chemicals? Both the method of testing and the reliability of the results are of concern. How do scientists communicate the findings once they have studied toxic effects — what is the language used to express relative toxicity? In the process of answering these questions, terminology unique to toxicology that industrial hygienists need to understand is introduced.

TOLERANCE TO TOXIC SUBSTANCES

The general public often believes that any exposure, however small, to a substance characterized as toxic is lethal. The body can tolerate exposure to toxic substances, as long as the level is reasonable. This has always been necessary to human survival, since contact with toxicants is unavoidable. Plant materials in our diet, for example, contain a variety of substances that are consumed along with the protein, starch, lipids, vitamins, and minerals that constitute nourishment. Many of these substances would seriously threaten health if taken in large quantities. After all, the chemistry of the plant serves the needs of the plant, not those who plan to kill and eat it. Some plant species produce toxic substances as an aspect of

[1] National Research Council, Committee on the Institutional Means for Assessment of Risks to Public Health, *Risk Assessment in the Federal Government: Managing the Process*, National Academy Press, Washington, D.C., 1983.

survival; for example, locoweed produces a compound neurotoxic to grazing animals. Safe food need not be completely free of toxicants — such food may not exist. It is necessary only that the toxicants are at a low enough level that they clear from the body and problems they caused are repaired before the person takes in any more such toxicants. Toxicologists determine what levels of exposure are safe for the target population, those who are exposed, then set standards to ensure that these levels are not exceeded.

THE TARGET POPULATION

In order to set realistic exposure guidelines, first consider the target population. Some branches of toxicology are concerned with hazards to fish, birds, or perhaps every living human being. For example, the target population when setting standards for the quality of urban air includes persons of every age and state of health. In this case, a level of carbon monoxide that can pose a threat to an elderly anemic individual with emphysema is of concern.

Industrial hygiene deals with workplace exposures. This translates into concern for a group of male and female adults roughly 18 to 70 years of age who are in reasonably good health. Problems relating to the health of a fetus are pertinent, since the workforce includes pregnant women. However, special problems associated with the newborn, the very old, or those with serious health disabilities are not a prime focus. The allowable levels of carbon monoxide in the plant atmosphere might be higher than would be recommended in urban air.

LENGTH OF EXPOSURE TIMES

The length of time the worker is in contact with the chemical is important. A dishwashing product that contains a detergent and ammonia might be perfectly safe for household use, given that exposure only occurs when the dishes are washed, but might be a serious health threat to a person whose skin is continuously exposed on the job. Urban air standards assume that residents breathe that air 24 hr/day 7 days/week. Exposure at the workstation occurs for an 8-hr day and a 40-hr week.

ANIMAL TESTING

How is toxicity tested in that target population? How would one determine how fast horses can run? First, one would obtain a device for measuring how fast an animal is running. Then one gets a horse and convinces it to run. Better yet, since the original horse is anywhere on a continuum from being a potential triple crown winner to being a lame and hopeless nag, the experiment would be repeated with a number of horses and the values obtained averaged. Would it make sense to conduct the experiment with mice instead, measuring how fast a series of mice ran? All sorts of arguments are immediately presented: mice have very short legs, mice move very quickly for their size (metabolic rate maybe?), mice don't have hooves, and so on. Even if the data were adjusted by saying a horse is 100,000 times larger than a mouse, it would not follow that it should run 100,000 times faster.

As solid as these arguments are against using mice to gather information about the running of horses, extending the concept — "predictive testing should not be done on dissimilar animals" — to all types of testing should be done cautiously. If the information sought concerns the potential of a new industrial chemical to harm workers, how does one obtain the answer?

> VOLUNTEERS FOR EXPOSURE TO THE NEW CHEMICAL WILL PLEASE FORM A LINE TO THE RIGHT AND LEAVE THEIR NAMES AND PHONE NUMBERS WITH THE SECRETARY.

Although it is true that data concerning toxicity to humans is most valid when exposure is to humans, people retreat from actually running such tests on themselves and seek an alternative procedure. Now maybe the mice can do the job for them.

In recent years the use of animals in research has been strongly, sometimes even violently, opposed by animal rights groups. Extremists in these groups have raided laboratories, destroying facilities and records that have set important research back years, and have caused other facilities to be turned into fortresses in an attempt to prevent further such destruction. The actions of such people give a movement that includes reasonable individuals with valid concerns a bad name.

Positive results of animal rights pressure groups include formulation and enforcement of standards for the care and housing of research animals that has eliminated some abuses of the past. It has stimulated thinking about alternatives to the use of test animals as surrogates or substitutes for humans, and as a result some new testing methods have been developed. For example, to study the effects of a chemical on a specific tissue, we may raise isolated cells of that tissue in a tissue culture medium. For example, a colleague of the author studies the effect of chemicals on sperm production, using cultured sertoli cells (the sperm producing cells) as a test system. Similarly, screening tests to determine if a compound is a mutagen are now done quickly and cheaply using special strains of bacteria in the Ames test.

However, for some testing, for example, an initial broad screening for any kind of systemic toxicity, the use of test animals is still necessary. A favorite cartoon depicts a physician sitting at the bedside of a patient. The physician has a cage containing a small animal in his lap and a hypodermic needle in his hand. The patient, in great agitation, is saying, "You mean you're just going to test it on a Guinea pig now?"

If the use of animal surrogates is necessary, an appropriate species must be selected. It is then important to obtain meaningful values for the toxicity of the substance to these test animals, and postpone concerns about using these data to predict toxicity to humans.

What Animal Shall Be Used?

Most toxicological data is collected using small rodents (rats, mice, Guinea pigs) as surrogates for human beings. There are numerous advantages to using such animals. They are small and relatively easy to maintain in good health in the laboratory. With a life expectancy of typically 2 years, one can see the effect of a lifetime of exposure in a relatively short time. They propagate rapidly. Because they have been used extensively in the past, scientists are familiar with their anatomy and physiology, their normal and abnormal states of health, and their response to a large number of other substances.

First, a species of animals is chosen, say Sprague-Dawley white rats. Notice that here the choice is more restrictive than using just "rats". Pure strains of laboratory rats have been developed and maintained for decades. They are bred to minimize differences between individuals. There are many varieties of rats, and experience teaches that there can be variation in response to a chemical among these varieties. In the hope of obtaining reproducible results, a single strain is selected that has been bred genetically pure for many years. To whatever degree it is possible for individuals to respond uniformly, it will happen with these rats. Furthermore, by using this variety repeatedly much can be learned about its characteristics, and scientists become practiced in the transfer of toxicity information into predictions of human response.

It is an important lesson that even in this population the exact dose to produce a particular effect may well be different for every rat. One would not expect the members of the human population, who are far from genetically identical, each to respond in the same fashion to a chemical. However, if a more reproducible response is obtained from the test group, and if they are studied well enough that one understands many of their differences from humans, the extrapolation of the animal data to humans is more dependable.

Extrapolating Animal Data to Humans: The Size Difference

In spite of all these advantages, using mice or rats to predict toxicity to humans involves many risks. In the estimation of the speed of a horse, the rat or mouse is not a miniature horse. Neither in measuring the toxicity of a chemical is a rat or mouse a miniature human. There are differences in metabolism, dietary patterns, basic anatomy, and size between mice and humans. As an example, if something is eaten by a human that then causes nausea, the human vomits it. Rats cannot vomit, so that same material would remain in the rat stomach. If the compound were toxic at the level given, it would be much less threatening to the human, who would remove it from the stomach automatically.

The difference in size between humans and small rodents is a major problem. The observation that 1 mg of a compound makes a white rat ill does not at all mean that the same 1 mg would make a human ill; in fact, it would be surprising if it did. Spreading that 1 mg around a 250-g rat is altogether different from diluting it into a 70,000-g human.

This difference is handled by employing units that indicate the weight of the compound given to the animal divided by the weight of the animal. Most commonly doses are listed in milligrams of compound given the animal per kilogram of animal body weight. Thus the dose of 1 mg per 250-g rat is a dose rate of 4 mg/kg. One assumes that the same dose rate should affect the human, and that 280 mg would have roughly the same effect on a 70-kg (154-lb) adult human. Because of differences between rats and humans this is almost certainly not exactly true, but it does provide a useful first approximation of toxicity to humans.

SAMPLE PROBLEM:

A. 25 mg of a narcotic substance under study causes a 200-g rat to sleep for 1 hour. What is this dose rate in mg/kg?

$$\frac{25 \text{ mg}}{200 \text{ g}} \times \frac{1000 \text{ g}}{1 \text{ kg}} = \frac{125 \text{ mg}}{1 \text{ kg}}$$

B. As a first approximation, what dose would be needed to produce the same effect on an 80-kg human?

$$\frac{125 \text{ mg}}{1 \text{ kg}} \times 80 \text{ kg} = 10,000 \text{ mg} \times \frac{1 \text{ g}}{1,000 \text{ mg}} = 10 \text{ g}$$

C. What dose would be predicted to put a 31-kg dog to sleep?

$$\frac{125 \text{ mg}}{1 \text{ kg}} \times 31 \text{ kg} = 3,875 \text{ mg} \times \frac{1 \text{ g}}{1000 \text{ mg}} = 3.9 \text{ g}$$

For reporting the toxicity of airborne toxicants, no adjustment is necessary for the ratio of animal to human size. One assumes that the rat and the human each breathe an amount of air that is in proportion to their size. The dose is therefore listed in terms of the concentration in the air, with no reference to animal size.

DESIGNING TOXICOLOGICAL EXPERIMENTS

Manner of Dosing

Next, a decision must be made regarding how to dose the animals. There are several pathways for dosing. Most commonly, the test substance is added to the food or water, placed directly into the stomach by gavage (a small tube inserted through the mouth into the stomach), injected into a vein or the peritoneal cavity, coated onto the skin, or added to the atmosphere the animal breathes. The final toxicity values obtained usually differ according to the path of entry into the body. Which path is chosen reflects the following concerns:

1. What will be the manner of exposure of workers in the plant? Whatever it is, it should be replicated as nearly as possible in the testing. A solvent to be used for degreasing parts may make contact with the skin of the worker or its vapors may be inhaled, but it is highly unlikely that the worker would ever swallow any. Testing should concentrate on exposure to the skin and lungs.
2. The technique should provide an accurate measure of the amount of compound taken into the test animal. If the food is dosed, then the amount of food consumed by the animal should be recorded. An advantage of the gavage tube or injection is the high level of accuracy with which the dose level is known.

The actual design of the experiment depends on how the dose is to be administered and what responses are to be studied (irritation of the skin, appearance of cancer, amount constituting a lethal dose, and so on).

Length of Testing Time

Testing time can vary. When the animal is dosed one time orally, by injection, or by coating on the skin, or when the dosing is for a limited time (as for a few hours one time by inhalation), the study is called acute. In such a test, the experimenter determines the level of the compound sufficient to achieve a monitored degree of toxicity. On the other hand, if the testing exposes the animal daily for months or perhaps the animal's lifetime, the study is termed chronic. Chronic testing reveals whether there is an accumulation of either the compound itself or the damage it causes. Knowledge of acute toxicity is always needed since people may have excessive contact with any chemical in their surroundings at any time. Chronic dosing is particularly important when workers contact a chemical every day, to evaluate whether damage accumulates.

Variability Among Test Animals

The toxicity of a compound is not a constant term for all animals. Differences are sometimes displayed between animals of even very closely related species.[2] Males and females sometimes display different degrees of sensitivity to a particular toxicant.[3] Some of the enzymes that change the chemical structure of a foreign substance in the body are produced at levels that depend on body levels of estrogens, female sex hormones. This results in variations in toxic response. Whether the variation is an increase or decrease in relative toxicity is hard to predict. These enzymes convert the foreign substance into a more readily eliminated compound, which would seem to predict that more enzyme activity equates to lower toxicity. However, the more readily eliminated product may also be more toxic.

[2] Comparing three strains of female rats exposed to the same dosage of 7,12-dimethylbenz(A)anthracene: 0% of the Marshal strain, 16% of the Long-Evans strain, and 100% of the Sprague-Dawley strain developed mammary tumors.
[3] For example, a dose of 7,12-dimethylbenz(A)anthracene that generates tumors in 79% of male rats causes tumors in only 18% of female rats.

Age is another important factor. Newborn (neonate) mammals are often more sensitive than adults, and animals of advanced age may show increased susceptibility as body defense mechanisms gradually lose effectiveness. For example, dosing at a constant rate in milligrams per kilogram with a barbiturate kills week-old rabbits, puts 2-week-old rabbits to sleep, and has no effect on 4-week-old rabbits.

Even if species, age, and sex are held constant, the effect of a compound on a group of test animals is still not precisely predictable. Wouldn't it be surprising if each of 10 animals was dosed with the same level of morphine, all fell asleep, then all woke up exactly 32 min later? Think of the differences between people: Joan can't eat raw vegetables, aspirin doesn't seem to work for Ralph, sour cream upsets Betty, I can't take certain antihistamines and drive safely, and those same antihistamines don't bother Bob. Individual test animals show variations in response, but the variations tend to cluster in a statistically predictable way around mean or average values. It would be surprising if 10 identically dosed animals awoke from their morphine-induced sleep in exactly 32 min, but not if the study was conducted twice and the group *averaged* 32 min of sleep each time. Variations can usually be explained in terms of differences in the activity of key metabolic enzymes used to chemically alter the toxicant, of the ease of crossing barriers into particular body compartments such as the brain, and of the effectiveness of the systems used to eliminate the compound from the body. These variations in the response of individual animals to a chemical do not make it impossible to do meaningful toxicity testing, but they need to be remembered when scientists design the experiments and interpret the results.

The experiment must be designed to consider these variations. To minimize complexities the animals used are as nearly identical as possible with respect to age and genetic makeup. Standard strains of test animals are available. Unless we want to focus on male/female differences, we average these out by using equal numbers of each sex. Thus the experimenter may determine reasonable toxicity values for that narrowly defined sample. Extrapolating the result to humans is a separate problem, one based on other parameters.

Now to perform the actual experiments. Suppose we want to determine what level of exposure to a narcotic puts the animal to sleep. Using a few animals and a broad range of dose levels constitutes a range-finding experiment that narrows the range of concentrations to study in a more comprehensive test. Once the approximate dose is estimated, a second experiment utilizes probably five or more concentrations in that range with at least four to five animals per dose level. If the experiment is properly designed, few become unconscious at lower levels, but an increasing percentage do as the dose level is increased.

REPORTING THE RESULTS — TOXICITY UNITS

Once we see how much compound does what degree of damage to what particular animal, the next problem is to communicate that information to all interested parties. For this purpose a whole dictionary of units has been devised. Most of these units are known by letter or letter and number abbreviations.

ACUTE LETHAL DOSE

Traditionally the most basic information, generally the first to be measured, is the value that describes how much compound taken in a single dose is likely to cause death. For compounds taken orally, injected, or absorbed through the skin, dosage is reported as weight of compound, either pure or dissolved in a noninterfering solvent, per unit animal weight. Recently, increasing concern about the use of animals in research has led to reassessment of the importance of this value. However, it is a traditional toxicity value, almost certainly listed for every compound that has been evaluated, and often it is the only value listed. It provides a valuable measure of relative toxicity, for example, allowing comparison of the relative hazard of a compound new to a process with that of a compound it might replace.

That all animals, no matter how carefully selected, do not respond identically to a given dosage now becomes a problem in communication of the findings. Do we select the lowest dose that caused a susceptible animal to die, the highest necessary to kill the most resistant animal, or an average dose? To resolve this, the first step is to prepare a dose–response curve from the data (Figure 4.1). The response, here percent mortality, is plotted on the y axis, and the dose, usually log dose in mg/kg, is plotted on the x axis. At the center of this curve is found the dose that is estimated to be fatal to half the recipient animals. This dose, the acute LD_{50}, is predicted to be lethal to 50% of the animals, and is the most common value used to describe the relative toxicity of a compound.Once the curve is established it is possible to describe other measures of toxicity such as LD_5, the dose lethal to 5% of the animal sample, or LD_{95}, the dose lethal to 95%. LD_{LO}, the lowest value to kill a single test animal, is often listed in summaries of toxicological information. Table 4.1 uses some familiar compounds to illustrate the range displayed by these values, and Table 4.2 describes in everyday terms the relative hazard associated with particular LD_{50} values.

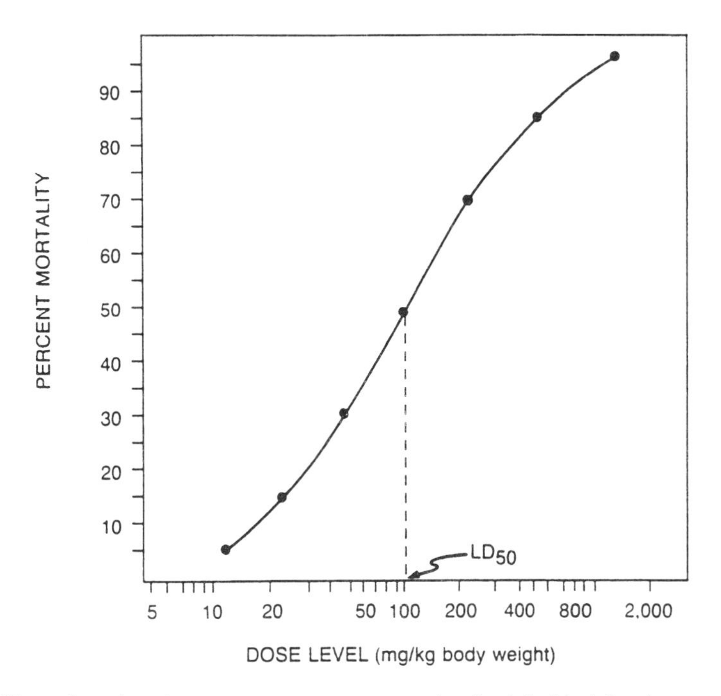

Figure 4.1. Shown here is a dose–response curve expressing the lethal toxicity of a compound. Notice that the dose units are mg/kg body weight. Although the data could be presented as dose or as log dose, a more compact and more easily interpreted plot results when the x axis is log dose, as it is here. The LD_{50} (see text) is determined by finding the intersection of the curve with 50% mortality on the y axis, then dropping a perpendicular line to the x axis from that intersection.

In cases where the dose taken by the animal does not have to be adjusted to the size of the animal (for example, when the intake is through the lungs), the standards are based on the concentration of the substance in the environment of the animal. The toxicity is expressed as the LC_{50}, or concentration lethal to 50% of the test group. The units in this case are usually either parts per million (ppm) or milligrams per cubic meter of air (mg/m^3). For a solid or liquid the "parts" in ppm normally are in terms of weight,[4] but when a gas is discussed the

[4] The milligrams per kilogram used in describing dosage levels is actually milligrams per 1,000,000 mg, a variation of ppm.

Table 4.1. Selected LD_{50} Values.

Compound	LD_{50} (mg/kg, rats, oral)
Glycerol	25,200
Ethanol	10,300
Ethylene glycol	8,500
Acrylic acid	2,600
Hydroquinone	320
Acrylamide	170
Acrylonitrile	93
Nicotine	1
Dioxin	0.001
Botulinus toxin	0.00001

Table 4.2. Toxicity Classes.

Toxicity Rating	Descriptive Term	LD_{50} wt/kg Single Oral Dose in Rats	4-hr Inhalation LC_{50}[a] in Rats (ppm)	Extrapolated Dose (g) for 70-kg Human
1	Extremely toxic	≤1 mg	<10	<0.07
2	Highly toxic	1–50 mg	10–100	0.07–3.5
3	Moderately toxic	50–500 mg	100–1,000	3.5–35
4	Slightly toxic	0.5–5 g	1,000–10,000	35–350
5	Practically nontoxic	5–15 g	10,000–100,000	350–1,000
6	Relatively harmless	≥15 g	>100,000	>1,000

[a] LC_{50} = concentration lethal to 50% of the test group.

general usage is volumes of gas per million volumes of air. Expressing a given gas concentration by weight and by volume does not produce the same number.

PROBIT PLOTS

The probit plot, an important graphical method for presenting dose–response information, is based on standard deviation calculations. Standard deviation is a statistical technique for describing the degree of deviation from the average value a set of data displays, and is described in Appendix A. In a large collection of data displaying a normal Gaussian distribution, 68.3% of the responses fall between one standard deviation (σ) less than and one standard deviation more than the average, 95.5% fall within $\pm 2\sigma$, and 99.7% fall within $\pm 3\sigma$. This translates into the following:

1. The dose that affects 50% of the animals is the LD_{50}.
2. The dose that affects 2.3% of the animals is 2σ below the LD_{50}.
3. The dose that affects 15.9% of the animals is 1σ below the LD_{50}.
4. The dose that affects 84.1% of the animals is 1σ above the LD_{50}.
5. The dose that affects 97.7% of the animals is 2σ above the LD_{50}.

Data may be considered in terms of how many standard deviations the value is from the average rather than in terms of its numerical value. This was first handled by calling the units NEDs (normal equivalent deviations), where zero is the average value and all values less than the average have negative numbers. In order to simplify handling the numbers by eliminating negative numbers, 5 was added to every NED value, so now the average is 5 and 1σ less is 4. These numbers are called probits.

The advantage of this system is that when log dose (x axis) is plotted against probits as the response unit, a straight line is produced (Figure 4.2). Visually estimating the best fit to such a line is relatively easy, and computer programs designed to optimize the fit of a straight line to data can be used. Using probits in dealing with toxicological data sounds complex, but in practice it is greatly simplified by the use of probit paper — special graph paper drawn so that when data is plotted according to the calibration of the axis, all the calculations are handled (just as the semilog paper takes the log of data plotted on the log axis).

THRESHOLD TOXICITY VALUES

The value one might wish to determine next is one that reflects the lowest level producing toxic effects (the threshold). This is needed in order to set standards for allowable maximum exposure. A word of caution is necessary before this discussion proceeds further. Whether considering rats, mice, or humans, there are a few individuals in any population whose response is extreme. These individuals exhibit behavior that statistically is highly unlikely. This could be an unusual display of resistance, and although this is interesting, it is of no value for setting standards. Unusually sensitive individuals also occur for a variety of reasons. Perhaps the animal was born with an abnormally high or low level of activity in a metabolic pathway which chemically alters the toxicant. If the pathway detoxifies the compound, low activity produces greater sensitivity. However, if the dangerous substance is one to which the original compound is converted by the pathway, toxic effects increase directly with the activity of this pathway.

Allergic Response

Most commonly, abnormally high sensitivity results when the individual develops an allergic response to the compound. This may take the form of the production of itchy welts on the skin, of an asthma response where the breathing passages become restricted, or of some other allergic symptom. This is not the normal toxic response observed for the chemical. Once a person is sensitized to the chemical, extremely low concentrations of the chemical produce the symptoms. A sensitized worker normally must be completely isolated from the chemical thereafter. Some chemicals are prone to produce allergy and are termed sensitizers.

Hypersensitive responses are difficult to predict and are a serious problem when attempting to set standards for exposure. In general, standards are based on the normal distribution of responses, rather than on a few hypersensitive reactions. For some chemicals, such as the isocyanates used to make polyurethanes, occurrence of sensitization is so commonplace that the allergic effect becomes the basis for standards.

Animal State of Health

The state of health of the animal is also a factor in response to a chemical on testing. A compound that produces edema (fluid) in the lungs is more deadly to an animal already suffering from a cold or pneumonia. Test animals must be in a good state of health at the start of a test, and symptoms appearing during a test should be the result of exposure to the chemical, not due to an entirely different health problem.

Terms Denoting Threshold Values

Several terms are used to describe a hypothetical dosage that is either the highest level to produce no response or the lowest level to produce a response. These values should be very similar, but often are not because of the variability of animal response and the design

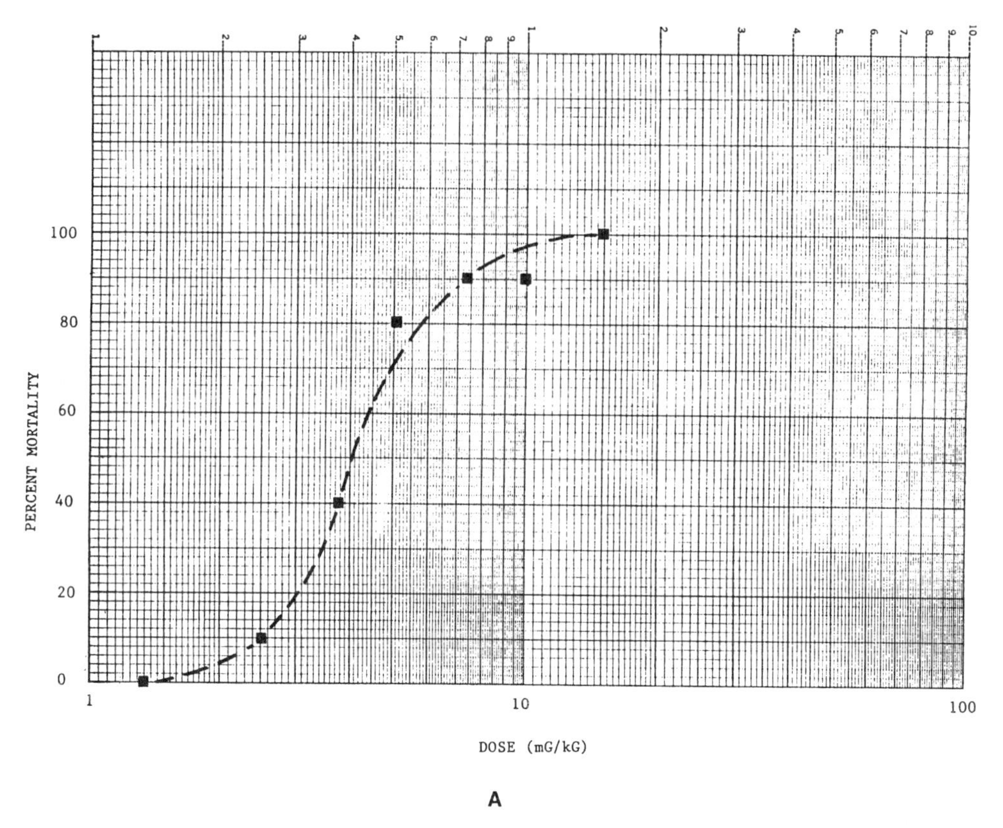

Figure 4.2. Here the same data is plotted two ways. In (A) semilog paper is used, producing the sigmoidal (S-shaped) curve often associated with dose–response studies. Locating a value for LD_{50} is not hard, especially since it is found at the steepest segment of the plot. However, drawing the best sigmoidal curve through the data may present problems. In (B) the same data is plotted on log vs. probit paper. Once again the LD_{50} is easy to locate, but drawing the best line through the data is much easier. Furthermore, it is very easy to extrapolate the line to the x axis to provide a threshold value for toxicity, something difficult to do in plot A.

of the experiment. The lowest dosage to produce a response is commonly referred to as a threshold value, or LD_{LO}. The highest dosage producing no symptoms in test animals generally is termed the NOEL, or no observed effect level. This may also be abbreviated NEL (no effect level) or NOAEL (no observed adverse effect level). Such a value is of great interest when establishing standards for exposure, but would not itself be used as a standard. People should not deliberately expose themselves to levels right at the brink of symptomatic response. It is possible that responses actually occurring at that level are not easily observed.

Multiple Exposures

Up to this point studies have been directed at determination of the relative toxicity of a single chemical. This compound is studied by dosing otherwise healthy animals in a carefully controlled environment that ideally poses no other threats. This is the only way such a study can be done. To assess simultaneously the effect of the chemical and a faulty animal diet or the chemical and a respiratory irritant in the atmosphere would very much complicate the study. One could not determine how much of the blame was actually due to the compound under study.

However, it is the real world that we wish to make safe. In the real world of a factory, we would expect the worker to be exposed to several different chemical or physical agents

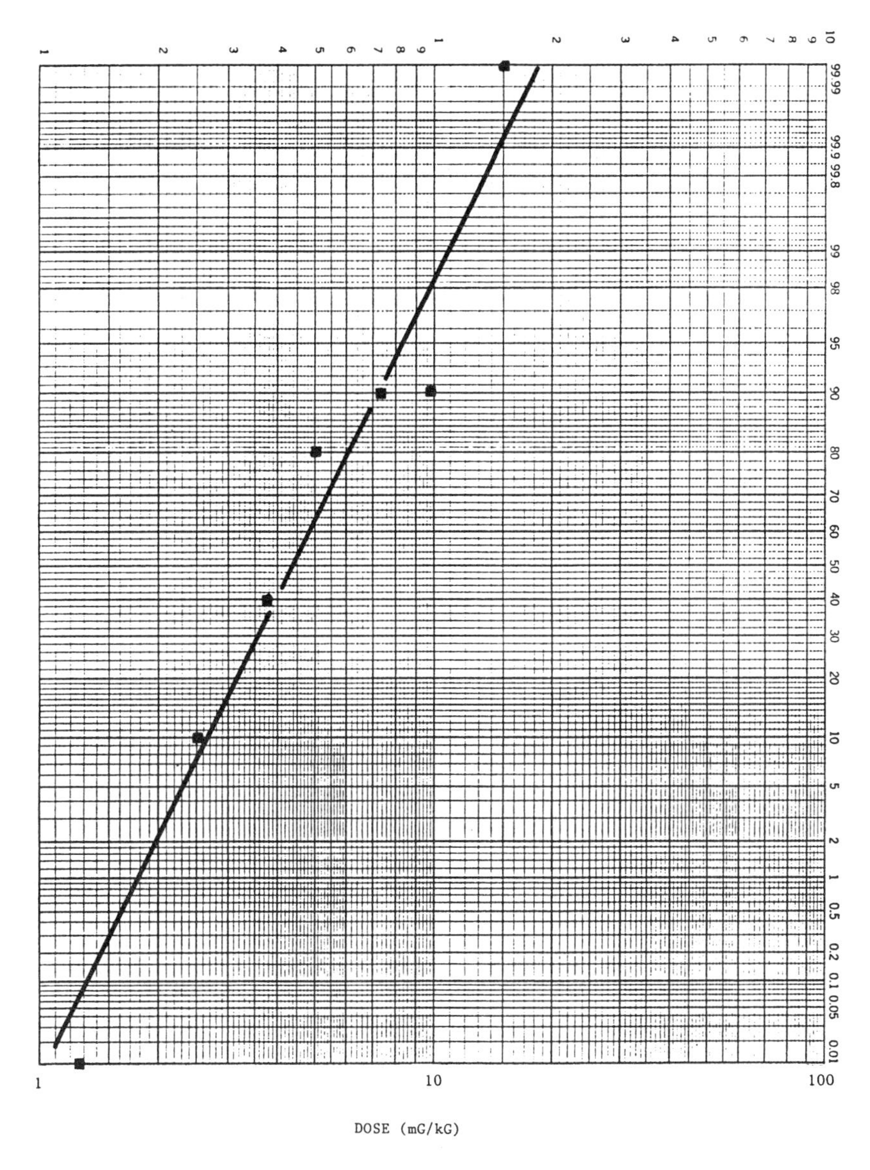

Figure 4.2B.

at once. Suppose we are testing a compound that damages the liver and our dosage is right at the threshold value. The liver, working valiantly to overcome the challenge, just manages to repair the damage as fast as the chemical causes it. The test animal would experience no difficulty. Now we expose the animal simultaneously to a low level of a second chemical that also damages the liver. The liver can no longer repair the damage as fast as it is caused, and starts to deteriorate. However, both chemicals are present below levels at which exposure should cause increasing damage.

If we dosed the animal concurrently at half the threshold concentration of each chemical, and we found this combination dose to be as high as the animal could tolerate without accumulating damage, we would describe the effects of those chemicals to be additive. Notice that now we are dosing with less than the amount tests indicate to be a threshold amount of each chemical. When the two chemicals are present at the same time, we need to revise our exposure standards. Commonly cited examples of chemicals with additive effects are parathion and malathion, members of the class of pesticides termed organophosphates. Organophosphates block motor nerve transmission, and each one added to a dose blocks its share. Similarly, the effect of combinations of certain barbiturates on certain reflexes is found to be additive.

Effects of toxic compound mixtures are displayed on a graph called an isobologram, and from this graph predictions are possible about the effects of mixtures of the two compounds. In Figure 4.3, the x axis is marked in dose units of compound A and the y axis is similarly marked for compound B. We mark the x axis with the ED_{50} (the effective dose or dose sufficient to cause some measure of response to 50% of a sample of animals) for compound A and the y axis with the ED_{50} for compound B. When we connect these two points with a straight line, the graph predicts the effect of mixtures of the two compounds whose toxic effects are additive. Pick a point along the straight line. Read the dose of compound A by dropping a perpendicular line to the x axis and the dose of compound B by drawing a line parallel to the x axis until it intersects the y axis. These two doses jointly administered should be the ED_{50} for the mixture.

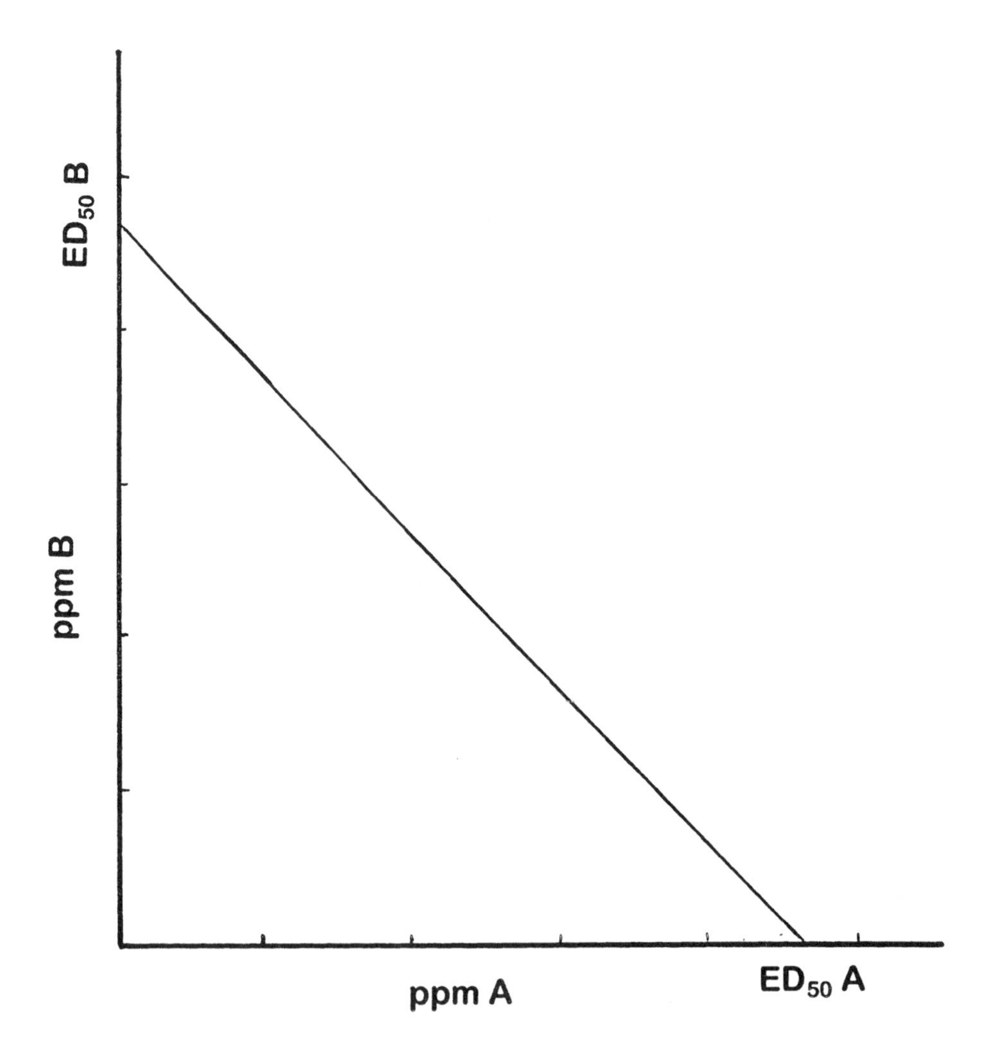

Figure 4.3. Shown here is the isobologram for two compounds (A and B) whose combined effects are additive. Concentration of A is on the x axis and concentration of B on the y axis. The line represents mixtures whose combined effects equal the LD_{50} or ED_{50} (see text for explanation).

When estimating the hazards of concurrent multiple exposures, health and safety specialists often assume additivity of the risks of the individual chemicals. A NIOSH study indicates that this is correct only about half of the time. About a quarter of the time additivity overestimates the true risk, in which case the compounds are termed antagonistic. For example, rats dosed with such chlorinated hydrocarbon insecticides as aldrin, chlordane, or lindane, then given the organophosphate insecticide parathion a few days later are more resistant to the parathion.[5] Mixtures of these two compounds would display toxicities that do not fall on the straight line of Figure 4.3. Experiments done with the mixtures would instead trace out a curved path like the one shown in Figure 4.4.

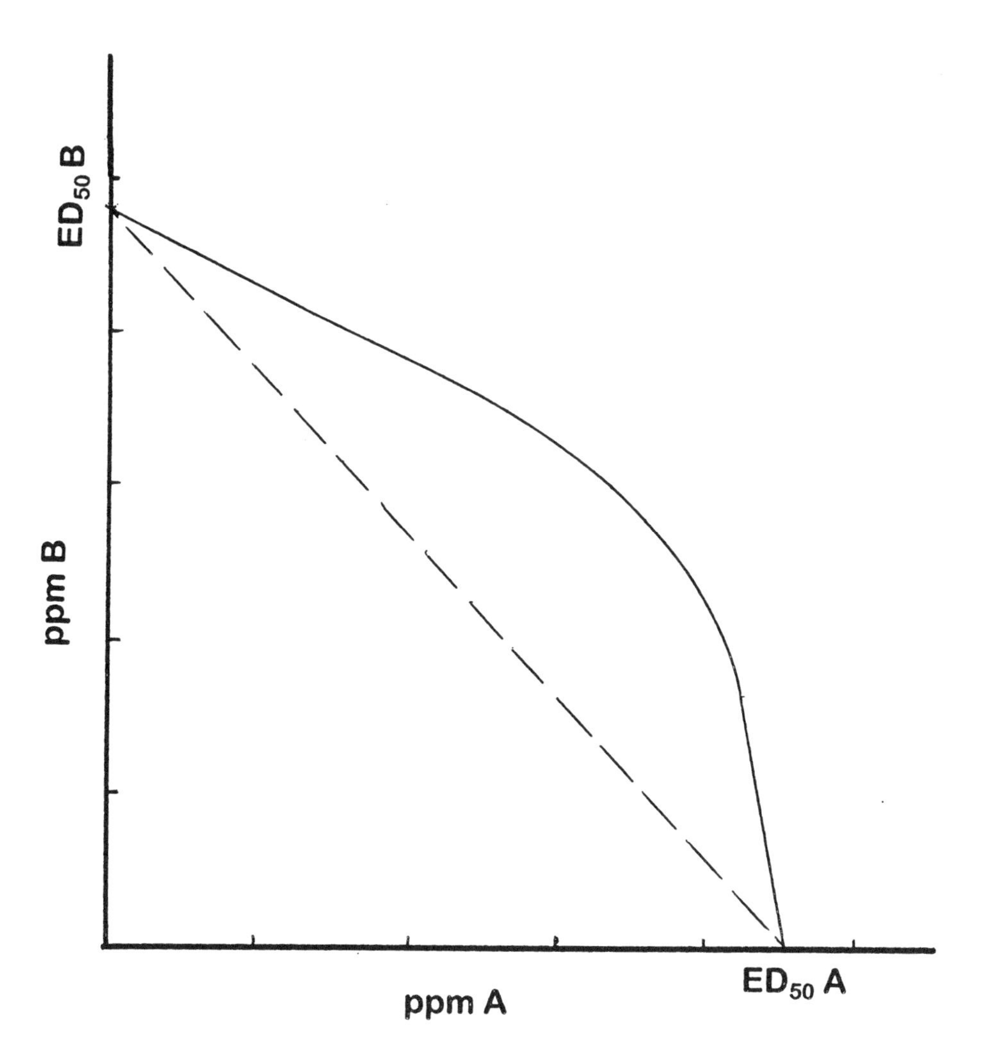

Figure 4.4. An isobologram of two antagonistic compounds. The layout is as in Figure 4.3.

Finally, and most significantly, in the other quarter of the cases a combination of two chemicals is more hazardous than addition of the individual toxicities would predict, an effect called synergism. An often quoted example of synergistic effect is the combination of ethanol and carbon tetrachloride, each of which is a liver toxin. In combination they cause the liver more damage than simple addition of their toxicities would predict. Some organophosphate

[5] W. L. Ball et al., *Can. J. Biochem. Physiol.*, 32, 440, 1954.

insecticides are synergistic in combination.[6] A plot of data collected with mixtures of two synergistic compounds would deviate from the straight line of simple additivity in the fashion shown in Figure 4.5. For any given mixture, the degree of synergism is calculated from the isobologram. Connect the EC_{50} points on the x and y axes with a straight line, as would occur with simple additivity. Locate the mixture of interest on the straight line and pass a line from the origin through that point. The distance along that line to the straight line divided by the distance from the origin to the isobole (the curved line representing experimental studies) measures the degree of synergism.

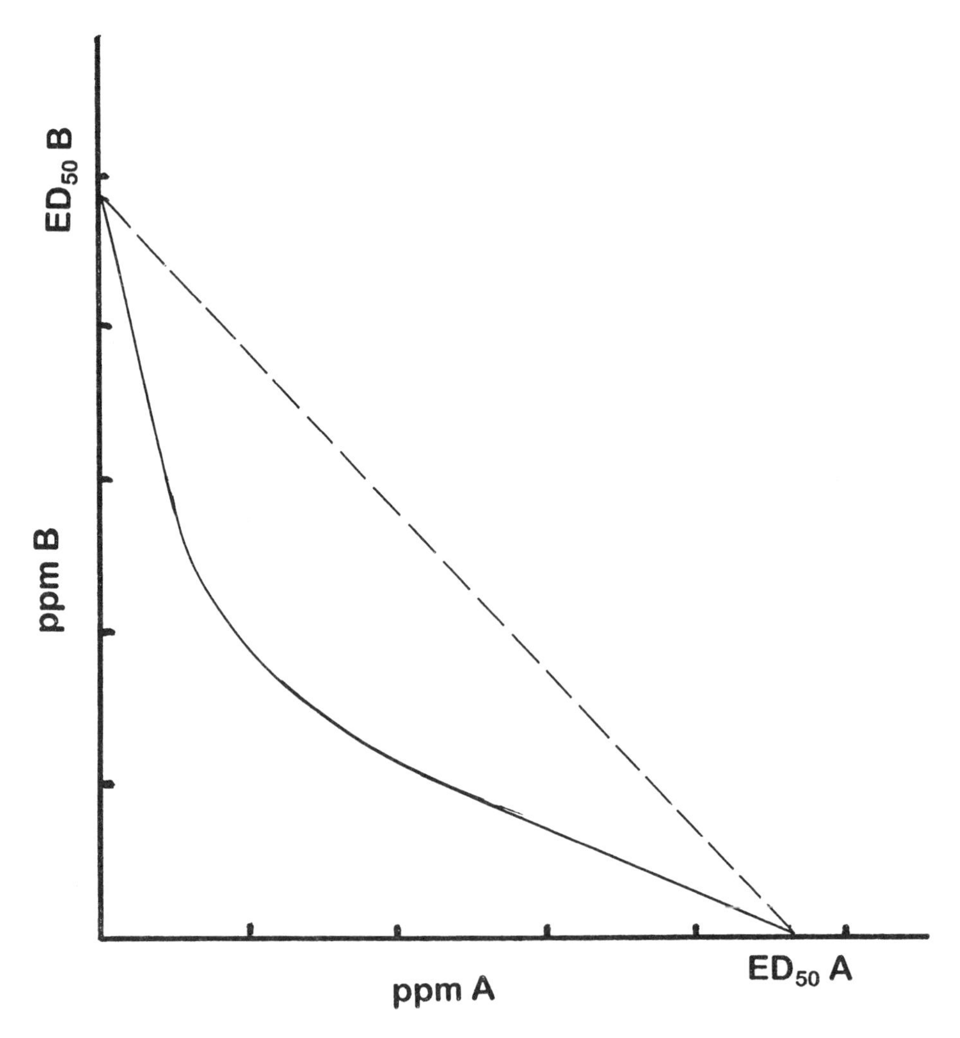

Figure 4.5. An isobologram of compounds whose effects are synergistic. The layout is as in Figure 4.3.

When two compounds are under study and A causes a toxic effect but B does not, and in combination the toxic effect of A is enhanced, B is a potentiator of A. This is also termed unisynergism. An example is provided by acrylonitrile, which causes duodenal ulcers in rats. Polychlorinated biphenyls (PCBs) do not cause duodenal ulcers, but when PCBs are given to the animals at the same time as acrylonitrile, the ulcer-generating activity of the latter is greatly enhanced.[7]

[6] J. Frawley et al., *J. Pharmacol. Exp. Ther.*, 121, 96, 1957.
[7] E. H. Silver et al., *Toxicol. Appl. Pharmacol.*, 64, 131, 1982.

In addition to dosing simultaneously with more than one chemical, combination effects can be seen from exposure to a chemical and a physical agent. Radiofrequency radiation sufficient to cause malformations in 30% of rat fetuses was combined with sufficient 2-methoxyethanol to cause malformations in 14% of rat fetuses. The combination induced malformations in 76% of the fetuses, a synergistic effect.[8]

THE SPECIAL PROBLEM OF CANCER

Assessing the carcinogenic potential of a compound is challenging. If a compound is hepatotoxic, animals exposed to it at high levels consistently show liver damage. By contrast, dosing with a carcinogen is like playing Russian roulette. Not every heavy cigarette smoker gets lung cancer, but the percentage of smokers that do is much higher than the percentage of nonsmokers. Of two carcinogens, one is more potent than another if, at the same concentration, it causes a higher percentage of animals to produce tumors. Studies must therefore be performed using a large enough test group that the result is statistically valid. Furthermore, tumors develop for reasons other than exposure to a particular compound, so there is a normal or baseline occurrence of cancer without dosing. To identify a carcinogen we must observe an increase in occurrence of tumors above this baseline level, so an undosed control group must be maintained for comparison. In summary, the percentage of animals lightly dosed with a carcinogen that develop cancer is higher than that of nondosed animals, more heavily dosed animals develop the disease even more frequently, and the frequency at a given level of exposure differs with the carcinogen used.

Response to a carcinogen is not necessarily observed until after an extended period of time. For example, in a society where women who smoked were less socially acceptable, movies portrayed glamorous women as being sophisticated when they lit a cigarette. The resultant increase in women smoking was followed by an increase in women with lung cancer 20 years later. The effect of some chemicals is to convert a normal cell to a precancerous cell. Perhaps months or years later, this precancerous cell may undergo other changes into a malignant cell. The transformation into a cancer cell may appear to be spontaneous, or it may be the response to a second chemical. Following such a long incubation, it is more difficult to establish unequivocally the cause of the original cell transformation. Studies to determine the carcinogenic potential of a chemical must extend for the life of the animal, so the introduction of carcinogenicity testing into the toxicological assessment of compounds has greatly increased both the effort and expense of such procedures.

In recent years the methods of assessing the carcinogenic potential of chemicals have been questioned. To learn whether there is likely to be one additional case of cancer per 10,000 population from use of a particular food additive, testing would have to involve many times 10,000 mice to be statistically valid, if exposure in the tests is at the levels of normal usage of the chemical. In order to do the testing with reasonable sizes of test groups, scientists must utilize the increased occurrence of cancer at higher dose levels. Testing is therefore done using extremely high concentrations of the chemical under study. At these concentrations many scientists claim that changes are caused that make the animal more susceptible to cancer, even cancer stimulated by other agents. For example, the immune system is an important defense against cancer because it destroys cells detected as different. High levels (but only high levels) of chemicals may suppress the immune system, and thus artificially raise the cancer rate. The EPA has developed a formal classification scheme for carcinogens that reflects some of these uncertainties (Table 4.3). The classifications of many compounds as carcinogens are now being reconsidered.

[8] B. K. Nelson, D. L. Conover, and W. G. Lotz, "Combined Chemical, Physical Hazards Make Exposures Harder to Calculate," *Occupational Health and Safety*, June 1994, p. 50.

Table 4.3. EPA Classification of Carcinogens.

Class A: Human Carcinogens
These compounds (less than 30) have been established to cause cancer in humans by epidemiological studies of human exposure at low levels.
Class B1: Probable Human Carcinogens
There is epidemiological support that these cause cancer in humans, but the data are not unambiguous.
Class B2: Probable Human Carcinogens
There is strong support that these compounds cause cancer in test animals, but no epidemiological data supporting there effect on humans.
Class C: Possible Human Carcinogens
There is support that these compounds cause cancer in test animals, but no epidemiological data supporting their effect on humans. The difference between Classes B2 and C lies in the quality of the results of the animal studies, and distinction between the groups is difficult to pinpoint.
Class D: Not Classifiable as to Human Carcinogenicity
There is no adequate evidence of carcinogenicity.
Class E: Evidence of Noncarcinogenicity for Humans
There is no evidence of carcinogenicity in two or more animal tests.

From U.S. EPA, "Guidelines for Carcinogen Risk Assessment," *Federal Register,* 51, 33993, 1986.

RISK ASSESSMENT

Finally, we must undertake the difficult task of risk assessment. By this we mean we must decide how much exposure to a given agent is safe and therefore should be allowable. Arthur Hayes, who served as commissioner of the Food and Drug Administration, which sets standards for exposure to or use of chemicals in foods, drugs, or cosmetics, defined *safe* as "A reasonable certainty of no significant risk, based on adequate scientific data."

Here we note the words *reasonable*, *significant*, and *adequate*. To the public, who wants to hear, "no risk", this sounds like hedging and waffling, but it is generally not possible to guarantee no risk to everyone, for reasons presented in this chapter. Furthermore, we are in transition from a time when there were no standards set for exposures toward one in which every risk is identified and regulated. We want to approach the task so as to make the most rapid possible improvements within the limits of our technology and resources. Tests are conducted, but they are limited of necessity because of the huge number of compounds of concern. Further, there are many aspects of toxicity that we need to understand much better to do the best possible job of regulation, for example, the causes of cancer. The objective today is to come up with the best recommendation possible, then remain alert to the need for modification of these standards as we gain more experience and understanding.

EXTRAPOLATING ANIMAL DATA TO HUMANS

There is considerable uncertainty involved in extrapolating animal test results to the prediction of toxicity to humans. The variability in response within the human population and the hypersensitive response must be considered. A rule of thumb applied when setting standards is to lower the recommended maximum exposure tenfold to protect against unforeseen higher toxicity in humans as compared to the test animals, then to lower it another tenfold to protect the hypersensitive members of the group of humans exposed. The resulting recommendation is dealt with as a permitted level of exposure. The level of the chemical in the workplace should be monitored, and the level found should be no higher than the permitted exposure level (PEL). Finally, the health records of the workers in that plant should be monitored in order to learn if the safety margin was sufficient, and if the chemical has an effect on humans of a type not anticipated by the animal testing.

PROBLEMS TESTING FOR TERATOGENIC PROPERTIES

Sometimes the testing process can be very difficult, as in testing for teratogenic potency. Such animal testing must be performed while the animal is pregnant. Traditional test animals, such as rats and mice, differ from humans in major aspects of the reproductive process. Fetal development in rodents moves forward much more rapidly. Small rodents bear litters rather than single offspring, and exposure varies with position in the uterus. The placentas of humans and small rodents differ in their effectiveness as barriers to the entry of foreign substances into fetal blood.

Extrapolating animal results to humans, always a problem, becomes even less dependable when assessing teratogenic potential. Aspirin, for example, is a potent teratogen in small rodents, whereas thalidomide is not. Had thalidomide been tested for teratogenicity in mice before it was released, it would have been evaluated as safe.

Exposure of a pregnant woman to a teratogen does not ensure birth defects. In addition, a very high level of malformed human offspring occur naturally. About 20% of all pregnancies result in spontaneous abortion. Little information is available to determine what percentage of these abortions is the result of malformation. Of the children carried to birth about 2% are physically malformed, and another 8% have a mental handicap. A chemical that caused 1% of the exposed population to develop cancer would not only be highly visible, but would be looked on as a serious threat. However, against such a high background level of childbirth abnormalities, an agent causing 1% teratogenic malformations might escape detection.

Regardless of their failings, animal teratogenic testing is all that is available before human exposure to a compound is allowed. After humans are exposed to the chemical, its teratogenic potency is likely to be detected only if it is a very effective teratogen.

UNITS FOR RECOMMENDED MAXIMUM EXPOSURE

The recommendations arrived at through animal testing are expressed in a variety of units. Because exposure by inhalation is the most likely route of exposure in a factory situation, particular attention focuses on it. The term MAC (maximum allowable concentration) was used for many years, but is not heard much today. Beginning in 1946 with the establishment of the Committee on Threshold Limits, the American Conference of Governmental Industrial Hygienists (ACGIH) has been devising standards for safe exposure termed TLVs (threshold limit values).[9] It is claimed that no worker has ever suffered harm when exposure was limited to the ACGIH recommendations.[10] Although ACGIH is a private organization and its exposure limits are not mandatory, TLV values have been sufficiently credible that they were adopted as government standards by West Germany, became the basis of the official U.S. standards, and have been since used as regulatory values in most industrialized countries.

Recommended maximum exposures are expressed on the basis of time-weighted average (TLV-TWA). Calculation of a TLV-TWA assumes an 8-hr day and 40-hr week. The exposure levels of the compound are measured at intervals. The TWA exposure is calculated by

[9] The ACGIH is a private group headquartered in Cincinnati, OH, that for many years has assembled recommendations for maximum exposure to chemicals and for other hazardous exposures in the workplace. Membership is composed of governmental and private industrial hygienists and academic investigators.

[10] M. E. Lanier, *Threshold Limit Values: Discussion and 35 Year Index with Recommendations*, ACGIH, Cincinnati, OH, 1984; H. E. Stokinger, "Threshold Limit Values: Part I," *Dangerous Properties of Industrial Materials Report*, Van Nostrand-Reinhold, New York, 1981.

multiplying the concentration of compound in each analysis by the length of time of exposure to that level. These are summed and divided by the total time to produce an average exposure level.

SAMPLE PROBLEM:

Given an exposure level of 2 ppm for 10 hr per week, 3 ppm for 20 hr per week, and 4 ppm for 10 hr per week, what is the TWA for this exposure?

Exposure (ppm)	Time (hr)	Product (ppm × hr)
2	10	20
3	20	60
4	10	40
	40	120

TWA = 120 ppm × hr/40 hr = 3 ppm

In such a standard, no importance is attached to the occurrence of a short period of high exposure, as long as it is counterbalanced by an adequate period of low exposure.

However, if a special hazard exists above a certain concentration in the air, the recommendation will include a ceiling value (TLV-C), a concentration the level must not exceed regardless of the time of that exposure. Limitations on high-level exposure may also be expressed as the acceptable maximum peak, which describes as the exposure limit both a concentration level and a period of time at that level. To use benzene vapor as the example, the TLV-TWA is 10 ppm. On top of this regulation, there is an acceptable ceiling of 25 ppm, so that during this 8-hr period levels higher than 25 ppm are not allowed for extended periods, even if they are countered by extended periods of low exposure. Finally, there is an acceptable maximum peak of 50 ppm for not more than 10 min. This means that the ceiling of 25 ppm may be exceeded, up to 50 ppm, if the exposure at the higher level is limited to 10 min. It must be remembered that if the level reaches 25 ppm for some period, or 50 ppm for under 10 min, these exposures must be balanced by extended periods of low exposure to allow the TWA to stay below 10 ppm.

Another term used to express a standard for intermittent high exposure is the short-term exposure limit (TLV-STEL). The STEL restricts worker exposure at the recommended maximum level to 15 min four times per day, where the exposures are separated by at least 1 hr.

Why is such a complex system used? If workers were exposed to relatively constant levels of chemicals in the plant atmosphere, the TLV-TWA would be a sufficient protection. However, it is not uncommon in a factory to have levels of a volatile material rise sharply for short periods of time. This would be the case when a process is carried out in a closed vat, but at the completion of the process, the vat is opened and the product removed. During that period, the levels of a chemical in the air rise sharply, then drop again when the vat is reclosed. TLV-Cs or STELs address the possibility that even if exposures are low on average, for some chemicals a high level of exposure is harmful even if sustained for only a short time.

CRITICISM OF TLVs

The adequacy of ACGIH recommendations has been questioned. Critics include scientists in European countries that adopted TLVs as standards. Also, in the U.S., people were upset that TLVs were used rather than NIOSH recommended exposure limits when OSHA attempted a sweeping update of PELs in the 1989 Air Contaminants standard. Common criticisms of TLV values include the following:[11]

- Some standards are determined without data concerning effects of long-term exposure.
- Data is lacking on reproductive effects.
- Carcinogenic risk is determined without human data.
- Standards for a number of chemicals consider only protection from eye and respiratory irritation.
- Some standards are based on data supplied by industry, leading to suspicion of bias.

Largely, criticisms center on faults in the data base used to establish the standard. As data bases improve, ACGIH adjusts the standards, publishing a new list each year. Furthermore, any attempt to establish a standard involves some level of risk. If a standard is used without problems, this verifies its relative safety.

EXPOSURE TO PHARMACEUTICALS

Workers in the pharmaceutical industry may be exposed to drugs in the workplace. All drugs are toxic, given a sufficient dose. The special knowledge of the behavior of pharmaceuticals that must be obtained as part of the process of gaining approval from the Food and Drug Administration for use of the compounds allows the industry to devise somewhat more sophisticated standards for safe levels of exposure. These exposure limit controls (ELCs) have the same goal as the TLVs of ACGIH, are similarly determined on a TWA basis, and express in mg/m^3 the maximum level that can safely be allowed in the workplace. They are calculated as follows:

$$\text{ECL} = \frac{(\text{NOEL})\,(\text{body mass})\,(\text{safety factor})\,(\text{fraction absorbed})}{(\text{inhalation flow rate})\,(\text{time to constant blood level})}$$

The safety factor is the term you wish to multiply your value by to be adequately low the threshold concentration for toxic effects. Inhalation flow rate is normally cubic meters per day and time to constant (steady state) blood level is proportional to biological half-life. For most drugs the fraction absorbed, the portion of the dose that reaches the blood, is 1.

BIOLOGICAL MONITORING

The TLV focuses on minimization of hazard by measuring and controlling the concentration of a hazardous chemical in the air workers are breathing. Another approach is to monitor the concentration of the chemical in the body. Since the purpose of air monitoring is to limit the amount of the chemical entering the body to safe levels, measuring levels actually in the body is more direct. It requires establishing a technique to assay the amount of chemical in the body, which is the most difficult aspect of employing such a standard.

In Chapter 2 the importance of the presence of the chemical in the blood was established, because in general the measure of a chemical entering the body by whatever pathway is its entry into the circulation, and the measure of success in removing it from the body is its rate of removal from the circulation. This suggests that blood testing is the best way to monitor exposure to a chemical, but clearly workers would object to frequent blood sampling, and other questions arise about the hazards of such an approach.

[11] J. Borak, "ACGIH's Threshold Limit Values Useful, but Formulas Are Still Controversial," *Occupational Health and Safety*, Aug. 1994, p. 26.

A less objectionable approach is to take urine samples from workers. These samples may contain the actual chemical, or they may contain instead a substance to which the body converted the original chemical (a metabolite). This approach is used in primary aluminum production, where exposure to fluorides is assessed by regular urine analysis. However, we really want to know how much of the chemical the worker has taken into his or her body, and without knowing the total volume of urine passed by the worker during the day, concentration of a single sample gives incomplete information. By selecting a compound whose excretion per day is relatively predictable, then expressing the concentration of the hazardous compound or its metabolite relative to the concentration of that urine component, we have a more meaningful measure of exposure.

The typical levels in a biological fluid that are found on exposure to the TLV-TWA level of a chemical are recorded as the biological exposure index (BEI). For example, exposure to 100 ppm xylene TWA typically produces 1.5 g methylhippuric acid (a metabolite) per gram of creatine (a reference compound in the urine) at the end of the workday. BEI values are published by the ACGIH in their annual *Threshold Limit Values for Chemical Substances and Biological Exposure Indices*.[12]

OSHA STANDARDS

When the Occupational Safety and Health Administration (OSHA) was established in 1970, they adopted TLVs as the best existing standards. OSHA uses the term PEL (permissible exposure levels) for their standards.[13] PEL values have been set for approximately 500 chemicals. Some of these standards have since been modified by OSHA on a one substance at a time basis.

Returning to the assumption of additivity in the case of multiple exposures, there is a simple formula that can be used to calculate compliance and safety when PELs are known for all components of the mixture. A fraction is made of the actual level of a chemical divided by the PEL, where both values must be in the same units. The sum of the fractions for the various components of the mixture are added, and if the total is more than 1, exposure goes beyond acceptable limits.

SAMPLE PROBLEM:

Given the following PEL values in ppm:

Benzene	10 ppm
Methylene chloride	50 ppm
Toluene	100 ppm

Waste solvent from a paint stripping operation is evaporating in a storage shed and the level of vapors is shown to be as follows:

Benzene	5 ppm
Methylene chloride	21 ppm
Toluene	15 ppm

$$\text{Exposure} = 5/10 + 21/50 + 15/100 = 107/100 = 1.07$$

The value is greater than 1 so the shed is in excess of recommended exposure levels.

[12] ACGIH, 6550 Glenway Ave., Bldg. D-7, Cincinnati, OH 45211-9438.

[13] A separate list of PEL values is maintained for standards in construction, which are not always the same as the general PEL values.

A recommendation for maximum worker exposure to a chemical is based on characteristics that go beyond the lethal effects. It includes consideration of adverse factors such as eye irritation, odor, skin irritation, or ability to produce narcosis. To the person in a plant charged with the responsibility for maintaining a safe environment, knowing the information upon which a recommendation is based is important.

OSHA or the chemical manufacturer provides other recommendations. NIOSH recommendations for exposure standards to replace PEL values are called recommended exposure limits (RELs). OSHA literature lists a value called immediately dangerous to life and health (IDLH). IDLH was developed as a part of the Standards Completion Program begun in 1974 by OSHA and NIOSH to establish complete occupational health standards. It represents the maximum level from which a person could escape in 30 min without any escape-impairing symptoms or irreversible health effects. The recommendations also include requirements for the use of eye protection, protective clothing, or respirators in handling a chemical.

OTHER STANDARDS

The primary sources of exposure recommendations for most countries in the world are the ACGIH TLVs and the maximum allowable concentration (MAK) values that are the basis of regulation in Germany. The operation of the MAK committee is similar to that of OSHA, and similarly, the final values published in the Bulletin of the Ministry of Labor become legal standards for Germany. By 1986, recommended safe levels of exposure had been established for only about 1700 of the more than 10,000 compounds used in industry. The limitations on time and finances in an organization such as the ACGIH or the MAK Commission require that they limit themselves to the study of compounds that are most likely to cause problems because of high toxicity, widespread use, or other factors.

Recently the American Industrial Hygiene Association (AIHA)[14] set up a WEEL (Workplace Environmental Exposure Level) Committee to perform a function similar to that of the ACGIH. By 1986 WEEL recommendations had been published for 46 chemicals. However, there are still many compounds for which guidelines are not available that are in common use by industry. Many companies have therefore undertaken to establish internal standards to serve their own safety program needs. At a meeting in 1983 of the American Industrial Hygiene Conference, these various standards were discussed (Paustenbach and Langner, 1986). In Table 4.4 we see the numbers of internal standards established by large companies.

Table 4.4. Numbers of Internal Corporate Safety Standards for Chemicals.

Company	Number of Special Standards
DuPont	~300
Exxon	50–100
Dow	~300
Rohm and Haas	~500
3M	15

Adapted from D. Paustenbach and R. Langner, *Am. Ind. Hyg. Assoc.*, 47, 809, 1986.

EPIDEMIOLOGY

Epidemiology is a valuable tool employed by toxicologists. Occupational epidemiologists gather data regarding the incidence of a particular health outcome from a population of workers who share exposure to a suspected hazard, and from another population, similar in

[14] The AIHA is similar to the ACGIH in many ways, but differs in encouraging corporate membership.

all other respects, who lack that exposure. A higher incidence in the exposed group must then be analyzed to determine if the result is statistically significant.

EPIDEMIOLOGICAL STUDIES OF SMOKING

A familiar example of an epidemiological study is the data indicating a higher frequency of lung cancer in smokers than in nonsmokers. What medicine now describes as our greatest public health threat was originally identified without use of laboratory studies. Further epidemiological studies established links between smoking and heart disease and between smoking and cancers in the mouth and throat or, in women, in the breast.

However, the epidemiological approach may or may not identify the agent causing the negative health effect. The tobacco industry exploited this fully, repeatedly arguing in the face of overwhelming statistical data that no one had ever *proven* a link between smoking and lung cancer, meaning no one had isolated a specific chemical from cigarette smoke that caused lung cancer in studies with laboratory animals. This sort of challenge to an epidemiological result is not valid. Given sound data, appropriate handling of the data, and statistically valid results, correlations obtained indicating a health problem must have an explanation.

First, one must choose to do an epidemiological study. What might lead to such a decision? Here are some examples:

1. Records of workers in a specific plant or specific industry seem to indicate a high incidence of days lost from work, or a suspiciously similar pattern of problems.
2. National data bases, such as the National Cancer Institute's *Surveillance, Epidemiology, End-Results* report, or the National Center for Health Statistics' *Monthly Vital Statistics* report, indicate a regionally high level of a disease, suggesting either an occupational or environmental cause.
3. Concern that a PEL for a chemical in use is not sufficiently low.
4. Concern that a chemical in use, but for which there is not a standard, is causing problems.

PROSPECTIVE AND RETROSPECTIVE STUDIES

Epidemiological studies may be prospective or retrospective. In prospective studies the medical records of workers presently being exposed to a chemical are examined to see if current exposure levels cause problems. When new standards are proposed for exposure to a chemical, they are based on our best toxicological test results. However, these standards are usually based on extrapolations of data obtained by experiences other than actual exposure of workers, i.e., tests with laboratory animals. Best laid plans may go astray, and effects of a chemical not observed in test animals may surface, or the chemical may prove to be much more toxic to humans than to animals for some reason. We therefore need to be alert to the occurrence of problems after the introduction of a standard. Prospective studies can be the best epidemiological studies, because the epidemiologist is in a position to gather whatever information is deemed useful, such as more complete worker health data or worker exposures to other chemicals.

Retrospective studies are employed after a problem has been noticed. Since exposures have already occurred, the epidemiologist has only whatever information has already been gathered. Medical records of workers displaying the problem are examined and compared to records of similar groups of workers not exhibiting symptoms of toxic exposure to see if the causative factor can be identified. Several chemicals have been implicated as carcinogens by observing elevated levels of a particular type of tumor in a specific population, then determining the chemical exposure shared by members of that population. This is how the carcinogenic nature of vinyl chloride was discovered.

DESIGNING AN EPIDEMIOLOGICAL STUDY

Epidemiological studies must be designed carefully. Badly formatted studies can seem to prove absurd correlations, like a linkage between eating apples and pregnancy among unmarried teens. Results of sloppy studies lead to cynical statements such as, "There are lies, damned lies, and statistics." The lung cancer and smoking study (see box) must have been well designed, because the tobacco lobby resorted to a "smoke" screen to convince the untutored public they need not face withdrawal from cigarettes.

A good study identifies two groups (exposed and not exposed), then restricts the suspected causative factor to one group and assures that the two groups are identical in other respects. In a study to determine if living at a higher altitude is a factor in lung problems, one should not choose the high-altitude group from hard-rock uranium miners and the low-altitude group from clean-room workers. Age, sex, and race all make a difference in response to a toxic agent. The groups should be the same in these regards, or the data may need to be normalized somehow for differences in the groups.

The result must be statistically valid. Usually this means that a sufficiently large population must be surveyed. There is no magic number of individuals that must be sampled. It depends on the frequency of occurrence of the health effect in the exposed population and on the selectivity of data. Suppose you were studying the effects of a spray paint solvent on worker health. Ideally, you would study just painters, and a relatively small sample of workers would work. However, you might find that the health data for the workers was not broken down by workstations, and you had to use the data for all workers in the plant. The subgroup of exposed workers is now diluted into a population of individuals not exposed. In another case, people may rotate to different jobs, and the sample of workers surveyed spend only a percentage of their time painting. In either case, a larger sample is necessary to obtain valid results. Statistical methods can assess the relative validity of a study based on frequency of occurrence of the negative effect and size of the sample surveyed, and can express the level of confidence to be placed on the results in quantitative terms.

Even a statistically high incidence of some problem may not be enough information to solve a problem. It may be clear that a problem exists, but not be clear what caused the negative effect. As an illustration, Table 4.5 shows a comparison of occurrence of phocomelia (birth defects of the limbs) in a German hospital in populations of mothers who gave birth before 1959 and after 1959. It was clear there was a problem, but based on that data alone there was no obvious course of action to minimize or eliminate the excess occurrence of phocomelia. It was a challenge, then, to study the health records of women who delivered babies before and after the appearance of high rates of phocomelia to identify use of the drug thalidomide as the causative agent. Those records had to include drug use of each woman, or the cause would have remained a mystery.

EPIDEMIOLOGY AND WORKPLACE HEALTH AND SAFETY

Several dangerous workplace chemicals have been pinpointed by epidemiology. The classical study establishing soot as a causative agent for scrotal cancer was essentially an

Table 4.5. Occurrence of Phocomelia.[a]

Time Period	Number of Cases
1949–1959	0
1959	1
1960	30
1961	154

[a] Birth defects in which arms and/or legs are malformed or missing. This is a dramatic example of epidemiological data serving to uncover the existence of an unsuspected teratogen. The next problem was to determine what substance had been used by these women that had not been used in the control period of 1949–1959. It was found to be the drug thalidomide.

Data from the University Pediatric Clinic, Hamburg, West Germany.

epidemiological study, with the chimney sweeps comprising the exposed group and displaying a high incidence of warts and tumors, and the rest of the population constituting the control group with little soot exposure and a low incidence of the problem. There may not have been a careful statistical analysis, but in that case it was sufficient.

The carcinogenic potentials of vinyl chloride and benzene were recognized in retrospective studies by higher frequencies of specific cancers in exposed workers, leading to rapid action to reduce worker exposures. The appearance of frequent male sterility at a single facility among workers handling bulk agricultural chemicals led to recognition that dibromochloropropane is a reproductive poison.

However, the full potential of epidemiology to uncover problems in chemical exposures is currently not utilized. Harmful exposures uncovered by looking at population data have usually involved an uncommon illness, for example, a rare form of cancer, or an exposure high enough that a relatively large percentage of those exposed were victims. Consider the present method of establishing safe limits of exposure. We collect data based on animal exposure, divide by factors we hope are adequate to cover the transfer to human exposure, and set an exposure limit. If the limit was marginally too high so that workers' health was compromised only infrequently or so that workers became only mildly ill, we might not notice. Epidemiological studies would test the success of the new standard.

Why do we not make more sophisticated use of epidemiology? OSHA presently requires the accumulation of medical information about individual workers, but there is no requirement that this data be assembled and examined with the goal of detecting deviations from the health data norm. Successfully applying epidemiological methods to such data would require first standardizing the methods of collecting and reporting health data. More detailed information would be needed on levels of exposure and more standardized methods of measuring exposure would be necessary. This clearly would increase the complexity and cost of monitoring the workplace. More awareness of trends in accumulated data as presently collected, however, could provide insights about exposures that need to be studied further, using more careful methods in the new studies.

KEY POINTS

1. Risk assessment based on toxicity studies establishes standards for workplace exposure.
2. Toxic substances enter the body constantly. Only when concentrations in the body rise to disruptive levels, or when damage occurs that is not repairable or accumulates because it is not repaired before the next exposure, do we experience problems.

3. When setting standards for exposure to a toxicant, the age, sex, and state of health of those likely to be exposed and the length of time they will be exposed must be considered.
4. Because of the risks involved, preliminary toxicity testing is performed on animals, usually small rodents. It is then necessary to extrapolate the data to predict risks to humans. There is less emphasis on testing with lethal levels of a toxicant, and alternatives to the use of animals are being developed.
5. When expressing results of toxicity studies, differences in size between humans and animal surrogates are accommodated by expressing data in milligrams (dose) per kilogram (body weight).
6. Dosing animals should be done so that (a) we know the intake accurately and (b) the manner of dosing resembles the manner of human exposure as closely as possible.
7. The length of exposure to a toxicant in testing varies with the information sought. Acute exposure studies immediate toxic effects and chronic exposure determines whether the compound or damage accumulates in the body.
8. Response to a toxicant varies according to sex and age. To a degree, the response of each animal or human is unique.
9. Testing is done using several small groups of subjects, each at a different dose level. Results are plotted to produce a dose–response curve. From this curve the LD_{50}, or dose lethal to 50% of the subjects, expressed in mg/kg, is determined. Testing airborne toxicants does not involve consideration of animal weight, is expressed in parts per million (ppm) and is termed the LC_{50}, or concentration lethal to 50% of the subjects.
10. The lowest dose to produce a response is the threshold value, and the highest to produce no effect is the NOEL (no observed effect level).
11. Combinations of chemical agents or chemical and physical agents may be additive, with a combined hazard equal to the sum of the individual hazards. When one agent reduces the hazard of the other the effects are antagonistic, and when one increases the hazard of the other they are synergistic.
12. Levels initially recommended as the upper limit for safe exposure are roughly NOEL/100. Such recommendations are termed TLVs by the ACGIH and PEL by OSHA. Such values are average exposures for 8-hr days and 40-hr weeks and are termed time-weighted averages (TWA). Ceiling (C) values may be listed regardless of the time of exposure or with a maximum length of time permitted at that level. A short-term exposure limit (STEL) is such a ceiling value used by OSHA.
13. Occupational epidemiology studies identify toxic compounds by comparing health statistics for an exposed group of workers with those of a similar but unexposed group.

BIBLIOGRAPHY

E. J. Calabrese, *Multiple Chemical Interactions,* Lewis Publishers, Chelsea, MI, 1991.

E. J. Calabrese, *Principles of Animal Extrapolation,* Lewis Publishers, Chelsea, MI, 1991.

D. Dollberg, *Analytical Techniques in Occupational Health Chemistry,* American Chemical Society, Washington, D.C., 1980.

V. A. Filov, A. A. Golubev, E. I. Liublina, and N. A. Tolokontsev, *Quantitative Toxicology,* Wiley, New York, 1979.

W. J. Hunter, Ed., *Evaluation of Toxicological Data for the Protection of Public Health,* Pergamon Press, New York, 1977.

C. Klaassen, M. Amdur, and J. Doull, *Toxicology,* 3rd ed., Macmillan, New York, 1986. (The first five chapters provide excellent basic coverage regarding the measurement of toxicity.)

F. W. Mackinson, R. S. Stricoff, and L. J. Partridge, *NIOSH/OSHA Pocket Guide to Chemical Hazards,* U.S. Government Printing Office, Washington, D.C., 1978. (A compact summary of PEL and IDLH values, physical properties, precautions in handling, and health hazards for 375 compounds used in industry.)

NIOSH, *Registry of Toxic Effects of Chemical Substances,* U.S. Government Printing Office, Washington, D.C.

D. Paustenbach and R. Langner, "Corporate Occupational Exposure Limits: The Current State of Affairs," *Am. Ind. Hyg. Assoc.,* 47, 809, 1986.

PROBLEMS

1. The recommended adult dose for an antibiotic is one 100-mg capsule three times a day. Would you give the same dose to a small child? If an average adult is 70 kg and a child who is to receive the antibiotic weighs 23 kg, how much antibiotic should the child be given per day, and in what pattern?
2. A chemical is used as a component in the solvent for automobile finishes sold to bump and paint shops. Many of these shops allow the paint to dry in the workplace air in a building that is closed in winter. Testing with rats established that 75 ppm is a safe level of exposure to this solvent. If the average rat in the test weighed 150 g, and a typical adult worker weighs 75 kg, what would your recommendation be for a maximum level of exposure to be allowed in a bump and paint shop?
3. In a facility that rebuilds used fuel system parts for automobiles including carburetors, fuel injectors, and fuel pumps, a solvent dissolves accumulated gums and fuel additives from the old units. Each unit is disassembled and the parts placed in a fine-mesh copper-screen basket. The container and contents are then shaken as they are sprayed with hot solvent in an enclosed machine. Workers then open each basket, remove and inspect all the parts, replace all worn and broken components, and reassemble the unit. When the basket is opened it is not uncommon to find some solvent still in depressions or chambers of the unit. A series of solvents has been proposed, all of which have satisfactory properties for the cleaning operation. You are asked, as the industrial hygiene consultant to the company, to choose the one that is safest for the workers. Here are measures of toxicity you have assembled for the solvents:

Solvent (boiling point)	LD_{50} (acute, oral, rat)	LD_{50} (acute, dermal, rabbit)	LD_{50} (90-day, oral, rat)	LC_{50} (24-hr, rat)
A (85°C)	920 mg/kg	3600 mg/kg	880 mg/kg	68 ppm
B (79°C)	870	3500	330	220
C (145°C)	125	2900	118	225
D (91°C)	780	850	699	256

As each solvent is considered for use, think of the pathway of worker exposure. Which would you recommend?

4. Table 4.6 displays oral and dermal LD_{50} values displayed by white rats dosed with a series of organophosphate insecticides. Using these values, answer the following questions:
   A. Which compound taken orally is known to be the most toxic to male rats? to female rats?
   B. Which compound taken orally is known to be the least toxic to male rats? to female rats?
   C. Name three compounds for which oral toxicity to male and female rats is very similar.
   D. Is it true that the relative oral toxicities, male vs. female, are closer in the case of phorate, where the difference in LD_{50} is only 1.2, than for trichlorofon, where the difference is 70?
   E. Name all the compounds that are more than twice as toxic orally to males than to females. to females than to males. As a generalization, does this family of compounds tend to be more toxic to males or females?
   F. Are these compounds typically more toxic by the oral or the dermal route?
   G. Look at the dermal toxicity of diazinon to males and females, of schradan for females, and of phosdrin for males. For each insecticide, what do the values tell us? What does high dermal toxicity infer?
   H. Is it a good generalization that compounds that are more toxic orally are also more toxic dermally?
5. Recall that 1 l is 1000 cm^3 and that 1 mol of a gas at STP (standard temperature, or 25°C, and standard pressure, or 1 atm) occupies 22.4 l.

Table 4.6. Acute Oral and Dermal LD_{50} Values of Organic Phosphorus Insecticides for Male and Female White Rats.

Compound	Oral LD_{50} (mg/kg)		Dermal LD_{50} (mg/kg)	
	Males	Females	Males	Females
Carbophenothion	30	10.0	54	27
Chlorthion	880	980	<4500	4100
Co-ral	41	15.5	860	—
DDVP	80	56	107	75
Delnav	43	23	235	63
Demeton	6.2	2.5	14	8.2
Diazinon	108	76	900	455
Dicapthon	400	130	790	1250
Dimethoate	215	—	400	—
Di-syston	6.8	2.3	15	6
EPN	36	7.7	230	25
Ethion	65	27	245	62
Fenthion	215	245	330	330
Guthion	13	11	220	220
Malathion	1375	1000	>4444	>4444
Methyl parathion	14	24	67	67
Methyl trithion	98	120	215	190
NPD	—	—	2100	1800
Parathion	13	3.6	21	6.8
Phorate	2.3	1.1	6.2	2.5
Phosdrin	6.1	3.7	4.7	4.2
Phosphomidon	23.5	23.5	143	107
Ronnel	1250	2630	—	—
Schradan	9.1	42	15	44
TEPP	1.05	—	2.4	—
Trichlorofon	630	560	>2000	>2000

A. How many liters are there in a cubic meter of air?
B. How many liters does 1 mol of gas occupy at 25°C and 1 atm pressure?
C. We sometimes see concentrations in air expressed as ppm (1 volume per 10^6 volumes) and sometimes see it as mg/m^3. How many mg/m^3 are there of a gas whose concentration is 1 ppm and whose molecular weight is 100 g/mol at 25°C?
D. We have a sample of air at 25°C that has two vapors in it whose concentrations are 2 and 4 ppm and whose molecular weights are 320 and 150 g/mol, respectively. Which has the highest concentration in mg/m^3?

6. You are on a committee trying to minimize the use of animals in toxicity testing. The suggestion is made that if you start by knowing the level of concentration in the air that a compound in question reaches in the workplace, you could simply expose an animal to that level to see if it is safe. You object to that approach because exposure recommendations should have a safety margin. (A) What is the safety margin described in the text, and (B) how would this "one animal, one level" test be modified to incorporate that margin? A toxicologist on the committee says, "Suppose we used that approach. Let us consider what the results would tell us. There are two possibilities: the animal lives and the animal dies. If the animal lives, we propose to say that this means the compound is safe at that level. (C) Is there any problem with that supposition?" Another participant then says, "I see where you are going with that. And of course the other side of it is (D) what information do we lack if the animal dies?"
7. A chemical used in cutting oils has been found to resist removal by preliminary treatment in industrial waste water and is going to the local sewage treatment facility. On checking, it is found that its oral acute toxicity is unknown, and tests are immediately undertaken. One proposed experimental design is to dose test animals by way of their drinking water and another is to inject the animals intraperitoneally (in the abdominal cavity where blood vessels readily absorb chemicals). (A) What are the advantages and disadvantages of each approach?

Injection is selected as the method of study, and 90 Sprague-Dawley rats are purchased. They are divided into 9 groups of 10 rats each, each group receiving a different dose level. Assume for simplicity that each of the rats is very close to 200 g in weight. The following data are collected:

Group Number	Concentration of Compound (g/ml)	Volume Injected (ml)	Number of Rats Surviving
1	0.0010	0.25	10
2	0.0010	0.50	9
3	0.0010	0.75	6
4	0.010	0.10	2
5	0.010	0.15	1
6	0.010	0.20	1
7	0.010	0.30	0
8	0.010	0.40	0
9	Pure water	0.20	10

(B) From this data what is the LD_{LO}? (C) What is a value for NOEL? Now you are going to plot the results. (D) Decide first what units the x axis should have in the dose–response plot and convert this data to those units. Then plot the data on three types of graph paper: conventional linear paper, semilog paper, and probit paper. Determine a value for LD_{50} from each. One of these plots is particularly good for estimating a threshold value. (E) Which plot is that and what is the value?

8. A series of chemicals are to be used together in the workplace and are being tested for their ability to cause male rat infertility. The results are displayed on the table below:

Compound	Male Rats Sterilized (%)
A	5
B	11
C	8
A + B	17
A + C	9
B + C	28

Assume the deviation expected on this data is about ±10%. Indicate whether each combination of compounds displays additive, antagonistic, or synergistic effects.

9. T. Inamasu and N. Ishinishi report (*Jpn. J. Hyg.*, 35, 179, 1980) that when animals are dosed with arsenite the LD_{50} is 31.4 mg/kg, but when these animals are pretreated with PCBs, the LD_{50} is 44 mg/kg. Are the effects of these compounds additive, antagonistic, or synergistic?
10. Lacquer is sprayed on some decorative screens in a booth, and the following levels of solvent components are found in the booth atmosphere:
    n-Butyl alcohol, 14 ppm
    Methyl cellosolve, 12 ppm
    Methyl alcohol, 25 ppm
    Would you recommend immediate replacement of the ventilation system of the booth? Use PEL values from Table 7.5.
11. An example of a difficult epidemiological experiment to design is the effect of a particular food on general health. Such studies are frequently done, and, given the huge differences in individual diets and lifestyles, require great care to produce a valid design. The author was recently at a talk where it was stated that the French have better health statistics regarding cancer and heart disease, because they drink wine, than the British, who drink beer. The premise was that antioxidants in the wine were responsible for the better health data. This question is not intended to debunk that statement, but to use it as an example of difficulties of experimental design. What questions would you want to see answered before you would accept this assertion?

CHAPTER 5

Toxicokinetics: Toxicants Into, Around, and Out of the Body

A chemical must contact a person in some fashion to cause injury. Some substances cause damage simply by contacting the skin or the most sensitive part of a person's outer surface, the eyes. In this chapter, however, we are concerned with substances that enter the body and are distributed by the blood stream. Once a toxicant is in the body, the concerns are how long it stays, what organs it damages, and how it is removed. The toxicologist discusses the rates of various events in this passage through the body (Figure 5.1) under the general heading *toxicokinetics*.

People who work in toxicokinetics sometimes refer to passage through the body by the abbreviation ADME (absorption, distribution, metabolism, elimination). Each of these steps has an effect on the blood level of the toxicant, and that blood level measures how much harm the toxicant is capable of doing. Plotting blood levels of toxicant against time, starting at the time of exposure, one sees a rising curve that peaks and gradually tapers again to zero (Figure 5.2). In that plot the highest level reached by the toxicant is termed C_{max} and the time required to reach that maximum is termed t_{max}. Because of individual differences, the plot that each identically exposed person generates is different. Further, a number of other factors must be considered, such as levels of physical activity, types and amounts of food consumed, and state of health.

ENTRY OF TOXICANTS INTO THE BODY

When we say a chemical has entered the body, we usually mean it has gained access to the circulatory system (the blood stream). Three important routes to the circulatory system are absorption through the stomach or intestines (the gastrointestinal [GI] tract), the skin, or the lungs.

BY THE GI TRACT

If a person is described as having been poisoned, immediately one thinks the person swallowed a toxic chemical; years of indoctrination by murder mysteries. Once swallowed, the chemical can cross the mucosal barrier of the stomach or intestine (the layer of cells lining the lumen of these organs) and enter the blood. This is the most important route of entry for toxicologists concerned with the safety of food, pharmaceuticals, or the water supply, so their studies focus on the hazards of specific ingested toxicants.

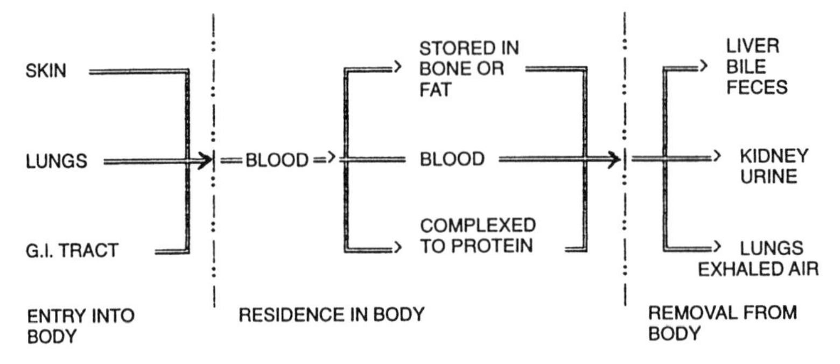

Figure 5.1. Flow diagram of a chemical into, around, and out of the body.

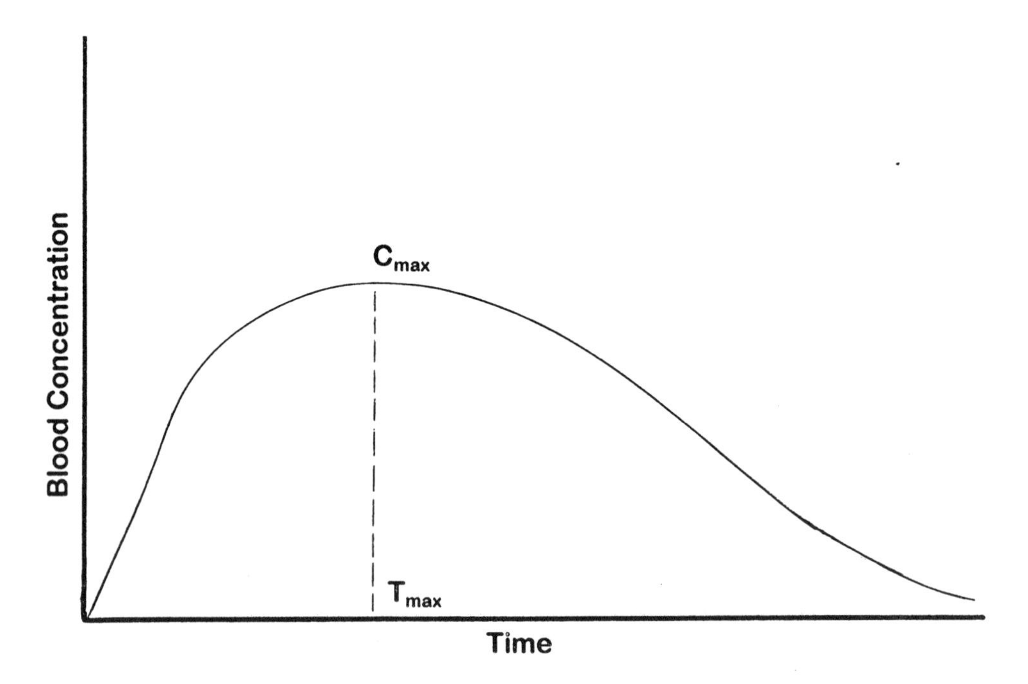

Figure 5.2. Time course of toxicants in body from a single dose. The blood level of the toxicant rises initially as it enters the blood. As the level rises, so does the rate of removal from the blood. Rate of entry is slowing, while rate of removal is rising. At C_{max}, the toxicant is entering and leaving the blood at the same rate. Thereafter, rate of leaving is greater, and the concentration steadily drops. The time at which C_{max} is reached is termed the T_{max}.

However, this pathway into the blood is the least threatening to the worker. Few workers would voluntarily place industrial chemicals in their mouths. Chemicals can inadvertently enter the GI tract if the worker does not carefully wash his or her hands before a meal break or at the end of the day, then transfers the chemical to food or a beverage. The prevention of such accidental intake is a matter of stressing personal hygiene.

THROUGH THE SKIN

Absorption through the skin is a more important workplace problem. Workers are likely to have contact with solvents, wash solutions, coolant solutions, lubricants, and other liquids in the course of performing their job. Beyond the question of irritation to the skin (Chapter 6), the hazard to the worker depends on the answers to two questions. First, how effective is the skin as a barrier to the entry of the compound into the body? Second, if the compound does enter the body, how much can the body tolerate without ill effects? Remember, the worker may also be inhaling vapors of that chemical, adding to the level of intake.

The skin effectively blocks entry of many chemicals. This is accomplished by an outer layer of tightly interlocked surface cells, the keratin layer of the epidermis. As these cells develop, they move farther away from the blood supply, replace normal cell contents with protein fiber, and cease to be living cells (Figure 5.3). A heavy keratin layer builds at points of high friction, such as the palms and soles of the feet.

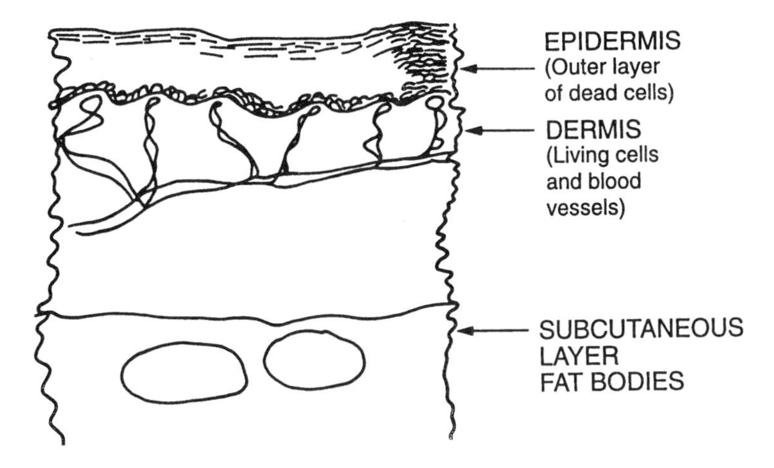

Figure 5.3. Cross-section of the skin.

However, compounds able to penetrate this keratin layer, most seriously a number of solvents, reach blood vessels below. Passage through the skin is easier where the skin is thinner, so it is more threatening to spill a chemical on the forearm or back of the hand than on the palm of the hand. Pores and hair follicles pass through the keratin layer, representing a potential alternative pathway for toxicant entry. However, pores and follicles are infrequent in occurrence, and so are considered to be of minor importance. Damage to skin, as by cuts, abrasions, or burns, increases the likelihood of a compound gaining entry. For this reason, when the potential of entry of a chemical through the skin is tested with animals, half the skin in the contact area is often lightly abraded. Warm water, especially water containing detergents, and some organic liquids loosen and soften the keratin layer, reducing its effectiveness as a barrier.

THROUGH THE LUNGS

Finally, compounds may enter the body by way of the lungs, the route of greatest concern in the workplace. Users of illegal drugs who smoke the drug (crack cocaine or ice amphetamines, for example) utilize this rapid route of entry to the blood, a route that bypasses metabolic mechanisms of protection in the liver. Most standards for permitted maximum exposure to a chemical (threshold limit values [TLVs], permissible exposure levels [PELs]) are standards for concentration of the chemical in the workplace atmosphere. Overall, the largest single effort in plant inspection for health hazards is directed at potential worker exposure to chemicals in the air.

Lungs consist of a highly branched set of air passages called the bronchial tree, terminating in little sacs, the alveoli (Figures 5.4 and 5.5). The barrier between blood and inhaled air is very thin in alveoli to facilitate the exchange of oxygen and carbon dioxide, and it is chiefly

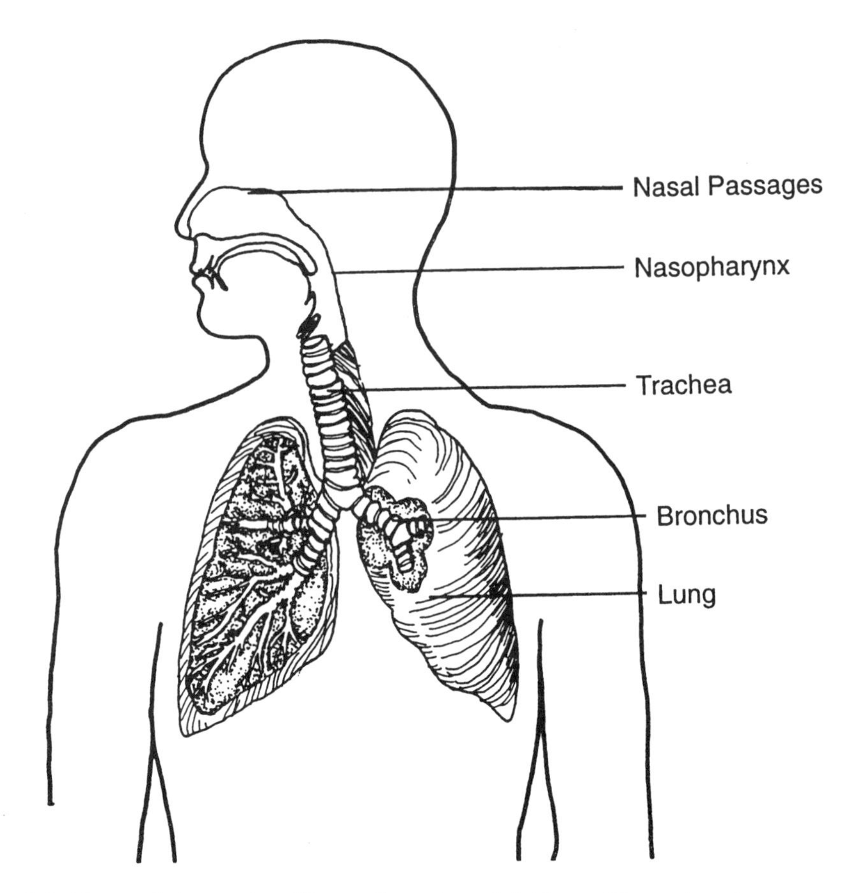

Figure 5.4. The respiratory system. We see the major subdivisions of the system: the nasal passages connected to the trachea and subdividing into the bronchial tubes. After many subdivisions the system terminates in closed sacs. All air passages up to, but not including, the terminal air sacs are covered with a sticky mucous surface that traps inhaled particles. Hair-like cilia move the mucus toward the nasal passages, carrying the trapped particles along. (From T. Godish, *Air Quality,* Lewis Publishers, Chelsea, MI, 1985. With permission.)

here that inhaled chemicals gain entry to the blood. More details about this route of entry are given in Chapter 7.

DISTRIBUTION OF TOXICANTS THROUGHOUT THE BODY

Compounds are transported by the blood to all parts of the body. Their immediate fate depends on two factors:

1. *Binding to protein.* Blood contains a variety of proteins, some of which bind to specific substances to transport them. Hormones, oxygen, and metal ions are examples of substances normally transported in this fashion. Some toxicants also bind to protein, which prevents them from passing through the capillary walls and entering the surrounding cells. Although this prevents them from causing damage, it also means they cannot enter the kidney tubular system to be removed from the body with the urine, so their stay in the body is extended. That is the good news and the bad news.
2. *Polarity of the compound.* Cell membranes are very nonpolar, essentially grease barriers. Chemicals that are very polar, such as most nutrients and minerals, would not be able to

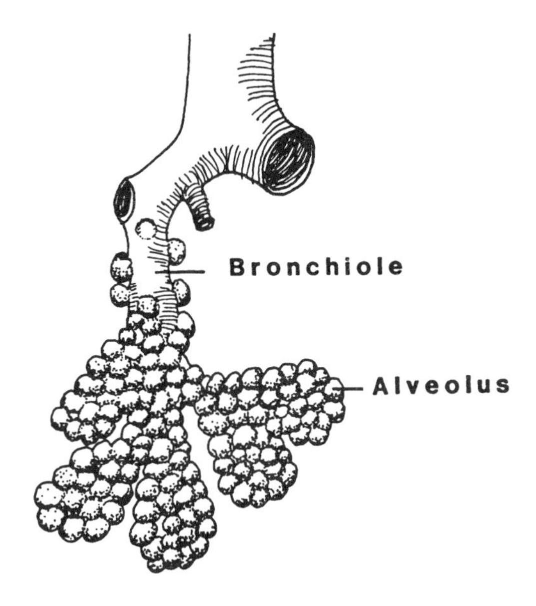

Figure 5.5. The alveoli. These thin-walled air sacs are the termination of the respiratory system. The barrier here between inspired air and the blood is very thin, allowing easy exchange of gases. (From T. Godish, *Air Quality,* Lewis Publishers, Chelsea, MI, 1985. With permission.)

cross these barriers into the body cells without specific transport systems built into membranes for that purpose. Polar toxicants are similarly blocked by membranes, although a few are moved by the nutrient transport systems because they resemble substances the cell needs. By contrast, nonpolar toxicants dissolve through the membranes and move relatively freely into body tissues. They may also enter body fat much as they would dissolve into a nonpolar solvent and remain stored there to be released later. For example, storage in body fat of nonpolar environmental toxicants such as chlorinated hydrocarbon insecticides is an important concern. Levels in the body can become relatively high when exposure to those compounds continues over a period of time, even when the overall level of exposure is relatively low.

METABOLISM

Throughout the body, but particularly in the liver, enzyme systems convert nonpolar foreign substances into different chemical species. Readers familiar with the enzymes of normal metabolism would find these systems surprising. Rather than being highly specialized, relatively few such enzymes collectively are able to chemically alter almost any organic structure. However, they catalyze changes at a much slower rate than do conventional enzymes.

Most of the changes promoted by these enzymes fall into two broad classes. One class is oxidation, and in oxidation reactions the substances become more polar (Figure 5.6). The other class, conjugation, involves coupling such foreign substances to very polar or charged structures (sulfate ions or charged derivatives of carbohydrates; Figure 5.7). Here too the effect is to increase the polarity of the compound. This increased polarity reduces access of the compound to cells and facilitates removal from the body by the kidneys.

This metabolism does not necessarily reduce toxicity at the same time. It may, but there are also cases where just the opposite occurs. For example, combustion produces polycyclic aromatic hydrocarbons (PAHs), which are relatively harmless compounds. However, they potentially could be retained in the body for a long time because they are nonpolar. Oxidation

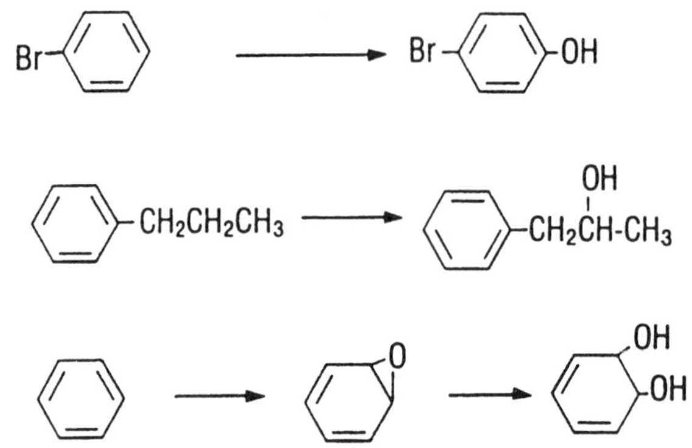

Figure 5.6. Oxidation of toxicants in the body. The products of foreign substance oxidation, which occurs largely in the liver, are more polar, making it easier for the kidney to remove them from the blood. Oxidized positions provide attachments for conjugation, shown in Figure 5.7.

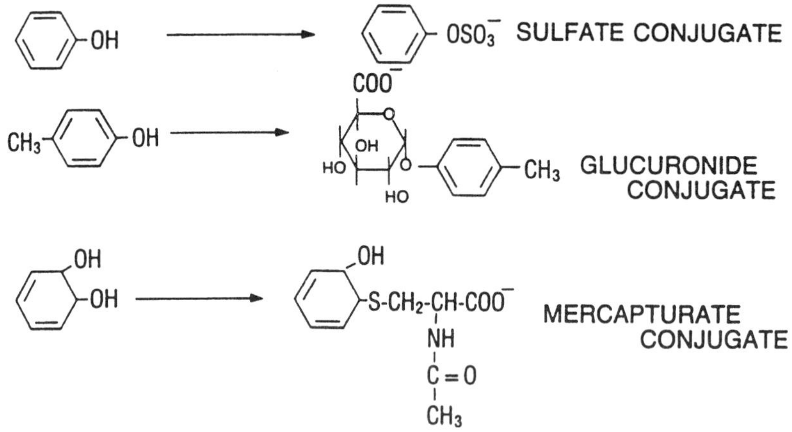

Figure 5.7. Conjugation of toxicants. The toxicant is atttached to another molecule. Notice in each case the attached molecule is charged, making the combination more water soluble. This facilitates removal from the blood by the kidney.

by metabolic enzymes converts them into products that are more polar, but these polar products are potent carcinogens. Once again we have a good news–bad news situation.

REMOVAL FROM THE BODY

KIDNEYS

The chief pathway for removal of toxicants from the body is to transfer them in the kidneys from the blood into the urine. Each time the heart contracts, about 25% of the blood pumped out to the body will pass through the kidneys. The basic functional unit of the kidney is the nephron, a structure that starts with the glomerulus where fluid is removed from the blood, extends through a long tubular system in which the content and concentration of the fluid is adjusted, and ends with the fluid (now urine) being passed into a collecting system that leads to the bladder.

In the glomerulus, blood is literally filtered through a porous membrane so that a water solution of small molecules enters the tubular system and the relatively large blood cells and proteins remain in the blood.[1] The filtrate passes down a tube, which has blood vessels closely associated with its walls. Selective transport proteins are built into this tube that bind desirable compounds such as nutrients and important minerals and move them back into the blood vessels. When the fluid reaches the end of this tube, glucose, amino acids, sodium ion, and about 99% of the water have been scavenged (Figure 5.8). As water returns to the blood, the concentration of solutes remaining in the fluid increases. Polar solutes are trapped in the absence of transport structures specific for them. However, nonpolar molecules are now much more concentrated in the urine than in the blood and, driven by this concentration gradient, dissolve through membranes of the tube walls and return to blood without the help of a transport structure.

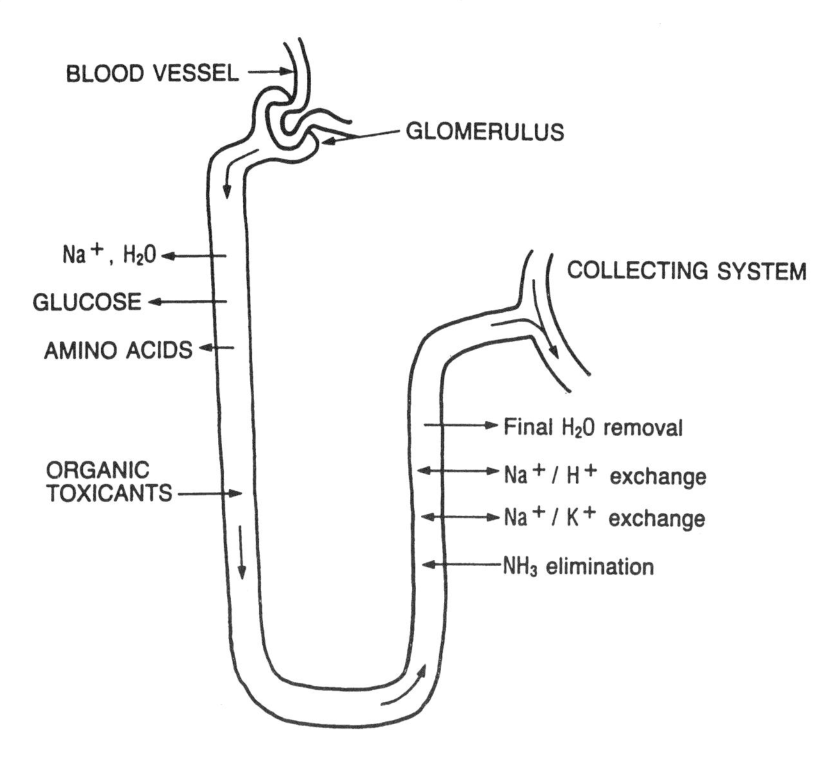

Figure 5.8. Kidney function. The fluid portion of blood is filtered into the renal tubular system at the glomerulus. As this fluid moves through the tubes, solutes the body wishes to retain (largely glucose, amino acids, and salts) are transported back into the blood, whereas the metabolic wastes and toxicants hopefully remain in the tubule. Nonpolar toxicants tend to dissolve through the tubule walls and reenter the blood, but polar toxicants remain. As sodium ion is returned to the blood, water follows to preserve the osmotic balance. Organic toxicants that passed the earlier filtration may be pumped into the blood later. Finally at the end of the system, exchanges are made to adjust blood pH and final recapture of water occurs.

The term *clearance* is used to describe the efficiency of the kidney at removing substances form blood. Clearance is the volume of blood completely cleared of a substance per unit time and is expressed as ml/min.[2] For a given compound (t), at a time when t is not entering the body, the amount leaving the blood equals the amount added to the urine:

[1] One of the symptoms of kidney damage is the appearance of blood or protein in urine, inferring that the glomerulus has been damaged and is allowing larger structures to pass into the nephron.
[2] The definition of clearance is misleading, because it seems to infer that the kidney takes a volume of blood and completely cleanses it of toxicant or waste product. In fact, compounds add to urine as the result of equilibria involving all of the blood filtrate, some equilibria being influenced by transport structures.

$$U_t V = Cl_t B_t$$

where U_t = concentration of t in urine
B_t = concentration of t in blood
V = volume of urine excreted per minute
Cl_t = clearance of t

$$Cl_t = \frac{U_t V}{B_t}$$

SAMPLE PROBLEM:

If 25 mg/ml of toxicant, t, are excreted in urine at a rate of 2.0 ml/min, and the blood concentration of t is 0.5 mg/ml, what is the clearance of t?

$$Cl_t = \frac{U_t V}{B_t}$$

Clt = (25 mg/ml)(2.0 ml/min)/0.5 mg/ml

Clt = 100 ml/min

A high clearance value is displayed by a substance that is cleared rapidly from the blood. A comparison of the expected clearance of a compound (such as creatinine) regularly appearing in urine with the present clearance in a particular subject provides a measure of the operating efficiency, and thus the health, of the kidney.

Basically, waste products and polar foreign substances are concentrated in the urine and are sent to the bladder. Other adjustments to urine are performed by the kidney, but those adjustments are of less importance to this discussion. When compounds in the urine are carcinogens, such concentrating raises the dosage of surrounding tissues and is the cause of kidney and bladder tumors.

LIVER

The liver removes a few foreign substances from the blood, transfers them into a fluid called bile, and sends them for storage to the gall bladder. Bile is primarily a solution of detergent-like molecules. When food enters the small intestine the gall bladder contracts and sends the bile into the intestine where the detergents of bile, called bile salts, stabilize the dietary fats as droplets (micelles). The toxicants placed in bile by the liver pass along the GI tract with food and eventually are eliminated with the feces, unless during transit they are reabsorbed through the intestinal wall. At best, this pathway eliminates only a small fraction of the amount and number of foreign substances removed by the kidney.

HALF-LIFE OF TOXICANTS

One of the studies performed on toxicants is the determination of their half-life in the body. By half-life we mean the length of time required when a known amount of a compound is taken into the body for half of it to be removed. Half-life depends on all the factors presented in this chapter. Compound polarity is an important factor. Polar compounds, restricted in their ability to cross membranes, tend to distribute less well into cells from blood, and remain trapped in urine, facilitating their removal from the body. They display shorter half-lives because of these factors. Nonpolar compounds are more difficult to eliminate because they distribute throughout the body cells, store in the fat, and transfer spontaneously back to the blood in the kidney by dissolving across membranes. Their half-life is generally much longer.

If a compound is bound to a blood protein its half-life is greatly extended. It does not enter cells readily, but it is not filtered out of the blood in the kidney either. Such binding is not a permanent attachment. During periods when the foreign molecule equilibrates off (is temporarily freed from) the protein, it acts like any similar unbound molecule and may then exert toxic effects or be eliminated.

Not mentioned earlier is the ability of a few toxic metal ions, because of their similarity to calcium ion, to be incorporated into bone structures. Cadmium and lead ions are significant examples. Bone is not a static tissue, but exchanges its ions with fluids surrounding the bone. There is a fluid bathing the bone surface that contains dissolved ions, and this fluid is separated by a membrane from surrounding body fluids. Transport structures to move ions either way are found in this membrane, and the regulation of blood calcium concentration is accomplished by transport at these sites under the control of a series of hormones. Ions resembling calcium are moved at the same time. Toxic ions incorporated into bone do not permanently remain there, but their entry into it greatly extends their half-life.

KEY POINTS

1. Toxicants enter the blood circulation to be distributed throughout the body through the GI tract, skin, and lungs. The GI tract is a minor route of entry in the workplace, because workers seldom swallow industrial chemicals. Passage through the skin, involving crossing the keratin layer of protective cells, is more likely. The lungs are the pathway of greatest concern, because they provide a very large surface across which airborne toxicants may transport.
2. Toxicants may bind to proteins in blood, which prevents them both from entering tissues to cause damage and from being removed from the blood by the kidneys.
3. Nonpolar toxicants cross membrane barriers easily, and thus enter cells readily to cause damage and are hard to remove from the body. They may also store in body fat.
4. Many cells, but particularly liver cells, metabolize toxicants to facilitate their removal from the body. Oxidation makes toxicants more polar and conjugation attaches them to polar structures.
5. The chief exit route from the body is transference to urine in the kidneys. A few compounds transfer to bile in liver and, when bile is added to the intestinal contents, may leave with the feces.
6. The time required to remove half the dose of a toxicant from the body is the toxicant half-life.

BIBLIOGRAPHY

E. J. Calabrese, Pollutants and High Risk Groups: The Biological Basis of Increased Human Susceptibility to Environmental and Occupational Pollutants, Wiley-Interscience, New York, 1978.

C. D. Klaassen, M. O. Amdur, and J. Doull, Toxicology: The Basic Science of Poisons, 3rd ed., Macmillan, New York, 1986. (A very thorough and complete presentation of the material in Chapters 1–3 of this volume is given in Unit 1.)

PROBLEMS

1. Workers get cutting oils on their hands during a gear-cutting operation in a transmission plant. A chemical used in cutting oils has an acute oral LD_{50} of 195 mg/kg. What is the likelihood of a worker being injured by eating an apple on break without washing his or her hands first? If the worker did this as a regular practice, what characteristics of the chemical could raise the level of hazard from this careless practice?
2. Workers in the repair and maintenance facility of a highly automated plant building two-cycle chain-saw engines and drive systems have borrowed some solvent from the painting operation to clean grease from parts before they work with them. A part is dipped briefly in the solvent, then carried to a bench. The solvent has a dermal LD_{50} of 90 mg/kg. When challenged that this is hazardous, the workers point out that no one has shown any ill effects from the practice, and the solvent does an excellent job of degreasing. You counter by saying, "You have been lucky so far. Here are circumstances that just haven't happened yet that would raise the toxic potential of skin exposure to this chemical: . . . "
3. On testing, the following compounds have shown potential as mold growth inhibitors to be added to water-based paints:

(a) 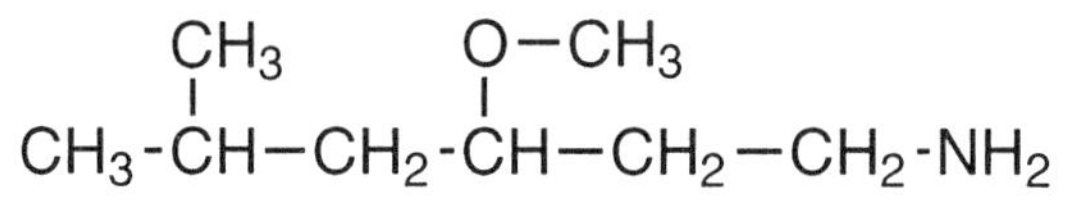
$$\begin{array}{rclcl} & CH_3 & & O{-}CH_3 & \\ & | & & | & \\ CH_3\text{-} & CH & {-}CH_2\text{-} & CH & {-}CH_2{-}CH_2\text{-}NH_2 \end{array}$$

(b) 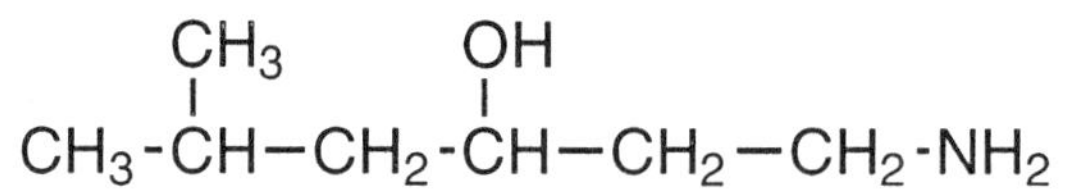
$$\begin{array}{rclcl} & CH_3 & & OH & \\ & | & & | & \\ CH_3\text{-} & CH & {-}CH_2\text{-} & CH & {-}CH_2{-}CH_2\text{-}NH_2 \end{array}$$

(c) 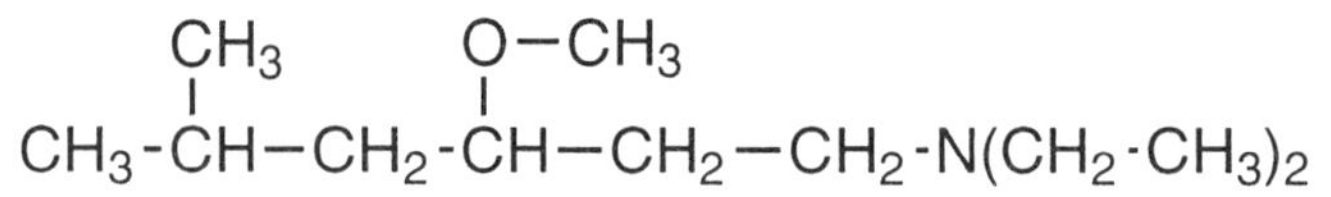
$$\begin{array}{rclcl} & CH_3 & & O{-}CH_3 & \\ & | & & | & \\ CH_3\text{-} & CH & {-}CH_2\text{-} & CH & {-}CH_2{-}CH_2\text{-}N(CH_2\text{-}CH_3)_2 \end{array}$$

(d) 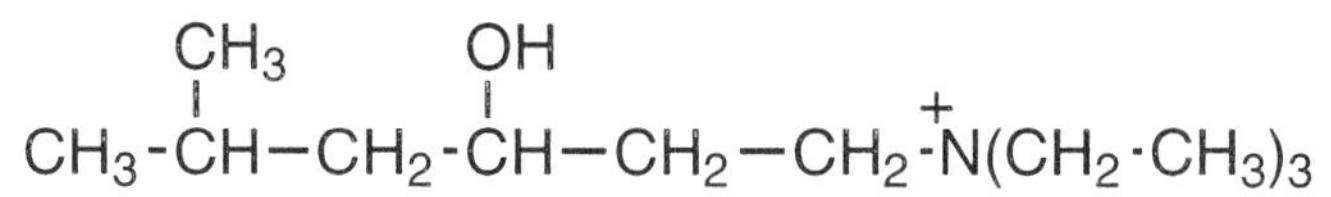
$$\begin{array}{rclcl} & CH_3 & & OH & \\ & | & & | & \\ CH_3\text{-} & CH & {-}CH_2\text{-} & CH & {-}CH_2{-}CH_2\text{-}\overset{+}{N}(CH_2\text{-}CH_3)_3 \end{array}$$

(e) 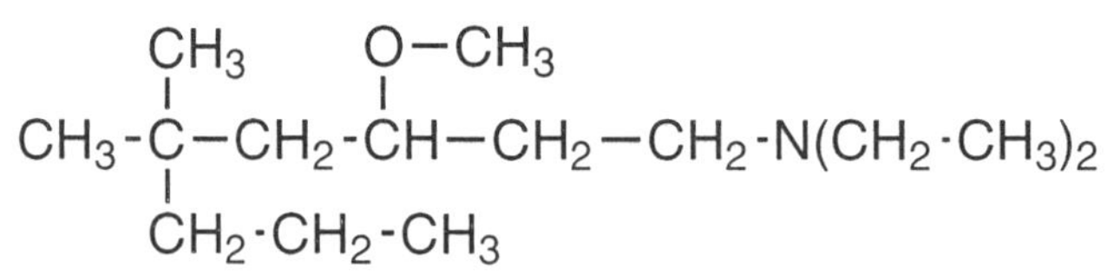
$$\begin{array}{rclcl} & CH_3 & & O{-}CH_3 & \\ & | & & | & \\ CH_3\text{-} & C & {-}CH_2\text{-} & CH & {-}CH_2{-}CH_2\text{-}N(CH_2\text{-}CH_3)_2 \\ & | & & & \\ & CH_2\text{-}CH_2\text{-}CH_3 & & & \end{array}$$

A. Predict the relative half-lives in the body that these would display based on relative polarity.
B. As a first assumption, would you expect values for clearance by the kidneys to be in the order you would have predicted for half-life?
C. Using test animals to study compounds (a) and (c), the following data were obtained:

Compound	Volume of Urine (ml/2 hr)	Conc. in Blood (μg/ml)	Conc. in Urine (μg/ml)
a	1.5	9.1	38
c	1.4	7.2	41

What are the clearance values?

D. Do these clearance values correspond to predictions based just on polarity?

E. What factor other than polarity affects half-life?

4. Refer again to the ADME described on the first page of the chapter. Based on your understanding of each of those subunits of the toxicokinetics process, discuss how each affects the C_{max} value and the rate at which it is reached.

Section III
Chemical Hazards in the Workplace

Chapters 6–11 present a detailed discussion of specific health and safety problems associated with chemicals. Hazards from chemicals used on the job are a major industrial concern, require extensive effort to monitor and control, and demand sophisticated training of the industrial hygienist and other health professionals to ensure competence in confronting them. Each chemical presents its own profile of low or high reactivity, toxicity, flammability, volatility, solubility, and other properties that affect the level of hazard encountered.

Discussion begins with chemical damage to skin, dermatosis, and its prevention. The next three chapters describe inhalation of chemicals from the toxicologic standpoint, outline methods of monitoring the plant atmosphere, and survey corrective and protective measures. Chapter 10 deals with fire and explosion hazards. Finally, Chapter 11 describes OSHA's special attention to workers in unpredictable circumstances: clearing hazardous waste sites, responding to unplanned chemical releases, and operating industrial facilities involving large amounts of hazardous chemicals.

CHAPTER 6

Occupational Dermatosis and Eye Hazard

To this point the focus of the text has been on damage caused by chemicals entering the body. Absorption across the skin is an important route of such entry,[1] and this aspect was presented in Chapter 5. Irritation of the skin or damage to the surface of the skin as a result of contact with chemicals is a different and important problem in occupational safety and health, and is now addressed.

OCCURRENCE OF INDUSTRIAL DERMATOSIS

Skin problems are the most prevalent job-related diseases. Of approximately 240,000 new occupational illness cases listed by the Bureau of Labor Statistics in 1988,[2] from 24 to 37% of these were skin diseases. These represent 70% of industrial claims paid by insurance companies. Furthermore, NIOSH believes skin problems are highly underreported. Incomplete reporting of dermatosis problems may be common because of a loophole in the requirements for reporting injury, which allows a company to omit a minor incident from the OSHA log if it is classified as an "injury" rather than an "illness" and no time is lost, especially if only first aid is required.[3] The rate of occurrence has dropped over the years as a result of programs to prevent skin contact with chemicals and of the changing nature of the workplace (separation of the worker from the work process, as by automation). The cost in the U.S. of occupational dermatosis, considering medical costs, compensation for lost wages, and loss of worker productivity, has been estimated by Mathias (1985) to be in the range of $220 million to $1 billion.

About a quarter of all workers utilize or are around some kind of skin irritant, and around 1% exhibit problems from this exposure. Workers in agriculture and manufacturing run the biggest risks. According to the Standards Advisory Committee on Cutaneous Hazards, very high rates of skin problems are found in leather tanning plants, poultry dressing plants, and meat packing plants. Industries involving contact with irritating chemicals such as plating and rubber fabrication also rank high on the list (Table 6.1).

[1] Skin absorption of chemicals is sometimes overlooked in the zeal for respiratory protection. Absorption through the skin can account for 25% of the uptake of chemicals. (S. L. Campbell, "Is Your Chemical Protective Clothing Quickly Becoming Landfill Material?" *Occupational Health and Safety*, June 1994, p. 33.)

[2] U.S. Department of Labor, Bureau of Labor Statistics, "Occupational Injuries and Illnesses in the United States by Industry, 1988," Bulletin 2366, U.S. Government Printing Office, Washington, D.C., 1990.

[3] 29 CFR 1904.12.

Table 6.1. Occupational Skin Diseases in High-Risk Industries.

Industry	Incidence Rate (per 1,000 workers)	Number of Workers (approximate)	Workdays Lost per Year
Leather tanning	21.2	22,900	1,392
Poultry dressing	16.4	89,800	4,405
Boat building and repair	11.1	48,000	854
Ophthalmic goods	8.5	38,000	1,390
Chemical preparations	8.3	36,700	855
Plating and polishing	8.3	61,400	1,270
Meat packing	7.2	164,300	1,561
Frozen fruits and vegetables	7.2	43,200	1,153

Adapted from Standards Advisory Committee on Cutaneous Hazards, OSHA.

SKIN ANATOMY

The skin is the largest organ in the body, including approximately 18,000 cm^2 (20 ft^2) of surface. The thickness of this covering varies from about 0.5 mm on the eyelids to about 4 mm on the back and such high abrasion areas as the palms of the hands and the soles of the feet. Skin is composed of two layers, epidermis and dermis, which are quite different in structure and function. A small percentage of the surface area includes pores and hair follicles, openings that lead into the interior of the skin.

EPIDERMIS

The epidermis, the outermost of the two skin layers, interfaces with the outside environment and must serve several protective functions. This layer must resist abrasion as the skin contacts other surfaces and objects, prevent unwanted loss of water from the body, resist the action of chemicals, and block the entry of microbes.

The epidermis is chiefly assembled from cells called keratinocytes. These originate at the epidermal inner layer with a group of basal or stem cells that can perform the usual cell functions, metabolizing nutrients that reach them by diffusion from the inner skin layer and dividing to form additional basal keratinocytes. As these cells divide, excess cells shift toward the outer surface of the epidermis. As they shift, they begin to produce fibrous structural proteins called keratin intermediate filaments. The filaments are organized and packed in the cell by interaction with a type of protein called filaggrin. The cell membrane is replaced by a rigid cell envelope composed of three layers of protein that are bonded together (cross linked) into a single tight layer, and that bind at the inner surface to the keratin inside the cell. Throughout this process the cell is losing its ability to perform normal cell functions and is steadily progressing toward the outer surface of the epidermis, driven by other cells from the basal cell layer below. Granules develop inside the cells that produce lipids (very nonpolar molecules) that are extruded from the cell and covalently bond to the outer surface to form a water repelling layer. The cells have now specialized to serve their protective functions, and form the outer stratum corneum or keratin layer. These outer cells are gradually rubbed off or otherwise lost (desquamation) and are replaced by new cells moving up from layers of living cells below. Humans are not aware of this loss of epidermis unless they accelerate it, as when they damage the skin by sunburn or if they are afflicted by the dread social disease, dandruff. The complete process from original basal cell division to desquamation takes about 28 days.

Melanocytes in the inner portion of the epidermis produce the pigment melanin to reduce damage to the skin by light, especially UV light. Humans who traditionally inhabited tropical regions of the world adapted to greater solar radiation by producing more melanin, and hence have darker complexions. A short-term increase in exposure to light stimulates increased production of melanin, resulting in suntan.

DERMIS

The inner skin layer, dermis, resists tearing or puncture. A meshwork of collagen fibers, rope-like protein structures that are strong without being rigid and unbending, is like a woven cloth, flexible but resistant to tearing. Should damage occur, specialized cells (fibroblasts) produce new collagen to heal the wound. Scars are the result of heavy collagen production.

Dermis contains blood and lymph vessels which supply the needs of the dermal and living epidermal cells. These blood vessels also play a role in body temperature regulation as blood circulating in the skin loses heat through the epidermis to the surrounding air. When body temperature rises, relaxation of muscle fibers in blood vessel walls causes an increase in dermal vessel diameter, increasing the amount of blood in the skin and consequently the rate of heat loss. Extra blood in the skin gives it a reddened or "flushed" appearance. Correspondingly, in cold weather the diameter of blood vessels shrinks to minimize heat loss.

Nerve endings in the dermis generate sophisticated sensory functions, allowing detection of physical contact, heat, cold, and texture. Potentially damaging contacts lead to a pain sensation, stimulating removal of the body from hazardous situations.

PORES AND HAIR FOLLICLES

Further temperature control is provided by the approximately 2,000,000 sweat glands of the dermis. These glands are coiled tubes deep in the dermis that extend through the epidermis to the surface of the skin, the opening being termed a pore. Sweat is simply water with dissolved salts collected in the coiled base of the gland, and it is produced at all times. Sweat coats the surface of the skin, and heat is lost as it evaporates. Production of sweat increases as the need to cool the body increases, either because of hot weather or the generation of excess body heat due to exercise or a fever. The rate of sweat production then exceeds the rate of evaporation and causes a person to become aware that he or she is perspiring. When the skin remains wet, the moisture loosens and speeds removal of the outer cells of the epidermal keratin layer. Continuous perspiration therefore reduces the protection against abrasion and the entry of chemicals provided by the outer epidermis. Skin that is constantly coated with sweat may also develop a "prickly heat" rash.

The surface of the skin grows inward to form a follicle or tube at the base of which the keratin-forming cells construct a shaft of hair. Hair follicles have their roots deep in the dermis. Sebaceous glands, generally associated with the hair follicle, produce sebum (skin oil), which travels up the follicle and coats the surface of the skin. A coating of skin oil repels water and water solutions of chemicals. Sebum accumulates in the follicle if the flow is blocked, causing a skin eruption (pimple, zit). Widespread blockage in follicles is termed acne.

Pores and hair follicles are passages into the dermis, and might therefore be suspect as major routes of entry of chemicals past the keratin layer. However, they represent only roughly 0.1% of the surface of the skin, so are not major factors in the passage of chemicals.

CONTACT DERMATITIS

More than 90% of dermatosis cases are caused by contact dermatitis, direct contact at a localized site with a chemical agent. Types of agents and the damage they cause to the skin fall into several categories, with overlap occurring between these classes. Effects occur at the site of contact with the chemicals, and in the workplace most often the site is the hands, wrists, or arms.[4]

Atopic people, people with a particular allergy problem producing hayfever or asthma and an atopic dermatitis, represent about a quarter of the population. However, they are the great majority of the victims of occupational dermatitis. Their risk is calculated at 13.5 times greater than that of the rest of the population.[5]

KERATIN LAYER AS A BARRIER

The keratin layer is inert and, aside from corrosives, must be penetrated by chemicals before their effect is observed. Abrasion to the skin damages or even crosses the keratin layer, reducing the barrier to chemicals provided by this layer. Chapping caused by cold weather or low humidity in an indoor work environment makes skin more susceptible to damage by chemicals. Physical exertion or a high workplace temperature causes sweating, which softens the keratin layer and makes it more permeable. Frequent immersion of the skin in hot water, such as dishwashing water, softens the keratin layer much as does sweating. In fact any frequent wetting of the skin such as occurs with health care workers, food processors, beauticians, and bartenders increases the likelihood of skin problems. Poor personal hygiene contributes to dermal problems by leaving irritants on the skin, but so also does overzealous cleansing. Sometimes the skin is exposed simultaneously to several harmful agents, each contributing to weakening the keratin layer barrier.

IRRITATION

Many classes of chemicals irritate the skin. The irritation may result from the action of the chemical itself or it may involve an allergic process. These two processes involve different elements and are considered separately. Mild irritation involves the arousal of inflammatory processes that result in fluid retention (swelling) and increased blood flow (reddening). Stimulation of the sensory nerves leads to itching, burning, or stinging sensations. As irritation worsens, cell death occurs along with further swelling and reddening, resulting in a rash. Damage to blood vessels eventually produces bleeding. Blisters are formed by fluids generated by inflammation in more severe cases.

IRRITANT CONTACT DERMATITIS

Irritation of the skin as the result of direct action on skin tissues is termed irritant contact dermatitis. Such irritation may be the result of direct chemical damage to cells, physical damage such as might be caused by sharp particles, the long-term effects of removing skin oils, or a combination of these causes. Irritating chemicals are estimated to be the cause of 75–80% of all work-related contact dermatitis (Shmunes, 1988).

The most common cause of irritation is a liquid chemical into which hands are immersed or which is splashed accidentally on the skin. Reports indicate that skin irritants are also carried to the worker through the air, and this source of contact should not be ignored. Airborne irritants can slip past even protective clothing, which seldom presents a continuous, airtight barrier.

Chemicals that do reach living cells cause irritation by a variety of individual toxic mechanisms. Testing programs express relative irritation by a chemical as the ID_{50}. Parallel to the LD_{50} or LC_{50}, the ID_{50} is the dose of irritating chemical that causes a skin reaction in

[4] J. E. Keil and E. Shmunes, "The Epidemiology of Work Related Skin Diseases in South Carolina," *Arch. Dermatol.*, 119, 650, 1983.

[5] E. Shmunes and J. E. Keil, "The Role of Atopy in Occupational Dermatoses," *Contact Dermatitis*, 11, 174, 1984.

half the individuals tested. This value provides a measure of relative ability to irritate when comparing a series of compounds. This allows anticipation of problems when a new chemical is to be introduced into the workplace.

One must consider other aspects of the work environment when estimating potential problems. For example, is skin wet frequently or at the time of contact with this chemical, or has earlier contact with solvents stripped protective oils from the skin?

Physical Damage

Solids with sharp particles can abrade the skin, directly causing irritation by physically damaging cells. This opens routes for other chemicals to penetrate the skin to cause further problems. The thin sharp fibers of fiberglass are an excellent example of this, and can be contacted both by handling and by an airborne route.

Accidents such as scrapes or falls cause abrasions. Inflammation can result with reddening and swelling. Once again the continuity of the skin as a protective organ is interrupted.

Solvents

Solvents mix intimately with — dissolve — other chemicals. Water is perhaps the most common vehicle used to dissolve substances, but in industry, the term *solvent* ordinarily refers to organic liquids (structures based on carbon) used to dissolve other organic substances. The chemical characteristics of such liquids vary, but all would primarily dissolve other organic substances and would not dissolve inorganic substances such as metallic salts. A few, such as acetone and smaller alcohols, mix with water, but most do not.

Irritation by solvents usually requires extensive and regular contact. Such contact removes skin oils from the surface of the skin and gradually loosens and removes keratin cells. Skin tissue loses water, producing drying and scaling. Cell membranes are disrupted, killing the cells. The effectiveness of skin as a barrier to entry into the body of the solvent itself or of other chemicals is reduced. With heavier exposure skin cracks or splits, producing open sores, further simplifying chemical entry.

Soaps and Detergents

Soaps and detergents (surfactants, surface active agents) are chemicals that have in one molecule both a nonpolar (water-insoluble) and a polar (water-soluble) segment. Detergents function at the interface between nonpolar and polar (usually aqueous) environments, and so have a large hydrocarbon component able to interact with the nonpolar environment and a charged, or very polar, structure attractive to water. Soap is the best known detergent:

$$\underbrace{CH_3CH_2CH_2CH_2CH_2CH_2CH_2CH_2CH_2CH_2CH_2CH_2CH_2CH_2CH_2CH_2}_{\text{Nonpolar hydrocarbon segment}}\underbrace{\overset{\overset{\displaystyle O}{\|}}{C}—O^-Na^+}_{\text{Polar segment}}$$

A Soap Molecule

Detergents belong to one of three chemical classes: anionic (generally quaternary amines), cationic (including soap, a carboxylic acid, and household detergents, often sulfonic acids), and nonionic (having a very polar structure such as a polyalcohol). When water is the medium used to clean objects contaminated by water-insoluble contaminants, the nonpolar part of dissolved surface active agents adheres to these contaminants, coating their surface. Nonpolar structures now have a polar surface and suspend in water.

Soaps and detergents are formulated differently for different functions, for example, for cleaning inert objects versus the grease-contaminated skin of workers. Commercial cleansing agents contain other chemicals, for example, to prevent hard water minerals from forming scum, to make the solution alkaline, or to preserve the agent from bacterial attack. If germicidal action is important, for example, in food handling or hospital environments, the agent may contain phenol or phenolic derivatives. Hand cleaners may contain abrasives to mechanically loosen the dirt. Waterless hand cleaners include both a detergent cream and petroleum distillate, the latter being able to dissolve the grease due to its nonpolar character.

Damage to skin by soap or detergent solutions requires extended or frequent contact. As with solvent damage, detergents remove skin oils and loosen keratin cells of the epidermis. Surface active agents are most effective when the water is hot, which accelerates these effects. Abrasives in mechanics' hand soaps do little damage to the thick skin of the palms, but may have more serious effects on the backs of the hands or the forearms. Damage to the skin by surface active agents makes skin more susceptible to irritation from other agents.

ALLERGIC CONTACT DERMATITIS

An allergic response to chemicals (allergic contact dermatitis) is an important cause of occupational dermatitis, although it occurs only about one-fourth as often as the irritant contact dermatitis and differs fundamentally in mechanism from it. Problems with irritation on contact with chemicals previously described affect most workers having such contact and depend on both the reactivity of the chemicals with components of skin and the amount of contact with the chemical. Allergic response may affect only a small proportion of workers, but those affected may show severe responses relative to amount of exposure, whereas those not affected show no response at all. Affected workers may not respond to a chemical on first contact, but become sensitized and respond at some later time. Thereafter that worker responds to very small quantities of the chemical, although the response may require several hours to appear. The resulting reddening, swelling, and rash are due to an inflammation of the tissue, and are hard to distinguish from irritant contact dermatitis.

Allergic response begins with the chemical, termed an antigen, being absorbed into the skin and reacting with a skin protein. This complex must then stimulate an immune response, the production of large numbers of antibody proteins that specifically bind the chemical–protein complex. This process may require about 1 to 3 weeks. The person is now sensitized.

Antigen–antibody complexes activate body mechanisms the purpose of which is to specifically bind and destroy invading bacteria or viruses and so prevent infections; however, the side effect is inflammation. The sensitized individual continues to produce antibody to the chemical. If the antigen had been a bacterium or virus, this aspect of the process would provide "immunity" to reinfection. However, in the case of a chemical antigen, future exposures result in these antibodies binding to the chemical (protein binding is now not necessary) and stimulating the reddening, swelling, itching, and other characteristics of inflammation, an allergic response. A dramatic allergic response occurs to even very small amounts of the chemical on reexposure.

Determining Allergen Identity

Diagnosis by occupational physicians of the cause of allergic contact dermatitis can be difficult. There is a time lag between exposure to the chemical and appearance of inflammation, complicating identification of the specific exposure that causes inflammation. The work environment is often a complex source of chemical exposures. Furthermore, the cause of the allergic response could be outside of the workplace. There are chemicals that are recognized as being potent allergens or sensitizers, and their use in the plant raises suspicions that they are the source of the problem. Shifting the worker to a workstation that eliminates contact with suspect chemicals may relieve the symptoms, but this is not assured, because only traces of the chemical may be necessary to evoke the response.

Patch testing is a useful diagnostic procedure for determining what chemical is causing allergic contact dermatitis. The most commonly used Finn chamber method is very similar to other variations of the method. Samples of suspect chemicals (often mixed into petroleum jelly) are placed in small aluminum cups, which are taped to the worker's back, usually for 48 hr. The exposed spots are evaluated for inflammation 1 or 2 days later, allowing time for any immediate irritation of nonallergenic origin to fade and for slow responses to appear. Tests must be done with controls to ensure that the worker's skin is not reacting just to tape or to being covered.[6]

Sources of Allergens

The most common allergic occupational skin problems are associated with the contact of outdoor workers with plants that produce potent allergens such as poison oak and poison ivy. Similar allergens are found in mango trees, Japanese lacquer trees, and cashew nut trees. Although contact with these latter plants in the U.S. is unlikely, products of those plants are found on the market. An oil is extracted from cashew nutshells and used in a range of plastics, lubricants, coatings, and other products. Lacquers and varnishes may be based on products of the lacquer trees. A person already sensitive to poison oak or ivy may respond to those products. Other plant components that cause problems to nursery workers include chemicals in a number of common plants, and forestry workers may be sensitized to lichens on bark or chemicals in sawdust.

Allergens are found throughout industry. In polymers, both plastics and rubbers, the monomers, hardeners, accelerators, antioxidants, and diluents of resins and rubbers include some allergens. For example, isocyanates, epichlorohydrin, bisphenol A, *p*-phenylenediamine, diethylenetriamine, and triethylenetriamine frequently cause allergic responses. A few metals are sensitizers. Nickel is the most common metallic allergen, and may be contacted in cuttings or solutions such as cutting oils or electroplating baths. The next most frequent metallic allergen is hexavalent chromium,[7] which is encountered in such industrial applications as tanning, electroplating, etching and photoengraving, and in pigments. Small amounts of hexavalent chromium are found in cement. Cobalt is also a potent allergen, but is less commonly encountered. However, cobalt is found in cement and in nickel products, and often workers sensitized to chromate or nickel are found also to be sensitized to cobalt.

Biocides, agents added to prevent deterioration due to growth of bacteria or fungi, are used in cutting oils, paints, adhesives, and hand soaps, and are sometimes allergens. Dyes, particularly azo dyes, may be allergens. Azo dyes are used in textiles, color photography, and hair coloring compounds. Agricultural workers can be sensitized to pesticides, particularly to organomercurials, carbamates, captans, triazines, and thiurams.

PHOTODERMATITIS

Some chemicals do not directly cause irritation on contact with skin, but symptoms of irritation appear later when the skin is exposed to sunlight. UV light converts them into different and more irritating chemicals.

[6] T. Fischer and H. I. Maibach, "Patch Testing in Allergic Contact Deramtitis," in H. I. Maibach, Ed., *Occupational Dermatoses,* 2nd ed., Year Book Medical Publishers, Chicago, 1987.

[7] Paradoxically, hexavalent chromium is a much more common allergen than trivalent chromium, but the actual sensitivity may be to the trivalent form. This is because the hexavalent form penetrates the skin more readily, then is converted to trivalent form and stimulates allergic response.

The most common phototoxic agent is coal tar (creosote, pitch). Workers who handle railroad ties often complain of "tar smarts". Symptoms include burning, stinging, and skin reddening.

A number of agricultural products produce the phototoxic agent furocoumarin, including celery, citrus fruits, parsley, dill, and carrots. In addition to irritation, skin may darken after exposure. Workers harvesting these crops are particularly susceptible, but anyone handling such produce extensively can be affected.

Less common causative agents include some dyes, drugs and other miscellaneous chemicals. Persons experiencing problems with phototoxic agents who cannot avoid exposure to the agent can obtain relief from the symptoms by placing sunscreen on exposed skin, thus avoiding the photoactivation.

HAIR PROBLEMS

Hair problems in the workplace include hair loss or color change. These represent a very small proportion of occupational dermal problems. Alopecia (loss of hair) is caused by agents that kill the living and growing cells at the hair root. Three chemicals have most frequently been the cause of such problems: thallium (which was actually used to remove unwanted hair), chloroprene dimers, and boric acid. Loss of hair occurs on exposure to ionizing radiation, commonly seen in patients undergoing radiation treatment for cancer.

Color change is most often caused by metals, copper and cobalt having been reported as producing green and blue colors, respectively. Picric acid and a few other chemicals also cause such problems.

ACNE-TYPE DERMATITIS

Acne, or skin eruptions, result when follicles are blocked so as to prevent the discharge of skin oils. The trapped sebum produces swelling and irritation. A number of working conditions stimulate this condition, particularly in individuals already prone to this problem.

Acne arises for reasons other than chemical exposure. Hot weather or high temperatures in the workplace induces perspiration, which results in swelling and softening of the keratin layer, possibly blocking the follicle. Skin friction, as from belts, hard seating, or face masks, also can cause pore blockage, and heat and friction together have an additive effect.

Hydrocarbon Agents

Acne commonly results from the handling of high-molecular-weight hydrocarbon compounds, particularly on fingers, hands, forearms, and skin in contact with oil-soaked clothing. Mechanics and machinists are particularly likely to have this problem, and lubricating greases and cutting oils are usually at fault. Roofers, road workers, and construction workers are likely to have contact with coal tar and pitch, which also block follicles.

The skin develops "blackheads", followed by swelling of the follicle with a burden of skin oil. This too is more serious in individuals more prone to acne, and improved personal hygiene reduces the problem.

Chloracne

Chloracne is similar in appearance but different in cause from the acne described above. The follicle is sealed with damaged keratin rather than oil. Chlorinated and brominated aromatic hydrocarbons cause this problem, with some chemical structures being much more potent than others. Chlorophenols and polychlorinated or polybrominated naphthalenes, biphenyls, and dibenzofurans are recognized causative agents. The problem may clear after contact with the chemical ceases, but improvement is usually slow.

Some halogenated aromatics are also significant systemic toxins. Most highly publicized has been 2,3,7,8-tetrachlorodibenzo-*O*-dioxin (TCDD, or just "dioxin"), whose toxic characteristics have been extensively investigated. Results have been sometimes controversial, and the waters have been greatly muddied by lawsuits or potential lawsuits brought by lawyers on behalf of Vietnam veterans exposed in the course of spraying defoliants. However, chlorinated hydrocarbons readily cross the skin as systemic toxins, and the appearance of chloracne is a sensitive indicator of systemic dosage for many exposed individuals.

New chloracne may continue to appear after exposure has ceased. It is hypothesized that this is due to chlorinated hydrocarbons being stored in fatty tissue, then gradually being released.[8]

ALTERATION OF SKIN PIGMENTATION

A few chemical agents cause pigmentation changes in skin either accompanied by irritation or in the absence of other symptoms. Melanin, the skin pigment, is a phenolic structure. Some phenolic chemicals interfere with melanin synthesis in the epidermis, resulting in temporary pigment loss. Heavier exposure may kill melanin-producing cells, inducing a more permanent loss of skin color. Other damage that kills skin cells, for example, exposure to corrosives described in the following section, may result in healing without the regeneration of the cells that produce melanin.

Following skin inflammation, there may be excessive pigment in the skin. Melanin may be deposited in the tissue by damaged melanin cells, and the occurrence is more marked in people with very dark skin, hence more melanin production. This may occur below the level of cells being carried outward to be shed, and so may persist for a long time.

CORROSIVES

Chemicals that cause massive tissue damage within short time intervals are termed corrosives. Corrosives destroy by direct chemical action, eroding the tissue and potentially producing deep wounds. As with serious burns, tissue damage can be extensive and the healing process produces heavy scarring. Corrosives have such potential for damage that skin contact must be avoided completely, and they should not be used without skin protection suitable to withstand the specific agent. Protective clothing contaminated with corrosives must be rinsed clean before garment removal. Corrosives are particularly threatening when used at high temperatures.

Acid

Most commonplace among industrial corrosives are concentrated strong acids, especially concentrated sulfuric acid. Sulfuric acid is the highest volume chemical produced in the U.S. and is used broadly throughout industry. Sulfur trioxide, a gaseous anhydride of sulfuric acid, is an intermediate in sulfuric acid production. Sulfur trioxide can react with moisture on the surface of the skin to produce concentrated sulfuric acid. By controlling the amount of water reacted with sulfur trioxide, sulfuric acid may be produced for use in a concentrated form that contains little or no excess water. Oleum, also called fuming sulfuric acid, has had less water added than the sulfur trioxide requires for complete reaction, and is therefore a mixture of sulfuric acid and sulfur trioxide. The lower the water content, the more hazardous the

[8] J. S. Taylor, "The Pilosebaceous Unit," in H. I. Maibach, Ed., *Occupational Dermatoses*, 2nd ed., Year Book Medical Publishers, Chicago, 1987.

sulfuric acid, with oleum being the most hazardous. Concentrated sulfuric acid causes more damage by rapidly dehydrating tissue than by its action as an acid. Skin contact might occur either with the liquid or with an airborne mist.

Concentrated nitric acid is used to make dyes, plastics, and fertilizer. It adds the ability to oxidize tissue to its effect as a strong acid. As an example of its effectiveness as a corrosive agent, hot nitric acid is used in the laboratory to completely digest tissue samples, destroying all carbon, hydrogen, and nitrogen compounds, for purposes of metals analysis. Skin contact results in ulcers and a reaction with keratin to produce a yellow colored product.

Phenol (carbolic acid) is a very weak acid that nonetheless seriously damages the skin. Phenol was the original antiseptic, and its derivatives are used in medicinals and a variety of agents designed to kill or prevent growth of microorganisms. Exposure leaves a thick layer of skin whitened and dead. It then absorbs rapidly into the body where further harm may occur.

Alkali

Concentrated solutions of strong bases (caustics) are corrosive in nature. The most common strong base in industry is sodium hydroxide (lye, caustic soda, white caustic). It is used in large quantities in textile, paper, soap, petroleum, and chemicals production. Once on the skin, caustic solutions are much more difficult to wash off than are acids, which heightens their ability to do harm.

Cement is an alkaline agent. It is prepared at very high temperature, which drives off even traces of water. When water is added, the components react with and incorporate that water, forming the rock-like product. Mixed with sand and rock, it is termed concrete. Typical cement contains aluminum oxide, silicates, calcium sulfate, none of which are corrosives, and calcium oxide. The latter reacts with water to produce calcium hydroxide, which is a strong base and is capable of skin damage. Cement is not a concentrated alkali like the lye described above, and damage requires long contact. Such contact eventually results in burns and ulceration, however, caused by a combination of the calcium hydroxide, abrasion, and removal of water from skin by the components of cement.

Testing Degree of Hazard

Knowing the degree of hazard associated with a corrosive generally translates into knowing its concentration. Thus the very hazardous concentrated sulfuric acid becomes relatively harmless in dilute solution. The Department of Transportation (DOT) is adopting the U.N. standards for packaging corrosives, which require testing to determine whether corrosives should be classed as I, II, or III for storage during transport. DOT is concerned about spills or releases during transport. There were 9000 such unintended hazardous material releases in 1992, about one-third involving corrosives. Corrosive spills caused one death and 23 major and 215 minor injuries.[9]

Traditional testing involved timing the destruction of the skin of a rabbit.[10] Alternatives to this method were sought. Concentrations of strong acids or alkalis can be estimated by measuring the pH of the solution or of a known dilution of the solution. The EPA has questioned the value of this method if applied broadly. DOT has approved the use of a patented CORROSITEX system. A vial of detection fluid is covered with a membrane that mimics skin. The time for a sample to break through the membrane is tested by timing the appearance of change in the detection fluid. The similar SKIN test uses cultured living

[9] D. E. Chenoweth, "The Acid Test," *Occupational Health and Safety,* Sept. 1995, p. 50.
[10] Three minutes, class I; 60 min, class II; 4 hr, class III.

human skin. A third method times depth of penetration of a metal block by the heated corrosive solution.

CANCER

When compared to other industrial skin problems, the development of skin tumors is a relatively uncommon but obviously very serious occurrence. Classes of skin tumors are associated with different classes of cells: squamous cell carcinoma, basal cell carcinoma, and melanoma.

For all cancer types, but particularly for squamous cell carcinoma, a sizable body of epidemiological evidence implicates UV light as a carcinogenic agent. Exposed skin of outdoor workers has a relatively high risk of tumor growth. Such tumors tend to appear in skin that already displays damage generated by the sun. Outdoor workers, particularly those with fair complexions, should wear protective clothing or sunscreen preparations.

Exposure to ionizing radiation also increases the likelihood of skin cancer, particularly squamous cell and basal cell carcinomas. Individuals that operate X-ray machines are at risk and must be careful to avoid exposure.

Polycyclic aromatic hydrocarbons are well-established carcinogens, particularly 3,4-benzpyrine and dimethylbenzanthracene. These are found in coal tar, asphalt, and creosote. Workers exposed to these materials may develop "tar warts" after years-long incubation periods, and a small proportion of these become malignant. Heavy exposure to UV light increases the frequency of malignancy.

Heavy and chronic exposure to arsenic results in warts (arsenical keratoses), which occasionally progress to become squamous cell carcinomas. The evidence linking arsenic and cancer is primarily epidemiological.

INFECTIONS

The skin generally provides good protection against infections. However, damage to the skin surface, as by abrasions or burns, chapping from weather, low humidity in the workplace, or poor personal hygiene, greatly increases the likelihood of these problems. Infections may be bacterial, fungal, or viral.

It may be difficult to determine whether a particular infection is work related or not, but some occupations carry an increased risk of a specific infection. Persons handling animals or animal products are at special risk. This includes then the original producer of the product (farmers and fishermen), transporters of the product (warehouse workers, dock workers), processors (meat, fish, and poultry packers, tanners, and furriers), and distributors (butchers).

Erysipeloid (fish handlers disease) is a bacterial infection that causes a spreading erythemia and is particularly common among those handling fish, but also is found among poultry and meat handlers. Anthrax used to be a serious threat, particularly from contact with goats or sheep. Inoculation programs have sharply reduced this problem in the U.S., but some danger still exists in handling hides imported from countries with less rigorous programs. Sheep or goats may also transmit orf, a viral disease, to ranchers and veterinarians.

PREVENTION OF SKIN IRRITATION OR DAMAGE

An important aspect of preventing dermatosis is the inclusion of this topic in the program of safety training provided by the company. Training programs often focus more heavily on accident prevention than on prevention of dermatosis.

PERSONAL HYGIENE

Cleaning the skin at breaks and at the end of the shift is an important part of dermatosis prevention. Cleansers selected depend on the nature of soil on the skin. Heavy-duty cleansers are appropriate where the skin is soiled with grease, paint, or other stubborn contaminants. However, the ingredients that make these cleaners "heavy duty" may also make them skin irritants. Such cleansers may contain a "grit" type of agent to help scrub off soil, and this agent can cause skin abrasions. Softer cornmeal or plastic grits are less likely to abrade skin than the harder pumice or sand abrasives. Abraded skin has reduced barrier properties. Cleansers, especially waterless cleansers, may also include organic solvents to dissolve grease, which can cause drying and cracking of skin due to stripping out of skin oils. Some solvents are more likely to cause these problems than others, so care in selecting cleansers can reduce this difficulty.

Where bacterial contamination is a problem, for example, in food processing or medical facilities, cleansers are required that contain special antibacterial agents. These range in effectiveness from agents that generally kill most bacteria and reduce chances of skin infections to agents that kill all bacteria, used when hazardous organisms may contaminate the skin.

Under most circumstances cleaning with soap and water is adequate. Employees whose skin is likely to become dried out by contact with solvents or detergents, and any using heavy-duty cleansers, may need to use moisturizing lotions after cleanup to prevent dry, cracked skin from developing.

PREDISPOSITION

Another route for reduction of the frequency of industrial dermatosis is to identify workers who are at greater risk of skin problems, and to avoid placing them at workstations that include potential harmful skin exposures. This requires careful screening, but can reduce the rate of incidents. Dry skin (low skin oil production) or older skin presents a less effective barrier to chemicals. People with an unusual tendency to sweat soften their keratin barrier, may dissolve some dry chemicals on their skin so as to increase irritating characteristics, and run greater risk of irritation from protective devices or clothing. Unusually hairy individuals and those who are already acne prone run a greater risk of occupationally induced acne problems.

Finally, persons with allergic tendencies (atopic individuals) are much more likely to experience industrial dermatosis. These can be identified as individuals with hay fever, asthma, eczema, or wool intolerance.[11] Further, once there is skin irritation, atopic individuals are slower to cure. Workers with psoriasis are at greater risk. Predisposing factors are discussed in more detail in Shmunes (1988).

FIRST AID — CORROSIVES

With the exception of phenol, which is not very water soluble, the immediate response to skin exposure to a corrosive should be thorough washing with water. Special agents such as polyethylene glycol are recommended for the removal of phenol, but in their absence soap and water is more effective than water alone. Because alkali is harder to remove from skin, washing of an alkali exposure should continue for at least 15 min. Persons without protective clothing who are exposed to a splash or spill of strong acid or alkali should remove their clothing, which may hold some of the corrosive agent, and shower immediately. Corrosives acting on skin can quickly destroy sensory nerves, reducing warning that damage is occurring. This may go undetected until serious harm is done if it takes place under clothing.

[11] E. Shmunes and J. Keil, "The Role of Atopy in Occupational Dermatoses," *Contact Dermatitis*, 11, 174, 1984.

In zones where the risk is high due to use of hazardous chemicals, emergency showers should be provided (Figure 6.1). The American National Standards Institute (ANSI) recommends that showers be located no more than 100 ft from a hazardous area.[12] Where acids or caustics are used, the recommendation is that the wash station be within 10–15 ft. Showers should be operated by a push handle or pull chain that does not have to be held after the flow starts. Such showers now commonly have shower curtains in case immediate removal of clothing is necessary.

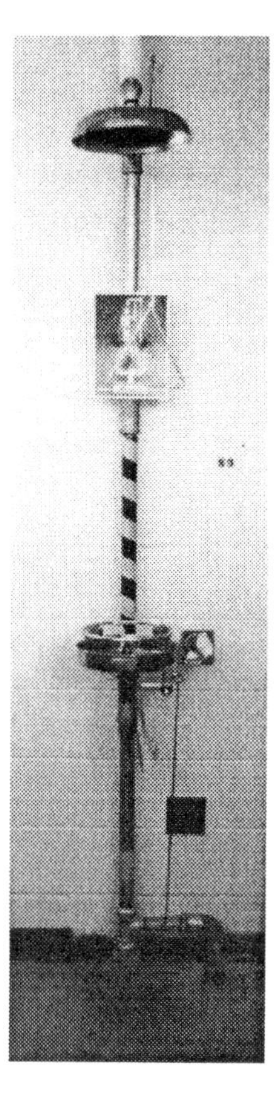

Figure 6.1. Safety shower. Pulling the chain causes a large volume of water to flow over the worker. This serves either to rinse off chemicals or to extinguish a fire. Some safety showers have enclosures, in case a chemical spill requires removal of clothing. Safety showers should be located close to areas where workers are in danger of chemical splashes. Notice this one also has an eyewash fountain.

PREVENTION OF SKIN CONTACT

Workers may be protected from skin irritation or damage by avoiding skin contact with the offending chemical or chemicals. In extreme cases, for example, in the case of a violent corrosive such as concentrated sulfuric acid, the worker is kept away from the chemical. In cases where extensive contact is necessary to cause a skin problem, for example, the problems created by detergents or by solvents, contact with the agents by the worker needs to be kept below a hazardous level. It is also possible for the worker to be physically protected from the chemical. Two common protections are barrier creams and protective clothing.

[12] The National Safety Council recommends 25 ft.

BARRIER CREAMS

Barrier creams block many chemicals from reaching the skin, but the level of protection is limited. They are coated on the skin at the start of the work day and as needed for replacement during work periods. They should be washed off and replaced at breaks, as for lunch, and removed at the end of the workday, just as you would remove safety gloves at those times. Removal is by washing with soap and water. Barrier creams are supplied in a variety of formulations, and care must be taken in selection of the type to use because some are better for particular applications than others.[13] Most contain a nonpolar "grease", either hydrocarbon or silicone, which is inert and repels water and many industrial chemicals. Acne-prone individuals may experience an outbreak of acne when using barrier creams. Further, this grease may transfer to the product being handled, with negative consequences.

Many questions arise: Is the coating thick enough overall? Was a spot missed? Is the cream removed by rubbing against other objects during work? If you developed a hole in safety gloves, you would know, but you cannot see that your barrier cream is gone from some area of your skin. Barrier creams should be used only in low-hazard situations, and if safety needs call for protective gloves, creams are not an equivalent replacement. However, use of barrier creams does not reduce manual dexterity as does wearing most gloves.

TYPES OF CHEMICAL PROTECTIVE CLOTHING

Protective clothing is used when contact with chemicals is unavoidable, as when manipulating objects freshly removed from a bath, or when there is a possibility of spills or splashes that could be harmful to the skin.[14] It may be essential to deal with emergency situations such as cleaning up a spill. Supplying protective clothing should not serve as justification for a job in which the worker is required to place his or her hands in continuous contact with a dangerous chemical. An engineering solution — the redesign of the task or process to eliminate such contact — is called for instead.

Compliance with OSHA regulations requires that where chemical protective clothing is necessary, there should be a written program, available to all affected personnel, that includes policies and procedures for the selection, use, decontamination, and maintenance of such clothing. This program should be reviewed at least annually. Worker training should be provided in the use, capabilities, and limitations of such equipment. Records should be kept of the hours of use of protective garments by workers.

There are two major decisions: (1) How much skin must be covered, and (2) From what material should the protective garment be constructed? Garments available range from just gloves to totally encapsulating chemically protective suits (Figure 6.2).

Less serious potential exposures are handled by some combination of safety glasses, face masks, gloves, splash aprons, raincoat-like garments, boots, or boot covers, used as is appropriate to provide adequate protection. These may be disposable, intended for a single use, or designed for long-term use.

MATERIALS USED IN PROTECTIVE CLOTHING

The second major choice is the material from which the clothing is to be constructed. There is no material that is uniformly resistant to all chemicals. In selecting the material, refer to tables provided by the manufacturer which list a series of common chemical hazards and indicate the performance of each material in blocking that chemical.

[13] A list of available products and usage recommendations is found in V. Meade, "What You Should Know about Barrier Creams," *Occupational Health and Safety*, June 1995, p. 51.

[14] A summary of available protective apparel, its characteristics, and the suppliers is found in *Occupational Health and Safety*, July 1996, starting on p. 25.

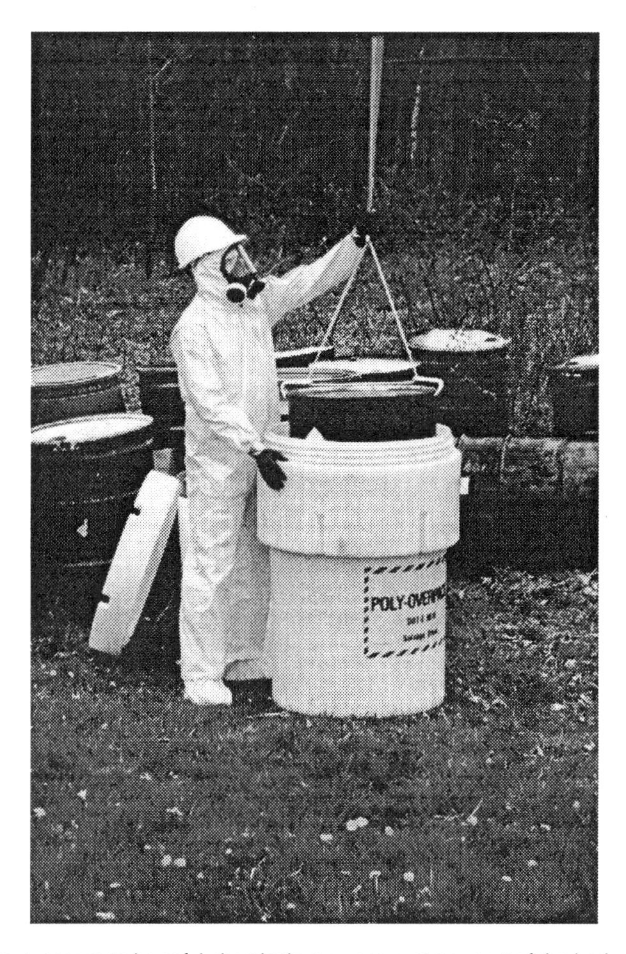

Figure 6.2. Where exposure to harmful chemicals may occur at any part of the body, protective clothing may be necessary. Here a worker clearing a hazardous waste site is completely protected, including face protection and the use of a respirator. (Photo courtesy of Lab Safety Supply, Janesville, WI. With permission.)

Informational charts are likely to characterize the protective garment according to resistance to degradation, penetration, and permeation.[15] Degradation means a physical breakdown or alteration of the material on exposure to chemicals, heat, UV light, or ozone. The material may weaken, harden, soften, swell, split, or discolor. Penetration refers to the passage of liquid through small holes in the material. These could be the result of stitching, natural pores, pinhole damage, and gaps at zippers or other fasteners. Finally, permeation is the diffusion of the liquid through the unpunctured material.

To protect against chemicals, the choice of material from which the garment is assembled is critical (Table 6.2). For example, hazardous water solutions of acids, alkalis, or strong oxidants are repelled by latex or latex-coated gloves, but latex is rapidly degraded by nonpolar solvents. Conversely, polyvinyl alcohol gloves are of value against nonpolar solvents, but become stiff and brittle on exposure to water, so readily that they must have a lining that absorbs perspiration.

The standard test of permeation for various materials is described in the American Society for Testing and Materials (ASTM) method F739, which is called the Standard Method for Resistance of Protective Clothing Materials to Permeation by Liquids and Gases. This test result indicates how long an item of protective clothing, for example, gloves, can be worn

[15] J. F. Rekus, "Fabrics that Repel," *Occupational Health and Safety*, Sept. 1994, p. 62; J. O. Stull, "Selecting Chemical Protective Clothing," *Occupational Health and Safety*, Dec. 1995, p. 20.

Table 6.2. Materials Used in Protective Clothing and Recommended Applications.

Material	Protects Well Against	Ineffective Against	Comments
Butyl rubber	Polar solvents	Nonpolar solvents	
Cotton	Many solids	Not for solvent protection	Mild abrasion protection
Ethylene vinyl alcohol	Most chemicals		Easily damaged; use with other stronger material
Leather		Not for solvent protection	Primarily abrasion protection
Natural rubber (latex)	Water solutions, polar solvents	Nonpolar solvents	Good cut and abrasion resistance
Neoprene rubber	Nonpolar solvents, acids, caustics		Structurally relatively weak
Nitrile rubber	Nonpolar solvents	Some polar solvents	Good cut and abrasion resistance
Polyvinyl alcohol	Nonpolar and many polar solvents	Water, alcohols	Good cut and abrasion resistance
Polyvinyl chloride	Water solutions, acids, caustics, some polar solvents	(See manufacturer's recommendations)	Good abrasion resistance
Vitron	Nonpolar solvents (see manufacturer's recommendations)	Polar solvents (see manufacturer's recommendations)	Structurally weak

Note: Nonpolar solvents include hydrocarbons and chlorinated hydrocarbons. Polar solvents include alcohols, ketones, and water. Some solvents are intermediate, such as high-molecular-weight alcohols, esters, and aldehydes.

immersed in the chemical before that chemical has passed through the glove and has begun to reach the hand. The material is tested by using it as one wall of a chamber into which the liquid is placed, and the time for the liquid to permeate is recorded. Results for a particular combination of liquid and material are not always the same because the analytical methods used to detect the liquid vary in sensitivity.

Another test result of interest is permeation rate. Once the chemical has permeated the protective material in the test described above, a measurement is taken of the constant rate at which the chemical crosses the barrier, usually expressed as $\mu g/cm^2/min$. Two materials might have the same breakthrough time, but quite different properties as a barrier once the chemical is passing through the material. ASTM has begun combining the two types of information to generate what is called a normalized breakthrough time. The test is run as described for determining a breakthrough time, but the time required to reach a penetration rate of 0.1 $\mu g/cm^2/min$ is measured. Differences due to varying sensitivity of detection of the chemical are eliminated, and the ability of the material to serve as a barrier is measured in a more meaningful fashion.

The ability of the material to resist degradation by the chemical is also of interest. The ASTM standard method (F903) measures the time until the material is degraded enough to allow bulk penetration. This test may be directed at measuring the resistance of seams in protective clothing or the durability of the material itself. Results are usually presented in broad qualitative terms (pass/fail; excellent/good/poor).

Strength is another important characteristic of protective clothing. Strength is measured as resistance to tears, puncture, cuts, abrasion, and fatigue (failure at a flex point). We are concerned both with the material and the seams. Most protective material is cloth based with a coating of the resistant polymer, although some is composed of polymer only. Protective materials are sold in different thicknesses, and it is reasonable to suppose that a thicker material lasts longer than a thin one. This reasoning must not be carried too far, however. Garments do not remain intact and leak free right up to some predictable time of failure, then completely dissolve. Failure occurs as the appearance of a small hole anywhere, often at a place where the garment is particularly stressed. Failure could occur at the thumb or finger surfaces of a glove where the worker contacts an object, at a flex point on the glove such as the crease at the base of the thumb, or at a defective place on a garment's coating. However, statistically one would predict longer useful life from a thicker material.

Flammability testing is done on protective clothing. There is a standard for this property (16 CFR 1610) that must be met, which involves measuring how rapidly flames spread on the material. ASTM method F1359 measures ease of ignition of the material and whether the flames self-extinguish after burning is initiated.

GLOVES

Gloves are the most frequently used protective clothing, and for good reason. Occupational dermatosis most often involves the forearms, hands, and fingers. In 1994 OSHA specified in 29 CFR 1910.138(a) that: "Employers shall select and require employees to use appropriate hand protection when employees' hands are exposed to hazards such as those from skin absorption of harmful substances, severe cuts or lacerations, severe abrasions, punctures, chemical burns, thermal burns, and harmful temperature extremes."

Gloves are available in a wide variety of materials, thicknesses, liners, lengths, and prices (Figure 6.3). There are various choices and compromises to be made when selecting gloves, depending on the particular work situation. The first consideration is type of hazard. The worker may need protection from sharp objects or abrasion, achieved by using cotton gloves with leather palms and finger pads. Hot objects require gloves that are heat resistant and are good insulators. However, neither of these types of gloves provide protection against chemicals.

Figure 6.3. Skin contact with chemicals may be prevented by gloves. It is important to select gloves of the proper thickness and material for a given chemical exposure (see Appendix E). (Photo courtesy of Lab Safety Supply, Janesville, WI. With permission.)

Gloves for chemical protection must block passage of the chemical to the employee's skin. Select a glove material that blocks the chemical by the criteria described above. Beyond that, the glove must be intact; even small holes are unacceptable. Because failure time is unpredictable and a percentage of even new gloves is defective, it is wise to spot check gloves each time before use. A quick, easy test for small holes is performed by folding shut the glove opening and rolling the material from the wrist toward the fingers to inflate it with trapped air. The appearance of the glove should also be inspected. A change in glove material (compared to a new glove) such that it becomes more tacky, slippery, discolored, or hardened, may predict forthcoming failure.

Another choice in glove selection is glove length. Gloves are available in lengths from those that end at the wrist to some that extend to the shoulder. The nature of the job determines what is appropriate.

A worker wearing gloves loses a degree of hand dexterity and ability to pick up objects. Workers will take gloves off because of this loss of dexterity. Of workers who experienced hand injuries, 70% should have been wearing gloves but were not, largely because of this loss of dexterity.[16] Correctness of glove fit is a factor. Glove materials vary in how well they grip an object. Thicker gloves make fine manipulation of objects more difficult, leading to more frequent spilling and dropping. Sometimes thinner, less expensive gloves should be used, then discarded frequently. Remember, a surgeon wears gloves, but they are very thin, fit tightly, and have excellent gripping characteristics.

In order to protect workers dealing with a variety of chemicals such that no single glove is adequate, the worker may "double glove" (wear one glove inside another). In such cases at least one of the gloves must be very flexible or the combination might immobilize the worker. Double gloving may also be done to improve the worker's ability to handle objects. If the protective glove provides a poor surface (perhaps slippery) for grasping the work object, a thin second glove pulled over it may provide a better surface for grasping.

Workers in cold environments may also require that the gloves be lined for warmth. This too adds to the clumsiness of the garment.

Protective clothing is also used as a guard against biological hazards (microorganisms). Recently there has been a turnaround in who is protected. Originally, surgeons and those assisting at an operation wore thin latex gloves to protect the patient from microorganisms carried by the surgical team. Now a large number of health care workers wear surgical gloves to protect themselves from organisms carried by the patients, so-called bloodborne pathogens.

However, some workers become sensitized to latex surgical gloves. Latex is a natural material, made from rubber tree sap. It is believed that the sap contains a protein that is the cause of the allergic response. Allergists estimate that about 2–6% of the public are allergic to latex. Many hospital workers now wear latex gloves continuously because of concerns about bloodborne pathogens. Frequent contact with the allergen raises the level of occurrence of allergic symptoms in these individuals, rising as high as 10–17%.

One solution involves wearing barrier creams under the gloves to protect the skin from the allergen in the gloves. Alternatively, the latex gloves may be replaced with gloves of another material, vinyl or nitrile. Such use of nonlatex materials then must extend to other hospital supplies, tubing, and a variety of seals. Finally, double gloving is possible, with the outer latex glove being changed between patients and the inner nonallergenic glove being worn continuously. Finally, alternatives to latex must be available for use with allergic patients.

MORE COMPREHENSIVE PROTECTION

The possibility of being splashed by chemicals adds the need for greater protection. More complete protection can be afforded in a number of ways, ranging up to complete body covering. Danger of splashing should lead to the use of safety goggles; eye protection is the highest priority. Transparent face shields protect the skin of the face from splashing. Under

[16] B. Gaither, "Matching Glove Utility and Dexterity," *Occupational Health and Safety*, July 1994, p. 26.

Figure 6.4. If full-face protection is necessary, a face mask may be worn. Notice, this is in addition to safety glasses, not instead of them. (Photo courtesy of Lab Safety Supply, Janesville, WI. With permission.)

no circumstances should a face shield be considered sufficient eye protection (Figure 6.4). Safety goggles should be worn under a face shield to ensure that chemicals do not enter from the sides, top, or bottom of the shield.

An apron provides some body splash protection. The protection of gloves can be extended by slip-on sleeves that fit under the gloves and extend up the arm. Such commercially available shields have elastic at each end, which should fit snugly, and come in a variety of materials. The seams up each side should be checked to see if they are sealed or stitched. Stitched seams may be satisfactory, but they do provide small access holes for chemicals to enter. Sleeves that seldom contact the chemical would not have to be discarded each time the gloves are changed, and they are less clumsy than long gloves.

When the nature of the hazard requires it, a raincoat-like full-body covering provides better coverage than sleeves and aprons. Foot protection in the form of pull-on boots worn over the shoes is then necessary. Foot covers must have a good grip on the floor so as to avoid causing slip-and-fall accidents.

ENCAPSULATION SUITS

The garments previously described protect against splashed liquids. If protection against vapors is required we need an encapsulating suit equipped with an independent air supply, either self-contained or supplied by a hose, thereby providing complete isolation from the chemicals. The totally encapsulating chemical protective suit is used in emergency spill cleanup or cleanup at a hazardous waste site when highly toxic chemicals are known to be present, or where the risk is unknown but is potentially very serious, such as first entry into a site where there is as yet undetermined potential for severe chemical exposure.

When a person is encapsulated, the usual mechanisms for controlling body temperature are frustrated. Extended wear requires a garment design that includes circulation of cooling air, an ice pack, or some other temperature-control device. The vision of the worker in the suit must not be dangerously impaired. Communication with an encapsulated worker is a problem that must be anticipated and handled. In recent years suit design has been improved. In particular, the weight of these garments has been sharply reduced.

REUSEABLE VERSUS SINGLE-USE PROTECTIVE CLOTHING

A protective garment that is intended to be discarded after a single use simplifies decision-making. The only serious problem remaining is deciding whether it can be disposed as regular trash or has been contaminated such that it needs special treatment as hazardous waste. Reusable clothing is usually more expensive to purchase, but generally carries a cost saving after reuse a few times, and may provide better protection because of superior construction.

Reuse of protective clothing requires that it be carefully inspected beforehand. A suit can be tested for leaks by inflation (ASTM F1052); then the leaks can be located by coating suspicious areas with soap solution and watching for bubbles. A leak in an otherwise sound garment can be fixed by the manufacturer.

DECONTAMINATION

Reuse also requires decontamination if exposure to toxic chemicals has occurred.[17] Decontaminating a single-use garment avoids the expense of disposal as hazardous waste. Washing down a contaminated encapsulating protective suit should be done before it is removed to ensure the worker is not contaminated during disrobing. Workers should never take contaminated protective clothing home to be cleaned, thereby taking the contaminant home also.

Contamination may be limited to the surface of the garment. Garments then may be aerated, washed with soap and water, or washed with solvent. Solvents used must not damage the material of the garment. Volatile liquids evaporate off the garment in a ventilated area. Heated air makes this method faster and more effective. If an organic liquid has started to penetrate the fabric — matrix contamination — time must be allowed for it to diffuse back out of the fabric, the reverse of permeation. There must be some method to ensure that decontamination was successful, but guidelines for this are not readily available.

The OSHA personal protective equipment standard and HAZWOPER[18] require that the employee be provided with clean equipment. They also require employee training in use of protective equipment, including cleaning and decontamination.

EYE HAZARDS

Eyes are irreplaceable and easily damaged. The surface of the eye at its most delicate point consists of the transparent cornea, across the outer surface of which is a layer of cells (Figure 6.5). The cornea is devoid of blood supply in order to improve its transparency, and it depends on diffusion of nutrients from underlying eye fluids. It is supplied with nerve endings, so damaging contact is painful. A person with a particle or some chemical in the eye is usually highly motivated to remove it. The surface layer of cells can be destroyed by chemicals. The water content of the cornea changes in the absence of these cells, causing it

[17] C. J. Kairys and Z. Mansdorf, "Decontaminating Protective Clothing Offers Challenge to Manufacturers," *Occupational Health and Safety,* May 1990, p. 36; T. R. Carroll, "Decontamination Matrix Helps Assess Reuse Potential for Protective Clothing," *Occupational Health and Safety,* March 1995, p. 54; decontamination is discussed in U.S. EPA, *Standard Operating Safety Guide.*

[18] 29 CFR 1910.132 and 29 CFR 1910.120.

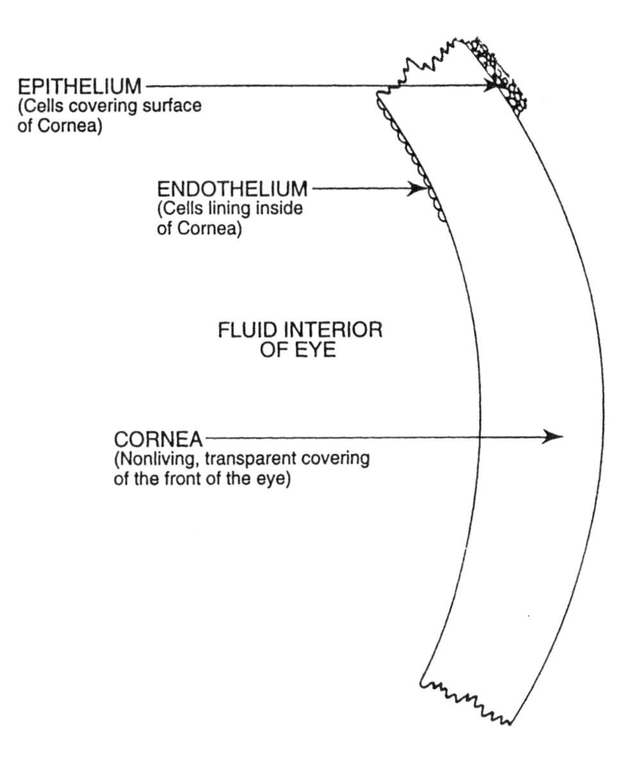

Figure 6.5. The cornea of the eye.

to become clouded. However, the cells frequently grow back, so, if the cornea itself is intact, the damage may be temporary.

There are three classes of concerns regarding eye protection: protection against physical damage from objects propelled at them by the work process (for example, metal or abrasive fragments from a grinder), protection from airborne dusts or liquid chemicals that may splash or be sprayed into the eyes, and protection against radiant energy (for example, intense UV light generated by arc welding). All eye protection must accommodate the vision correction needs of the worker, either by fitting over prescription glasses or by including corrective lenses in their construction.

DAMAGE BY CHEMICALS

Although any reactive chemical presents a threat, most eye accidents involve acids, alkalis, solvents, and detergents. Acids and alkalis kill the surface layer of cells and can cause corneal structural damage leading to scarring and blindness. Alkalis present the greatest problem because they are more difficult to remove, and during the extended time required to remove them, they can cross the cornea and do damage inside the eye. Nonpolar acid or base anhydrides are also hard to remove, and they react with water to produce heat. Sulfur dioxide and ammonia, the most common examples of such compounds, are used in a variety of ways in factories. They are gases and are often used under pressure, for example, in a refrigeration system. A leak in a line can result in the gas entering the eye under pressure. Lime (calcium oxide) also presents a special problem, because particles from a dusty environment may penetrate the surface of the eye. Once inside, lime reacts slowly with water to produce calcium hydroxide, a very strong alkali.

Organic solvents are generally chemically nonreactive and largely nonpolar chemicals. They damage the surface cells of the eye on contact, and, although this process is painful, it is much less threatening than acid or alkali exposure. The pain on contact drives the victim to remove the solvent (or any other chemical) from the eye as quickly as possible, minimizing the damage. Solvents are unlikely to do structural damage to the cornea.

All detergents are irritating to the eye, but the three classes — anionic, cationic, and nonionic — vary greatly in the degree of irritation. Anionic detergents, which are used for their bactericidal properties in food processing, hospitals, and other institutional settings, are by far the most irritating, whereas nonionic detergents are the mildest to the eye.

Airborne particulates, dusts, can be eye irritants. Mining, woodworking, and steel making or metal forming facilities are places where such dust may be a problem.

FIRST AID — CHEMICALS IN THE EYES

If any chemicals contact the eye, immediately flush the eye with quantities of water. This is not the time to display your cleverness as a chemist by using alkali to neutralize an acid spill — such treatments usually cause more damage. Eyes should be opened as widely as possible to ensure that border areas are well flushed. Flushing should continue for at least 15 min; some experts suggest even longer.

Where risk of exposure is high, emergency eyewashes should be provided; plant safety inspections should include checking for their presence (Figure 6.6). Workers should be trained in the use of emergency eyewash equipment. Emergency eyewashes are turned on by an obvious hand or foot handle that starts the flow, and need not be continuously pushed for the flow to continue. The ANSI standard (ANSI Z358.1) calls for a minimum flow of 0.4 gal/min at a low velocity. Self-contained units — units that use stored water rather than being plumbed into the water supply — should have enough water to operate continuously for 15 min.

Figure 6.6. Eyewash fountain. This device delivers a large flow of water at low pressure directed at the eyes.

It avoids confusion if all emergency eyewash units in a facility operate in the same fashion, for example, all utilize a foot-controlled valve. Both an emergency shower and an eyewash station are likely to be called for in the same location, and suppliers provide units that combine these features.

DAMAGE BY RADIANT ENERGY

Excessive UV exposure causes irritation to the surface of the eye, called conjunctivitis. The victim has a sense of a foreign object in the eye, experiences lacrimation (tearing), and may want to avoid exposure to light. The severity of the symptoms are in proportion to the intensity of exposure. Symptoms often clear in a few days. Arc welding is an intense source of UV light, and even short-term exposure causes eye problems.

Just as when driving on a sunny day or sitting on the beach, reflected light — glare — at levels that are not dangerous can be fatiguing. If the glare is a necessary part of the job environment, for example, from a hot furnace, reducing glare improves job comfort.

Lasers

Lasers present a serious eye hazard. The danger arises from the intensity of the source. Although the laser may be a tiny beam, lasers concentrate high levels of light energy in that beam. Lasers have a variety of characteristics, including the wavelength and intensity of the source. Even low-energy sources, such as pen-like laser pointers, do damage when exposure is extended. When lasers are used in the workplace, shielding and screening should block stray radiation.

Such intense energy may damage the cornea and is likely also to do internal damage to the eyes, injuring the lens or the retina, the nerve endings at the back of the eye that transfer light images to the brain. Consider that when blood vessels leak in the retina, a problem experienced by individuals with diabetes, lasers are used to seal the vessels by burning the tissue. Powerful lasers are used to cut steel.

ANSI has published a series of recommendations concerning laser use, the ANSI Z136 series.[19] Laser safety experts have devised limits for laser exposure termed maximum permissible exposures, or MPEs. These are based on classifying lasers into four groups according to relative risk.

PREVENTION

When safety recommendations call for the wearing of eye protection, the proper safety glasses should be used and the wearing of this protection should be enforced.[20] OSHA regulations for face and eye protection are found in 29 CFR 1910.133, and they specify that any protective devices purchased after July 5, 1994, shall comply with ANSI standard Z87.1-1989.[21]

Goggles or face masks that guard against flying objects must be shatterproof, provide adequate coverage regarding the direction from which objects are propelled, and be comfortable and easily cleaned. ANSI uses standard impact tests to rate the suitability of glasses.

Safety glasses designed to protect against chemicals should completely enclose the eyes to prevent entry of chemical into the eye from any angle: sides, top, and bottom (Figure 6.7). Venting provided to minimize steaming of totally enclosing glasses should have a baffle design that prevents them from leaking chemicals that strike the glasses at some particular angle. If fogging is a problem, anti-fog lenses are available from some manufacturers. Spectacle-style safety glasses are not acceptable protection against potential chemical splashes. Face masks are often used to protect the entire face, and these are valuable as skin protection, but they do not substitute for safety glasses because they allow chemical entry from the sides.

[19] The source of these standards is the Laser Institute of America, 12424 Research Parkway, Suite 125, Orlando, FL 32826.

[20] A summary of protective devices for the eyes and face, listed by manufacturer, is supplied in the September 1996 *Occupational Health and Safety.*

[21] A comprehensive list of products available for various hazards is found in G. Arrotti (1995) (see Bibliography).

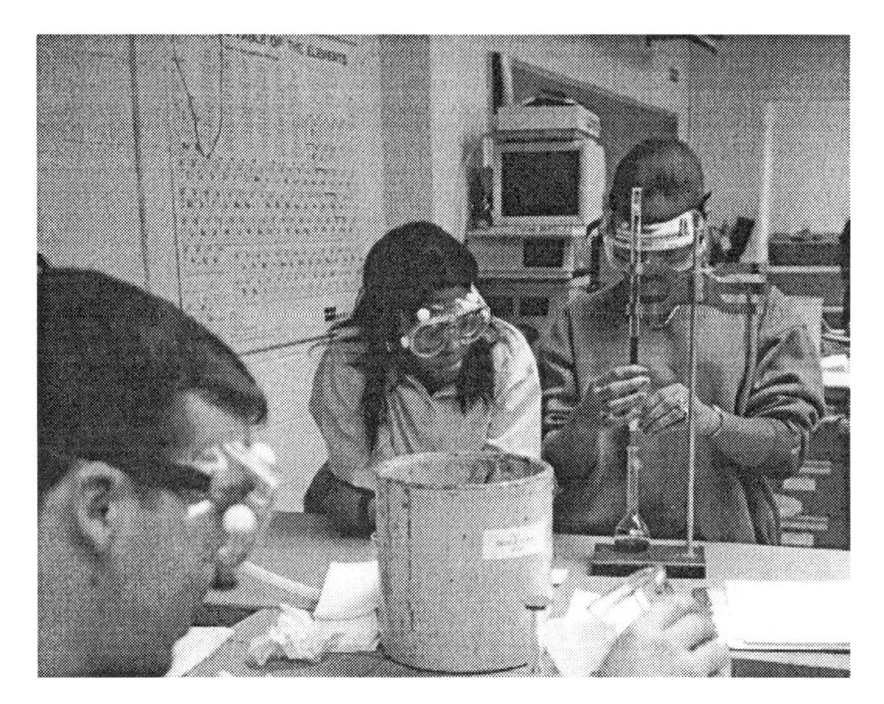

Figure 6.7. Safety glasses. Here beginning chemistry students practice eye safety. To provide complete protection, safety glasses must protect the eyes from the top, bottom, and sides, as well as from directly in front. Such glasses have baffled vents to reduce steaming in hot, humid conditions.

Airborne particulates are going to be excluded only by safety wear that fits tightly around the eye, such as goggles. If dust also requires respiratory protection, a full-face respirator handles both problems.

An obvious but sometimes overlooked point is that no style of safety glass can protect the eyes from harmful gases or vapors. For example, on contact with the moist eye surface hydrogen chloride gas dissolves to form hydrochloric acid, a damaging strong acid. The only protection in this situation is a full-face respirator.

Protection against UV or intense visible radiation is provided by tinting. Approved tinted eyeware is marked with a shade number indicating the darkness of tint. Welding or cutting operations are a special concern, and specific standards for tinting and recommendations regarding deciding which shade number to use are shown in Table 6.3. Another problem situation arises when workers must look into a hot furnace, which may produce high levels of radiation. Having tinted safety glasses that must then be used in normal light conditions may be a problem. A reflective or tinted face mask that can be dropped in place over untinted safety glasses when looking at the furnace may be useful.

KEY POINTS

1. Skin has two layers, the outer epidermis and the underlying dermis.
2. The epidermis has a surface keratin layer of dead, protein-filled cells. These are constantly replaced by cells modifying into keratin layer cells as they grow outward.
3. The dermis contains collagen fibers, nerves, and blood vessels. Blood vessels expand when the body is overheated, cooling the blood.
4. Skin is pierced by pores and hair follicles. Evaporating sweat from pores lowers the skin temperature. Sebaceous glands in pores coat the skin with nonpolar sebum.

Table 6.3. Filter Lenses for Protection Against Radiant Energy.

Operations	Electrode Size 1/32 in.	Arc Current	Minimum[a] Protective Shade
Shielded metal arc welding	Less than 3	Less than 60	7
	3–5	60–160	8
	5–8	160–250	10
	More than 8	250–550	11
Gas metal arc welding and flux cored arc welding		Less than 60	7
		60–160	10
		160–250	10
		250–550	10
Gas tungsten arc welding		Less than 50	8
		50–150	8
		150–500	10
Air carbon	(Light)	Less than 500	10
Arc cutting	(Heavy)	500–1000	11
Plasma arc welding		Less than 20	6
		20–100	8
		100–400	10
		400–800	11
Plasma arc cutting	(Light)[b]	Less than 300	8
	(Medium)[b]	300–400	9
	(Heavy)[b]	400–800	10
Torch brazing			3
Torch soldering			2
Carbon arc welding			14
Gas welding			
Light	Less than 1/8	Less than 3.2	4
Medium	1/8–1/2	3.2–12.7	5
Heavy	Greater than 1/2	Greater than 12.7	6
Oxygen cutting			
Light	Less than 1	Less than 25	3
Medium	1–6	25–150	4
Heavy	Greater than 6	Greater than 150	5

[a] As a rule of thumb, start with a shade that is too dark to show the weld zone. Then go to a lighter shade that gives sufficient view of the weld zone without going below the minimum. In oxyfuel gas welding or cutting where the torch produces a high yellow light, it is desirable to use a filter lens that absorbs the yellow or sodium line in the visible light of the (spectrum) operation.
[b] These values apply where the actual arc is clearly seen. Experience has shown that lighter filters may be used when the arc is hidden by the workpiece.
From 29 CFR 1910.133(a)(5).

5. Contact dermatitis is damage caused to the skin by chemicals. Abrasion, chapping, solvents, or extensive contact with water, especially water containing detergents, reduces the effectiveness of the keratin layer as a barrier.
6. Inflammation involves reddening and swelling of skin due to contact with an irritant. This produces an itching or burning sensation.
7. A few agents cause loss of or color change in hair.
8. Allergic contact dermatitis is an inflammation of skin due to allergic response to a chemical.
9. In photodermatitis, innocuous agents on skin are converted to irritants on exposure to UV rays in sunlight.
10. Blocking pores either with grease (acne) or by keratin destruction by chlorinated hydrocarbons (chloracne) produces pustules when sebum is trapped in pores.
11. Some agents reduce or intensify skin pigmentation.
12. Corrosives chemically destroy skin tissue, generally producing scarring. Corrosives are strong acids or alkalis.
13. Direct contact with carcinogens can initiate skin cancer.

14. Workers dealing with animals or animal products are particularly susceptible to skin infections.
15. Protection for skin can vary from simple nonpolar barrier creams, to gloves, to full protective body covering.
16. The material chosen for protective clothing should be appropriate to the chemicals encountered.
17. Decontamination of protective clothing requires knowledge of both the chemistry of the contaminants and the properties of the garment.
18. Eyes present a special problem for protection because the surface is easily damaged, and damage may lead to permanent blindness.
19. First aid for chemicals in the eyes involves extended rinsing with water at eyewash stations.
20. Eye protection against impact by flying objects should meet ANSI standards.
21. Eye protection against dust or splashed chemicals must fit the face closely.
22. Eye protection against radiant energy comes in a range of shade numbers, depending on the degree of protection required.

BIBLIOGRAPHY

R. M. Adams, *Occupational Skin Disease,* 2nd ed., W. B. Saunders, New York, 1990.
G. Arrotti, "The Eye Protection Program," *Occupational Health and Safety,* Aug. 1995, p. 28.
D. J. Birmingham, "Occupational Dermatoses," in *Patty's Industrial Hygiene and Toxicology,* 4th ed., Vol. I, Part A, John Wiley & Sons, New York, 1991.
D. J. Birmingham, *The Prevention of Occupational Skin Diseases,* Soap and Detergent Association, New York, 1982.
G. Dupis and C. Benezra, *Allergic Contact Dermatitis to Simple Chemicals,* Basel Dekker, New York, 1982.
M. P. Esposito, T. O. Tiernan, and F. E. Dryden, "Dioxins," EPA-600/2-80-197, EPA, Cincinnati, OH, 1980.
A. A. Fisher, *Contact Dermatitis,* 3rd ed., Lea & Febiger, Philadelphia, 1986.
K. Forsberg, *Chemical Protective Clothing Permeation and Degradation Database,* Lewis Publishers, Boca Raton, FL, 1993.
K. Forsberg and S. Z. Mansdorf, *Quick Selection Guide to Chemical Protective Clothing,* Van Nostrand-Reinhold, New York, 1993.
J. S. Johnson and K. J. Anderson, *Chemical Protective Clothing,* Vols. 1 and 2, American Industrial Hygiene Association, 1990.
H. I. Maibach, *Occupational Dermatoses,* 2nd ed., Year Book Medical Publishers, Chicago, 1987.
C. G. T. Mathias, "The Cost of Occupational Skin Disease," *Arch. Dermatol.,* 121, 332, 1985.
C. G. T. Mathias, "Occupational Dermatoses," in Carl Zenz, Ed., *Occupational Medicine: Principles and Practical Applications,* 2nd ed., Year Book Medical Publishers, Chicago, 1988.
S. A. Ness, *Surface and Dermal Monitoring for Toxic Exposures,* Van Nostrand-Reinhold, New York, 1994.
E. Shmunes, "Predisposing Factors in Occupational Skin Diseases," *Occupational Dermatoses,* 6, 7, 1988.
E. Shmunes, "The Role of Atopy in Occupational Skin Disease," *Occupational Medicine — State of Art Review,* Vol. 1, Hanley and Belfus, Philadelphia, 1986.
L. A. Stewart, G. J. Engelken, and N. H. Nicol, "Essentials of Occupational Contact Dermatitis," *Dermatol. Nursing,* 4, 175, 1992.
J. P. Tindall, "Chloracne and Chloracnegens," *J. Am. Acad. Derm.,* 13, 539, 1985.

PROBLEMS

1. Consider the structure of the skin and how it varies with location. If the entire hand is submerged in a chemical, assuming the skin is completely intact on that hand, where will transport of the chemical into the blood be likely to occur first?

2. List conditions that reduce the effectiveness of the keratin layer as a protective barrier.
3. A group of 35 workers in a textile plant are exposed to an azo derivative dye component, and 2 become sensitized to it.
   A. Of various options (providing protective gloves, etc.), what is the best course to follow with these two sensitized workers?
   B. Why do only these two respond, and not the entire exposed group?
   C. There is no other chemical, of some 15 different chemicals used in that section of the plant, that sensitized a worker. What characteristic of that chemical might explain that?
4. You are newly employed as the safety consultant to a small industrial plant manufacturing sheet metal warning signs to be mounted in high-hazard areas of factories. You are asked to recommend protective gloves to be used by workers in the plant and stay within a set budget. You identify these worker tasks:
   There is a small press where worker 1 loads precut sheets of metal, which often have sharp edges, into a feeder for a machine that stamps sheet metal into signs with raised letters. These signs are moved on a conveyor to a station where worker 2 places them individually in a machine that rounds the edges. Worker 3 then places the signs in a jig and holes are drilled by a machine. Worker 3 removes them from the jig, hangs them on a rack, and mounts the rack on a conveyor system that slides them into an acid bath to clean them before painting. The rack is transferred by worker 4 to a water rinse, then into a chamber where blowing hot air dries the signs. They are placed on a holder by worker 5 to be spray painted. They are turned over after one side is painted by worker 5 who handles them by the holder, which is coated with some paint overspray, and placed back into the paint machine to paint the other side. Worker 6 places the sign in its holder on the conveyor into a paint drying oven. The hot sign emerges from the other side. Worker 7 removes it from the holder and places it on a stand where the paint for the raised letters is rolled on. The finished sign rolls under heat lamps on a conveyor to the packing room, where worker 8 packages signs for shipment to distributors.
   A. Which workers (by number) must have hand protection?
   B. What about the use of gloves by the other workers?
   C. Discuss how appropriate each of these statements is to this situation?
      (1) A distributor gives a price break on purchases of 200 pairs of gloves at a time. By getting all required gloves alike, taking advantage of the price break is feasible, and a better grade of gloves can be afforded.
      (2) The tasks require a variety of glove types, all of which must be replaced every day, so only the cheapest grade gloves fit the budget.
      (3) Some gloves can be kept in use for a long time, others need to be replaced more frequently.
      (4) If heavy-duty gloves are purchased for worker 4, they need to be inspected before use each day.
      (5) Heavy-duty gloves are a good investment for workers 5 and 6.
   D. How will the gloves be chosen for worker 4?
   E. Is there any worker who should have protective clothing in addition to gloves?
   F. The plant owner asks you about eye protection regulations. Where are design standards for eyeware found? Does OSHA provide specific or general guidelines?
5. Heavy-gauge gloves are purchased by an employer to be used several days before being discarded by workers in a plant where solvents and moderately strong acids are in use. You are asked to outline a procedure for glove use and daily glove inspection. What should be done?
6. Medical terminology is commonplace in references frequently used by industrial hygienists, and you need to be able to decipher it. As stated previously, a medical dictionary is a good investment for a person planning a career in the field. The following are from an excellent advanced toxicology book.[22] Translate them into laymen's terms.

[22] E. E. Emmett, "Toxic Responses of the Skin," in C. D. Klaassen, M. O. Amdur, and J. Doull, Eds., *Toxicology: The Basic Science of Poisons*, 3rd ed., Macmillan, New York, 1986.

"The dermis has substantial vascular plexuses, unlike the epidermis, which is avascular. . . . The dermis has a plexus of lymphatics, which drain to the regional lymph nodes and the thoracic duct. The dermis has abundant sensory and sensorimotor nerves."

"The skin and particularly the epidermis is an actively metabolizing organ that is capable of significant biotransformation of xenobiotics."

"Contact dermatitis is manifest by signs of erythema and edema in experimental test animals. In humans more varied responses are seen, and erythema and edema frequently progress to vesiculation, scaling, and thickening of the epidermis. Histologically the hallmark is spongiosis or extracellular edema of the epidermis.

"A number of physiologic and pathologic changes occur in the skin as a result of exposure to the ultraviolet (UV) component of sunlight. These include erythema (sunburn); thickening of the epidermis; darkening of the existing pigment (immediate pigment darkening); new pigment formation (delayed tanning); actinic elastosis (premature skin aging); proliferative and other changes in epidermal cells; suppression of T lymphocytes; actinic keratosis, a precancerous condition; and the development of squamous cell cancers, basal cell cancers, and probably melanomas."

7. Where in HAZWOPER does OSHA discuss the use of totally encapsulating chemical protective suits? Read the section. Now locate the discussion of decontamination of protective clothing. Are specific procedures outlined?
8. You are the safety officer in a plant producing plastic bumpers for the auto industry. The bumpers are compression molded from heavy plastic sheets in hot dies. Bumpers are then trimmed, some edges ground smooth, and the finished bumpers are spray painted with primer. An employee develops a skin rash. How would you determine the likely cause of the rash?

CHAPTER 7

Inhalation Toxicology

In Chapter 3 we briefly considered the lung as a route of entry into the body. In the industrial setting this is the most important such route. Consequently, control of the workplace atmosphere is the objective of the majority of industrial hygiene studies and government regulations. This chapter presents more detail on lung structure and function and describes interaction of the lung with specific atmospheric contaminants.

Inhaled toxicants may consist of dispersed molecules (gases or vapors), aerosols, or particles. By vapors, we mean evaporated liquids. Aerosols are droplets of liquid small enough to stay suspended in the air.

LUNG STRUCTURE AND THE ENTRY OF TOXICANTS — A MORE DETAILED DESCRIPTION OF THE RESPIRATORY SYSTEM

THE NASAL CAVITY

The first segment of the respiratory system is the nasal cavity, which is lined with a moist mucus layer. This compartment warms and moistens the air before it enters deeper portions of the system. The sticky mucus surface traps particles, particularly larger particles. Strong irritants initiate sneezing, preventing penetration farther into the system and clearing out contaminants.

Entry into the lungs normally is by way of the nose and nasal cavity. However, when assessing the hazard of a chemical to the complete work force, such trapping must be discounted as a means of protection because some workers are constant mouth breathers. Other workers experiencing respiratory congestion, as from a cold, become temporary mouth breathers. Finally, a person exercising strenuously resorts to mouth breathing as the most rapid pathway to get more air into the lungs. The alternative path through the mouth is far less effective at blocking progress of toxicants farther into the lungs.

THE PHARYNX AND BRONCHIAL TUBES

Beyond the nasal cavity is the pharynx, which subdivides more than 20 times to form the bronchial system. The larger bronchial tubes have cartilaginous structures in their walls to prevent collapse of the tubes without causing them to be rigid. Smooth muscle is found in these walls, and its contraction narrows and relaxation widens the diameter of the passages. The diameter of the passages is regulated in response to physical demands on the body by the brain through the sympathetic and parasympathetic nervous systems and the hormone epinephrine. The final subdivisions, the bronchiole, narrow to about 1 mm in diameter and lack the cartilaginous wall found in the larger branches.

A thin lining of mucus, perhaps 5–10 μm thick, is excreted by special cells and completely coats this bronchial "tree". When the lungs are irritated, the rate of mucus production increases. As in the nasal cavity, the mucus serves to trap particles. The inertia of particles gives them a tendency to travel in a straight line, so they "spin out" onto the mucus when the airways bend at a branch point. Larger particles have greater inertia, so are more likely to be trapped.

Particles settle out of still air, but are held in suspension in moving air. Faster moving air can suspend larger particles. At each bronchial branch the total cross-sectional area of the system increases because the sum of the areas of the branches is greater than was the area before branching. Visualize a simple model where a duct carrying moving air splits into two branches each the diameter of the original duct. Each branch is now carrying half the original air and therefore has half the original flow rate. As the total cross-sectional area of the bronchial system increases, the flow rate correspondingly decreases and larger particles settle out onto the mucus.

Beneath the mucus layer is a forest of hair-like cilia which extend perhaps 4 μm outward from the cell. These cilia move like little whips as rapidly as 20 strokes per second, shifting the mucus layer at a rate of about 1 cm/min upward toward the nasal cavity. Trapped particles are carried up this mucus "escalator", and eventually the mucus and its particulate burden enter the throat and are swallowed. Hopefully the chemicals of the particulate will now be removed from the body with the feces, but some components may be absorbed from the GI tract.

Gases and vapors pass completely through the bronchial system unless they are water soluble and dissolve into the moist lining. Those that dissolve and are irritants stimulate coughing, clearing the air out of the lung.

ALVEOLI

At the end of the bronchiole are the alveoli, little sacs in which oxygen and carbon dioxide are exchanged with the blood. Before age 10 human lungs generate the adult population of about 300,000,000 alveoli, each about 250 μm in diameter. The surface area for gas exchange ranges from about 35 m^2 after completely exhaling to about 100 m^2 on taking in a deep breath. There is no mucus barrier on the alveolar walls.

The outside surface of the alveolus is 90–95% covered with capillaries, and the thickness of tissue separating air in the alveolus from blood in the capillary is in the micron range. Gases and vapors that enter the alveolus transport only a short distance across this barrier to enter the blood. Chemicals from particles that reach the alveolus may also cross this barrier and enter the blood.

Phagocytic Cells of the Alveolus

Phagocytic cells populate the inside surface of the alveoli and are thought to be responsible for depositing a detergent-like chemical mixture that helps prevent alveoli from collapsing. Phagocytic cells engulf foreign substances and bacteria and cleanse the alveoli of inhaled particles. They then migrate toward the bronchi to be moved out of the lungs on the mucus. Exposure to dust stimulates expansion of this cell population. If components of the particles are toxic to phagocytic cells or if the population grows very large so more are dying at any given time, the dead cells release digestive enzymes that damage lung tissue.

Unfortunately, components of cigarette smoke narcotize phagocytic cells, with the result that both particulate and cells accumulate in the alveoli. For this reason cigarette smokers experience greater risk from exposure to airborne toxicants than nonsmokers.

THE NATURE OF ATMOSPHERIC CONTAMINANTS AFFECTS THE LIKELIHOOD OF ENTRY INTO THE BLOOD

GASES OR VAPORS

Gases or vapors are not trapped on sticky mucus as are particles, so they can penetrate with the inhaled air all the way to the alveoli, then readily cross the thin barrier into the blood. Gases and vapors may also do damage to some degree anywhere in the lungs. A large fraction of more polar, hence more water soluble, compounds dissolve into the watery mucus lining before they reach the alveolus, causing them to remain longer in the lung. Once dissolved, the gas or vapor must pass through the mucus lining in order to reach the wall of the lung. This lining is relatively thick in the upper airways, but decreases in thickness toward the alveoli. Once reactive chemicals penetrate to the wall of the lung, they may do structural damage to the lung.

If the toxicant is a strong irritant, the individual ceases to inhale, either voluntarily because of distress or involuntarily because of bronchial contractions or spasms. Irritants stimulate the production of mucus, increasing the thickness of the layer and its effectiveness as a barrier. Subsequent coughing up of the excess mucus and its removal from the lung also carries out the irritant. Reluctance of the worker to inhale strong irritants lessens the hazard of these compounds.

PARTICLES

Throughout evolutionary history humans have been exposed to dusty air. The bronchial lining of mucus and its movement out of the lungs described earlier is the anatomical response to this environmental problem. However, this system can be saturated or toxic particles may be inhaled in the workplace. Pneumoconiosis is a general term used to describe the harm done by inhalation of solids as the result of occupational exposure. Common symptoms include dyspnea (shortness of breath), chest pains, fatigue on exertion, and cyanosis (shortage of oxygen). Accompanying damage to the lung circulatory system can cause strain to the right side of the heart, which is responsible for pumping blood to the lungs. Chemicals transfer from particles in the lung into the blood stream, traveling to other organs and causing damage.

Airborne particulates are classically subdivided into groups based on their source and character. Solid particles, for example, may be dusts if they are produced by agricultural processing or a grinding type of operation, fumes if they arise from high-temperature metal processing (generally metal oxides), and smokes if their source is combustion of organic substances. Dusts are generally at the high end of the size range, perhaps from 1 to 150 μm in diameter. Fumes are smaller, running 0.2 to 1 μm in diameter, and smokes usually have a diameter of less than 0.3 μm. Aerosols are mists or fogs of water droplets condensed on smaller solid particles, or droplets of organic material such as result from a spraying operation.

Although toxic hazard primarily relates to the chemical nature of the particulate, the size of the particle is the second most important consideration. Larger particles are more likely to be trapped early on the mucus and removed as the mucus is moved out of the lung. The greater penetration by smaller particles lengthens their stay in the lung, increasing the chance of harm. The greatest potential for problems results when particles reach the alveoli.

A significant percentage of particles a few microns or less in diameter reach the alveoli. For purposes of classification a boundary has been set at 5 μm, and particles that size or less in diameter are termed respirable. Larger particles are called nonrespirable. The ACGIH has recently subdivided the nonrespirable class into the very large particles that are largely trapped in the nose and nasal cavity, inhalable particles, and those that are stopped in the bronchi, thoracic particles. The logic of this subdivision is that inhalable particles are often cleared directly from the body, as by blowing the nose. Those trapped on the bronchial mucus are carried with the mucus out of the lung and are generally swallowed, becoming GI tract toxicants rather than respiratory tract toxicants.

Various problems result from particulates entering the lungs. Dyspnea results from obstruction of the airways. Chemicals in particles may stimulate some contraction of the smooth muscle lining the bronchial tubes. Inhaling dusty air stimulates excess mucus production, narrowing the passages. Finally, the mucous glands and cells themselves may enlarge on long-term stimulation (hypertrophy).

Chemicals in particles may produce local irritation. Such irritation can cause fluid to collect in the lungs, a type of pneumonia. Certain particulates generate fibrosis, a scarring of the lung accompanied by a permanent loss of functional capacity. Physical activity is limited by the combined capacity of the lungs to provide oxygen to the blood, and the heart and circulatory system to deliver the oxygen to the tissues, and there is excess capacity of lungs over circulatory system in this process. Therefore, loss of lung capacity may not limit capability for physical exertion for a time, and so go unnoticed.

SPECIFIC EXAMPLES OF HAZARDOUS PARTICULATES

SILICA

Silicosis is a fibrosis of the lungs caused by inhaling respirable particles (>5 μm) of silica (SiO_2) over a long time period. Phagocytic cells in the alveolus engulf these particles and die, silica being toxic to them. The digestive enzymes released by these cells on death damage lung tissue, leading to scarring (fibrosis). Victims of silicosis have few symptoms in early stages, but display silicotic nodules in chest X-rays (simple nodular silicosis). In advanced cases where there is loss of functional lung mass, the victim experiences shortness of breath and coughs sputum. Now X-rays reveal large areas of scarred tissue (conglomerate silicosis). The symptoms develop slowly and may therefore be missed. If the victim is a smoker, the symptoms may be misinterpreted as being the result of tobacco use. Infections of the lung are common, and the victim is more likely to develop tuberculosis.

The TLV for dusts containing silica depends on knowing the percentage of silica in the dusts to which the worker is exposed:

$$\text{TLV - TWA} = \frac{10\ \text{mg}/\text{m}^3}{\%\ \text{silica}} + 2$$

Silicosis was the first occupational lung disease to be reported. It may result from the inhalation of any uncombined form of silicon dioxide. Grinding with silica abrasives, sandblasting, pottery making, or working with sand in a foundry are possible industrial exposures. Sandstone quarrying, "hard rock" mining, or tunneling also provide serious occupational contact. Fibrosis sometimes found in the lungs of coal miners probably results from silica released into the mine air during tunneling and inhaled, rather than from any effect of the coal dust itself.

ASBESTOS

Asbestosis, lung disease caused by inhalation of asbestos fibers, is perhaps the most publicized of the lung fibrosis diseases. What is termed asbestos is actually a number of different mineral materials that share the property of being convertible to the familiar fibrous product. They are silicate compounds of calcium, sodium, iron, chromium, nickel, or magnesium. Not all varieties of asbestos present the same degree of hazard. Asbestos strands readily break into short lengths and enter the air as very small particles, often less than 1 μm in length.

The thin fibers lodge in the smaller subdivisions of the air passages, causing irritation and edema. Continued exposure leads to permanent scarring of the lung with a resulting loss of lung capacity and tissue flexibility. Victims often have a cough, and lung infections are very common.

In addition, some types of asbestos are carcinogenic. Bronchial carcinoma (malignant mesothelioma) occurs in association with asbestos exposure. The type of asbestos called crocidolite is a particularly potent causative agent for malignant mesothelioma. Combining the inhalation of asbestos fibers with the smoking of tobacco sharply increases the risk. It is hypothesized that the surface of the fiber, as it lies embedded in the lung, serves to collect and concentrate carcinogens from the tobacco smoke.

Asbestos has found a broad range of applications largely based on its fibrous character and resistance to high temperatures. These include pipe insulation, floor tile, roofing shingles, cement, clutch pads, brake linings, wall board and plaster, and heat- and fire-resistant garments. It has been used in factories for insulation of steam pipes and boilers, as an air filter medium, in board form for partitions, as fireproofing on beams, and as insulation under the roof to prevent heat loss.

Substitutes have been found for asbestos in many applications since the discovery that cancer can result from exposure, so the level of worker exposure has dropped. However, OSHA estimates that 3.9 million U.S. workers are still at risk of asbestos exposure. Although care is taken in new installations not to use asbestos in a hazardous fashion, many older facilities have asbestos present in locations such that workers are at risk of exposure. Particular care must be exercised in removal of old asbestos from such sites; an expert with the proper equipment and experience is necessary. There is less emphasis now on very hazardous removal of asbestos from old sites, and more focus on managing it in place in a safe fashion.

In 1994, OSHA issued rules (29 CFR 1910.1001) designed to protect workers at risk of asbestos exposure. The rules lower the PEL to 0.1 fibers per cubic centimeter and spell out detailed work practices to lower exposure of workers. Special attention is paid to asbestos removal procedures.

TALC

Talc is hydrated magnesium silicate, and is usually found in association with deposits of dolomite. It is a component of tile, paper, and paint and it is used as a lubricating powder with rubber products such as tires and gloves or as talcum powder. It does not cause the degree of fibrosis that we see with asbestos, but occasionally it is a cause of pneumoconiosis.

MICA

The mineral form of mica is called muscovite. It is found as plates that separate into thin sheets, or as a powdery product. It serves well as a heat or electrical insulator, so it is used in electrical devices and appliances. The ground material is added to roofing, and it has served as a mold release agent in the rubber industry. It infrequently is responsible for pneumoconiosis.

FIBERGLASS

Fiberglass or glass wool is simply very long, thin fibers of glass. It is used as insulation and as a reinforcing material in plastics. Tiny fragments of glass wool carried as a dust in the atmosphere can enter the respiratory tract and cause irritation, much as skin contact with glass wool can cause skin irritation. Recent studies based on both epidemiological studies of fiberglass workers and laboratory studies with animals indicate there is no correlation between exposure to airborne fiberglass and lung cancer.[1]

METALS AND METAL OXIDES

Metals and their compounds may appear in the workplace as the result of grinding, sawing, milling, or otherwise shaping metal parts. When metal is melted or formed at high temperature, metal oxide fumes may be released. Several of these fumes are irritating or damaging to the lungs. A set of symptoms very much like influenza is encountered, termed metal fume fever. These problems are discussed in more detail in Chapter 19.

COAL

The mining of coal has caused a great deal of lung damage over the years. Coal particulate produces characteristic symptoms called coal miner's pneumoconiosis (black lung disease). Coal dust deposited in the lungs leads to breathlessness. Fibrosis is not a direct symptom. The introduction of wet drilling techniques to minimize dust production was a major advance in reducing the magnitude of this problem.

COTTON FIBER

Another dangerous class of dusts are the natural product particulates. An example is the dust arising from cotton in the course of making cloth. Fibrous material in the air in cotton mills carries an agent that causes brown lung disease. Other problems have been experienced by farmers inhaling molds and organic particulates, but these are outside the scope of this discussion.

OTHER DUSTS

Not all dusts are as damaging as those already presented. When deposition in the lung does not cause inflammation or fibrosis, health problems are significantly reduced. For example, calcium carbonate (limestone or marble) and calcium sulfate (gypsum) dusts are not as threatening. Similarly, the iron dust and iron oxide fume that welders encounter is not as damaging as many other metal oxides. Iron compounds may accumulate in the lungs, a problem termed siderosis, but they do not stimulate fibrosis. However, welders need to take precautions because many of the items being welded, as well as welding rods themselves, include more dangerous metals in their composition.

HAZARDS OF GASES

Gases generally cause one of two types of problems for workers. Some are asphyxiants, meaning that ultimately they result in tissues having an inadequate supply of oxygen. Others are irritants that damage or kill cells lining the nose, throat, and lungs, resulting in discomfort and pain, constriction or spasmatic contraction of breathing passages, or filling of lung passages with fluid.

[1] J. Johnson, "Studies Support the Safety of Fiberglass," *Occupational Health and Safety*, Dec. 1992, p. 7.

ASPHYXIANTS

In living organisms the energy needed to perform biological work, processes such as contracting muscle, transmitting nerve impulses, and the complex chemical synthesis required for cell repair and growth, is supplied by adenosine triphosphate (ATP). This molecule has three adjacent phosphate groups. When a phosphate–phosphate bond is hydrolyzed (almost always the bond attaching the end phosphate), energy is released to do biological work. By the laws of thermodynamics (these are not in the *Federal Register*), that same amount of energy must be supplied to reestablish the phosphate–phosphate bond. ATP is reassembled in conjunction with the oxidation of dietary molecules (carbohydrates, fats, and proteins), using the energy released by this oxidation to drive ATP formation.

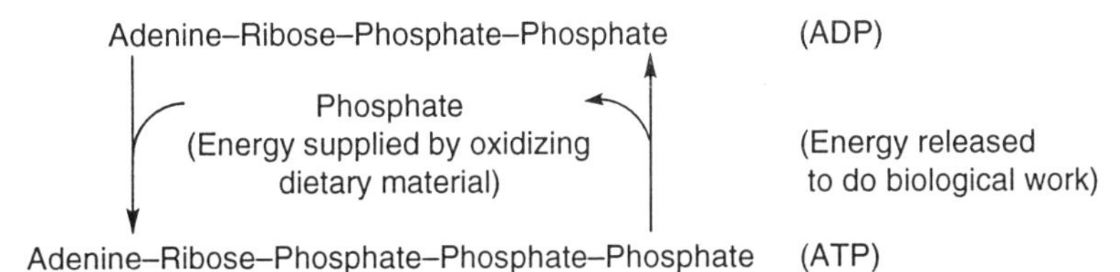

When the supply of oxygen to the body is reduced, these metabolic oxidations slow or cease, ATP supplies dwindle, and vital processes driven by ATP are halted.

The immediate danger from asphyxiation lies in reducing the supply of oxygen to the brain. Brain cells are more susceptible to the loss of their oxygen supply than are the cells of other tissues. Furthermore, once dead, brain cells are not replaced. Within 5 to 8 min of a complete loss of the oxygen supply, serious, possibly irreversible brain damage is taking place. In gradual asphyxiation the early symptoms are a feeling of euphoria, followed by an increase in the respiratory rate. Later there is fatigue and headache, then nausea and vomiting. Finally, the victim collapses.

Hypoxia is a general term describing the reduction in oxygen supply to tissues. Low levels of oxygen in the air or interference with respiration, the original intake of oxygen into the blood, is termed anoxic hypoxia. If respiration is normal and the air has sufficient oxygen, but the capacity of blood to transport oxygen is impaired, the problem is termed anemic hypoxia. If the blood flow is inadequate due to heart problems or a blocked blood vessel, it is termed hypokinetic hypoxia. Finally, the tissues may be poisoned so as to be unable to use the oxygen, a problem termed histotoxic hypoxia. Asphyxiants are subdivided into simple and chemical subgroups.

Simple Asphyxiants

Simple asphyxiants dilute the oxygen supply to levels that no longer support life. This requires that the diluting gas be present in extremely high concentration. Small concentrations prove harmless, because simple asphyxiants are chemically inert. Nitrogen, helium, argon, and such hydrocarbons as methane belong to this class. These are not detectable by odor. The natural gas smell associated with methane is actually trace amounts of a very strong smelling mercaptan added to facilitate recognition of a leak. Carbon dioxide is a borderline member of this group. It creates problems primarily by displacing oxygen from the environment. However, as an acid anhydride it exerts a physiological effect on the person inhaling it, for example, stimulating deeper, more rapid breathing. Carbon dioxide is a dense gas, so it may accumulate at higher concentrations at floor level or at low points in tanks or compartments that workers might enter. Awareness of the hazards of carbon dioxide are important because of the widespread use of this compound in industry.

Sometimes oxygen-free atmospheres are deliberately produced, as in some special welding techniques. More dangerous is the situation where the worker enters an area unaware that the oxygen supply is dangerously low. This happens in once-sealed compartments in which an oxidative process has been occurring, or in tanks where evaporation of residues of the liquid once stored in the tank has displaced the oxygen.

Chemical Asphyxiants — Carbon Monoxide

Chemical asphyxiants (Table 7.1) block the transportation or use of oxygen by some chemical reaction. Important examples include carbon monoxide, hydrogen cyanide, and hydrogen sulfide. Carbon monoxide is an odorless gas. It is found in coal mines and anywhere combustion is occurring, such as around internal combustion engines and furnaces. An operation of particular concern is the heating of vats or tanks with a gas flame. As the burning gas strikes the cold surface, the gas temperature is lowered, leading to incomplete combustion and the generation of carbon monoxide.

Table 7.1. Asphyxiants.

Compound	PEL[a] (ppm)	IDLH[b] (ppm)
HCN	10	50
CO	50	1,500
H_2S	20 (ceiling[c])	300
CO_2	5,000	50,000

[a] PEL = permissible exposure level, time-weighted average, 8-hr day.
[b] IDLH = immediately dangerous to life or health.
[c] 50 ppm; 10-min peak.

In the blood, carbon monoxide combines with hemoglobin, the oxygen transport compound, making hemoglobin incapable of binding oxygen. The bond with carbon monoxide is at the same location on hemoglobin as the bond with oxygen, but is more than 200 times as strong. Hemoglobin with carbon monoxide attached is unavailable for oxygen transport, producing a temporary, anemia-like condition. The greater the carbon monoxide exposure, the higher is the percentage of hemoglobin tied up in this fashion (Table 7.2). A portion of the hemoglobin of smokers is already incapacitated, because carbon monoxide is found in cigarette smoke, and they are at greater risk. Because of the tight carbon monoxide–hemoglobin bond, simply removing a victim from the source of the gas does not quickly correct the problem. Hollywood movies portray the situation to the contrary where the person attempting suicide is taken out of the garage and wakes up good as new. Emergency services often supply pure oxygen to a victim, not only to make optimum use of unbound hemoglobin, but also to speed equilibration of carbon monoxide off hemoglobin by use of the high oxygen concentration.

Table 7.2. Effects of Binding of CO by Hemoglobin.

Percentage of Hemoglobin as CO Complex	Cause/Effect
5	Caused by one pack of cigarettes per day
10	Caused by 50 ppm CO in air (the PEL[a])
15–25	Headaches, nausea
40	Collapse
60	Death

[a] PEL = permissible exposure level.

The Haldane equation, devised in 1912, allows calculation of the degree of saturation of blood with carbon monoxide at various concentrations of CO in the air:

$$\frac{Hb_{CO}}{Hb_{O_2}} = M\frac{p_{CO}}{p_{O_2}}$$

where M is about 245 at normal blood pH

SAMPLE PROBLEM:

Suppose p_{CO} were 10 torr and p_{O2} was approximately normal (160 torr). What fraction of the hemoglobin is nonfunctional because it is bound to CO?

$$\frac{Hb_{CO}}{Hb_{O_2}} = M\frac{p_{CO}}{p_{O_2}}$$

$$\frac{Hb_{CO}}{Hb_{O_2}} = 245\frac{10\text{ torr}}{160\text{ torr}} = 1.6$$

$$\frac{Hb_{CO}}{\text{total Hb}} = \frac{1.6}{1.0+1.6} = 0.62\text{ (or 62\%)}$$

CASE STUDY: CO IN THE WAREHOUSE

Two employees of a Michigan plastics company become ill from inhalation of CO. Investigation traced the CO to ten propane-fueled forklifts, which had been in service around the clock 10–12 years. Five units were replaced, and all units were equipped with electronic devices to adjust the air/fuel ratio to maintain optimal combustion. Concentrations of CO in the exhaust stream dropped from as high as 1000 ppm in old, unregulated forklifts to 10–15 ppm in new forklifts and 40 ppm in old forklifts fitted with the controllers. A bonus to the company is a 10% reduction in propane consumption. Increased ventilation and installation of CO monitors provided the rest of the fix.

Adapted from D. Riley, "Creative Controls for Carbon Monoxide Emissions," *Occupational Health and Safety,* July 1995, p. 22.

Hydrogen Cyanide

Hydrogen cyanide, famous as the lethal gas used in the gas chamber, has a faint bitter almond odor. Salts of this compound are used in some ore extractions, in electroplating, in preparing some plastics monomers, and in a few other industrial processes.[2] In the body, cyanide binds to the compounds that are the prime users of the oxygen in the tissues, blocking the process of energy production in the body. Most often the hazard from hydrogen cyanide

[2] The salts are very toxic if ingested, but, as salts, are not volatile.

arises when a solution containing such salts as sodium or potassium cyanide is accidentally mixed with acid. This converts the cyanide ion in solution into gaseous hydrogen cyanide.

Hydrogen Sulfide

There are no direct uses of hydrogen sulfide, but it does occur as a byproduct in refineries, some plastic and rubber processes, and tanneries. Sewers also contain hydrogen sulfide. It has a distinctive rotten egg odor at very low concentrations, but at dangerous levels hydrogen sulfide blocks the sense of smell. In high concentrations it causes respiratory paralysis.

IRRITANTS

Irritant gases either cause direct damage to the linings of the air passages or in that moist environment they are converted to molecules that do. They are often referred to as upper respiratory tract and lower respiratory tract irritants, a distinction based on how far into the lung they are likely to penetrate.

Upper respiratory tract irritants (Table 7.3) are unlikely to cause serious injury if the worker can leave the area, because they are so unpleasant to encounter that workers hold their breath and quickly escape from contact. This limits entry to the nose, throat, and larger passages of the respiratory tract — the upper respiratory tract. Upper respiratory tract irritants are more water soluble and are irritant as water solutions. A burning sensation and possible spasms of the bronchial tubes are the result of these compounds contacting the lung surface. Important examples of upper respiratory tract irritants include hydrogen chloride (the gas form of hydrochloric acid), sulfur dioxide, ammonia, and chlorine (a borderline case).

Table 7.3. Upper Respiratory Tract Irritants.

Compound	PEL[a] (ppm)	IDLH[b] (ppm)
HCl	5	100
HBr	3	50
HF	3	20
SO_2	5	100
NH_3	50	100

[a] PEL = permissible exposure level, time-weighted average, 8-hr day.
[b] IDLH = immediately dangerous to life or health.

Lower respiratory tract or whole-lung irritants (Table 7.4) pose a significantly greater threat. Generally they are less water soluble, so they dissolve into and penetrate the mucus more slowly. The full potential for damage follows a slow reaction with water, converting them into more damaging compounds. For both these reasons the irritation of the lung on first being inhaled is not as extreme as with upper respiratory tract irritants, so victims allow a damaging dose to move through the entire lung.

For some of the more dangerous compounds such as phosgene or nitrogen dioxide there is a long time lag, even hours, between the inhalation of the gas and the full impact of the compound on the lungs. The irritation causes the lungs to fill with fluid, producing a pneumonia-like condition. In its extreme form a person is drowned in his or her own body fluids. Beyond the immediate crisis, an extreme invasion of the lungs by an irritant can leave the lung permanently scarred, resulting in a reduction of lung capacity.[3] Such paralysis of

[3] Lower respiratory tract irritants such as nitrogen dioxide and ozone are low-concentration components of photochemical smog and cause lung damage to residents of areas afflicted with this problem.

Table 7.4. Lower Respiratory Tract Irritants.

Compound	PEL[a] (ppm)	IDLH[b] (ppm)
Cl_2	1.0	25
Br_2	0.1	10
F_2	0.1	25
NO_2	5.0	50
O_3	0.1	10
$COCl_2$ (phosgene)	0.1	2

[a] PEL = permissible exposure level, time-weighted average, 8-hr day.
[b] IDLH = immediately dangerous to life or health.

breathing and fluid in the lungs were the prime cause of death in the tragic escape of methyl isocyanate (actually an irritant vapor) in Bhopal, India, in 1984. Reduced lung capacity is a common complaint among the survivors.

HAZARDS OF VAPORS

The evaporation of organic liquids into the workplace atmosphere represents a major regulatory concern. The mechanisms by which individual compounds threaten worker health span the entire spectrum of toxicology from mild irritation, through systemic damage as the chemical enters the blood, to the causing of cancer. The threat varies with the concentration of the vapor in the workplace atmosphere, and this in turn depends on how the compound is used, ventilation and other safeguards, and the volatility of the liquid that is the source of the vapor. Volatility reflects the rapidity and degree to which the liquid evaporates. Two physical constants measure volatility: the vapor pressure and the boiling point. Vapor pressure measures the amount of a compound that can enter the atmosphere at a given temperature. For example, if the atmospheric pressure were 750 mmHg (slightly below 1 atm) and the vapor pressure of a compound at the existing temperature were 75 mmHg, that compound would evaporate until 75/750 or 10% of the molecules in the air above the container of liquid were molecules of that liquid. Vapor pressure increases with temperature, and the boiling point is the temperature at which the vapor pressure is as high as the existing atmospheric pressure. Compounds with low boiling points must therefore have high vapor pressures, and would be described as very volatile.

The importance of volatility in industrial toxicology can be demonstrated very clearly by considering again the Bhopal tragedy in India. The compound that caused the deaths and injuries was methyl isocyanate, used to make carbamate insecticides. Toxicologically, it bears a strong family resemblance to other isocyanates, such as toluene diisocyanate, used in the plastics industry. However, methyl isocyanate is very volatile (boiling point, 39°C). With a rise in the temperature of the storage tank, it rapidly converted to vapor, blowing the safety valve and blanketing the town of Bhopal. Had everything been the same, except that the tank had contained toluene diisocyanate (boiling point, 251°C), the amount of vapor escaping would have been much smaller and the major tragedy would not have occurred.

To list all the compounds with hazardous vapors would be a gigantic task, but the recommended exposure maxima of some more common organic solvents are listed in Table 7.5. For any liquid to be used in the plant, referring to the label on the container, or better, to the MSDS gives warning of any hazard. A danger exists whenever hazardous liquids are used or stored in open containers, are sprayed onto surfaces, evaporate from coatings or from dipped parts, or are used in any other fashion that allows evaporation into the atmosphere. Furthermore, hotter liquids have greater vapor pressures, and so escape more rapidly.

Table 7.5. Recommended Maximum Exposures to Solvents.

Solvent	PEL[a] (ppm)	TLV-TWA[b] (ppm)	STEL[c] (ppm)
Acetaldehyde	200	100	150
Acetone	1000	750	1000
Acetonitrile	40	40	60
n-Amyl acetate	100	100	150
Anthracene (coal tar pitch volatiles)	0.2[d]	0.2[d]	—
Benzene	10 (25c)	10	25
n-Butyl acetate	150	150	200
n-Butyl alcohol	100	50	—
Carbon disulfide	20	10	—
Cellosolve	200	—	—
Cellosolve acetate	100	—	—
Chlorobenzene	75	75	—
Cyclohexane	300	300	375
Dibutylphthalate	0.1	1	2
1,2-Dichloroethylene	200	200	250
Diethyl ether	400	400	500
Dimethylformamide	10	10	20
Dioxane	1	—	—
Ethyl acetate	400	400	—
Ethyl alcohol	1000	1000	—
Ethyl chloride	1000	1000	1250
Ethylene glycol	—	50	—
Ethyl formate	100	100	150
Furfural	5	2	10
Furfuryl alcohol	50	10	15
n-Heptane	85	—	—
n-Hexane	100	—	—
2-Hexanone (methyl butyl ketone)	100	5	—
Isobutyl alcohol	100	50	75
Isopropyl acetate	250	250	310
Isopropyl alcohol	400	400	500
Methyl acetate	200	250	—
Methyl alcohol	200	200	250
Methyl cellosolve	25	25	35
Methyl cellosolve acetate	25	25	35
Methyl chloride	100	50	100
Methyl cyclohexanone	100	50	75
Methylene chloride	25	125	500
Methyl ethyl ketone	200	200	300
Naphtha (coal tar)	100	—	—
Naphthalene	10	10	15
Octane	75	—	—
n-Propyl acetate	200	200	250
n-Propyl alcohol	200	200	250
1,1,2,2-Tetrachloroethane	5	1	5
Tetrachloroethylene	100 (200c)	50	200
Toluene	200	100	150
1,1,1-Trichloroethane (methyl chloroform)	350	350	450
1,1,2-Trichloroethane	10	10	20
Trichloroethylene	100	50	150
Xylene (mixed isomers)	—	100	150

[a] PEL = permissible exposure level.
[b] TLV-TWA = threshold limit value–time-weighted average.
[c] STEL = short-term exposure limit.
[d] mg/m^3.
From W. H. Lederer, NIOSH Registry of Toxic Effects of Chemical Substances, 1981–1982, U.S. Department of Health and Human Services. Regulatory Chemicals of Health and Environmental Concern, Van Nostrand-Reinhold, New York, 1985.

WARNING PROPERTIES OF GASES AND VAPORS

A part of the woodsy lore we pick up as children and retain, unless it is challenged, is that poisons taste nasty and harmful chemicals in the air smell bad. Workers often assume that if the air has no unpleasant odor it is safe, and if they smell something, they are at risk.

Gases or vapors that have a strong odor or are irritants of the respiratory passages are said to have good warning properties. Escape of such compounds is readily noticed, permitting corrective measures to be taken before dangerous levels are reached. However, many compounds are odorless or will already have accumulated to dangerous levels before an odor is detected. A good example is provided by three chlorine-containing gases: hydrogen chloride, chlorine, and phosgene. Hydrogen chloride is a strong acid when dissolved in water, and is a severe irritant to the eyes and respiratory tract. It has excellent warning properties. The irritation is so intense that a person exposed to even low concentrations of this gas will struggle to avoid inhaling it, even at levels that would do little harm. Chlorine is not an acid, but does react with water to produce acid. It is therefore less irritating, so that a person exposed to chlorine might accept dangerous levels into the lungs in an emergency situation. Phosgene is the most dangerous of the three. It reacts only very slowly in the lungs to produce acid and does not have a strongly repugnant odor. Dangerous levels could easily be inhaled inadvertently, the damage happening some time after inhalation.

The lowest airborne concentration that a person can smell is termed the odor threshold. This very subjective value is usually determined as the concentration at which 50% of individuals surveyed first detect the odor. In Table 7.6 a comparison is made of the odor threshold, hazardous properties, and PEL of a few common hazardous chemicals. See for yourself the lack of correlation between hazard and odor.

Table 7.6. Vapor Warning Properties.

Vapor	Odor Threshold (ppm)	PEL-TWA[a] (ppm)	Hazard
Acetone	13	1000	Irritant
Acrylonitrile	17	2	Mutagen, neurotoxin
Carbon disulfide	0.11	4	Neurotoxin
1,2-Dichloroethane	88	1	Mutagen, renotoxin, hepatotoxin
Ethylene oxide	430	1	Possible carcinogen; broad effects
Hydrogen sulfide	0.9	10	Respiratory paralysis
Tetrachloroethylene	27	25	Hepatotoxin, renotoxin

[a] PEL-TWA = permissible exposure level–time-weighted average.

Compounds with an unpleasant odor at levels that are low enough to be safe may still need to be controlled to prevent the annoying odor, to reduce worker concerns, and thus to provide a comfortable workplace. Hazard should not be equated with repugnant odor or irritation to the lung, but that unpleasant awareness of a compound is a safety feature.

OCCUPATIONAL ASTHMA

Asthma is the inflammatory response of lungs to the inhalation of an allergen, a substance to which the individual is allergic. The difficulty is common, with approximately 3% of the population diagnosed as having an asthma problem. Asthmatic response usually begins with a heavy exposure to a sensitizing agent (the allergen). As with other allergies, once he or she is sensitized the individual responds to even very small amounts of allergen. Occupational asthma is allergic response to an agent inhaled in the workplace. The response may be delayed, making it hard to draw a clear association between the workplace exposure and the symptoms. Skin tests may be needed to identify the allergen (Chapter 6).

Response is mediated by antibodies in the lung (often IgE antibodies) and includes contraction of the bronchial smooth muscle, which produces a constriction of the air passages. The victim, now suffering an asthma attack, is short of breath, wheezes, coughs, or generally has difficulty taking in sufficient air, and can be severely distressed. Symptoms of an asthma attack are treated with β-adrenergic drugs and derivatives of theophylline, which relax the bronchial smooth muscle. Antiinflammatory steroids may be used in extreme cases. If airborne allergens cannot be removed adequately from the atmosphere, sensitized workers should be removed from the vicinity of the allergen.

Some chemicals are unusually effective sensitizing agents. Higher levels of asthma occur among workers who have contact with animals or who work with specific plant products. Diisocyanates, discussed above, are excellent examples. Methylene bisphenyl isocyanate and toluene diisocyanate are used in the plastics industry and are currently regulated at a PEL of 0.02 ppm. Because of the sensitizing potency of these chemicals, NIOSH recommends a PEL of 0.005 ppm. Experience indicates that if levels of potent sensitizers are kept low, fewer workers are sensitized. Stricter regulation of these compounds was part of the package of new regulations OSHA passed that was reversed by the courts in 1989.

KEY POINTS

1. Inhaled toxicants may be gases, vapors, particles or aerosols.
2. The mucus-lined nasal cavity can trap particles, but many workers are mouth breathers.
3. The pharynx and bronchial tube are lined with mucus. The bends in air passages and the increasing cross-sectional area and consequent slowing of air flow with lung penetration favor particles depositing on the mucus.
4. Mucus in the lungs is continually moved, carrying the trapped particles, toward the nasal cavity.
5. Only a thin wall separates air in the alveolus from blood in capillaries. Alveoli have no mucus, but do have phagocytic cells to engulf particles.
6. Polar gases or vapors dissolve in the moist lung lining. Here irritants generate coughing.
7. Nonpolar gases or vapors are not retained long in the lung, but they are more capable of crossing the alveolar wall and entering the blood.
8. Particles include dust, fumes, and smoke, according to origin.
9. Only particles smaller than a few microns penetrate well as far as the alveolus.
10. Irritation causes inflammation, resulting in fluid collecting in the lungs.
11. Some particles generate scarring (fibrosis) in the lungs.
12. Silicosis is caused by long-term inhalation of SiO_2. In advanced cases there is fibrosis and loss of function.
13. Asbestosis is a lung fibrosis problem caused by asbestos fibers. Some varieties of asbestos also cause malignant mesothelioma.
14. Talc and mica occasionally cause fibrosis.
15. Fiberglass is a lung irritant.
16. Many metal and metal oxide particles cause metal fume fever, which has symptoms similar to influenza.
17. Coal dust deposits reduce lung capacity in a problem called black lung disease.
18. Cotton fibers cause a fibrosis called brown lung disease.
19. Simple asphyxiants are gases that dilute oxygen in the air.
20. Chemical asphyxiants block oxygen either from reaching or from being used in the tissues. These include CO, HCN, and H_2S.
21. Upper respiratory tract irritants are water soluble and cause immediate irritation on entering the lungs. This limits the willingness of the victim to allow them to go deeply into the lung. They include HCl, SO_2, Cl_2, and NH_3.

22. Lower respiratory tract irritants react slowly with water to produce irritants, causing damage some time after inhalation. These include phosgene and NO_2.
23. Vapors are evaporated liquids. Lower boiling liquids produce vapor in larger concentrations.
24. Irritation and odor are warning properties of gases and vapors that make us aware of their presence.
25. Compounds that sensitize the lungs cause occupational asthma.

BIBLIOGRAPHY

E. J. Calabrese and E. M. Kenyon, *Air Toxics and Risk Assessment,* Lewis Publishers, Chelsea, MI, 1991. (Includes a large number of assessments of specific compounds.)

D. B. Menzel and M. O. Amdur, "Toxic Responses of the Respiratory System," in C. D. Klaassen, M. O. Amdur and J. Doull, Eds., *Toxicology: The Basic Science of Poisons,* 3rd ed., Macmillan, New York, 1986.

R. F. Phalen, *Inhalation Studies,* CRC Press, Boca Raton, FL, 1984.

R. P. Smith, "Toxic Responses of the Blood" in C. D. Klaassen, M. O. Amdur, and J. Doull, Eds., *Toxicology: The Basic Science of Poisons,* 3rd ed., Macmillan, New York, 1986.

H. Witschi and P. Netthesheim, *Mechanisms in Respiratory Toxicology,* CRC Press, Boca Raton, FL, 1982.

G. W. Wright, "The Pulmonary Effects of Inhaled Inorganic Dust," in G. D. Clayton and F. E. Clayton, Eds., *Patty's Industrial Hygiene and Toxicology,* Volume I, Part A, 4th ed., John Wiley & Sons, New York, 1991.

PROBLEMS

1. In this problem the objective is to visualize the environment of chemicals entering the respiratory tract and consider the likelihood of that chemical being trapped in each region.
   A. Consider a particle entering the nose. Describe changes in the surroundings of the particle as it moves from the nose to the alveolus, and the effect of these surroundings and changes on removal of the particle from the air.
   B. What changes in the particle itself can affect the likelihood of its entrapment, including chemical composition, degree of adsorption of water, size, and density?
   C. Contrast the movement of a molecule of a nonpolar gas such as methane in terms of the likelihood it will be stopped.
   D. Now contrast the movement of a molecule of a polar and irritating gas such as HCl.
2. Before opening a bottle of CF_3–CH_2–OH you read the warnings on the label and learn it is somewhat acidic and is a respiratory irritant. What effect does a quantity of an irritating gas have in the lungs that increases the likelihood of its removal?
3. You are discussing fibrosis of the lung with a friend, who says, "Why should fibrosis increase the burden of the right side of the heart, the side pumping blood to the lungs? Doesn't loss of lung tissue simply mean there is less lung tissue the heart needs to supply with blood?" You reply, "No, ...!"
4. You are an industrial hygienist inspecting a plant where the following operations are being performed:

   (1) Materials and parts are delivered to the plant at an open-air loading dock and (2) are shifted into the plant to sites along the assembly line by gasoline-powered forklift trucks. (3) Relatively low melting aluminum alloys are heated in an open cauldron using gas flames. The molten alloy is poured into molds to form garden tractor transmission housings. (4) The dies are cooled by circulating water, housings are removed from the molds, and the housings are clamped in a jig. (5) Holes are drilled for press-fit bearings and to insert studs to assemble the components. (6) A grinding wheel grinds one face flat for attachment of the cover. (7) Stud holes are tapped (threaded) and compressed air is used to blow the housing clean. (8) The assembly is dipped in a chlorinated hydrocarbon solvent to degrease the surfaces, is

blown dry with compressed air, and is placed in a basket that is the cathode of an electroplating system. (9) It is passed through a tank of cupric cyanide, sodium hydroxide, and sodium cyanide where the housing is copper plated. (10) The basket then passes through a water rinse tank and goes into a bath of nickel sulfate, nickel chloride, and boric acid where the housing is nickel plated, and finally through a second water rinse bath. All baths are electrically heated. (11) The part is again placed in a jig and the outer surface is polished with buffing wheels and a fine abrasive. (12) The surface is covered with a template and is spray painted with the company logo. (13) Heat lamps rapidly dry the paint. (14) Studs are screwed into the threaded holes with power drivers and bearings are pressed into their housings. (15) Finished parts are packed in Styrofoam pellets in shipping crates for transfer to the main assembly operation in another state.

A. Identify each step in the operation during which you need to be alert for airborne contaminants being generated.
B. Consider the plating operation. Tanks lose water over time as a result of heating and solution being carried out on parts. Concentrations of plating bath solutions in rinse tanks rise as they are used. You propose that instead of adding water to the plating solution, you add water from the rinse tank, then replenish the rinse tank. How does this reduce waste and what dangerous air contaminant may be prevented by this procedure?

5. The ACGIH TLV for asbestos is complex and reveals a great deal about the nature of the problem. It reads as follows:
   Amosite, 0.5 fibers > 5 $\mu m/cm^3$ TWA
   Chrysotile, 2 fibers > 5 $\mu m/cm^3$ TWA (95% of asbestos used)
   Crocidolite, 0.2 fibers > 5 $\mu m/cm^3$ TWA
   Others, 2 fibers > 5 $\mu m/cm^3$ TWA
   A. For most chemicals, standards describe mg/m^3, ppm, and mg/kg, and depend on measuring the weight of the chemical in the dose volume. How must the dosage be measured here?
   B. Why is it specified that the dose must be of fibers of dimensions less than 5 µm?
   C. Why is the standard for crocidolite so low?
6. Compare the values for PEL and TLV in Table 7.5. Do they more often agree or disagree? Where they differ, is OSHA or ACGIH more conservative?

CHAPTER 8

Protecting the Worker I: Monitoring the Plant Atmosphere

The totally safe workplace is an unreachable goal. However, the efforts of many researchers, government workers, legislators, labor union officers, company managers and concerned members of the voting public are bearing fruit (Figure 1.1). Many dangerous practices of the past have been eliminated. The right-to-know laws give the worker access to information needed for self-protection. Regulatory efforts have moved us toward common recognition of what constitutes a good working environment, and they serve as a lever to move less concerned managers toward providing those conditions. Heightened awareness of the potential for harm has caused many to be more thoughtful and favorably disposed toward efforts to provide and maintain a safe workplace, and to exert greater effort toward anticipating dangerous situations, rather than to react to a problem only after harm has already occurred. The importance of this progress is heightened by the rapid appearance of new technologies and the need to change processes more frequently to meet mounting competition in a world marketplace.

Chapters 8 and 9 outline steps necessary to maximize the safety of the workplace with respect to airborne pollutants. Before a health and safety program can be established, it is necessary first to identify the hazards. Once identified, the best measures to protect workers from these hazards must be devised. Finally, a program of monitoring of the workplace ensures that protective measures are working.

IDENTIFYING HAZARDS

Assume the role of the compliance officer. Identifying hazards present for the worker in an existing or planned facility is a monumental task, even if one restricts consideration to airborne contaminants. Start by identifying all chemicals entering the facility for whatever purpose. Using this inventory, assess the hazards of each chemical. The toxicity and volatility of each chemical can be obtained from a number of sources, the most obvious and readily available of which is the MSDS provided by the supplier of the chemical.

Once you decide what chemicals are of concern, there is a three-step process: collecting samples, analyzing samples, and deciding if a hazardous situation exists.

Toxicity and hazard are not the same thing. Toxicity data indicate levels of exposure to the chemical that cause harm and the type of harm likely to occur. Hazard includes additionally the likelihood of injury occurring, given the manner of use of the compound. It is necessary to consider how each compound is used in the plant. NIOSH recommends preparation of a flow sheet that lays out the process being performed in a schematic fashion. With the knowledge of where each chemical is used, a walking tour of the plant is taken. At that time, the most likely sites for exposure can be identified.

A large dose of good common sense is important when evaluating plant hazards. A compliance officer is a detective, looking for clues. Candy bar wrappers on the floor indicate workers are eating on the job and may add chemicals on their hands to the snack. One must see objects not for their function, but for their potential to release chemicals. A forklift is not a device to lift heavy objects, but instead is a carbon monoxide generator. A valve is not a device to control flow of liquids in pipes, it is a potential site for leaking. Particular attention should be paid to enclosed processes such as open vats, degreasing tanks, and spraying or drying operations. Where bags of dry chemicals are opened and dumped into a hopper is a high-risk site. In fact, any place where the chemicals are handled or transported deserves special attention. Receiving and shipping departments are places of special concern. Notice how the workers interface with the plant machinery, paying special attention to whether workers are at locations with a high risk of chemical exposure. What is the type and layout of plant ventilation? Sampling air in high-risk areas is useful to determine if suspected problems are real problems. A summary at this point serves as the foundation for a plan to control hazards to the workers.

Where there are legally established limits, noncompliance can involve fines and court orders. Given that the health of the workers is at stake, the control of levels of contaminants in the plant air is a project to be taken seriously. How does one measure levels of dangerous chemicals in factory air on a day-to-day basis?

SAMPLING

Two approaches are used for sampling air:[1] the instantaneous or grab sample and the continuous or integrated sample.

GRAB SAMPLING

In grab sampling, a single aliquot (sample) of air is captured to be analyzed for concentration of the contaminant. The method of capturing the sample is usually quite unsophisticated. One may simply open an evacuated bottle in the appropriate location, then seal it immediately. Or draw a sample of air into an evacuated syringe body, then cap the end of the syringe. In either case, the container is empty at the outset, so there are no previous contents to contaminate the sample. Opening the evacuated bottle can be done remotely to allow sampling where a person might prefer not to go, sampling mine or sewer gas, for example. It is necessary to know the temperature and pressure of air being sampled to allow calculation by the gas laws of the exact amount of air that has been collected.

Such grab sampling limits the analysis to a relatively small quantity of air, so it is most useful in situations where either the analytical method is very sensitive or the contaminant is relatively concentrated. A larger sample is taken by pumping the air into a plastic bag (Figure 8.1). Such bags are lightweight, are easily stored, and can hold several liters. Any sample taken in the course of a few minutes is considered a grab sample.

INTEGRATED SAMPLING

Integrated sampling involves extended pumping of a stream of air through a trap, which captures the contaminant. Such sampling involves a much larger volume of air, allowing accumulation of a quantity of the contaminant. This permits the use of less sophisticated

[1] A list of suppliers of air sampling equipment by manufacturer is in *Occupational Health and Safety*, May 1996, pp. 46–48.

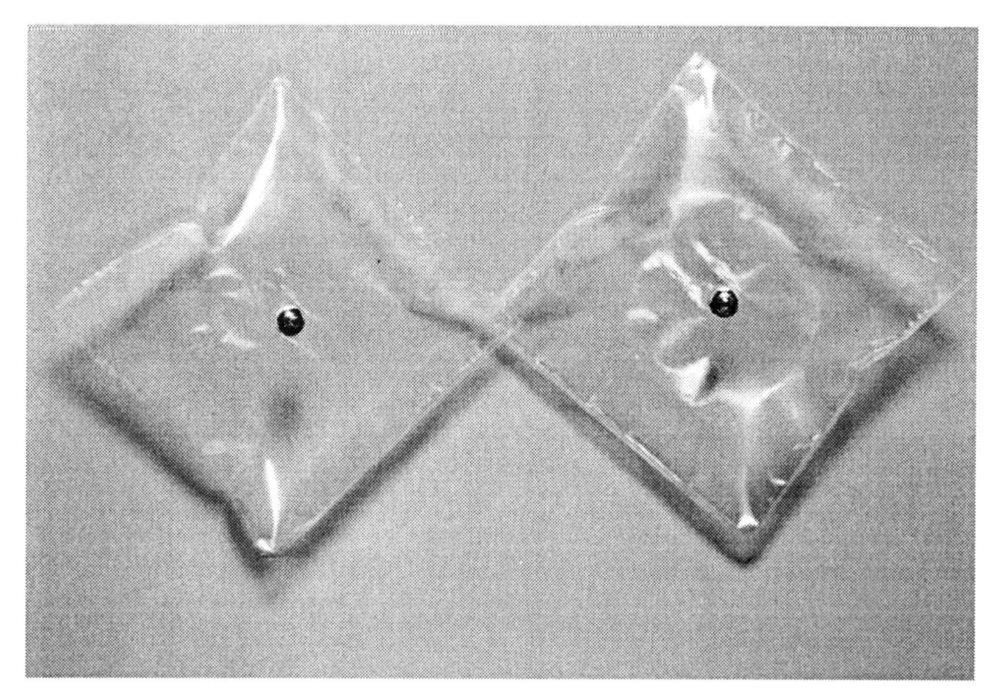

Figure 8.1. Grab sampling is performed here by pumping a sample of air into these bags. Such bags must be completely evacuated at the start, free of leaks, and constructed of materials that do not release contaminants into the sample.

methods of analysis and/or the detection of smaller concentrations of airborne contaminant. Integrated sampling determines the average level of the airborne contaminant over the testing period and does not distinguish a constant moderate level from a low overall level with a series of very high peaks. The indicated moderate level might be in compliance, whereas the peaks in a relatively continuous low level might exceed ceiling value limits.

Calibration of equipment is important in integrated sampling. The pump that draws the air through the trap must have a constant flow rate. There must be a device to calibrate the pump by measuring that rate of flow. Finally, a trap is needed that effectively and reproducibly removes contaminant from the air, then releases it all at the time of analysis. The captured sample must be stable (capable of being stored until analysis is performed without significant change) and in a suitable form for analysis, the system should require a minimum of manipulation in the field, and, if possible, the trap should not use corrosive or hazardous materials.

GENERAL AIR SAMPLING

There are various locations at which one may wish to take an integrated sample of a chemical in the plant air. A general plant air sample is useful to give an overall measure of plant contamination. One might also be concerned with escape of chemical at a known or suspected point source, such as an open vat, a spraying operation, or a valve. Measurements made at a source of contaminant escape should not be used as values representing overall contamination of plant air. Air collected at a point source will later be diluted by plant air or may be removed effectively by the ventilation system. However, such a reading indicates hazard to a worker at that location and estimates the effectiveness of systems that clear the air.

A variety of stationary devices are available that either collect a sample for later analysis or give a direct reading of the contamination of the air at that location. Such devices may depend on appearance of a specific absorption of infrared light, change in the transparency of a filter, change in the pressure drop across a filter, scattering of light by airborne particulate, or a variety of other techniques. Devices are available to take samples automatically at timed intervals. It is important to understand the sampling method used by a monitor of this sort to be sure that it is appropriate for the contaminant under study. Installation of any such monitoring equipment should consider other chemicals present that could interfere with the analysis.

PERSONAL AIR SAMPLERS

The most important air to sample is the air inhaled by the individual worker. Such air must be collected near the face. Unless we wish to attach the worker by a tube to a large stationary device, which would restrict the free movement of the worker and thereby distort the results of the study, the entire apparatus must be small and lightweight enough to be carried about conveniently by the worker. This device, in spite of its small size, must meet adequate standards for analysis.

Such personal air samplers are available and are in common use. They consist of a small, battery-powered air pump that can be worn on the belt (Figure 8.2), to which a trapping device is attached. A tube pinned to the clothing near the face carries the air to the trapping device (Figure 8.3).

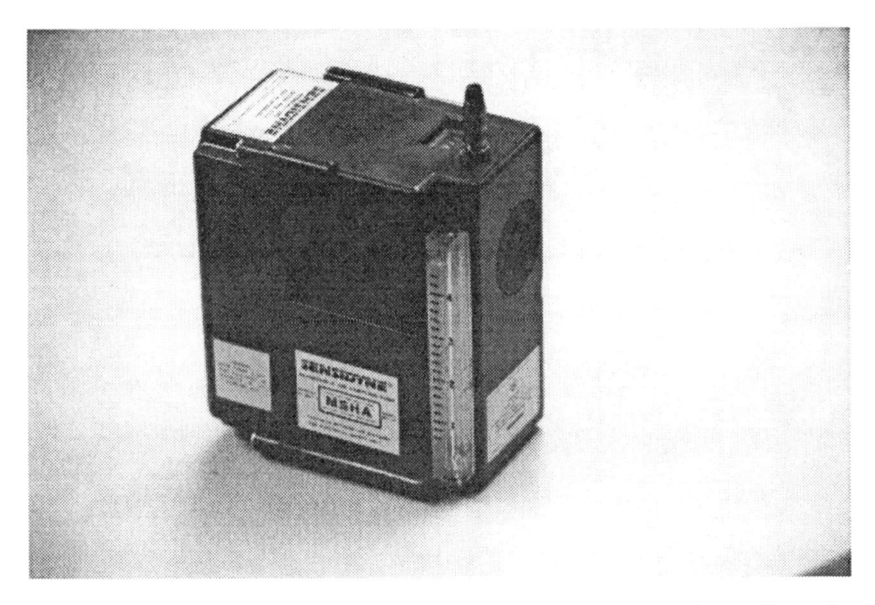

Figure 8.2. This battery-operated portable air sampling pump is worn by the worker and is used to collect sample from the worker's breathing zone. The rate of air flow is adjustable, must be determined before each usage, and can be roughly checked on the rotometer on the edge of the housing.

TRAPS FOR COLLECTING CONTAMINANTS

Three types of traps are used on these devices: filters, adsorbent tubes, and liquid traps. The nature of the contaminant determines the best trap to employ. Particulate contaminants are collected on filters, whereas gases and vapors utilize adsorbent tubes or liquid traps.

Filters

A good filter strikes a balance between stopping the particles and allowing an adequate rate of air flow. Thus, large filter holes allow good air flow, but fail to trap small particles.

Figure 8.3. This worker is wearing a personal air sampling device. Four sampling tubes (traps) are mounted to the air intake. (Photo courtesy of Mine Safety Appliances Company, Pittsburgh, PA. With permission.)

Filters are made of a variety of materials, and the composition of filter chosen depends on the method of analysis to be used.

If the total trapped particulate were simply to be weighed, that requires a filter that does not absorb water from the air, altering its weight independently of particulate capture. Polyvinylchloride or polyvinylchloride-acrylonitrile filters are effective for this purpose.

Cellulose esters, such as cellulose acetate, are useful when the contaminant is to be analyzed in solution. To accomplish this, the filter must be ashed and the residue dissolved or the filter with the residue on it must be dissolved. Dissolving is accomplished by the use of nitric or acetic acids, sodium hydroxide, or solvents such as acetone or dioxane. This is the procedure if metals are to be measured by atomic absorption.

Where the filter is to be studied microscopically, as would be the case for a study of asbestos in air, polycarbonate filters are superior because of their smooth surface. They can also be dissolved in base or dioxane.

If inertness to the chemicals being trapped is important, glass fibers can be used. They trap particles effectively with a low pressure drop in the flow line, withstand higher temperatures, vary little in weight due to absorption of water from the air, and withstand the use of acids or organic solvents to solubilize trapped material. However, if metals are to be analyzed, especially by atomic absorption, glass may contribute metallic impurities. Teflon filters are also inert and are nonpolar, so that nonpolar organics are retained.

Paper filters are available in a large range of porosities and weights and are very inexpensive, but there are some important disadvantages. These include the variability of their ash content if the filter is to be ashed or otherwise broken down during analysis. They change in weight due to the absorption of water from the air sampled, and even change weight without drawing sample through them as the humidity changes, so that accurate determination of the weight of particulate deposited on the paper is impossible.

Respirable Particulate Samples

If the contaminant is in particulate form one must consider the effect of the size of the particles in assessing the toxic threat. Penetration of particles into the lung is a function of particle size; the smaller the particle the greater the penetration (Chapter 4). Because the larger particles are likely to be carried out of the lung, they do not present the same level of threat as do the smaller particles, which reach the alveolus where removal is less effective.

When total particulate content of air is measured by noting the increase in weight of a filter after passage of a known air volume, the method does not distinguish small and large particles. If one compares two samples of air with identical weights of particulate contamination per unit volume — one contaminated with particles of all sizes and the other with largely respirable particles — the latter sample might have from tens to thousands of times higher numbers or concentrations of the dangerous respirable particles. Measuring the content only of small particles assesses hazard more accurately. Techniques have been devised to collect the particulate in at least two stages. Commonly, the first is a cyclone, a centrifugal device that spins out the high-inertia large particles, and the second is a filter to collect the respirable dust (Figure 8.4).

Adsorption Tubes

The second class of traps is adsorption tubes. The term adsorption describes the binding of substances to the surface of a solid resulting from attractive forces between the solid and the collected substance. Adsorption tubes are small glass tubes packed with adsorbent and sealed at each end. The adsorbent surface of such a sealed tube remains clean until the tube is opened, which should be done only immediately before use. At that time the ends are broken off and the tube is inserted into a system that draws an air sample through it. After the desired volume of air has passed through the tube, it is removed from the apparatus, the ends are capped, and it is sent to a lab for analysis. To be effective, adsorbent should have a high collection efficiency (trap a high fraction of the contaminant to which it is exposed), have a high desorption efficiency (release a high fraction of the sample at the time of analysis), and have a high breakthrough capacity (adsorb a large amount of contaminant before 5% passes through untrapped and appears on a backup tube).

Breakthrough capacity depends on the humidity of the air being sampled. Adsorbents have sites on their surface that are particularly effective at binding contaminants due to the presence of charged groups, and, to a lesser degree, to the shape of the site. Water binds well to these sites, often better than the contaminant itself. As humidity rises, an increasing fraction of sites bind water, and so are unavailable to bind contaminant. As a direct result, the breakthrough capacity drops. The NIOSH standard requires that breakthrough capacity be twice the OSHA recommended maximum level measured at 80% humidity. Tubes are sometimes designed with a drying agent in front of the adsorbent to remove humidity from the air. This may sound like an ideal solution to the problem of humidity, but an absorbent used to remove water must not also remove contaminant from the air, a more difficult requirement to meet.

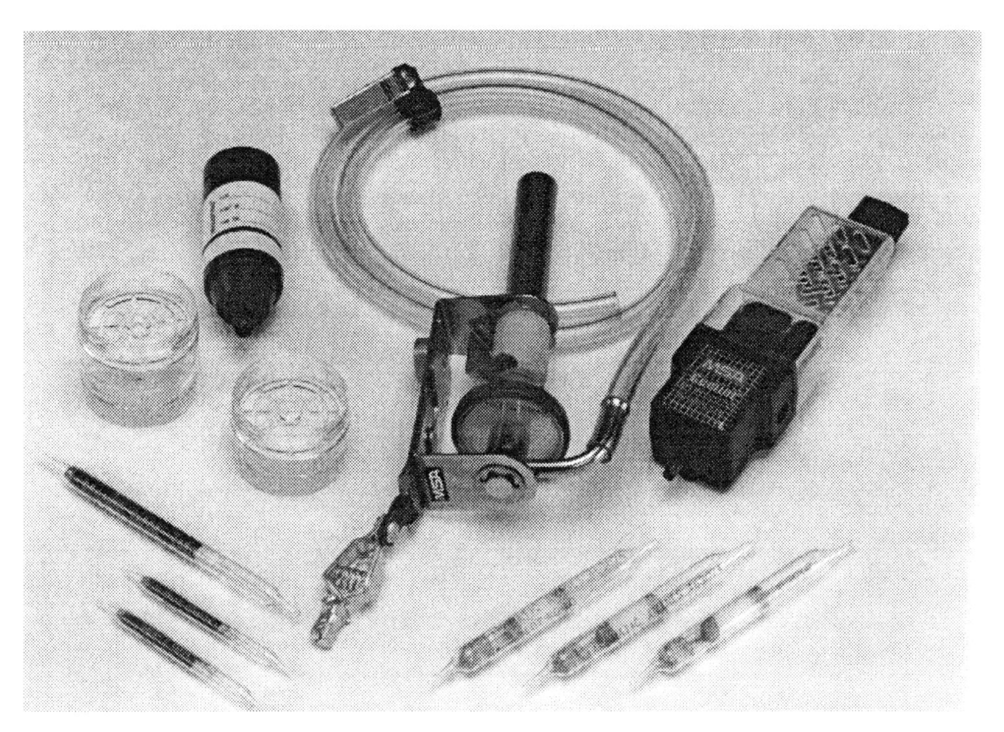

Figure 8.4. Shown here are the accessories used with a personal air sampling pump. Filters used to collect particulate are mounted in canisters or clear plastic mounts (upper left). The latter accommodate a variety of kinds of filter medium. Air enters and leaves through holes in the centers of the holders. These would be sealed around the outside and labeled before use. Once the sample is collected, the entry and exit holes are capped, and the sample is ready for analysis. In the center is a cyclone unit. Only respirable air reaches the filter when this is used. The unit on the right is a holder for adsorption tubes, this assembly capable of mounting two tubes. Most vapor adsorption is done with either charcoal tubes (lower left) or silica gel tubes (lower right). (Photo courtesy of Mine Safety Appliances Company, Pittsburgh, PA. With permission.)

Once adsorbed, the contaminant must then be removed for analysis. The adsorbent binds gases and vapors to the degree that they have polar character. They can then be removed either by using a polar solvent or by driving them off by heating the tube. Once off the tube, one of the techniques for analysis is used to measure the quantity of the contaminant.

Types of Adsorbent

The packing for adsorbent tubes is usually either activated charcoal or silica gel, although other agents have been used. Activated charcoal is the most frequently employed adsorbent (Figure 8.4). Because the particles are very small, charcoal has a large surface area, consequently a large capacity to adsorb airborne organic vapors. It is particularly effective when the organic vapor is chemically relatively nonpolar (for example, hydrocarbon, chlorinated hydrocarbon, ester, ether, or ketone). Charcoal is less useful with more polar compounds, such as alcohols, amines, and phenols. These compounds are collected effectively by the charcoal, but bind so tightly to the charcoal surface that they are difficult to remove quantitatively for analysis.

Silica gel has advantages in collecting polar vapors because the surface is less strongly adsorbent (Figure 8.4). It binds vapor adequately and later releases it successfully. A major disadvantage to silica gel is that it also adsorbs water strongly. In fact, it is frequently used commercially as a drying agent. When sampled air is humid, quantities of water are adsorbed and the remaining capacity of the surface may be inadequate to trap contaminant vapors.

Liquid Traps

Bubblers or impingers at one time were the primary technique for entrapment of organic compounds, and, although they are less commonly used today, they still have value. A major disadvantage is the requirement that a worker wearing a bubbler must keep the tube upright.

Air is bubbled through a liquid which dissolves or reacts with the contaminant (Figure 8.5). If air flow through the bubbler is too fast, the contaminant may not be collected effectively, so the person sampling the air must understand the limitations of the system.[2]

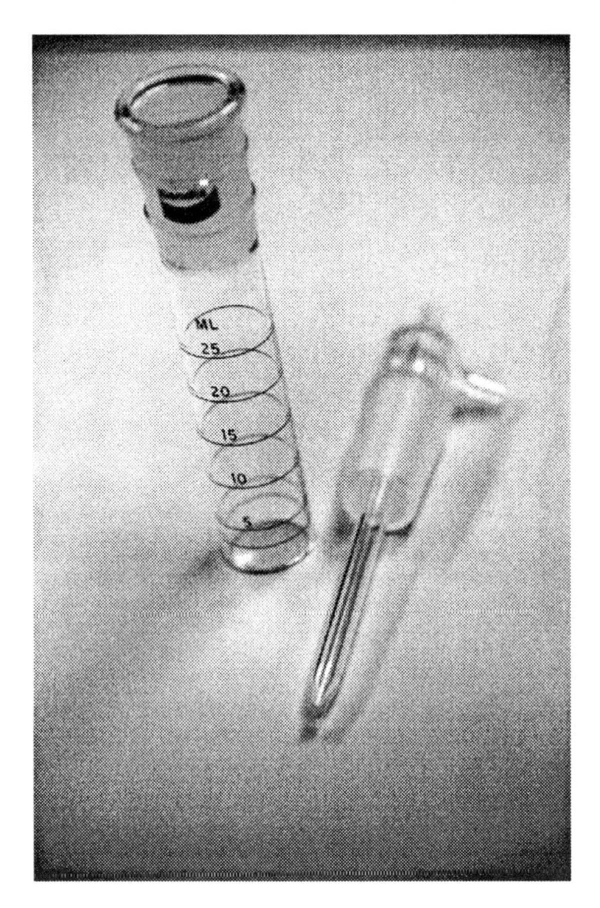

Figure 8.5. Air may be sampled by attaching a bubbler or impinger to the intake of the air sampling pump. The liquid in the trap dissolves or reacts with the contaminant of interest.

Bubbler techniques are improved by adding reagents to the solvent that react with the contaminant to form less volatile products. Analysis is then performed on the reaction products. The chemistry taking place in the trap can be complex, and on occasion side reactions interfere with the analysis. By the same token, the fact that many chemicals not of interest might be collected, but do not react with the chemical reagent and are therefore ignored, can simplify the analysis.

[2] Help in using liquid traps is available from the *NIOSH Manual of Analytical Methods* or the *OSHA Analytical Methods Manual.*

DIFFUSIONAL SAMPLING

The preceding collection devices depend on pumping air through a collecting device. Tubes are marketed where the manner of operation is simply to open the tube and place it in an area where information about air contamination is desired. Air diffuses through the tube opening and is trapped on the adsorbent packing. There is normally also a chemical agent on the adsorbent that reacts with the specific contaminant to produce a colored product. The contaminant binds and reacts with the first adsorbent encountered, so the color appears first at the open end and proceeds along the packing with further exposure. The length of color band measures the concentration of contaminant. The tube has numbered markings along the length of the adsorbent packing. When placed on site for a standard length of time, the numbered markings read air concentration, usually in parts per million, directly (Figure 8.6).

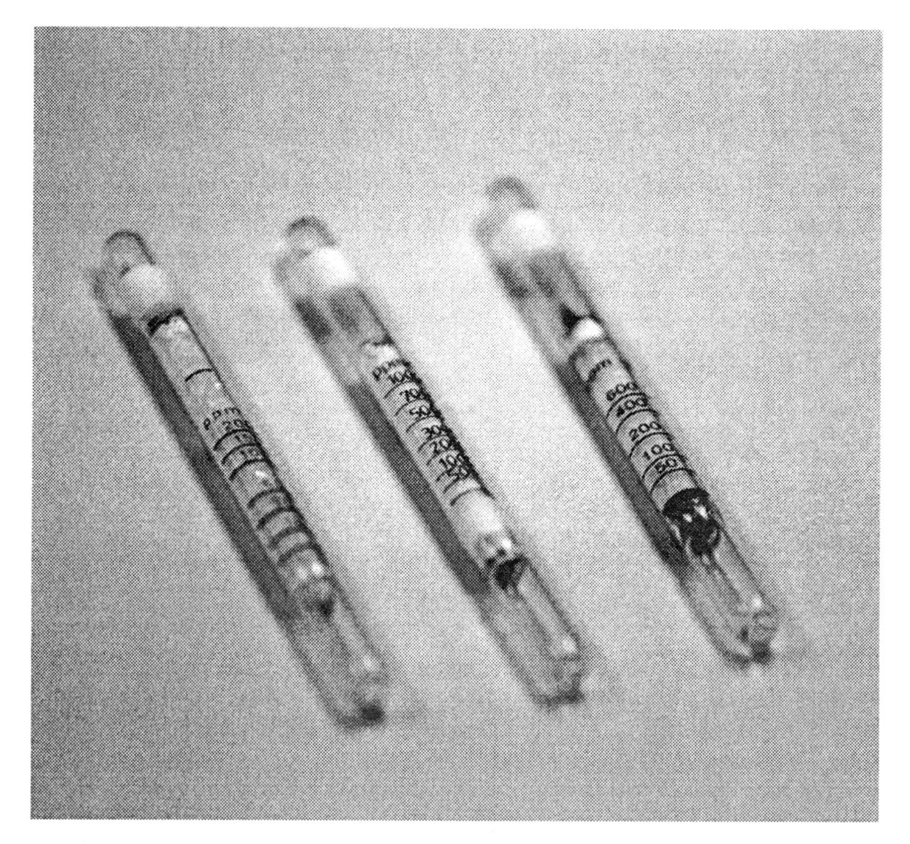

Figure 8.6. Passive diffusion tubes. These tubes are opened by snapping off the glass lower end. Air diffuses into the tube and the contaminant of interest reacts with a chemical agent on the solid filling of the tube to form a colored product. When the tube is hung in the plant for a specified time period, the concentration of contaminant can be read directly from the farthest penetration of the stain on the calibration of the tube.

Diffusion tubes are inexpensive, do not require the purchase and maintenance of personal air pumps, and are not as bothersome to workers as the wearing of a pump and collection device. The concentrations indicated by these tubes must be regarded as approximations, and are of greatest value in establishing that air levels of a specific contaminant are low enough that more careful analysis is unnecessary. A high reading on a diffusion tube should trigger the use of a more accurate analytic method.

ANALYTICAL CHEMISTRY

Before attempting analysis we need to know what we are analyzing. Those readers not familiar with analytical chemistry must accept that to be given a completely unknown mixture for analysis is to be handed a very difficult task. Gas chromatography and high-performance liquid chromatography (HPLC) can do some remarkable separations of complex mixtures. Even so, there is no guarantee that a single peak displayed during such a separation does not represent more than one component. Furthermore, without a list of probable components and a set of standards for comparison, such a separation tells us only the degree of complexity of the mixture, but says nothing about the identity of each component. Coupling the separation technique with one for identification (for example, mass spectrometry), then interfacing the whole with a computer that has a library of identified mass spectrometry patterns might do the job. However, the cost of such an instrument makes its routine use for spot checks impractical. Fortunately, when analyzing the air of a particular workplace, we normally know what chemicals are present.

ERRORS IN ANALYSIS

How do we assess the quality of available analytical methods? There are two kinds of error in such measurements. One is termed determinate error or bias, and describes a problem that causes the system to give an incorrect answer whose error is always on the same side of the correct value, either higher or lower. Suppose we are pumping air through a trap, then removing it from the trap for analysis. Bias results from a flaw in the system, such as incomplete removal of the contaminant from the air, failure to remove the contaminant completely from the trap, or from incorrect calibration of the flow rate of the air pump. If the contaminant were incompletely trapped or released, all the readings would be too low, and if the pump were assumed to be moving 2 l of air per minute through the trap, and it were actually moving 2.5 l, the calculated value would always be too high.

Bias is detected by careful calibration of all parts of the system with standards. If bias is found, the system may be adjusted to run properly. If, however, bias is unavoidably built into the system, as if the recovery from the trap using known samples is always at the level of 95%, making all the readings 5% too low, one could then correct by increasing all the values obtained by 5%.

The other problem is indeterminate or random error, and is the scattering of the values on both sides of the true value. Random error occurs when it is difficult to reproduce some part of the procedure exactly. For example, in pipeting the delivery volume varies slightly or a lab worker is reading a meter and guessing which calibration on the dial is nearest to the needle. Sometimes the value will be too low and sometimes too high.

Random error is dealt with by running several samples. A single sample could be a low value, a high value, or possibly right on the mark, but several samples would vary randomly on all sides of the correct answer. By running several samples, then averaging the results, a much more dependable value is obtained. Furthermore, by dealing statistically with variability in the results, the relative importance of random error in the method can be evaluated. Usually either standard deviation or 95% confidence limit is calculated (Appendix A). The more tightly values cluster around the average value, or the larger the number of samples analyzed, the smaller is the value for these error terms.

NIOSH has published standards for the validation of sampling and analytical methods. If one is measuring a sample that is at the level of the OSHA recommendation for maximum exposure, the value obtained should have a 95% confidence limit of ±25% where there is no bias error, and less if there is a bias correction. Bias should not exceed 10%. The sample should be capable of being stored for 1 week at room temperature with no more than 10% degradation. Other standards are included in appropriate discussions later in the chapter.

MINIMIZING ERROR

The industrial hygienist is in charge of determining contaminant levels and is responsible for the quality of answers obtained. Where devices analyze workplace air directly, the industrial hygienist must understand their use, calibration, and limitations. When the industrial hygienist collects the sample and sends it to a lab for analysis, it is necessary to understand all aspects of sample collection. Further, the limitations of the analytical method employed by the lab should be understood. As every student of analytical chemistry knows, every measuring technique includes some error. Some methods are more error prone than others, so the industrial hygienist must understand how accurate the desired answer must be and how dependable analytical results are, and select a method of analysis capable of producing the desired level of accuracy.

Understanding the analytical method allows anticipation of possible interferences. We may wish to know if levels of *n*-butanol are safe and choose a technique calibrated in *n*-butanol levels that is based on oxidation of the *n*-butanol. However, if this technique responds to every oxidizable vapor, it is of no value in determining the level of a single component in an atmosphere including a complex mixture of such vapors.

The instrument used for collection and/or analysis must have been carefully maintained and calibrated. Instruments used in the workplace are the bread and butter of the industrial hygienist, and should be well understood and carefully pampered. Standards are available that allow accurate calibration of an instrument before use.

ANALYSIS OF SAMPLES

ON-SITE ANALYSIS

Stain Tubes

It is very useful to be able to spot check air in suspicious locations of a plant, making an approximate determination of levels of known contaminants in the air during a plant inspection. A variation of the adsorption tube called the stain tube or detection tube is designed to give a direct reading, providing the inspector with an instant assay of the levels of particular chemicals without the delay or expense involved in a laboratory analysis.

In a stain tube, a chemical reagent is added to a stationary support (silica gel, polymer, glass, etc.) that reacts with the contaminant of interest to produce a color. The treated adsorbent is packed in a thin glass tube which is sealed at the ends. Stain tubes are used most frequently in conjunction with a large syringe device to analyze a grab sample. The ends of the tube are broken off, the tube is attached to the syringe, and a fixed volume of air is drawn through the tube (Figure 8.7). In the earliest tubes, the degree of color produced in the tube was compared to a chart of color standards to determine the concentration of contaminant in the air. Today instead we use a reagent that reacts rapidly with contaminant and is at a carefully adjusted level in the tube. The contaminant reacts immediately with the reagent at the inlet of the tube, leaving the reagent changed in color. The next contaminant to enter the tube must move farther along to unused reagent, extending the stain. The tube is premarked with calibrations, and the length of stain measures concentration of contaminant in the air (Figure 8.8).

Such a tube can be used in two ways: to measure the length of stain from a fixed volume of air or to measure the volume of air necessary to produce a standard length of stain. Most tubes sold today are calibrated to read parts per million in the air directly from the markings on the tube when a standard amount of air, perhaps 10 full strokes of the plunger, is drawn through the tube. The number of strokes needed so as to use the tube calibration directly is provided in the literature supplied by the manufacturer.

Figure 8.7. Air sampling with a detector tube. A sample of air is drawn through the detector (stain) tube using a hand pump. Generally, several pump strokes are required to sample the correct volume of air. (Photo courtesy of Mine Safety Appliances Company, Pittsburgh, PA. With permission.)

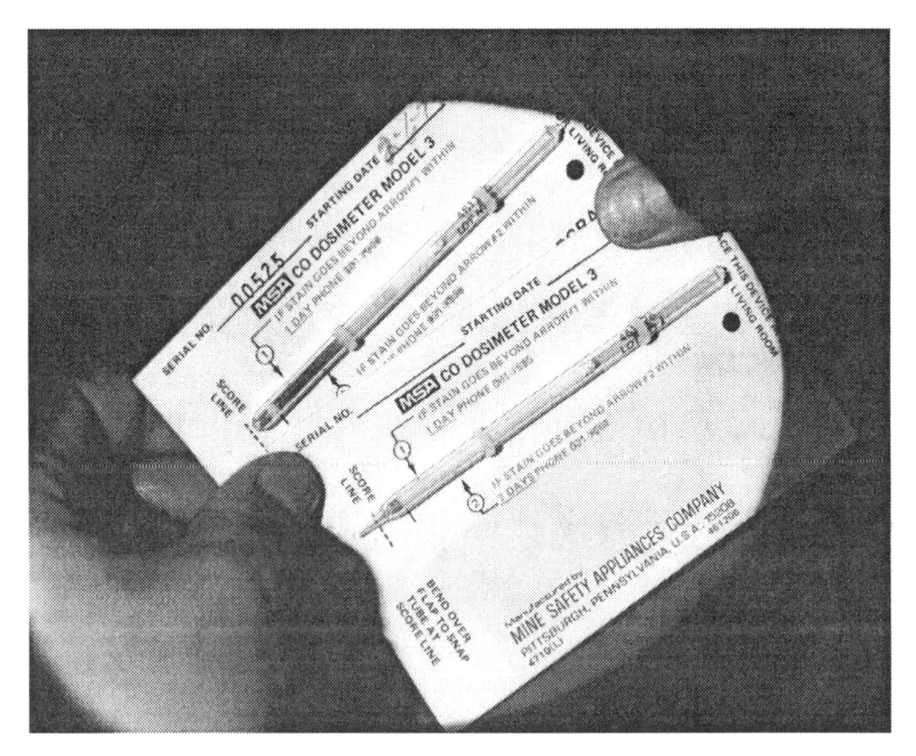

Figure 8.8. Detector tubes. Shown here are two detector (stain) tubes. The upper tube is stained due to the presence of carbon monoxide in the air sampled. To use these the ends are broken off and they are attached to a hand pump (Figure 8.7). The concentration can be read directly from the tube calibration if the correct volume of air is drawn through the tube with the pump. (Photo courtesy of Mine Safety Appliances Company, Pittsburgh, PA. With permission.)

SAMPLE PROBLEM:

The tube manufacturer states that the n number for the tube is 10, meaning if you draw 10 volumes through the tube the calibration of the tube is read directly. After four strokes the tube reads 250 ppm, and it is clear that further volumes will exceed the tube capacity. Estimate the concentration of the contaminant.

$$250 \text{ ppm} \times \frac{10}{4} = 625 \text{ ppm}$$

Tubes are available for a wide range of compounds. The use of these tubes requires that the operator have some sophistication about the situation under study. The tube is labeled to be used for the analysis of a specific compound.

However, on examination, the packings of some tubes are found to be the same for the analysis of several compounds. One supplier lists stain tubes for 169 different analyses. However, if one eliminates those entries where the tube is also listed for another compound, and those for which the difference from one tube to another is just a matter of range, the list is narrowed to 79 different packings. Many of these may be variations in the treatment of a common packing. As an example of the possible problems, the stain may be the color change in an oxidizing agent such as a chromium salt, and several organic compounds are oxidized by this salt. Clearly there is the possibility of misinterpretation of the results where more than one such oxidizable contaminant is present in the air. Some tubes take this type of problem into account in their design. The portion of the tube where the reaction occurs is preceded by segments that react with likely interfering substances. It is necessary for the person assaying the air sample to be aware of the contaminants likely to be present in the air under study and the method employed in the tube being used.

Two other points must be kept in mind when using stain tubes. First, these systems are good only for an approximate measure of concentration. A very low reading ensures the inspector that an area is safe, and a high reading alerts the inspector that a problem may exist, indicating the need for more careful sampling and laboratory analysis. Estimates indicate an error range of perhaps ±25%. Second, stain tubes can deteriorate with time, and are always dated. Expired tubes should never be used when data is collected for official records. In fact, the expiration date of a tube used should be included as part of the record.

Portable Direct Reading Detectors

There are a number of examples of instruments that detect concentrations of a gas or vapor directly. For example, a very useful detector is one that sounds an alarm when oxygen levels are too low (Figure 8.9). Workers entering a confined space are therefore immediately warned if a portable air supply is needed. Another common detector measures levels of combustible gases or vapors. Detectors for a series of specific air contaminants are manufactured. Some portable devices mount a few, often four, of these in a single, easily carried unit (Figure 8.10).[3]

[3] A list of detection devices by manufacturer is found in *Occupational Health and Safety*, May 1996, pp. 34–38.

Figure 8.9. This battery-powered monitor sounds an alarm if oxygen levels drop below 19.5% or rise above 23%. This is particularly useful to workers entering confined spaces where oxygen levels may have been depleted. (Photo courtesy of Lab Safety Supply, Janesville, WI. With permission.)

LABORATORY ANALYSIS OF SAMPLES

The trend over the years has been for laboratory analysis of environmental samples to be done increasingly by instrumental methods. The advantages include greater speed of analysis, possible automation of heavily used procedures (as has happened extensively in the field of clinical analysis), and an elimination of the kind of random error associated with a chemist working at the bench. The need for analyses of industrial and environmental samples has grown rapidly, and a number of individuals have purchased analytical instruments and started businesses specializing in specific types of analysis.

One must guard against the temptation to assume that because an analytical instrument cost a substantial amount of money, it cannot produce a wrong answer. All instruments should be used under the supervision of well-trained and informed chemists, and should be carefully maintained and calibrated. No instrumental method is any better than the standardization techniques used to check its accuracy.

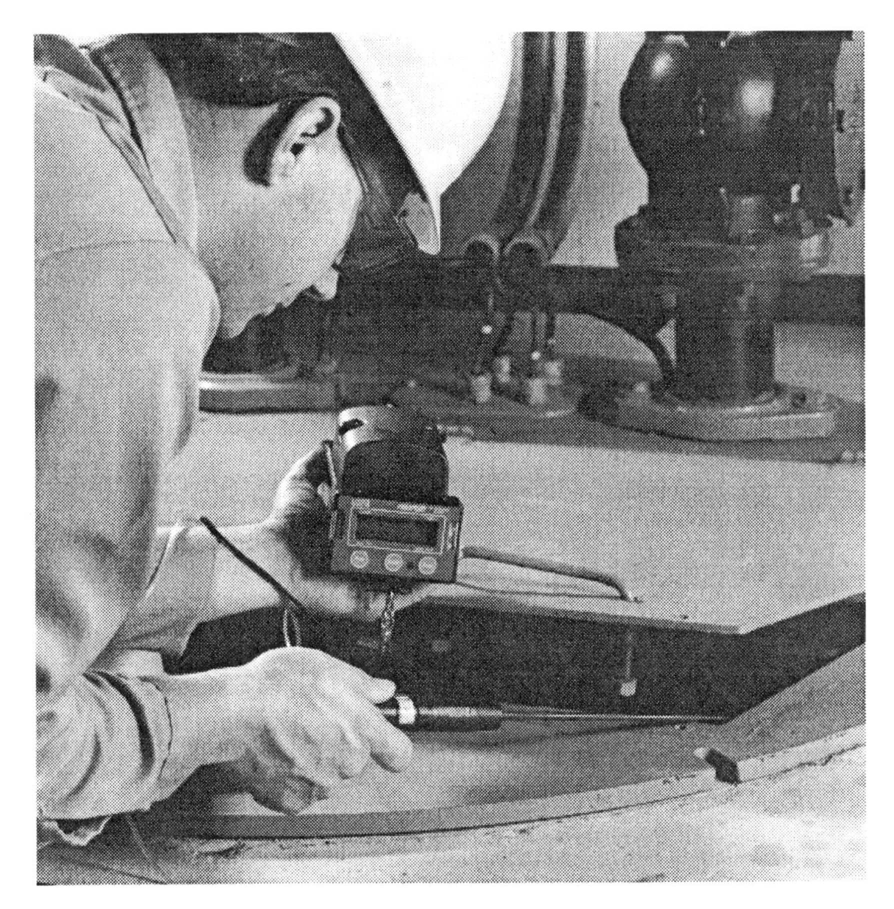

Figure 8.10. Here we see a gas detector equipped to measure four different gases, and presently reading percent oxygen. It can also detect and measure Cl_2, CO, and H_2S. Such a device is commonly used in conjunction with entry into confined spaces. (Photo courtesy of Mine Safety Appliances Company, Pittsburgh, PA. With permission.)

Gas Chromatography

Peter Eller, in his chapter in *Patty's Industrial Hygiene and Toxicology* (Volume IIIA, p. 526), summarizes the analytical methods used in the NIOSH-OSHA standards, and the number of kinds of assay done by each method. Of 216 analytical methods, 165 utilize gas chromatography. This is not surprising, since the technique is capable of sophisticated separations and the results can be converted easily to quantitative information. The instrumentation is relatively simple, is comparatively inexpensive, and has been heavily used since the 1950s (Figure 8.11). Portable gas chromatographs have been developed that can be used on site.

A gas chromatograph consists of a packed tube, the column, through which an inert carrier gas is flowing, and which is mounted in an oven to control temperature (Figures 8.12 and 8.13). A large variety of column packings (stationary phases) are available, and selection of the correct column depends on the chemicals to be analyzed. Advice is available from companies that supply columns.

A constant temperature is necessary for analytical reproducibility. There is a heated chamber into which the sample is injected either as a small volume of liquid, which immediately volatilizes, or as a compressed gas. Carrier gas sweeps the sample onto the column, where compounds are retarded to varying degrees by attraction to the packing of the column.

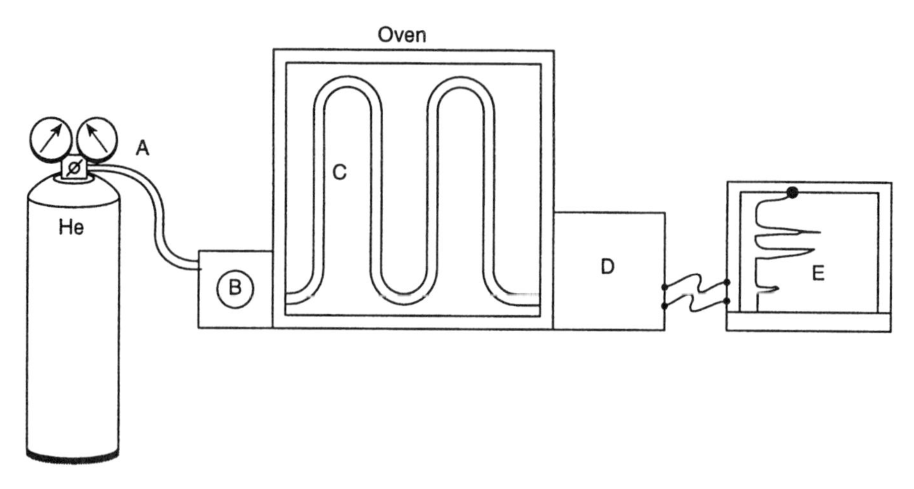

Figure 8.11. The components of a gas chromatograph are shown in this diagram. The moving phase is an inert gas — helium or nitrogen are used commonly — fed into the instrument at a controlled pressure to ensure a constant flow rate (A). Sample, usually a solution, is injected with a very low-volume syringe (1–10 μl) into the hot injection port (B), where it is vaporized. It is swept through the column (C) by the moving phase gas, where the components of the injected mixture are held back according to their individual attraction for the stationary phase of the column. Separated components are sensed by one of several methods in the detector (D), and a record of their position and relative quantity appears as peaks on the moving paper of a recorder (E).

They emerge from the column as a series of separated bands. These pass through a detector, which responds to their presence by producing a small electrical current proportional to the amount of the compound. The current is amplified, and information is presented to the operator as a series of peaks above the baseline on chart paper. Under constant conditions of temperature and carrier gas flow rate in a given column, the length of time required to move a particular compound through the column (retention time) is a constant and can be used to identify the compound. The area under the peak on the chart paper is proportional to the amount of the compound. Conventional recorders are being replaced with computerized devices that determine and print directly onto the output graph the retention time and relative area of each peak.

Since any column packing and carrier gas can be used on any instrument, the biggest fixed difference from one instrument to another is the design of the detector. Three detectors have had long usage in gas chromatography. The simplest and easiest to use is called a thermal conductivity (TC) detector. It operates on the basis of the change in resistance of a hot metal wire when the temperature is changed by the passage of a separated compound across it in the midst of a steady flow of carrier gas. Instruments with TC detectors are stable and easy to run and maintain. However, they are not used much for the analyses of interest in this text because the sensitivity of this method is relatively low.

Flame ionization detectors, on the other hand, are used a great deal in analysis of contaminants. They consist of a very clean hydrogen flame burning between two plates with a large voltage drop between them. The output of the column is added to the fuel for the flame, and when a separated compound enters the flame, it burns with the production of a quantity of ions. These ions allow a flow of current between the plates, and this current is amplified. The flame ionization detector is capable of detecting compounds across a wide concentration range, down to quite small amounts. It is the most favored technique, being specified in 144 of the 165 NIOSH-OSHA gas chromatographic methods.

Finally, the electron capture detector depends on the ability of molecules containing highly electronegative atoms (fluorine, chlorine, oxygen, etc.) to deflect the electrons being emitted

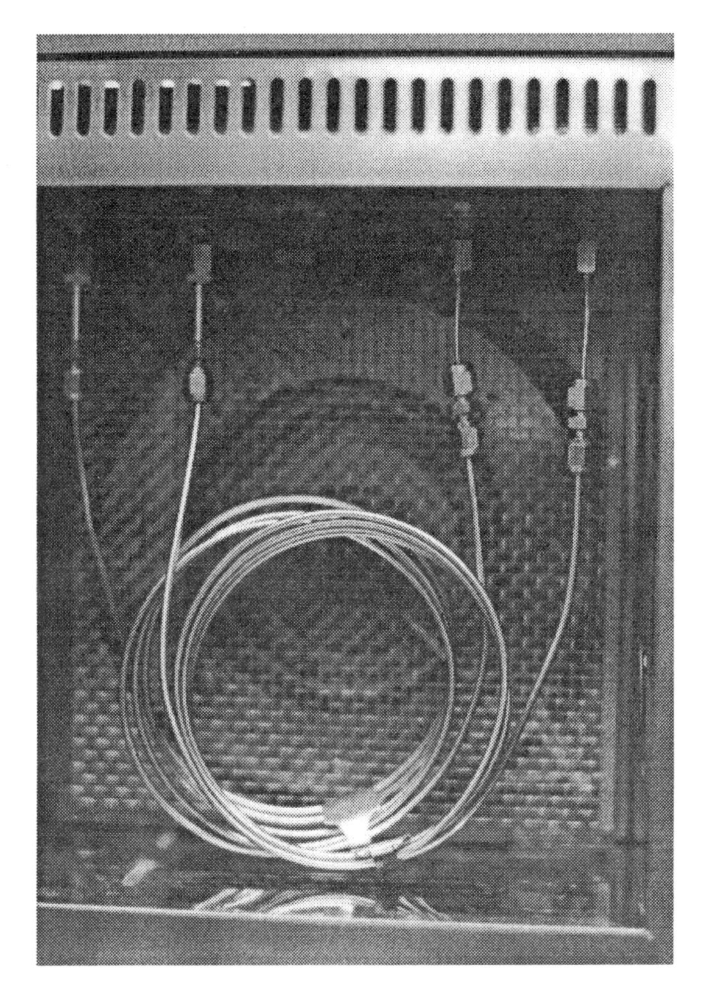

Figure 8.12. Shown is the interior of the oven of a gas chromatograph. The compartment is well insulated and has a heater and blower in back to maintain constant temperature. In gas chromatography the stationary phase is packed into a column and the moving phase (carrier gas) carries the sample through the column toward the detector. Here the stationary phase is a liquid mounted on a solid support and packed in a metal tube. This instrument has two columns mounted in a single oven, each with a different stationary phase to allow flexibility in choice of separation technique used.

by a source of beta radiation. It is able to detect tiny amounts of molecules containing electronegative atoms. However, its disadvantages are that it cannot differentiate relative amounts in large samples, it is not sensitive to molecules without electronegative atoms, and it is more trouble and expense to maintain than the other two detectors.

Atomic Absorption Methods

For measuring metals in a sample, the method of choice is atomic absorption spectrometry. Metals in solution are drawn into a flame or stream of hot air. The light of a wavelength that excites electrons of the metal of interest is passed through the stream of rising gas, and a photocell measures the light reaching the other side of the gases (Figure 8.14). Light used to excite the metal atoms is now missing, and the concentration of metal is calculated from the degree of this absorption of the light. In principle, a separate light source specialized in producing the correct wavelength is used with each metal studied, but lamps are

Figure 8.13. The view is similar to that in Figure 8.12. Here we have a capillary column, a very long, narrow-diameter glass tube wound onto a rack. The stationary phase is a liquid coated onto the walls of the tubing. Because of the great length of the column, much better separations are achieved by a capillary column.

available to be used for more than one metal (Figure 8.15). The flame method, which reads down to the microgram range, is the least sensitive of the atomic absorption techniques. The flameless method and the graphite rod technique extend the sensitivity by several powers of ten.

Spectrophotometric Methods

Visible and ultraviolet spectrophotometry is useful to assay a few selected compounds. This method also operates on the basis of the measurement of the amount of light used to excite electrons, but the electrons are in molecules rather than around single atoms. It is necessary that the molecules absorb light in the range covered by the particular instrument. Some molecules do this naturally, whereas others can be induced to do so by reaction with the appropriate reagent. The method requires only a few milliliters of solution and can detect milligrams to micrograms of the compound.

Figure 8.14. At the heart of the atomic absorption spectrometer is the burner. The gas supply to this burner is coming from the lower left (A). A wide flame is produced at the top of the burner (B), and the light enters through the window (C) and passes through the length of the flame. The sample is drawn up the fine tube from the flask (D) and is fed into the flame.

Infrared spectrometry is used for environmental analysis. A gas sample can be read directly, but a long light path must be used because of the relative insensitivity of the method. The lower energy infrared light is used, and the energy of this light increases the rate of internal vibration in organic molecules. Structures such as carbonyl or alcohol groups each respond to specific wavelengths of infrared light (Figure 8.16). Often a dedicated instrument is used, which is set to read at a single wavelength, a wavelength absorbed by a chemical group in a compound of concern in the particular plant (Figure 8.17). When using infrared spectrometry, it is important to realize that every organic molecule containing a particular structure absorbs light of the same frequency, so, for example, the method would not differentiate between two ketones when the frequency is set to detect the carbonyl group absorption.

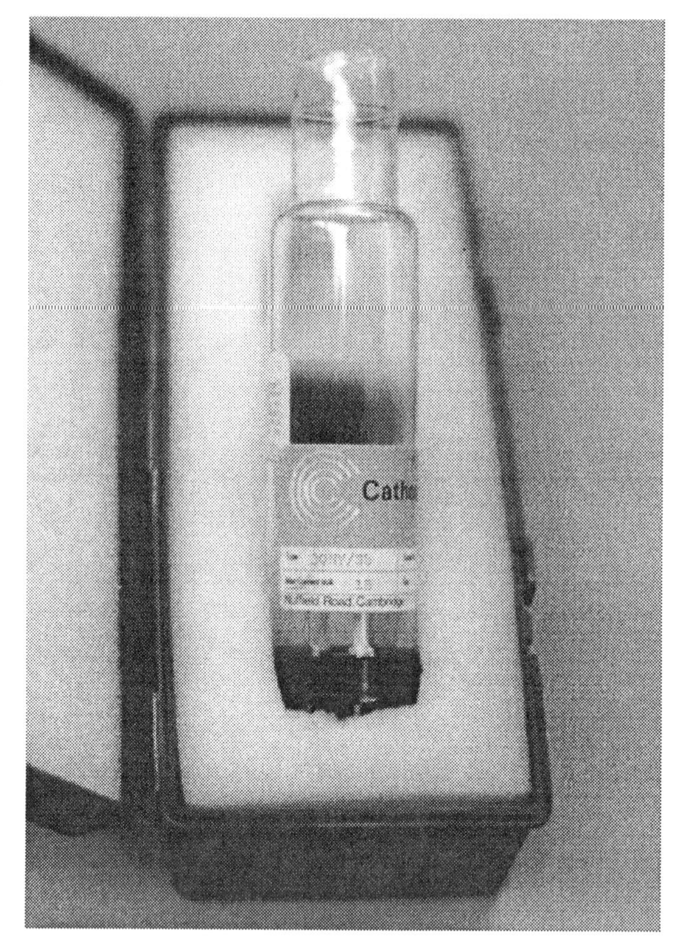

Figure 8.15. Specialized lamps are needed to allow analysis of a variety of metals. Some, like the one shown here, are used for a single element (antimony), whereas other lamps can be used for more than one metal.

Other Methods

Other techniques are used in a few cases for specific assays. X-ray diffraction is used occasionally, particularly for crystalline silica. Specific ion pH meter electrodes are sensitive to certain ions in solution, are easily used, and have a few applications at present. This method can be expected to expand as new electrodes are developed, particularly since the basic instrument is inexpensive. HPLC is a technique that is increasing popular, and will certainly expand its applications in the field. New column packings are being developed, and less complex and expensive instruments are available. It is similar in principle to gas chromatography, but analyzes solutions rather than vapor and gas mixtures.

Some very sophisticated analyses can result from combining methods. Analysis of individual gas chromatographic peaks by mass spectrometry has the potential of identifying the compound in each peak after an unknown mixture has been separated. Mass spectrometry fragments molecules into charged structures, which are then separated by mass. Each compound produces a distinctive pattern of fragments, which can be stored in a computer. A given analysis can then be compared with items in the computer memory, and so identified. In a similar fashion, infrared spectra can be stored in a computer. A new rapid infrared method, fourier transform infrared (FTIR), can scan the peak from the gas chromatograph, and can identify the molecule from a search of the computer files. These instruments are complex

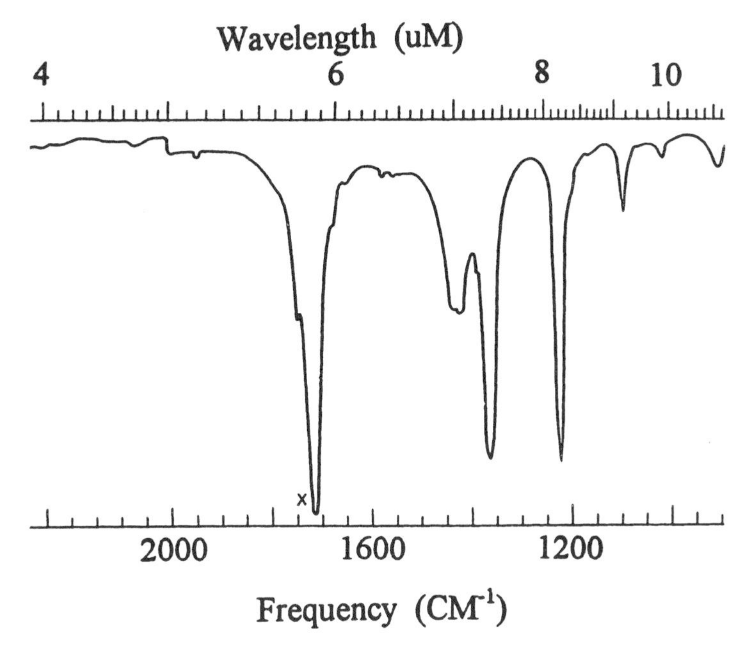

Figure 8.16. Shown is a portion of the infrared spectrum of acetone. The strong absorbance bands each represent a single type of vibration of a specific structure in the molecule. Here, the band to the far left is generated by the carbonyl group. An instrument designed to detect a single wavelength can provide continuous monitoring for compounds in the air absorbing that wavelength.

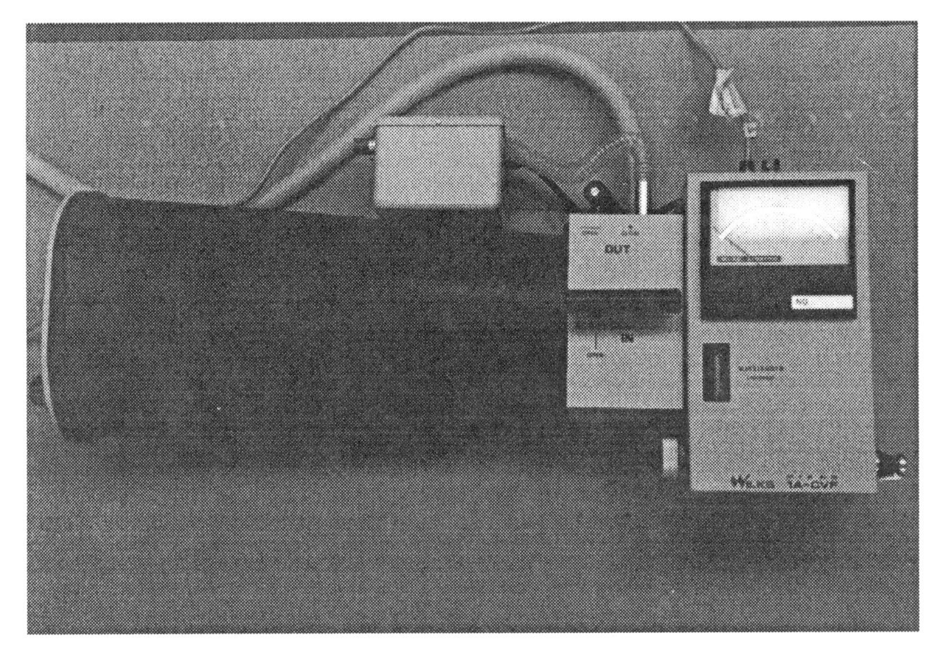

Figure 8.17. Portable infrared spectrophotometer. This instrument can run on batteries and be taken into the workplace. Adjusting the instrument to a specific wavelength makes it selective for compounds with a particular functional group. The meter (right) indicates the degree to which light is absorbed at the wavelength for which the instrument is set, and the signal can also be sent to a recorder for continuous monitoring. A pump (top) draws air to be tested into the black horn-like structure on the left. This contains a series of movable mirrors that allow changing the length of the path through which the infrared light travels, altering the sensitivity of the instrument.

and expensive and are used largely when the demands go beyond the need for a routine survey, particularly when the identity of the contaminants is unknown.

CHAIN OF CUSTODY

The practicing industrial hygienist may run instruments that do analysis on site, but is less likely to actually do the final laboratory analysis where the sample is taken out of the worksite. It is important to understand how these analyses are done, so the above discussion is important. A more practical question is to understand how these samples that are to be analyzed elsewhere must be handled. The legal terminology is the establishment of a valid chain of custody.

First, samples must be carefully labeled as they are taken, identifying the person doing the sampling and the time, date, and location for each sample. These labels should correlate with records in the person's log book. The sample stays in the possession of the industrial hygienist or in a secure place of storage under the control of the industrial hygienist until it is shipped to the laboratory. Shipping should be by a secure shipping vendor. Most important, complete records of all aspects should be kept. Computerized database systems are available.

DECIDING ABOUT THE SAFETY OF THE WORKPLACE

The final stage in the three-part approach to determining the relative safety of the plant air is to make a judgment from the results of analytical work as to whether the level of a contaminant is too high. Here some of the judgment of the industrial hygienist is replaced by government standards. These standards are the listed PEL values of OSHA,[4] of the Mine Safety and Health Administration, or of a state agency that has taken responsibility for the task. Other recommended standards exist that do not have legal status, but can serve as guidelines in the absence of a specific PEL. These include recommendations of NIOSH, the research arm of the federal government in such matters. The ACGIH-TLVs and the American National Standards Institute (ANSI) Z.37 committee report are published values that, in part, were the basis of the PEL values originally set by OSHA. The ACGIH updates their list of values every year, providing for almost all the exposures an industrial worker is likely to experience. The National Academy of Sciences and the National Research Council (NAS-NRC) have published short-term public limit (STPL) values.

The management of a plant must be familiar with the regulations and their intent. If this intent is not clear, the appropriate government office should be contacted for clarification. The analytical methods used should be the same as or similar to those used by OSHA in establishing the PEL. It is desirable that laboratory work be done at a laboratory accredited by the American Industrial Hygiene Association (AIHA). Recommendations about the number of samples to be run are provided by NIOSH.[5] Good quality control should be incorporated into the analytical plan, with documentation of the analytical sequence, careful maintenance and calibration of instruments, and cross-checking of samples with another laboratory. NIOSH operates a proficiency analytical testing (PAT) program, which evaluates the quality of work done by service laboratories. Careful record keeping is essential to meet the requirements of regulatory agencies, to record the effectiveness of the control systems, and to be used in case a chemical involved in the plant operation becomes part of an epidemiological study.

[4] See 29 CFR 1910.1000.

[5] N. A. Leidel and K. A. Bush, Department of Health, Education, and Welfare Publication (NIOSH) 75-159, Superintendent of Documents, Washington, D.C.

KEY POINTS

1. Grab sampling involves taking a single instantaneous sample.
2. Integrated sampling requires pumping a quantity of air through a trap over an extended time period.
3. Sampling may be done of the general plant atmosphere, of suspected trouble zones, or of the air in the workers' breathing zone.
4. A variety of automatic air samplers may be used for continuous monitoring.
5. Personal air samplers pump air from the workers' breathing zone through a trap.
6. Particulate is collected on a filter. The choice of filter material depends on the analysis proposed.
7. Stain tubes have a color-producing reagent coated onto a solid support. The length of the color zone produced measures the concentration of contaminant.
8. Adsorption tubes adsorb contaminants onto a solid medium.
9. In liquid traps, air is bubbled through a solvent, perhaps with dissolved reagent, to capture the contaminant.
10. Determinate error (bias) happens when values cluster to one side of the correct values. It is corrected, if it is consistent, by multiplying answers by a correction factor.
11. Indeterminate (random) error results from random variation in some aspect of the analytical procedure, producing a scattering of the answers. Random error is handled by averaging a large number of observations.
12. Gas chromatography, the most used standard occupational analytical method, separates vapors as they are passed mixed with a gas carrier across a column packing that variably attracts components of the mixture.
13. Metals in particulates are most often vaporized and selectively assayed by light absorption in the gas stream of an atomic absorption spectrometer.
14. Other analytical methods used include visible, UV, and infrared spectrophotometry, liquid chromatography, and mass spectrometers or infrared spectrophotometers coupled to gas chromatographs.
15. Once the concentrations of contaminants are known, the values are compared to prescribed limits.

BIBLIOGRAPHY

ACGIH, *Advances in Air Sampling,* Lewis Publishers, Chelsea, MI, 1988.

ACGIH, *Air Sampling Instruments for Evaluation of Atmospheric Contaminants,* 6th ed., Cincinnati, OH, 1983.

ACGIH, *The Documentation of Threshold Limit Values for Substances in the Workroom Air,* 5th ed., American Conference of Governmental Industrial Hygienists, Cincinnati, OH, 1986.

ACGIH, *TLV Values for Chemical Substances and Physical Agents in the Workroom Environment with Intended Changes for 19XX,* American Conference of Governmental Industrial Hygienists, Cincinnati, OH (latest edition, revised annually).

L. J. Cralley and L. V. Cralley, *Patty's Industrial Hygiene and Toxicology,* Vols. IIIA and IIIB, John Wiley & Sons, New York, 1985.

P. M. Eller, Ed., *NIOSH Manual of Analytical Methods,* 3rd ed., U.S. Department of Health and Human Services, Cincinnati, OH, 1984.

T. Godish, *Air Quality,* 2nd ed., Lewis Publishers, Chelsea, MI, 1991.

M. Katz, *Methods of Air Sampling and Analysis,* APHA Intersociety Committee, American Public Health Association, Washington, D.C., 1977.

L. J. Keith, *Identification and Analysis of Organic Pollutants in Air,* Ann Arbor Science Publishers, Ann Arbor, MI, 1984.

S. D. Lee, T. Schneider, L. D. Grant, and P. J. Verkerk, Eds., *Aerosols,* Lewis Publishers, Chelsea, MI, 1986.

Occupational Safety and Health Act, Public Law 91-596 S2193, 91st Congress, December 29, 1970, Superintendent of Documents, Washington, D.C.

J. B. Olishifski, *Fundamentals of Industrial Hygiene,* 3rd ed., National Safety Council, Chicago, 1988.
OSHA Analytical Methods Manual, 2nd ed., U.S. Department of Labor, Salt Lake City, UT, 1990.
C. H. Powell and A. D. Hosey, *The Industrial Environment — Its Evaluation and Control,* 2nd ed., Public Health Service Publication 614, Superintendent of Documents, Washington, D.C., 1965.
A. C. Stern, *Air Pollution, Vol. III. Measuring, Monitoring and Surveillance of Air Pollution,* 3rd ed., Academic Press, New York, 1977.

PROBLEMS

1. Air in a metal-working facility contains large and small particles of a metal oxide. A grinding operation is the source of the large particles, whose average size is 24 μm, and a welding operation is producing a fume whose average particle size is 3.2 μm. Air testing in the shop is performed by weighing a filter, pumping a known volume of air through it, and weighing it again. About 80% of the particles on the filter collected in a sampling of the total room air are of the small size. The density of the oxide is 5.6 g/cm^3.
   A. What is the percent by weight of small particles?
      A ventilating device is placed at the grinding operation that traps most of the particulate from that source. Filter weight increases are now only 25% of the previous values on daily testing.
   B. Comment on the relative improvement in safety, assuming the metal oxide is a health hazard.
   C. If the ventilation had been placed by the welding operation, what would have happened to the test results for air contamination?
2. In each of the following scenarios, indicate whether grab sampling or integrated sampling is better, and name the method of chemical analysis that would probably be used on the collected sample.
   A. A large underground tank that had stored tetrachloroethylene at a plant has been out of use since the company stopped using that chemical 2 years ago. The tank is going to be entered by a worker to check its physical condition with the plan of using the tank to store a different solvent. How will the safety be checked before the worker enters?
   B. A welder is assembling custom air ducting for commercial building projects out of galvanized sheet steel. The zinc coating usually has traces of cadmium, which is significantly toxic. The OSHA standard for cadmium (TWA) is 0.1 mg/m^3 with a 0.3 mg/m^3 ceiling, and recommendations by both NIOSH and ACGIH are for a lower standard. Two things are to be assayed: the average daily exposure and the peak level during actual welding.
   C. A plant builds small (dorm room) refrigerators. The finished unit without its door is spray painted as it moves along a line by automatic equipment in a ventilated enclosure and is dried by heat lamps. Doors are similarly finished on another line. Workers assemble the doors onto the unit immediately afterward. The units are then inspected for quality of finish, and those with flaws go to a station to the side where paint touch-up is performed by hand-held spray guns. We are concerned with overall worker exposure to the paint solvent, particularly by those workers mounting the doors and by workers doing occasional hand spraying.
3. In each of these analytical procedures indicate whether the indicated problem produces determinate or indeterminate error. If determinate, will resulting analytical results be too high or too low?
   A. Air in a closed coal mine is being sampled prior to workers reentering the facility. An evacuated bottle of known volume is lowered into the mine and the bottle is remotely opened, then closed again. (1) The pump used to evacuate the bottles is worn so that the bottles to be used were not completely evacuated, all having 0.065 atm of air remaining. (2) Oxygen in the sample is determined by noting the volume drop in the air sample at constant temperature and pressure as the oxygen reacts with an excess of hydroquinone. Volume can be read to ±3%. (3) The combustible gases (here primarily CO and CH_4) are estimated by passing air samples through a combustion apparatus and measuring the heat

produced. The apparatus is not insulated well enough so that a percentage of the heat is lost to surroundings.

B. A personal air sampling pump draws air through a charcoal trap to sample volatile organic material. In the laboratory a known volume of CS_2 is drawn through the charcoal to extract the trapped organic compounds, and the resulting solution is injected into a gas chromatograph for analysis. (1) Calibration of the air sampling pump was improperly done so that the actual volume of air pumped per minute is 7% lower than the calibration indicates. (2) Where concentrations of volatile organics in the air are high, the capacity of the charcoal is slightly exceeded. (3) Samples of 1 μl are injected into the gas chromatograph by a technician by use of a 5-μl syringe. (4) The entrance to the column in the gas chromatograph has particulate deposits, which reduces the flow rate of carrier gas through the column. (5) The technician opens the sample vials and lets them stand with the CS_2 evaporating for a period of time before injecting the sample into the gas chromatograph.

4. A plant produces polybutadiene, and at regular intervals at times when the plant is in full operation the level of butadiene in plant air is measured. For one period this data was obtained:

Volume of Air Sampled (l)	Level of Butadiene (ml)
200	1.91
150	1.47
255	2.35
105	0.76
430	5.75

A. What is the average concentration in ml/l?
B. What is the standard deviation for this data?
C. Given the student t table values, what is the 95% confidence limit?

5. A spot check is made for solvent in the air near a degreaser using a stain tube at the time parts are being cleaned. The ends of the stain tube are broken and it is inserted into the syringe-type air pump. Instructions say to pump the syringe 10 times, then read the concentration in ppm directly from the tube calibration. The calibration stops at a maximum value of 100 ppm, and the stain reaches 88 ppm after six strokes. From that information, what is the concentration of solvent in the air?

6. You are visiting a plant to do a safety inspection and you bring along a case of personal air sampling pumps with the following accessories:
Charcoal adsorption tubes
Cellulose acetate filters
Polyvinylchloride filters
Polycarbonate filters
What would you select to test for each of the following air contaminants at specific plant locations?
Asbestos
Fe_2O_3 fume
Benzene vapor
Total mass of particulates

CHAPTER 9

Protecting the Worker II: Providing Clean Air

The worker must be provided with clean air. Where exposure problems are revealed, the next step is corrective action. In a plant we are usually dealing with a point source, and, as those involved with environmental problems will readily indicate, this is a relatively easy problem to solve. Several courses of action are possible.

REDUCING THE LEVEL OF CHEMICALS

Various remedies reduce levels of chemicals in plant air.

1. Be sure the plant is operating as designed by instituting a program of thorough plant maintenance. Inspections of valves and seals to catch leaks as soon as they occur prevent needless chemical release. Belts running blowers should be functioning properly and the ventilation system should not be blocked by deposits in ducts.
2. Good plant housekeeping prevents accumulation of dusts or spilled volatiles that can gradually raise levels of toxic substances in air. Housekeeping must not expose the janitorial person; i.e., a person pushing a broom can be heavily exposed to dusts.
3. Minimize air contamination from unplanned releases. These include accidental spills or escapes during such plant maintenance as valve repairs or tank cleaning. Careful planning can reduce frequency of spills, but accidents are going to happen. If containers of chemicals are moved with a forklift, for example, it is likely that one day a worker will spear the chemical container rather than the pallet. OSHA requires a written plan for removal of spilled chemicals (29 CFR 1910.1200), and further requires that workers are trained in the correct response to such an occurrence. Know the properties of the chemicals in use. Proper disposal of any rags or other absorbents used in such a cleanup is important. It does no good to stop the evaporation of a spill by soaking it up with absorbent rags, then allowing the chemical to evaporate from the rags. During disposal, one must avoid combining chemicals that react with each other. Such chemical incompatibilities are indicated on the warning label of the chemical container and on the MSDS.
4. Entry of chemicals into the air is more likely when chemicals are handled or transferred. Air contamination often is reduced by simple, common-sense changes. As an example, the loading of powdered solids into drums is a potential major source of particulate contamination of the air. If the powder is delivered down an overhead tube that is positioned over the open drum, escape of material into the air is greatly reduced by fitting the end of the delivery tube with a flat sheet that serves as a cover over the drum during filling.

REMOVING THE WORKER FROM THE CHEMICAL

Sometimes the safety of the worker can be improved by relocating the job to a spot distant from the source of the chemical. Many chemical processes are readily adapted to remote operations, but this is not always a useful option.

CHANGING THE PROCESS IN THE PLANT

Once a process is set up to function in a particular way and is running successfully, there is reluctance to change it. Driven by the requirements for greater safety, equipment manufacturers and process designers invent new technologies to reduce amounts of chemicals required. Spray painting, where solvents are a serious source of air contamination, provides an example. New methods use reduced amounts of solvent in the paint, substitute a less toxic solvent, or even virtually eliminate solvent entirely.

Replacement of a chemical by a less toxic one is a measure that improves the safety of the plant in many ways. Handling the chemical is less dangerous, spills are less threatening, waste handling is less critical, and levels of air contamination do not have to be controlled as closely. Sometimes such a change is opposed because the substitute chemical is more expensive. If the alternative is an expensive engineering change, however, substitution of another chemical often yields savings when the price of protecting workers adequately against the more toxic chemical is added to the cost sheet.

VENTILATION

One can substitute low-hazard for high-hazard chemicals or redesign processes to prevent the entry of chemicals into the air, but ultimately one must accept that hazardous or otherwise undesirable substances will enter plant air. This air must be removed from the vicinity of the worker, which is accomplished by ventilation, the planned supply and removal of air from a defined space or volume. Ventilation must be employed when chemicals present a health, fire, or explosion risk. Ventilation is the most important tool for bringing a workplace into compliance. Besides increasing worker safety, ventilation benefits include improvements in ease of housekeeping in the plant, employee comfort, and machine maintenance.

OSHA does not specify methods of ventilation or volumes of air moved, but rather requires that the employer operate the facility such that levels of these chemicals in the air are safe, regardless of the methods employed to achieve this goal. However, for certain processes, such as grinding, buffing, and woodworking, local laws may specify ventilation practices.

This chapter presents ventilation design features, but not with the goal of having the industrial hygienist or occupational safety specialist design ventilation systems. Initial design is complex, and is the task of a ventilation engineer. However, people concerned with worker safety must understand what comprises a good system and what is needed to ensure that a system in place meets the intended operating parameters.

PARAMETERS OF A VENTILATION SYSTEM

Several terms are used to describe ventilation characteristics. Air pressure is the force exerted by air molecules randomly colliding with a surface. The reference pressure is atmospheric pressure, the same pressure a weather reporter discusses on the evening news. Air pressure in a duct connected to an operating fan differs from atmospheric pressure. This static pressure is above atmospheric pressure (positive pressure) when the fan is forcing air into the duct (a blower), and is below atmospheric pressure (negative pressure) in the case of an exhaust fan (Figure 9.1). In home heating systems, heated air propelled through the heating ducts to the rooms of the house has positive pressure, and air being brought back to the furnace through the cold air returns has negative pressure. Positive pressure moves air from heating ducts into the rooms, and negative pressure causes room air to move into cold air returns.

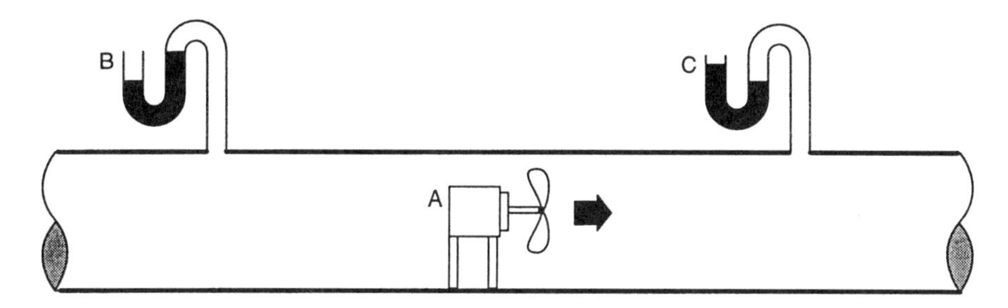

Figure 9.1. Static pressure can be compared to atmospheric pressure using a water manometer. Such a device consists of a U tube containing water that connects to the duct at one end and is open to the atmosphere at the other. Difference in water height in the U tube measures difference in pressure. When ducts carry air forced through by a fan (A) at the entrance to the duct, static pressure is higher than atmospheric pressure (C), and when air is exhausted from a duct, the static pressure is lower than atmospheric pressure (B).

Air moves through or into ducts at a certain velocity, usually measured in meters or feet per minute (fpm).[1] Greater velocity means greater friction between the moving air and the duct walls. Overcoming this frictional resistance requires the expenditure of more energy at the blower.

As air moves in the ducts, velocity pressure is generated. The moving air particles are not moving randomly. An object placed in the duct is struck by air particles more frequently on the side from which the air is coming than on the side toward which it is going (Figure 9.2). This pressure is what ultimately moves particles of contaminant through the duct along with the air.

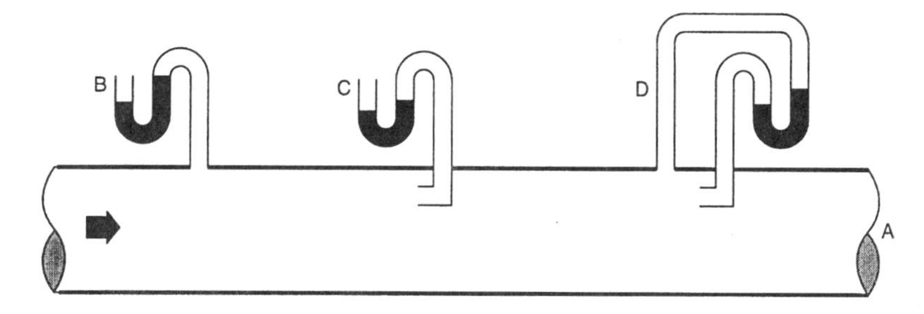

Figure 9.2. In this duct the fan is at the end marked A and is moving the air to the right. Manometer B measures the difference between atmospheric pressure and static pressure, as did the manometers in Figure 9.1. By directing the inlet of the manometer tube toward the flow of air, manometer C displays the difference between atmospheric pressure and the sum of static pressure and velocity pressure. Since manometer D responds to static pressure at one end and the sum of static pressure and velocity pressure at the other, the reading measures velocity pressure directly.

For a given size and mass of particle, a specific velocity of air has sufficient velocity pressure to keep the particle suspended and moving with the air. The term *capture velocity* is used when the concern is to move particles outside the ventilation system through an

[1] At the time of writing, U.S. engineers still prefer units in the English system, and it is in these units that most calculations are performed. Both English and metric units are used in this text and the reader is encouraged to learn to convert from one system to the other.

opening into a duct, whereas *transport velocity* describes the velocity needed to keep the particle suspended in the duct until a suitable collector or exit is reached.

Finally, there is the air flow, the quantity of air moved per unit time through the ducts of the system and measured in cubic meters or cubic feet per minute (m^3/min or ft^3/min [cfm]). To increase the air flow in a system of specified capacity, the velocity must be increased. If the air flow is through an exhaust system leaving the building, it is an obvious but sometimes overlooked fact that there must be an equal flow of replacement air into the building.

GENERAL EXHAUST VENTILATION

Ventilation systems fall into two major classes: general and local exhaust ventilation. General exhaust ventilation (GEV, dilution ventilation) means removal of contaminants by the movement of the entire air mass into, around, and out of the workplace. An indoor facility employs ventilation for heating the plant in cold weather, and for controlling humidity and supplying cooler air in hot weather. If the system that heats or cools the air blows the air into the plant, then draws it back in for retreatment (as home heating and cooling systems do), this is not a GEV system. The GEV system must involve air exhaust facilities and corresponding air intake. In a GEV system, chemicals are simply diluted into the workplace air, which is exhausted at a rate that prevents chemical concentrations from rising to a harmful level. GEV is most satisfactory when the contaminants are gases or vapors and have a low toxic, fire, or explosion risk.

Ventilation engineers sometimes describe the total capacity of a ventilation system in terms of room changes of air per hour, the number of times an amount of air equal to the volume of the room is exhausted per hour. Careful design is required to ensure that this volume of air flow is adequate, moving air through all parts of the facility and moving it rapidly enough to prevent hazardous materials from rising to harmful levels. In simple terms:

$$E = \frac{Q}{V} \tag{1}$$

where E = rate of air exchange in room changes per hour (h^{-1})
Q = volumetric air flow generated by GEV in m^3/min or cfm
V = the volume of air in the room in m^3 or ft^3

Some assumptions include:

- The generation rate of the contaminant (G) is constant.
- Entry of the contaminant is by a simple, known source.
- Exit is by exhausting air only.
- Air passing through the room is perfectly mixed.

The assumption of perfect mixing is obviously an ideal situation, but one that is central to the calculations about to be described. Serious questions arise when a GEV system is selected for a facility and there is large variation in contaminant concentrations in the room.

Knowing the rate of generation (G, in units of mg/min), the concentration of contaminant in the exhaust air (C, mg/m^3), the rate of air flow (Q), and total mass of contaminant in the room (M), one may calculate change in amount of contaminant in the room (ΔM) for a period of time (Δt).

$$\Delta M = G\Delta t - QC\Delta t \tag{2}$$

Concentration of contaminant in room air (C), the more significant information, is M/V, where V is the volume of the room. Equation 2 can be divided by V to obtain:

$$\frac{\Delta M}{V} = \Delta C = \frac{G\Delta t}{V} - \frac{QC\Delta t}{V} \tag{3}$$

The two terms in this equation are the rate of generation of contaminant (first term, GΔt/V) and the rate of removal (second term, QCΔt/V). If concentration in the air is rising ($\Delta C > 0$), contaminant is not expelled as fast as it is generated. Concentration does not rise linearly under these conditions, because as concentration rises, the rate of removal also increases, although not as rapidly. Instead concentration approaches a maximum limit concentration (C_{max}) where the increased rate of removal due to higher room concentration finally matches the rate of generation. The equation relating concentration to time (derived in Burgess, 1989) is

$$C = \frac{G}{Q}\left[1 - \exp\frac{-Qt}{V}\right] \tag{4}$$

The system approaches a maximum concentration ($t \gg 0$), which is

$$C_{max} = \frac{G}{Q} \tag{5}$$

Not surprising, the equation tells us that the maximum concentration reached depends on the rate of generation of contaminant and the rate at which air is exhausted. Once this relationship has been experimentally worked out, we can estimate the effect on maximum level of the contaminant in the workplace air resulting from change in either rate of generation of contaminant or rate of removal of air.

If generation of contaminant ceases, it is useful to predict how rapidly concentration of contaminant is reduced. This is determined as follows:

$$C_2 = C_1 \exp\left[\frac{-Q}{V}(t_2 - t_1)\right] \tag{6}$$

Here C_1 is the original concentration and C_2 is the new concentration of contaminant after exhausting air at rate Q for a time period of t_2–t_1. It can be derived (again, Burgess, 1989) that the half-life of the contaminant ($t_{1/2}$) — the time required to remove half the contaminant — is estimated from rate of air exhaustion and volume of the room, once generation ceases:

$$t_{1/2} = 0.693\frac{V}{Q} \tag{7}$$

With these equations, simple predictions can be made about variations in concentration of a contaminant as conditions change.

These equations were based on assumptions, including that of perfect mixing of air in the room. This would produce ideal transfer of contaminant generated into air leaving the room. In practice, air moving toward the exhaust point mixes with air containing contaminant in a less than ideal fashion, resulting in a buildup of contaminant above predicted values in the vicinity of the source. To include this factor in the equation, Q is multiplied by a constant, K (distribution factor):

$$Q_a = KQ_i$$

where Q_a = an actual Q
Q_i = an ideal Q

K has a value greater than 1, with less effective mixing generating a larger value. K would be very hard to determine, and is actually increased arbitrarily to insert a safety factor into the calculations. The value of K is not a fixed characteristic of general ventilation but can be influenced by simple design considerations. Most important here is the location of air inlets and outlets.

SAMPLE PROBLEM:

A solvent is spilled in a room with dimensions 20 m × 50 m with a 5-m ceiling. The GEV system exhausts 100 m³/min. The spilled solvent is cleaned up, but not before it reaches 25 mg/m³ vapor in the air. (A) How much time is required to reduce the level to half the initial level, assuming perfect mixing? (B) How long would it be before the level is down to the PEL of 10 mg/m³?

(A)

$$t_{1/2} = 0.693\frac{V}{Q}$$

$$= 0.693\frac{(20\text{ m})(50\text{ m})(5\text{ m})}{100\text{ m}^3/\text{min}}$$

$$= 34.7\text{ min}$$

(B)

$$C_2 = C_1 \exp\left[\frac{-Q}{V}(t_2 - t_1)\right]$$

$$10\text{ mg/m}^3 = 25\text{ mg/m}^3 \exp\left[\frac{-100\text{ m}^3/\text{min}}{5000\text{ m}^3}(t_2 - t_1)\right]$$

$$\ln 0.4 = -(0.02/\text{min})(t_2 - t_1)$$

$$-0.916 = -(0.02/\text{min})(t_2 - t_1)$$

$$(t_2 - t_1) = 45.8\text{ min}$$

Engineers use a simple formula to estimate the air flow necessary to control a known contaminant escaping into plant air at a known rate:

$$\text{cfm} = \frac{G \cdot SG \cdot K \cdot 6{,}720{,}000}{MW \cdot TLV} \quad (8)$$

where cfm = needed flow rate
G = rate of release in pt/hr
SG = specific gravity of contaminant
K = distribution factor
MW= mol wt of contaminant
TLV= TLV or PEL of contaminant (ppm)

Notice that this formula does not require room volume and that units are canceled by the numerical constant, as long as the units for each entry are as shown here.

SAMPLE PROBLEM:

Given Equation 8, determine the flow rate necessary to control the air concentration of acetone (MW = 58.1, SG = 0.797) escaping at a rate of 8 pt/hr when the TLV is 1000 ppm. Assume K = 5.

$$\text{cfm} = \frac{\text{G} \cdot \text{SG} \cdot \text{K} \cdot 6{,}720{,}000}{\text{MW} \cdot \text{TLV}}$$

$$= \frac{(8)\,(0.797)\,(5)\,(6{,}720{,}000)}{(58.1)\,(1{,}000)} = 3{,}680$$

LIMITATIONS OF GENERAL VENTILATION

The difficulties of dependence on general ventilation as the only system to maintain a safe environment are several:

1. Exhausting a volume of air equal to the capacity of the building does not ensure that all air in the building has been changed once, because air in the center of the pathway between inlets and exhaust vents moves well and is changed more than once, whereas air in corners and off to the side may be relatively stagnant. The source of a hazardous contaminant may be away from the path of direct and optimal flow of air through the facility.
2. Air carrying a hazardous material may be moved toward the worker or past several other workstations on its way to be exhausted, increasing general exposure to the contaminant.
3. Processes that release larger amounts of material for a short time, and very little between times demand a high level of air movement during the short time of release. Unfortunately, the time contaminant is being released usually coincides with a worker being at the site. GEV systems do not adapt well to such operations.
4. Seasonal reductions in the ventilation rate may reduce air flow in hazardous areas below desirable rates.
5. A loss of efficiency may occur as a system ages or between times of routine maintenance as ducts or filters become blocked with dust or fan belts slip.
6. The calculations assume the entering air is clean, high-quality air and so does not contribute to contamination of the room air.

These problems are overcome by good design and careful maintenance. The location of inlet and outlet vents is important in preventing stagnant zones in hazardous locations and in minimizing the degree of contact workers away from that location have with the contaminant. Air should move from the cleanest toward the dirtiest locations in the workroom. Standards must be established for normal operation of the system, and a pattern of routine inspection should be instituted to ensure continued adherence to these standards.

As energy costs rise, interest increases in avoiding the discharge of large volumes of heated air during cold times of year. This requires either installation of heat exchangers that will transfer some heat from outgoing air to incoming air or installation of systems to remove impurities from the air, allowing it to be recirculated. The latter course has obvious monitoring requirements.

Another approach to setting adequate ventilation standards involves basing the rate of air flow on the type of activity in the room and either the number of persons in the room or the room area. For example, the American Society of Heating, Refrigerating, and Air Conditioning Engineers (ASHRAE) recommends 20 cfm/person for an office, 30 cfm/person in a dry-cleaning facility, and 0.5 cfm/ft^2 in a warehouse.[2] Assumptions made about good mixing of the air when basing safety on room changes of air apply here too.

Finally, one must always be alert to problems created when there is a change in process in a facility. The introduction of new or larger sources of chemicals or particulates may create hazards the previously satisfactory system cannot now handle. A new process may include a substance whose airborne concentrations must be held to lower levels than those previously in use. Even without a change in chemicals or in total level of output into plant air, problems arise if the process change abandons a steady moderate rate of emission in favor of intermittent larger releases. Rearranging the locations of workstations could result in exposure to harmful materials by people previously out of their path. Management concern focuses on how changes affect the cost and efficiency of operating the facility, and it is the job of safety personnel to be alert to how the changes impact safety.

LOCAL EXHAUST VENTILATION

Local exhaust ventilation (LEV) systems remove air at the point the hazard is generated. In many states certain operations, such as grinding, must employ local ventilation. LEV has clear advantages, especially when dealing with particularly hazardous substances or an unusually dirty individual operation.

1. The hazard can be removed by moving relatively small volumes of air, thus saving energy.
2. The problem of exposing a large segment of the workforce to a locally generated hazard is reduced.
3. If contaminants must be removed from the air before it is discharged from the plant, the task is now restricted to a relatively small air volume. Air purifying equipment can be much smaller in scale and cost.

The employer is rewarded for investment in LEV machinery and ductwork by lower costs for compliance.

DESIGN OF A LOCAL EXHAUST VENTILATION SYSTEM

An understanding of LEV system components helps the compliance officer spot flaws and possible bad design features, alerting this person to the need for more detailed analysis of air contamination at particular workstations. A local ventilation system has four components:

1. At the workstation there must be a hood, the air intake device designed to capture the hazardous material.
2. Air is then carried through ducts away from the site.
3. An air cleaning device may be inserted into the ductwork system to reduce the levels of the contaminants in the exhaust air.
4. Finally, there is a motor-driven fan that drives air from the inlet(s) to the exhaust.

[2] American Society of Heating, Refrigerating, and Air Conditioning Engineers, Inc., *Ventilation for Acceptable Indoor Air Quality*, ASHRAE Standard 62-1989, 1791 Tullie Circle NE, Atlanta, GA 30329.

Variations in the design of an LEV system include the volume of air exhausted at the hood, the design of the hood, and the location of the hood with respect to the source of contamination.

HOODS

The hood can be anything from the open end of a duct placed near the source of contamination (a plain inlet) to a sophisticated enclosure surrounding the entire operation. The system fan is moving air through the ducts away from the hood such that air pressure is lower at the hood opening. Room air moving toward this low pressure zone at the opening carries contaminants into the duct.

Hood Design

Hood design should be appropriate to the contaminant being trapped, the geometry of the work area, the location of the worker, and the need of the worker to have access to the work in progress. It is an obvious but very important point that in order for the rest of the system to have any value, the contaminant must be captured at the hood, so hood design is critical.

Room air entering the hood opening needs to move with enough velocity to capture the contaminants. The term *capture velocity*, the velocity of air flow needed to move specific contaminants into a hood, varies with the type of contaminant, ranging from approximately 20 to 650 m/min. Large particulate material requires a greater capture velocity than vapors or gases.[3] The velocity must also be sufficient to overcome usually variable cross-drafts between the source and the hood opening.

Velocity of air entering a hood is measured with a velometer. Velometers may be very simple, depending on the deflection of a vane by moving air. A more sophisticated design converts the rate of rotation of a propeller placed in the air stream or the degree of cooling of a hot wire by the moving air to air velocity. Like any tool used by an industrial hygienist, a velometer must periodically be recalibrated.

A velometer is used to calculate the air flow through a hood. A bit of experimentation with a velometer at a hood opening establishes that the air velocity differs with where it is measured. It is highest at the center and drops as the edges are approached. This is the result of friction of the air with duct walls. The average flow rate is needed for an accurate calculation of total air flow. The dimensions of the hood are measured and the hood area is calculated. Then a grid is laid out from the dimensions establishing a series of uniformly spaced imaginary points. Flow rate is measured and recorded at each of these points, and the values are simply averaged. Multiplying the average flow rate by the hood area establishes the air flow through the hood:

$$Q = vA \tag{9}$$

where Q = air flow of the hood
v = average flow rate
A = area of the hood

Hood Reach

Away from the hood opening, air velocity drops. The distance at which the hood still provides sufficient capture velocity for the specific contaminant is called the reach of the

[3] Engineers are more likely to use feet per minute (fpm).

SAMPLE PROBLEM:

Calculate the air flow in a hood 2 × 4 ft in size with the following flow rate measured on an imaginary grid:

1. 31 fpm	4. 42 fpm	7. 33 fpm
2. 34 fpm	5. 45 fpm	8. 36 fpm
3. 32 fpm	6. 40 fpm	9. 31 fpm

The average air flow is 324/9, or 36 fpm. The area of the hood is 4 ft × 2 ft, or 8 ft^2.

$$\begin{aligned} Q &= vA \\ &= (36\text{ fpm})(8\text{ ft}^2) \\ &= 288\text{ cfm} \end{aligned}$$

hood. Equations are derived to predict the reach of particular hood designs,[4] but given the variables it is hard to design a system with a reasonable capture velocity value without resorting to trial and error.

The distance between the hood opening and the source of contamination is very important. A simplified calculation of the effect of distance from the opening requires use of this equation:

$$v_x = \frac{kQ}{x^2 + kA} \tag{10}$$

where v_x = velocity of air flow at distance x

Q = volume of air passing through the hood opening per minute (This is v_0A, where v_0 is the air velocity at the opening.)

k = a constant dependent on the shape of the opening. (For square to round openings, k is near 0.1.)

A = the area of the opening

The volume of air entering the hood per minute is the velocity of air at the opening times the area of the opening.

Using this relationship one can calculate the effect of distance from the opening. Suppose the average velocity of air at a square hood opening is 200 fpm and the area of the opening is 4 ft^2. What is the velocity of air flow 2 ft from the opening?

$$\begin{aligned} v_2 &= \frac{kQ}{x^2 + kA} = \frac{kv_0A}{x^2 + kA} \\ &= \frac{(0.1)(200\text{ fpm})(4\text{ ft}^2)}{(2\text{ ft})^2 + (0.1)(4\text{ ft}^2)} \\ &= 18\text{ fpm} \end{aligned}$$

[4] Chapter 6 in Heinsohn (1991) (see Bibliography).

A distance of only 2 ft is enough to reduce the rate of air flow to less than one-tenth of its original value.

Types of Hoods

There are three basic types of hoods: capture hoods (Figures 9.3, 9.4, and 9.5), enclosures, and receiving hoods. A capture hood is placed near the source of contamination and the flow of air is sufficient to carry the contaminant into its opening. When designing a capture hood these points are important:

1. The distance from the hood opening to the source of contamination must be as small as possible. A short distance allows capture with a lower air flow into the hood.
2. The geometry of the source must be considered. For example, a hood opening at one side of a large open vat is close to one side of the vat but is far from the other side.
3. Movement of air into the hood should be away from the worker.
4. If the contaminant is given an initial velocity in one direction by the process, for example, particulate coming from a grinding wheel, the hood opening should be placed on that side of the process to take advantage of, rather than having to overcome, that velocity.
5. The hood must not interfere with the process or the worker.
6. Cross-drafts of air from passing people or vehicles, doors opening, and the like can seriously interfere with effective capture of contaminants. It is hard to anticipate cross-draft occurrence and difficult to calculate its impact quantitatively; it thus requires some overdesign to compensate safely.

Design features may be added to minimize some of these problems. A round hood opening at the side of a vat draws air more effectively at the center than at the edges, so it is not uniformly effective across the vat surface. Replacing the round hood with a wide slot results in a system that draws relatively uniformly across the entire end of the vat. This modification is not without cost, because a slot generates turbulence in the air inside the duct, increasing its resistance to flow. The effect on Equation 10 is to decrease the value of K, thereby reducing the velocity of the air at a given distance outside the hood opening.

When dealing with a wide vat at which there are restrictions due to access to the vat and location of the workers such that only one side may have a hood opening, air under pressure may be introduced on the opposite side of the vat and directed across the vat to the hood opening. This is termed a push–pull system (Figure 9.5). The velocity of air from the air source must be adjusted to move the contaminants up to but not beyond the hood opening, or else this becomes just another cross-draft. This can be done without spreading the contaminant in every direction because a blower can direct air successfully, unlike an air intake which draws from every direction. To illustrate this principle, hold your hand about a foot from your mouth and blow a small volume of air at the hand. Now repeat this, but this time draw in sharply about the same volume of air. Did you feel the moving air in each case?

The simplest case of local ventilation would be to construct a duct that ends at the source of contamination. If a hole is cut in a metal sheet the shape of the duct end, and the duct is mounted to that sheet so the air comes through the hole in the sheet, the sheet is called a flange. Simply putting a flange on the opening of the hood increases its effectiveness significantly. This is because air entering a simple opening is partially drawn from behind the opening, and the flange blocks this. The approximate effect of a flange about as wide as the opening itself is to change Equation 10 to the following:

$$v = \frac{kQ}{0.75\,(x^2 + kA)}$$

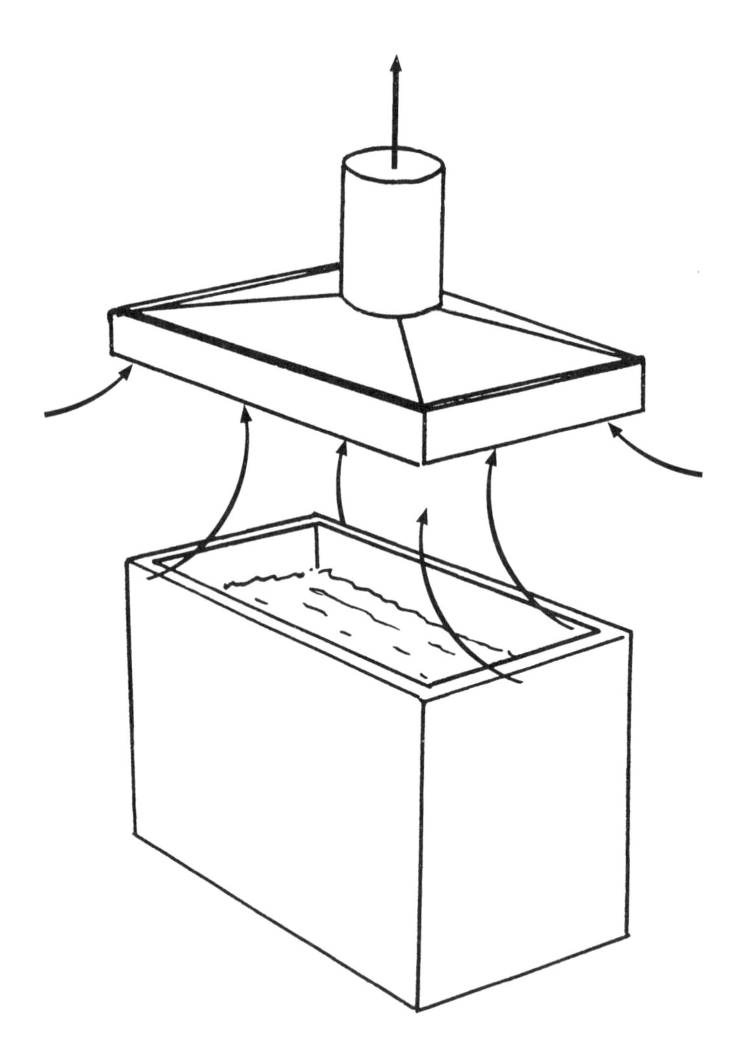

Figure 9.3. A canopy hood is suspended over the vat or other source of contaminant. Such a design allows access to the workstation from all sides. However, the capture of dense vapors is very inefficient. Light vapors and hot air rise toward the canopy to be captured.

Once a flange is added to the hood, the first step has been taken toward building an enclosure. Adding sides blocks cross-drafts and increases the portion of intake air that is carrying contaminants. Once we have built a box around the source (including an air inlet somewhere), contamination control is achieved with much lower air flows (Q) and the system is less subject to interference in its operation by external factors (Figure 9.6). This is only possible when it is not necessary to have access to the operation.

Finally, a receiving hood takes advantage of velocity given to the contaminant in a specific direction. An air intake hose attached to a saw or grinder so that the particles are directed right into the opening can be very effective. Some local ordinances require such a system to be installed. Another type of receiving hood is a canopy intake placed over a hot process. A canopy hood is a tempting design because it places the hood out of the way of workers and machinery, but it is an ineffective design for a cold process. However, in a hot process the contaminant rises due to thermal upcurrents of air and is conveniently collected. Such an arrangement may move contaminants toward workers at the edge of the vat, and if so should not be used.

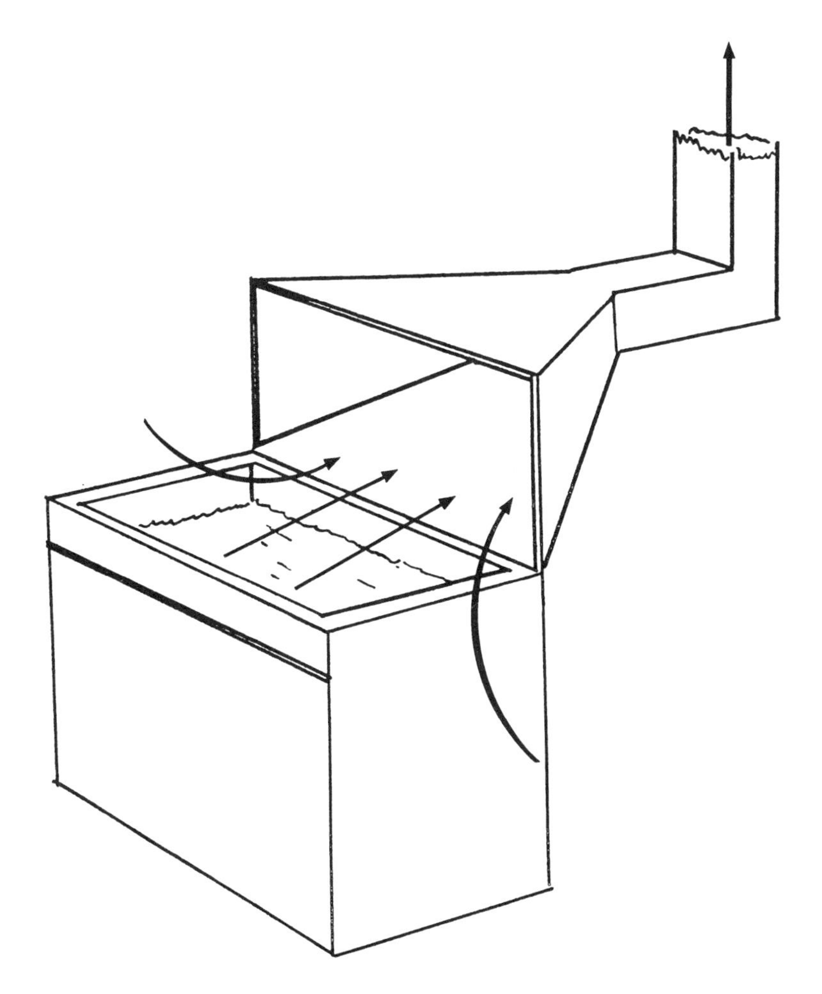

Figure 9.4. In this lateral hood air moves across the vat into the hood. In such an arrangement air enters the hood from locations other than over the vat. It is important that the air have high enough velocity or that the vat be sufficiently narrow that vapors from the opposite side of the vat enter the hood.

DESIGN OF VENTILATION SYSTEM COMPONENTS

The design of hood features unique to LEV systems has been discussed in some detail. Now we turn to the design and operation of the rest of the ventilation system.

DUCTS

First recognize that ducts are not unique to LEV systems. A GEV system could be as simple as a fan exhausting air at one end of a room and an inlet connected to the outdoors at the other end. However, incoming air may be distributed to several rooms, may be transported down walls to be released at floor level, or may enter a room at well-spaced points to obtain better air distribution (higher K values). Such sophistication in air delivery requires ductwork.

Two problems are associated with the design of the duct system to move air: power requirements to move the air and proper distribution of the air in a multi-intake system.

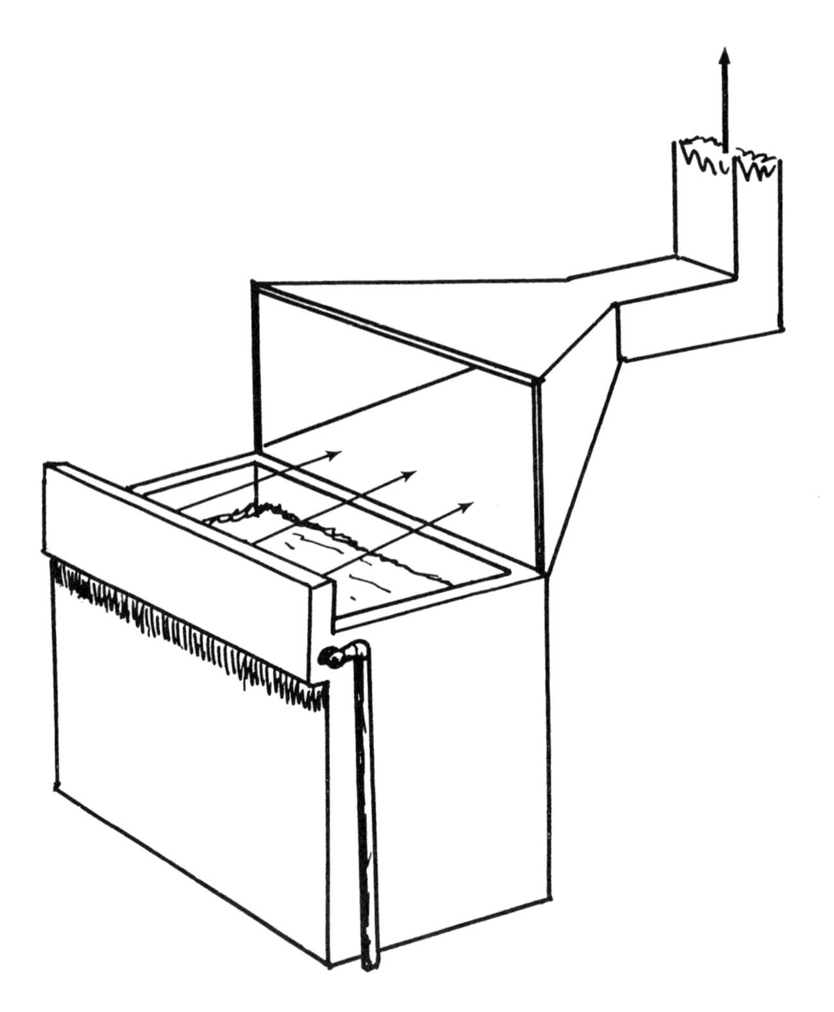

Figure 9.5. This push–pull system differs from the simpler lateral hood in that nozzles direct compressed air across the vat toward the hood. Such a system ensures efficient capture of vapors from the side opposite the hood.

Beyond air flow design, it is necessary to have a means of cleaning the system and checking its performance.

The difficulty of moving air through the duct system is called the resistance of the system. Resistance is the result of friction of the moving air with duct walls. High resistance must be overcome by more powerful fans, and therefore has a price in energy needed to run the system. Round ducts have less duct surface for a given cross-sectional area than any other shape, and therefore provide lower resistance than ducts of other shapes. The higher the velocity of air through the duct, the greater the resistance. To move the same amount of air through a smaller diameter duct requires that the air move at a higher velocity, so smaller ducts have higher resistance. Now larger fans may be needed. Money saved in initial installation by fabricating smaller ducts may be lost many times over in operating costs. Resistance is also increased by designing sharp bends into the system. Just as it takes less energy to move a new Taurus than a Model T Ford down the highway at the same speed, smooth curves in ductwork lower the energy requirement to move air (Figure 9.7). A rule of thumb for duct curvature is that the radius of a turn should be at least 2.5 times the duct diameter.

There is a minimum velocity required in the duct when the contaminant is particulate. Heavier particles require greater velocity, with minimum velocities ranging from approxi-

Figure 9.6. An enclosure. A laboratory hood provides an example of an almost complete enclosure of the source of contamination. The window goes up to allow manipulation of the operation in the hood, then closes down so that all air entering the slot at the bottom of the window goes up the stack.

mately 600 m/min for fumes from welding or soldering to 1700 m/min for dust from grinding metals.[5] Too low a velocity results in solids being deposited in the ducts, eventually blocking them. Vapors and gases do not settle out, and an air velocity of around 650 to 1000 m/min is recommended.

The second, and more difficult, problem in duct design is providing proper distribution of air coming from each of several hoods in a complex ventilation system. One possibility is to design the duct diameters so that sufficient air flow is maintained at each hood for the contaminants being captured (Figure 9.6). Another is to place gates, adjustable barriers, in the duct to adjust the flow. A good design based on balancing duct diameters is more efficient,

[5] Recommendations are given in the ACGIH manual.

Figure 9.7. Here a small plant has a general exhaust ventilation system. The blower to the right removes air exiting the plant at several points on the ceiling. Notice the duct avoids sharp bends, allowing air to be moved with less friction. Notice too that as more air enters the main duct, its diameter increases. Done properly, this ensures roughly equal flow at each entrance to the duct. The blower on the left serves a local exhaust ventilation installation.

and gates are places to deposit particulates. However, it is easy to change a gated system and rebalance it by adjusting the gates, whereas a system with fixed duct diameters is inflexible.

FANS

Capacity, efficiency, noise, and maintenance problems are issues in the selection of a fan. Relative capacity of a fan can be indicated by its manufacturer, but how much air is actually moved in a given installation depends on the resistance of the system. Efficiency determines the cost of running the fan. Noise is more important in some installations than others, and its relative importance must be judged when a system is designed or modified. In general, of two fans moving the same volume of air per minute, the larger, slower running fan is quieter. Maintenance procedures vary with operating conditions, but must be understood so that a periodic inspection and cleaning plan can be designed. Fans must not be installed and forgotten. No amount of ductwork will move air without a functional fan.

Fan design is divided into two groups: axial and centrifugal (Figure 9.8). Axial fans are like the familiar ceiling fan or room fan in a house, with propeller-like blades attached to an axle. Air is drawn from one side of the spinning blades and is propelled toward the other. Axial fans can be direct drive or belt driven with the motor offset. They are often used in simple GEV systems. Blade shape is a factor, but regardless, axial fans tend to be noisy.

In centrifugal fans a housing encloses the blades. Air enters the housing at the drive shaft, and blades force air out of the housing like a paddle wheel. Again they may have a direct drive or be belt driven. Centrifugal fans with straight blades direct particulate out toward the housing. Because dust does not accumulate on the blades, these fans are sometimes characterized as "self-cleaning". Some centrifugal fans have curved blades. These have higher efficiency and are quieter than straight blades. However, particulate is trapped in these blades, so they require periodic cleaning.

Selection of the correct fan for a system, GEV or LEV, requires a ventilation engineer. The safety professional ensures that the fan in place is providing correct air flow by checking

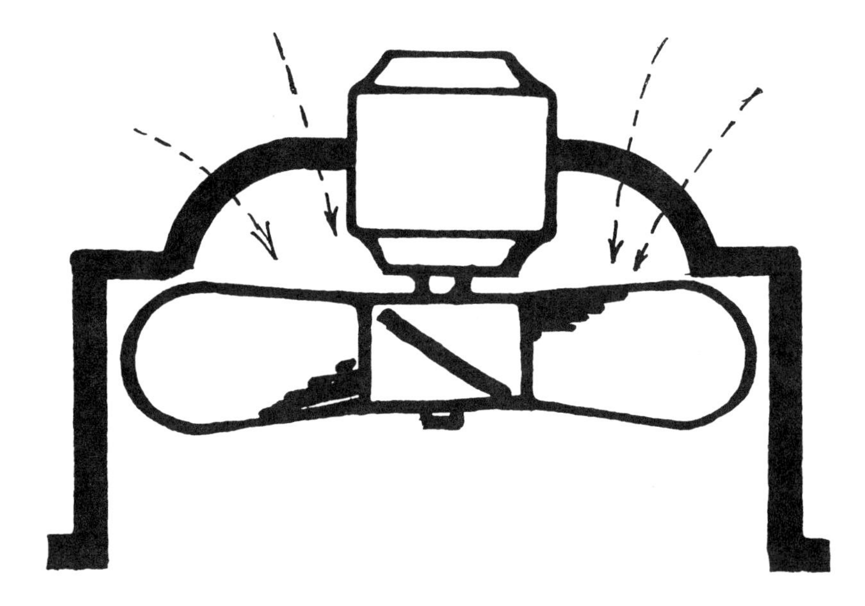

Figure 9.8A. Axial and centrifugal fans. (A) An axial fan. Such fans have propeller-like blades. The motor may be attached to the blades, or drive may be by a belt from the motor at one side. The latter allows lower speeds on larger diameter fans, a quieter arrangement. (B) A centrifugal fan. Air enters at the center of the housing and is "spun out" of the system at the perimeter by a "squirrel-cage" fan. The roof fans in Figure 9.7 are centrifugal fans.

velocities at hoods and comparing pressures in ducts with design specifications by use of water manometers at access points. Fans should have scheduled maintenance, especially if a belt drive is employed.

Fans that exhaust air from ducts are often seen on the building roof. The fan exhaust is directed upward into a chimney-like stack. The stack is intended to carry the exhaust air high enough that it is carried away, not pulled back into the fresh air intake. The exit velocity should be sufficiently high that if colored smoke were added to the exhaust air, a plume would rise perpendicularly above the building. The stack must be sufficiently high. A rule of thumb suggests at least 10 ft above the roof line, if the roof line is within 50 ft of the stack.

Directing stacks vertically permits rain to enter the air outlet. Various designs to handle this problem are employed. The least successful of these is to install a cone shaped cap, with point upward. Although this blocks the rain, it also redirects the air sideways, possibly into the air intake, and may create back pressure in the duct by restricting the exit of the air. More successful is the installation of a larger diameter sleeve around the top of the stack extending above the stack opening. Rain normally falls at an angle, strikes the side of the sleeve before reaching the stack opening, and runs down the inside of the sleeve onto the roof. In another useful design the fan is offset at an angle to the stack. The stack then extends below the entry point of the fan. Rain runs down the stack past the blower entry and exits from a hole at the bottom.

CLEANERS

A cleaner is used if the exhaust air is hazardous or a nuisance to the general public, or if the air is to be recirculated. Cleaner design depends on the nature of the contaminants. Removal of particulate requires a quite different approach from removal of gases or vapors.

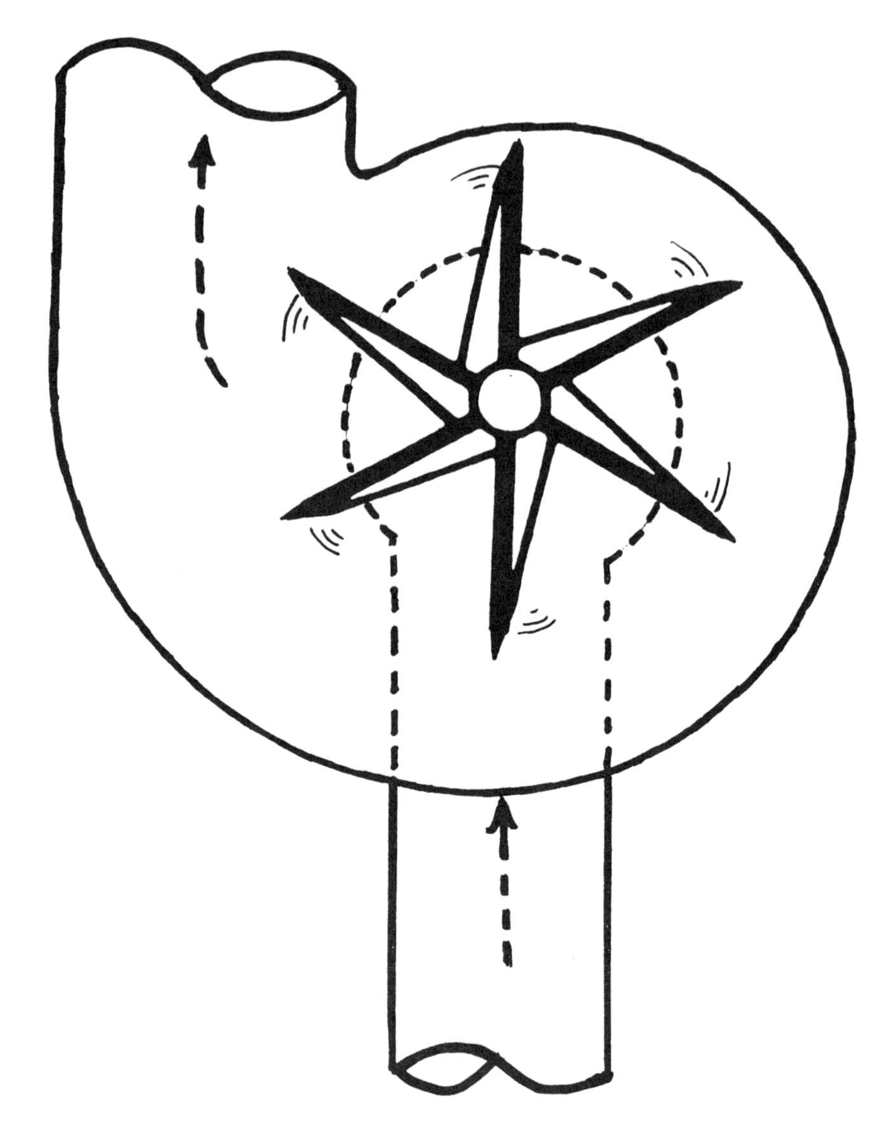

Figure 9.8B.

Particulate Removal

Most cleaners are designed to remove particulate from exhaust air. Recall the relationship between particle size and the ability of the particle to penetrate deep into the lungs. As a rule, particles under 5 μm are classed as respirable and are therefore more hazardous. Unfortunately, the easiest particle removal methods are effective only against larger particles and are useful primarily for removal of nuisance dusts.

The efficiency of particulate removal (amount usually being expressed in weight terms) is often calculated as follows:

$$\text{Efficiency} = \frac{\text{Amount collected}}{\text{Amount entering cleaner}}$$

This number can be deceptive. Respirable particulate may not be removed at all in a system rated as 98% efficient, because the respirable dust particles individually weigh so little.

ESTIMATING AIR FLOW

It is sometimes useful when assessing the adequacy of a ventilating system to be able to estimate air flow. Some new systems include sensors, which provide the information directly. If not, an estimate is possible from data about the fan, given some basic information. The formula is

$$Q = \frac{(8.5)\,(V)\,(A)\,(ME_{fan})\,(E_{motor})}{(FTP)\,(K_{dl})\,(d)}$$

where Q = air flow through the fan (cfm)
V = voltage at the fan
A = amperage of fan motor
ME_{fan} = mechanical efficiency of the fan (0.5–0.7; available from supplier)
E_{motor} = efficiency of the motor (0.90–0.99; available from supplier)
FTP = the total static pressure drop across fan (inches on the water gauge)
K_{dl} = drive loss factor of fan (1.05–1.25; available from supplier)
d = a correction for altitude (1.0 at sea level, less 0.03 per 1000-ft increase)

An estimate can also be made if data about the heating or cooling system is available. If the rate of heat transfer at the heat transfer coils is known and the temperature change of the air is measured, the calculation is relatively easy:

$$Q = HT/(1.08\Delta T)$$

where HT = the rate of heat transfer (BTU/hr)
ΔT = temperature change in air (°F)

Taken from D. J. Burton in *Occupational Health and Safety,* June 1994, p. 65.

Removing Large Particulate

The simplest system passes the exhaust air through a large settling chamber. Velocity drops sharply in such a chamber, just as water in a fast moving stream slows as it enters a pond. As velocity drops below the capture velocity of the larger particles, these settle to the cone-shaped bottom of the chamber from which they can be removed periodically through access at the bottom of the cone. Baffles added to the chamber cause air to change direction sharply. These remove some particles as they impact the baffle.

In centrifugal or cyclone collectors air is swirled around a cone-shaped chamber, moving faster as the diameter becomes smaller (Figures 9.9 and 9.10). The particles spin out against the sides of the chamber and drop to the bottom, much as the bends of the bronchial tubes remove particles from the air in the lungs. Once again larger particles are removed much more effectively. A high-efficiency cyclone has a smaller diameter cone, spins the air faster, and so is more effective at removing smaller particles. Such a cyclone has a much higher resistance and may require installation of a larger capacity fan.

A dynamic precipitator adds a motor-driven impeller to a cyclone to drive the air against the sides of the chamber, which raises efficiency to the level of a high-efficiency cyclone. Respirable particles escape collection even in high-efficiency cyclone and dynamic precipitator designs.

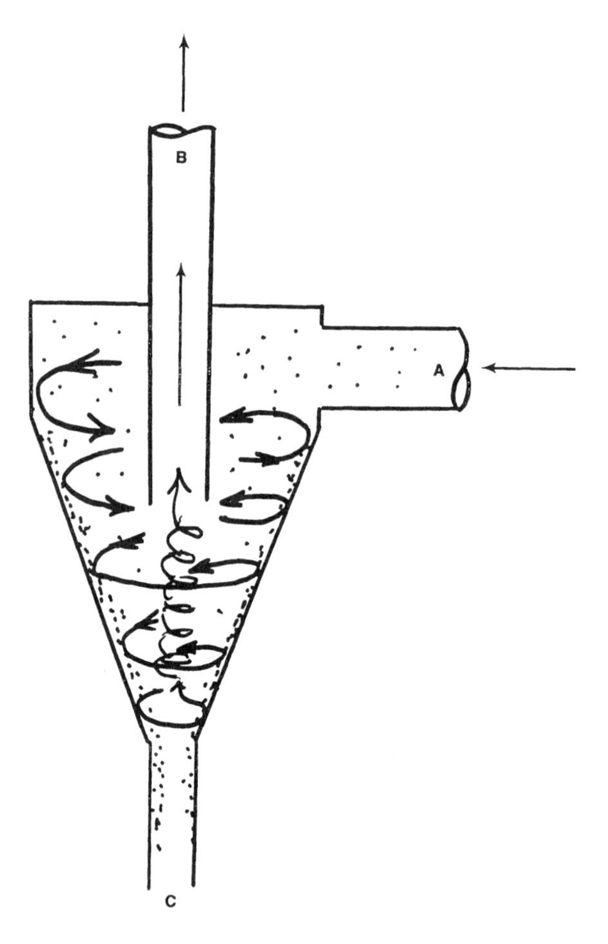

Figure 9.9. In a cyclone collector, particulate-contaminated air is directed into a conical chamber (A), forcing it to spin around the chamber. As it moves down the cone (toward smaller diameter), the spinning becomes faster. Particles are driven by centrifugal force against the walls of the chamber and deposited there. Faster spin rates deposit smaller particles. Treated air is drawn out of the center (B), and deposited particles are collected at the bottom of the chamber (C) and transferred into a container for disposal.

Removing Respirable Particulate

Cloth filters are frequently employed as cleaners. A bag house has a large chamber in which air enters at the bottom and has to pass through cloth bags to exit at the top (Figure 9.11). Although respirable particles pass through the pores in new cloth, as the bag is used the larger pores are blocked and respirable particles are removed. Periodically the bags are shaken, so that accumulated particulate is dropped to the bottom of the chamber for removal. This design is restricted to removal of dry particulate, because oily or wet droplets seal openings in the cloth.

Water can be used in a variety of ways to remove both particulate and water-soluble gases or vapors. Water can be sprayed in a chamber or dynamic precipitator, or air can be passed up a scrubber tower in which water is sprayed or flows over baffles or a packing material. If used water is placed in a tank, allowing particulate to settle out, the water sometimes may be reused.

Electrostatic precipitators utilize a discharge electrode and a grounded collecting electrode (Figure 9.12). Particles are given a charge with the negative discharge electrode, then move toward the positive plate of the collecting electrode. Periodically the accumulated particulate

Figure 9.10. A cyclone collector on a small industrial building.

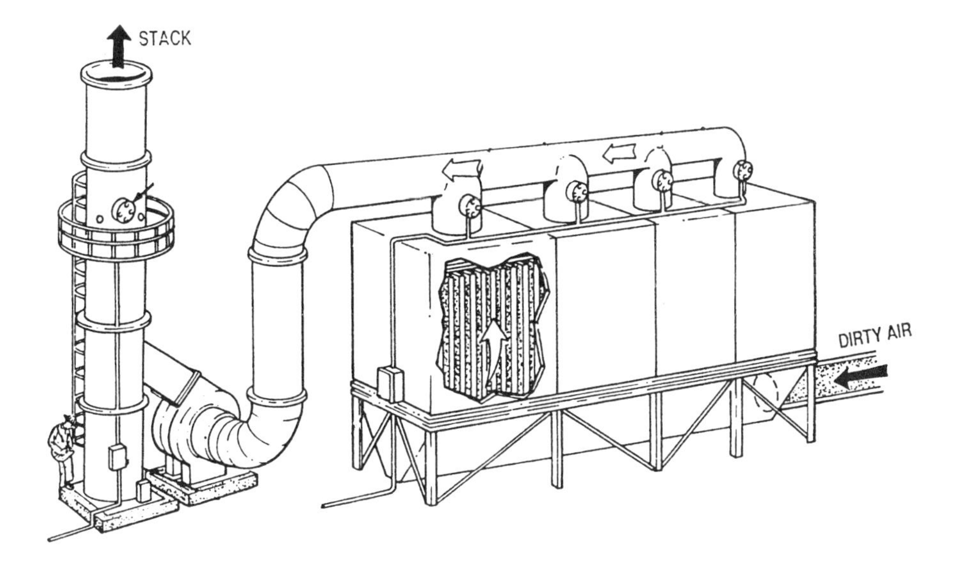

Figure 9.11. In a baghouse, particulate-contaminated air enters at the lower end of the chamber. In order to exit, it must pass through hanging cloth filters, which remove particulate from the air. Periodically the filters are shaken, causing collected particulate to fall to the bottom where it can be collected for disposal. Notice here the centrifugal fan is offset, so that rain entering the stack runs straight to the bottom rather than down into the fan. (Photo courtesy of Auburn International, Danvers, MA. With permission.)

is removed from the plates. Such a system is efficient and adds little resistance to the system. Preliminary cleaning of contaminated air is normally done, perhaps with a settling chamber or cyclone, to reduce the load on the system.

Vapor and Gas Removal

Removal of gases or vapors requires a different approach. Wet scrubbers work unless the contaminant is not water soluble. Activated charcoal filters remove most contaminants effec-

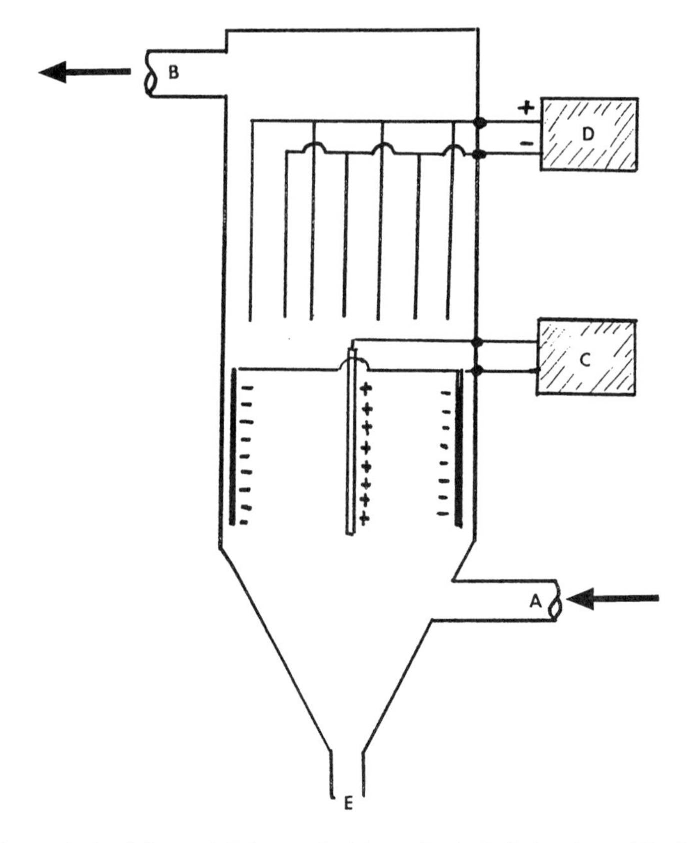

Figure 9.12. In an electrostatic precipitator, particulate-contaminated air enters at the lower end of the chamber (A). In order to exit (B), it must pass through an electrical field generated by power supply C, which gives the particles a charge. They then pass by plates given a charge by power supply D, and particles attach to the oppositely charged plates. Particulate then falls to the bottom where it can be collected for disposal (E).

tively. Cooling exhaust air may cause the vapor to condense. Passing exhaust air over burners, possibly in the presence of a combustion catalyst, incinerates combustible contaminants.

WALL LOSSES

Airborne contaminants may be adsorbed on solid surfaces in a room, and so removed from the air. Walls in the kitchen of a house or restaurant where frying is done have a grease residue. Later, some residues may desorb from these surfaces and reenter the air. Tobacco smoke is a familiar example — and is why hotels have nonsmoking rooms. You know when you enter a room that is occupied regularly by smokers from the stale tobacco smell in the air even if no one is there. This is evidence of the desorption of the smoke components from the walls. This may well work with chemicals in the workplace. This could explain anomalies in measurements of volatiles in a room when ventilation rates and generation rates do not explain the observed levels of a chemical in the air at a given time.

Equations have been derived to quantify this effect, but a great many characteristics of the contaminant and the walls must be known to employ them. Clearly, though, the magnitude of the effect is proportional to the surface area in the room. Both adsorption and desorption are first-order effects, directly proportional to the concentration of the contaminant. The strength of adsorbance to the surface can be expressed as a constant characteristic for the surface and the contaminant. Other factors such as the degree of mixing of the air, which determines how frequently contaminant collides with wall surface, enter into the calculation.

We will not pursue these calculations, but rather discuss the phenomenon qualitatively so that it can become part of the thinking of a person whose task may involve dealing with contamination in a room. A quantitative discussion is provided in Heinsohn (1991).

CHEMICAL PERSONAL PROTECTIVE EQUIPMENT — RESPIRATORS

It is time to broaden the objective of the chapter — protecting the worker from airborne contaminants — and make some general statements about personal protective equipment (PPE). In deciding whether PPE is needed, first the level of hazard faced by the worker must be assessed. Two aspects are important in protective equipment: (1) protection from exposure of the skin to damaging or toxic agents and (2) provision of safe breathing air. The first of these concerns was introduced in some detail in Chapter 6.

Once the hazard is identified and its magnitude understood, it is useful to consider the four levels of protection defined by the EPA.

Level A. Here the worker must be isolated from his or her environment as completely as possible. The PPE consists of a totally encapsulating suit including boots and gloves and constructed of materials resistant to the chemical hazard (a "moon suit"). This suit is equipped with a positive-pressure, self-contained breathing apparatus (SCBA). Such an outfit might be worn by emergency response personnel entering a spill area or a worker at a hazardous waste site involving highly toxic chemicals.

Level B. The positive-pressure, self-contained respiratory system is also employed at this level, but the skin protection is less critical. Chemically resistant clothing is employed, but the worker is not encapsulated.

Level C. The respiratory system provides air purification, and chemically resistant clothing is worn.

Level D. Regular work clothing with no respiratory protection is sufficient. The hazard is low.

It is desirable to employ an "engineering solution" to problems of airborne contaminants, reducing levels by ventilation. However, there are times when the use of respirators is appropriate. Clearing a hazardous waste site requires full-time respiratory protection. Certainly in plants where chemicals are handled, respirators should be available for emergencies such as spills, equipment malfunctions, or special maintenance procedures. An adequate emergency preparedness plan ensures that respirators of the correct design are on hand.

Respirator recommendations are included on the MSDS for the chemical. These recommendations are usually based on the level of the contaminant in the air, and in an emergency there is seldom time to do an analysis. It is best then to err on the side of caution.

RESPIRATOR DESIGN

Two styles of face masks are available: a half-face mask covers the nose and mouth and a full-face mask covers the entire face, providing eye protection too. The mask must be flexible to fit snugly at the edges without gaps and leaks. The body of the mask is made of natural rubber, silicone, or polyvinylchloride. Rubber masks have been used for many years and are still available. Rubber is more susceptible to damage by chemicals such as hydrocarbon solvents and ozone than other face mask structural materials. Very flexible silicone rubber bodies are the most comfortable to wear, but become slippery if the wearer is sweating. Polyvinylchloride masks are satisfactory and have a cost advantage, but tend to be inflexible in the cold.

There are two classes of face mask type for respiratory protection: negative-pressure and positive-pressure masks. In a negative-pressure mask air is drawn into the mask when the wearer inhales. To enter the mask, air must pass through a purifying device. In a positive-pressure system breathing air is supplied from a source external to the face mask.

NEGATIVE-PRESSURE SYSTEMS

The simplest situation is protection against particulate in the air. Air entering the mask passes through a filter to remove particulate. In a low-hazard situation, masks can simply be made of heavy paper to be discarded at the end of the day or when they become blocked. Better protection is provided by a face mask to which removable filters are mounted (Figure 9.13). Filters vary in mesh size, and therefore in the size of the smallest particle they allow to pass through. It is important to recognize that a filter mask traps only particulate material. If the problem is a gas or vapor, a filter style of respirator does not serve. Many workers do not understand this, so clear labeling and instruction is important.

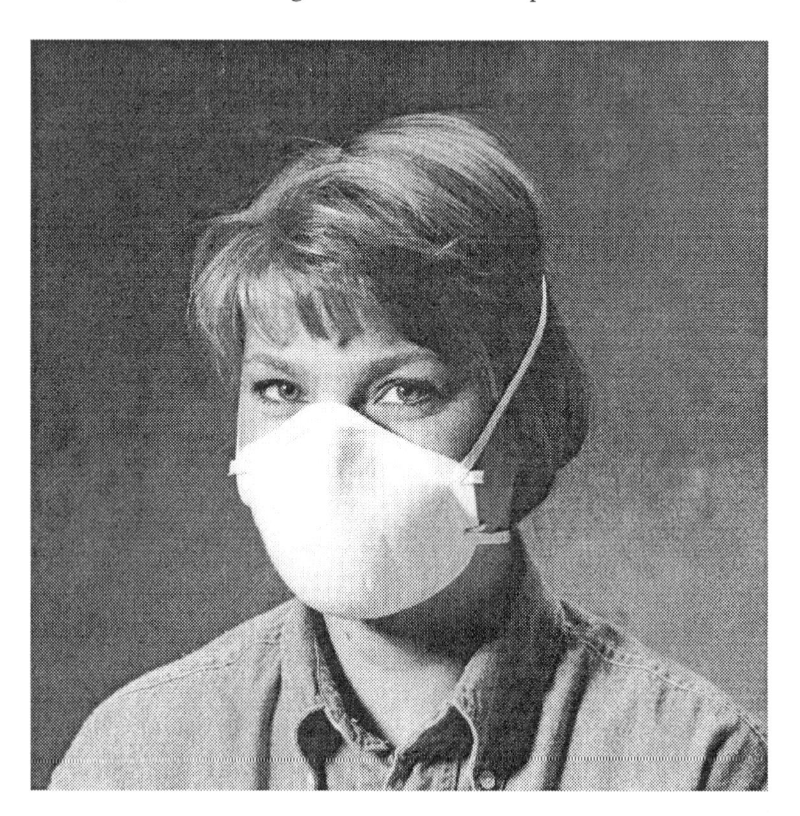

Figure 9.13. Simple filter respirator. Good fit to the face is essential; otherwise the dirty air will preferentially go through the leaks rather than the filter. (Photo courtesy of Mine Safety Appliances Company, Pittsburgh, PA. With permission.)

To remove a gas or vapor, air must pass through an adsorbent, such as activated charcoal, or a substance that reacts with the gas or vapor. Respirators with this feature have removable cartridges containing an active agent that is replaced when the adsorbent is spent. Manufacturers provide a number of different cartridges, each color coded (Table 9.1). The cartridges used must be correct for the hazardous gas or vapor involved, and the recommendations of the manufacturer should be followed. Once again, many workers feel that they are safe when wearing a respirator, even the wrong one, and must be instructed which is the correct cartridge to employ. Adding a filter outside the cartridge also removes particulate and may extend the life of the cartridge.

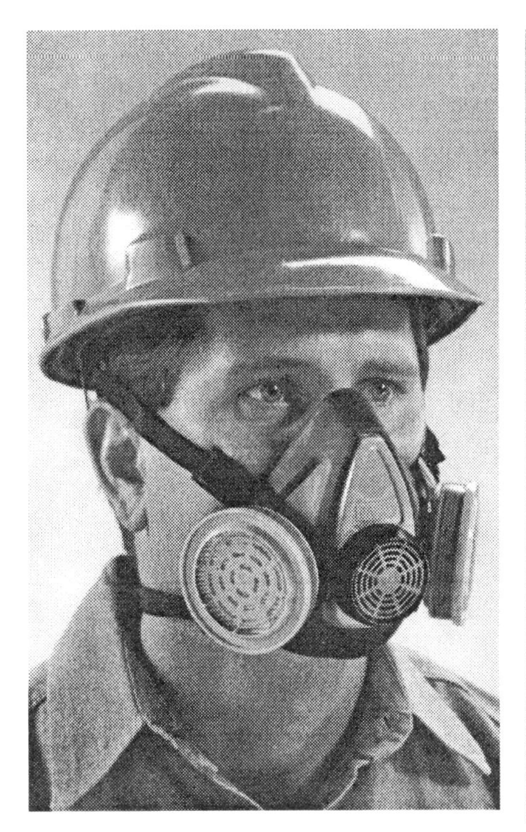

Figure 9.14. Shown here are examples of face masks that utilize replaceable cartridges to treat the worker's breathing air. A range of cartridges are available to deal with particulate, vapor, and gas hazards. The photo on the left shows a half-face respirator and the one on the right a full-face respirator. The full-face respirator provides the bonus of eye protection and is intended for more hazardous atmospheres than the one shown at left. It provides the option of breathing through the filter or connecting to an external air supply. (Photos courtesy of Mine Safety Appliances Company, Pittsburgh, PA. With permission.)

Table 9.1. Color Coding for Respirator Cartridges.

Contaminant	Color
Acid gases	White
Organic vapors	Black
Ammonia gas	Green
Carbon monoxide	Blue
Acid gases and organic vapors	Yellow
Acid gases, ammonia, and organic vapors	Brown
Acid gases, ammonia, carbon monoxide, and organic vapors	Red
Radioactive materials (not tritium or noble gases)	Purple
Dusts, fumes, and mists	Orange

Note: An orange or purple stripe is added to an otherwise coded canister to indicate blocking of dusts and mists or radioactivity, respectively.

CHECKING RESPIRATOR FIT

The first requirement of a negative-pressure respirator is that it fit closely to the face, preventing leakage at the edges. The respirator issued to an employee must generally fit or adapt correctly to facial contours. Whatever cartridge is installed, it resists the movement of air through it. If the respirator leaks at the edges, there will be less resistance to air entering through the leak than by the correct route, and the effectiveness of the device is sharply reduced.

Testing for correct fit when the respirator is issued can be done by a qualitative fit test. This is done by exposing the employee with respirator to a harmless chemical. One protocol accepted by OSHA uses isoamyl acetate (banana oil), which has a strong odor. Smelling bananas indicates the presence of a leak. Other approved tests use saccharin, depending on tasting the sweetness, and stannic chloride, which is irritating. All these chemicals are readily detected at low concentrations (Figure 9.15).

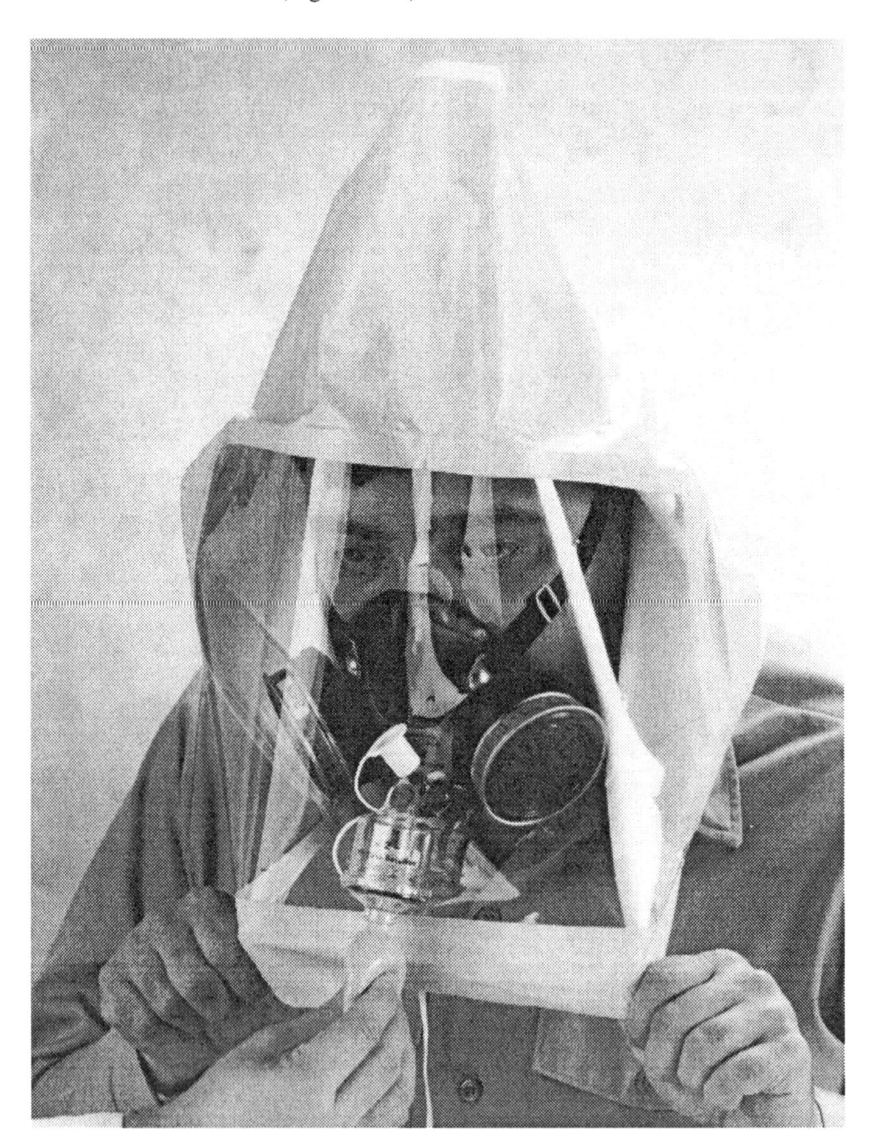

Figure 9.15. Fit testing a respirator. With the worker in the respirator, the head is covered with a hood. Squeezing the bulb adds a particulate such as saccharin powder or a harmless volatile liquid such as isoamyl acetate (banana oil) to the air under the hood. The worker readily detects a leak by taste (saccharin) or smell. (Photo courtesy of Mine Safety Appliances Company, Pittsburgh, PA. With permission.)

A quantitative fit test is done by actually sampling and analyzing the level of some agent inside the respirator while it is being worn. The respirator is equipped with a high-efficiency particulate filter and the wearer is exposed to a standard source of particulate. The fit factor is the ratio of the concentrations of particulate outside and inside the mask. Sampling the air inside the respirator is a difficult problem; the equipment is expensive, and other factors demand that experienced experts do this study.

The respirator should be spot checked by a positive-pressure or a negative-pressure test each time the employee puts it on. In the positive-pressure test the worker blocks the valve through which exhaled air exits, then exhales. The mask should inflate slightly and remain inflated for a few seconds. In a negative-pressure test the employee blocks the air intakes on the cartridges and inhales. The mask should press against the face. The negative test also checks operation of the air exit valve. When the worker inhales, this valve should close, preventing untreated air from entering the mask.

POSITIVE-PRESSURE SYSTEMS

Where contaminant levels are very high, removal by the respirator is inefficient, or oxygen levels are low, a breathing apparatus is used to supply breathing air. A portable or self-contained breathing apparatus includes a tank of air strapped to the back of the worker and connected to the face mask. Newer equipment includes an adapter that allows the cylinder to be refilled during use, eliminating the need for the worker to leave the work area to change bottles. Alternatively, the face mask may be connected by a flexible hose to tanks or an air pump located outside the danger zone (Figure 9.16).

Where compressed air from tanks is used, OSHA uses the standard of the Compressed Gas Association for breathing air, titled G-7.1. In its most recent revision, this standard calls for:

Oxygen, 19.5–23.5%
CO, <10 ppm
Hydrocarbons, <5 mg/m^3
CO_2, <1000 ppm
Free of odors, OSHA refers to this as Grade D air.

OSHA also has standards for air compressors used for supplying breathing air. These include filters, absorbents, and some warning systems.[6]

Supplied air may be constantly flowing into the mask, a continuous-flow system. In such a system the respirator may fit loosely. With positive air pressure in the respirator, air leaks out of the mask, not into it. A loose fitting respirator can be less expensive in construction, and in fact, may be disposable. In jobs where the respirator may become contaminated, such as a spray painting operation, a disposable respirator is highly desirable.

Demand-flow systems supply air only when the wearer inhales. The pressure drop in the mask activates a valve that allows air to enter. Similarly, the pressure increase during exhalation closes the intake valve and opens the exhaust valves. Such a respirator must fit the face tightly, otherwise outside air might enter leaks during inhalation.

Finally, there is the pressure-demand system. In this system the intake valve never completely closes. As a result, the interior of the mask is always under at least slight positive pressure, eliminating the chance of outside air leaking into the breathing zone.[7]

In respirators with a loose face mask, air flow rates must be between 6 and 15 cfm, and for tight-fitting masks, a flow rate of 4–15 cfm is necessary. Excessive flow rates of dry compressed air dry out the respiratory tract and eyes. Measuring the volume of air passing through a respiratory system to ensure compliance is difficult. However, flow rates depend on the line pressure, hose length, and other design features. Respiratory equipment manufacturers recommend a line pressure that provides a correct flow rate.

[6] 29 CFR 1910.134 (d)(2)(ii).
[7] J. F. Rekus, "A Breath of Fresh Air," *Occupational Health and Safety*, May 1991, pp. 22–28.

Figure 9.16. Supplied-air systems. The worker in the above photo is wearing an SCBA respirator. Air is provided for a limited time period by the tank carried on the worker's back. The worker in the photo on the next page is connected to a system that supplies air through a line from a pump or tank at a distant location. In an emergency, the worker can disconnect from the primary air supply and leave, breathing air from the small bottle. (Photos courtesy of Mine Safety Appliances Company, Pittsburgh, PA. With permission.)

NIOSH rates various types of respirators by something called the protection factor. This factor is derived from the ratio of the contaminant in the air outside the mask by the concentration inside the mask. Thus, if benzene is at a level of 500 ppm outside the mask and 2 ppm inside the mask, the protection ratio is 500/2, or 250. The types of equipment described above have been assigned protection factors for comparison to the systems described here (Table 9.2).

In cases where the atmosphere is contaminated badly enough that it would be classed as IDLH, a system with a remote air supply should be supplemented with a small portable air tank worn by the worker and attached to the mask to allow escape if the main system should fail, a combination breathing apparatus. Asbestos removal or hazardous waste remediation are examples of times such a system would be employed. In such an atmosphere, a standby worker should be on hand to help if the respiratory system fails. The worker also should be wearing a harness to facilitate removing him or her from the contaminated area.

RESPIRATOR MAINTENANCE

Written safety plans should include regular inspection of respirators. Valves and lines must be in working order, rubber parts must be flexible and unbroken, viewplates must provide good vision, and straps and mounts should function well.

All respirators should be cleaned at regular intervals, and those that are shared by users must be cleaned and disinfected after each use to avoid the spread of disease. There is no

Figure 9.16. (Continued).

Table 9.2. NIOSH Protection Values for Respirator Systems.

Type of Respirator	Protection Value
Continuous-flow	
Loose-fitting respirator	25
Tight-fitting respirator	50
Demand-flow	
Half-face mask	10
Full-face mask	50
Pressure-demand	
Half-face mask	1000
Full-face mask	2000

standard format for respirator cleaning, and even equipment manufacturers differ in their recommendations. Disinfectant chemicals that may damage parts of the equipment should be avoided, but those used must do a good job. Diluted bleach, which some manufacturers recommend, can attack and slowly degrade some respirator fabrics. Alcohol reduces the life of rubber parts such as latex or neoprene, resulting in loss of good fit or damage to valves.

High temperatures (above 120°F) also damage rubber components. Ordinary detergents often do not adequately disinfect the mask, but cationic detergents of the sort used by hospital laundries are useful. These should be rinsed from the equipment thoroughly, because they are more likely to irritate skin and eyes than mild detergents.[8]

THE SPECIAL CASE OF ASBESTOS

In writing the 1994 Asbestos Standard (29 CFR 1910.1001), OSHA established a PEL of 0.1 f/cc and designated types of respirators to be used at levels in excess of this. The following table is drawn directly from the standard:

Respiratory Protection for Asbestos Fibers.

Airborne Concentration of Asbestos or Conditions of Use	Required Respirator
Not in excess of 1 f/cc (10 × PEL)	Half-mask air purifying respirator other than a disposable respirator, equipped with high-efficiency filters.
Not in excess of 5 f/cc (50 × PEL)	Full-facepiece air purifying respirator equipped with high-efficiency filters.
Not in excess of 10 f/cc (100 × PEL)	Any powered air purifying respirator equipped with high-efficiency filters or any supplied-air respirator operated in continuous flow mode.
Not in excess of 100 f/cc (1000 × PEL)	Full-facepiece supplied-air respirator operated in pressure demand mode.
Greater than 100 f/cc (1000 × PEL) or unknown concentration	Full-facepiece supplied-air respirator operated in pressure demand mode, equipped with an auxiliary positive-pressure self-contained breathing apparatus.

Note: a. Respirators assigned for high environmental concentrations may be used at lower concentrations, or when required respirator use is independent of concentration. b. A high-efficiency filter means a filter that is at least 99.97% efficient against mono-dispersed particles of 0.3 μm in diameter or larger.

From 29 CFR 1910.1001, Table 1.

EVALUATING THE PROGRAM

Once a program to control airborne hazards is in place, evaluation is necessary. Good evaluation has three aspects. First, it is necessary to monitor the atmosphere of the plant to see if the levels of toxic chemicals are safe and in compliance. Second, the condition of the ventilating system should be checked periodically to ensure it is meeting design specifications. Third, employees should be trained regarding the hazards and the use of such emergency devices as respirators. Finally, good employee health records must be maintained.

MONITORING THE AIR

Monitoring the air requires several types of measurement (Chapter 8). Concentrations of chemicals in the general plant air should be measured to ensure that overall conditions in the plant are satisfactory. Levels in high-hazard parts of the plant should be separately measured. In some cases, it is worthwhile to invest in automatic monitors that record atmospheric levels of chemicals likely to be in high-hazard areas. These could be connected to an alarm system, so that warning is given of unexpectedly high levels of escaping chemical.

[8] P. Eisenberg, "Do You Make These Common Mistakes in Maintaining Your Respirator Masks?" *Occupational Health and Safety*, Feb. 1992.

When the level of contaminant in plant air is normally high, for example, above the action level, more careful continuous monitoring is necessary. The air in the workers' breathing zone can be sampled occasionally by having the workers wear individual monitoring devices. Where there are ceiling limits imposed for the particular chemical, a series of individual readings are necessary to be certain that the ceiling is never exceeded.

TESTING VENTILATION SYSTEMS

It is no more realistic to install a ventilation system and assume it never needs to be touched again than it is to buy a new car and assume it never needs service. At the time of installation the operating parameters of a ventilating system should be measured, first to ensure it accomplishes the task for which it was designed and, second, to establish baseline values for later comparison to determine if parts of the system need service. Such retesting should be done at periodic intervals.

If changes are made in a plant, such as an expansion or shifting of an operation or the introduction of a new process, the ventilation system should be analyzed to ensure it is adequate for the new situation. A given fan moves a given amount of air, hence less air per station as new hoods and ducts are added. Furthermore, a new outlet changes the distribution of air intake and may require rebalancing the system.

An early concern in analysis of a system is determination of general patterns of air movement. Devices that generate a fine particulate smoke, often of titanium tetrachloride, are useful. Generating smoke near a workstation determines if air is actually moving away from the worker. A hood may draw well at its center, but allow contaminants to escape at the edges of its intended reach. Air turbulence resulting from hood design may divert contaminants from capture by the hood. Smoke reveals this spillage of contaminant. The effects of cross-drafts of air on hood performance can be assessed. A large amount of smoke released in a room helps estimate how quickly air is changed by a general ventilation system, and may identify regions of stagnant air.

Testing whether a system is operating according to original design parameters involves testing pressures and comparing them to values measured at the time of installation. Atmospheric pressure is expressed in a variety of units: mmHg, pascals, pounds per square inch (psi). Static pressure measurements focus on the difference between pressure inside and outside the duct and are relatively small values compared to total atmospheric pressure. It is common to use a water manometer, in its simplest form a U tube containing water, one end of which is inserted into a hole in the duct. Negative pressure in the duct results in the water level dropping on the atmosphere side of the tube. The negative pressure is reported as this difference in level in inches of water.

Velocity pressure in a duct is measured by a Pitot tube, a variation of the water manometer having a right angle bend in the tube (Figure 9.17). When inserted into the duct with the tube pointing directly at the air flow, it provides a pressure reading that is the sum of the static pressure and the velocity pressure. A second tube surrounds the first and is not open in the direction of the air flow, but rather has openings perpendicular to the air flow. This measures static pressure. If one side of the water manometer is connected to each tube, the manometer reads the difference between the total pressure and the static pressure, which is the velocity pressure.

Either a deflecting vane velometer or a rotating vane anemometer are most often used for measuring air velocity at large hood openings. In the deflecting vane velometer, air strikes a vane, which deflects in proportion to the air velocity, and the angle of deflection is read on a scale directly in feet per minute. It can be adapted to different velocity ranges by changing probes, the low end of the low-velocity probe reading 30 fpm. A rotating vane anemometer has a multivaned fan blade that is spun by the moving air, and the speed of rotation is read directly in feet per minute. A hand-held unit reads down to about 100 fpm. With either of

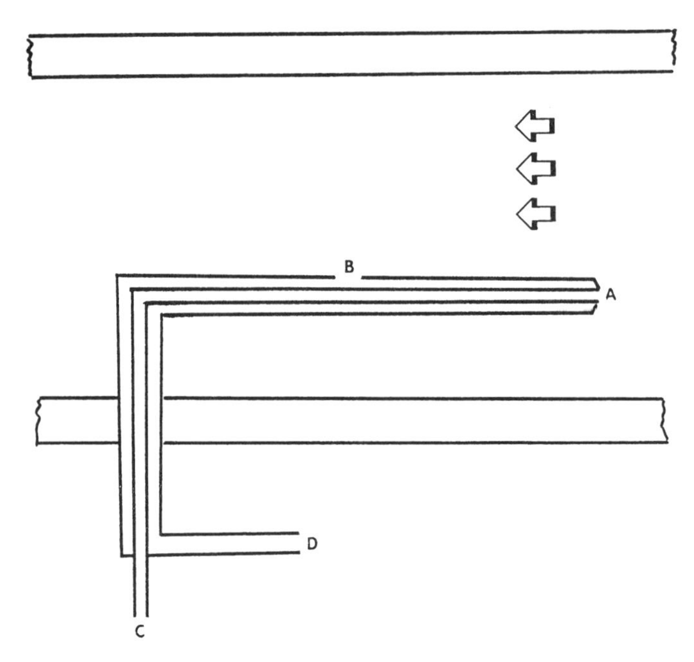

Figure 9.17. A pitot tube allows one to measure both static pressure and velocity pressure in a duct. In this sketch air is moving from right to left. Opening A in the tube is directed toward the air flow, and a manometer connected to tube C measures the sum of static pressure and velocity pressure. The opening in the outer tube (B) is directed perpendicularly to the air flow, and a manometer connected to tube D measures only static pressure. A manometer connected at one side to tube C and at the other to tube D reads velocity pressure.

these instruments error is introduced if air velocity at small hood openings is measured, because the blockage of the hood opening by the instrument distorts the value obtained.

Other devices are used. The resistance of a heated wire changes with temperature, and if the wire is placed in the air flow, it is cooled in proportion to the rate of flow. Such a heated wire anemometer is particularly useful at low air velocities, reading well below the lowest range of either the deflecting vane velometer or the rotating vane anemometer.

When measuring air velocity at a hood opening several readings must be taken at different locations in the opening, because the velocity changes from the center to the edges of the opening. In a long straight duct the centerline velocity is highest, dropping down to zero at the surface of the duct wall. Measurements in a duct should be done 7–8 duct diameters away from a bend or constriction to avoid distortions in the flow from these features.

A basic understanding of the design and parameters of a ventilation system is important to the safety professional. Good ventilation is one of the most important protections for workers when hazardous materials are in use, and systems should be checked to ensure they are performing as is necessary to achieve this end.

EMPLOYEE HEALTH RECORDS

By law, good employee records of health and exposure must be maintained.[9] If a problem arises involving either an individual worker or the general health of all employees in a sector of the plant, the records are a means of learning the cause of the problem, so the situation will not be repeated. Many times, an epidemiological study of company health records have revealed a previously unsuspected hazardous condition. This is particularly true of types of

[9] See 29 CFR 1904 and 29 CFR 1910.440.

toxicity for which animal surrogate testing is less reliable, for example, in the cases of carcinogens or reproductive toxins.

ENTRY INTO CONFINED SPACES

More than 300 workers die per year as the result of entry into confined spaces. A confined space has limited access ports and inadequate or nonexistent ventilation, a space not intended for continuous human occupation. There may be a bad atmosphere, it may potentially trap the worker, or it may present the danger of engulfing the worker. Storage tanks, silos, bins, utility chambers, and tunnels for pipelines or wiring are examples. These spaces are referred to as permit-required confined space, or simply permit space. Virtually every industrial site has examples of confined spaces, and such areas may need to be entered for maintenance, cleaning, or service of machinery. In 1993, when OSHA wrote confined space regulations, it was estimated that "about 224,000 establishments have permit spaces; 7.2 million production workers are employed at these establishments, and about 2.1 million workers enter permit spaces annually. . . . compliance with these regulations will avoid 53 worker deaths and injuries, 4900 lost-workday cases and 5700 nonlost-time accidents annually."[10]

CONFINED SPACES WITH BAD ATMOSPHERES

Three classes of problems are found with the air in confined spaces: improper oxygen concentrations, flammable or explosive vapors or gases, and toxic vapors or gases. More than one of these problems could be found in a given chamber. Simple portable meters are available that measure oxygen level and level of combustible gases or vapors. Most include one or two other detectors, often CO and H_2S.[11] Recognize that satisfactory readings on such a monitor does not answer questions about specific toxic gases or vapors beyond CO and H_2S, so if chemicals have been associated with this space, additional testing is needed. Workers must be trained on the correct use of these monitors, the manufacturer's recommendations about calibration should be followed, and a calibration log should be maintained.

When testing remotely before entry by drawing the sample through a long tube, the tube should be made of teflon. Other kinds of tubing may absorb some of the toxic gases and so produce a falsely low reading. Testing should be done at several levels, because stagnant air is not necessarily uniform in composition. For example, air directly above an evaporating liquid may have a higher concentration of the vapor. The monitor should accompany the worker, in case of a change in conditions inside the space. OSHA requires that retesting of the atmosphere be done each 4 ft of travel, moving slowly enough to allow the monitor to respond.

Blowers attached to flexible ducts may be used to ventilate an area before attempting to enter. This may correct the problem, but the worker must wear appropriate respiratory protection if it does not. When safe air composition has not been established with certainty, the worker should use a respirator. In any case, emergency respiratory devices such as SCBAs must be available if an emergency arises.

Incorrect Oxygen Levels

The most common problem with air in a closed compartment and the leading cause of death from entering a confined space is that the air has too little oxygen. Oxygen levels are normally 20–21% in the atmosphere. Many people have had the experience of traveling to

[10] OSHA booklet 3138, Occupational Safety and Health Administration, 1993.
[11] 29 CFR 1910.146(c)(5)(ii)(C).

OPEN PITS AS CONFINED SPACES

Hydraulic oil was spilled in a 15-ft deep pit in a machine shop. A worker dumped in a quantity of methyl chloroform (1,1,1-trichloroethane) to dissolve the oil and went down a ladder to mop up the spill. In a short time he collapsed. He died of cardiac arrest on the way to the hospital. Simulating the circumstances later, investigators found that the level of methyl chloroform at the floor of the pit was over 50,000 ppm, perhaps 10,000 ppm at face height with the worker standing on the floor of the pit, but only about 200 ppm at the rim of the pit. 30,000 ppm is considered to be lethal in 5–6 min.

Adapted from W. A. Burgess, *Recognition of Health Hazards in Industry,* 2nd ed., Chapter 5, John Wiley & Sons, New York, 1995.

a high altitude and experiencing fatigue on performing normally simple activities such as climbing stairs. The percentage of oxygen in the air may be normal in those locations, but there is less oxygen because there is less air, so one experiences a problem with inadequate oxygen supply. The point is, physical efficiency is impaired even at levels of oxygen that are lower but not life-threatening. Difficulty breathing is experienced at 14% levels and mental confusion sets in at 12%. A 10% level leads to unconsciousness and 8% to death. The OSHA requirement is for at least 19.5% oxygen for worker entry without a breathing apparatus.

Oxygen can be removed from the atmosphere of a confined space in a number of ways. Oxidative processes such as rusting inside a closed metal tank use up the oxygen supply. Gases entering or generated in a confined space displace oxygen. For example, a liquid residue remaining in a tank can evaporate, displacing the tank air with vapor. The density of the displacing gas is important. A dense gas initially concentrates at the bottom of the space, shifting the original air upward. When sampling the air in a confined space before worker entry, do not assume that a satisfactory oxygen concentration measured only high or low in the compartment establishes safety at all other levels.

A less common problem, oxygen levels of 23% or greater, creates an unusual fire or explosion risk. Substances slow to burn under normal conditions become readily flammable. The OSHA upper oxygen limit is 23.5%.

Flammable Gases or Vapors

Flammable gases, vapors, or dusts constitute the second class of hazard. Tanks or vats that once stored flammable substances and are being cleaned or serviced may well have enough of the substance remaining, particularly residues of a volatile liquid, to create the potential for fire or explosion. The lower flammable limit (LFL) or lower explosive limit (LEL) should be obtained, perhaps from the MSDS, before measuring levels of the material in the space. Before entry such a space may be flooded with a gas that does not support combustion, such as nitrogen, a process termed *inerting*. A possible problem in inerting is to test for oxygen level and decide there is no fire or explosion hazard, then to leak oxygen back into the space during entry.

People associate combustion with liquids, so combustible dusts are less likely to generate concern, but they can constitute a serious threat. Bins containing agricultural products have been known to explode violently on ignition. As a rule of thumb, dust levels sufficient to obscure visibility at 5 ft should be considered dangerous.

Toxic Gases or Vapors

Finally, toxic vapors or gases are a concern. Assaying for the known contents or former contents of a confined space is helpful in selecting the test method, but concern should not be limited to these chemicals. Carbon monoxide and hydrogen sulfide are toxic gases found commonly, and a survey for these and other common contaminants is a wise precaution. Remember that many toxic substances have poor warning properties.

OTHER CONFINED SPACE HAZARDS

Any bin, tank, or silo that contains liquid or flowable solid presents, in addition to questions of the suitability of the air, a possibility of engulfment by the liquid or flowing solid. This should be a consideration before the worker enters the space, and plans for preventing engulfment and for dealing with it if it happens should be in place. Stored grain is a common engulfment hazard. A study by Purdue University documented 232 fatal engulfments in a 30-year period. Their data indicated that about 80% of the incidents led to the death of the victim.

Further, a worker could require removal from a confined space because of heat stress or because of health problems such as a stroke or heart attack.

PERMIT ENTRY SPACES

In January 1993, OSHA defined a systematic procedure for entering a space defined as a permit-required confined space. It is therefore a key decision whether or not this procedure is necessary (Appendix F). OSHA defines such a space[12] as one that:

1. Contains or has a potential to contain a hazardous atmosphere
2. Contains a material that has the potential for engulfing an entrant
3. Has an internal configuration such that the entrant could be trapped or asphyxiated by inwardly converging walls or by a floor that slopes downward and tapers to a smaller cross-section
4. Contains any other recognized serious safety or health hazard

The employer is required to have a written procedure for employee entry into permit spaces, including permit forms, and must provide training to employees who may either enter or supervise the entry of another into a hazardous confined space. Procedures and decisions regarding entry are outlined in a flow diagram (Figure 9.18). All such spaces must be posted with a danger sign indicating that it is a permit space into which unauthorized entry is prohibited.

There is a step-wise procedure when such a space is to be entered. A permit must be filled out and signed by an entry supervisor, who is responsible for ensuring that conditions are acceptable for entry and for terminating the job if they become unsuitable. The employer ensures that the entry supervisor understands the hazards and symptoms displayed by a worker being injured or overcome by those hazards. The supervisor is responsible for correctly fulfilling the points covered on the entry permit and for canceling the permit when the job is completed or terminated for some other reason. The supervisor removes unauthorized individuals who enter or attempt to enter the space during the time the work is in progress.

Blanket permits are not written such that one permit covers entry into a series of confined spaces. Each entry must have its own permit. The permit includes the following points:

[12] 29 CFR 1910.146(b).

APPENDIX A TO § 1910.146—PERMIT-REQUIRED CONFINED SPACE DECISION FLOW CHART

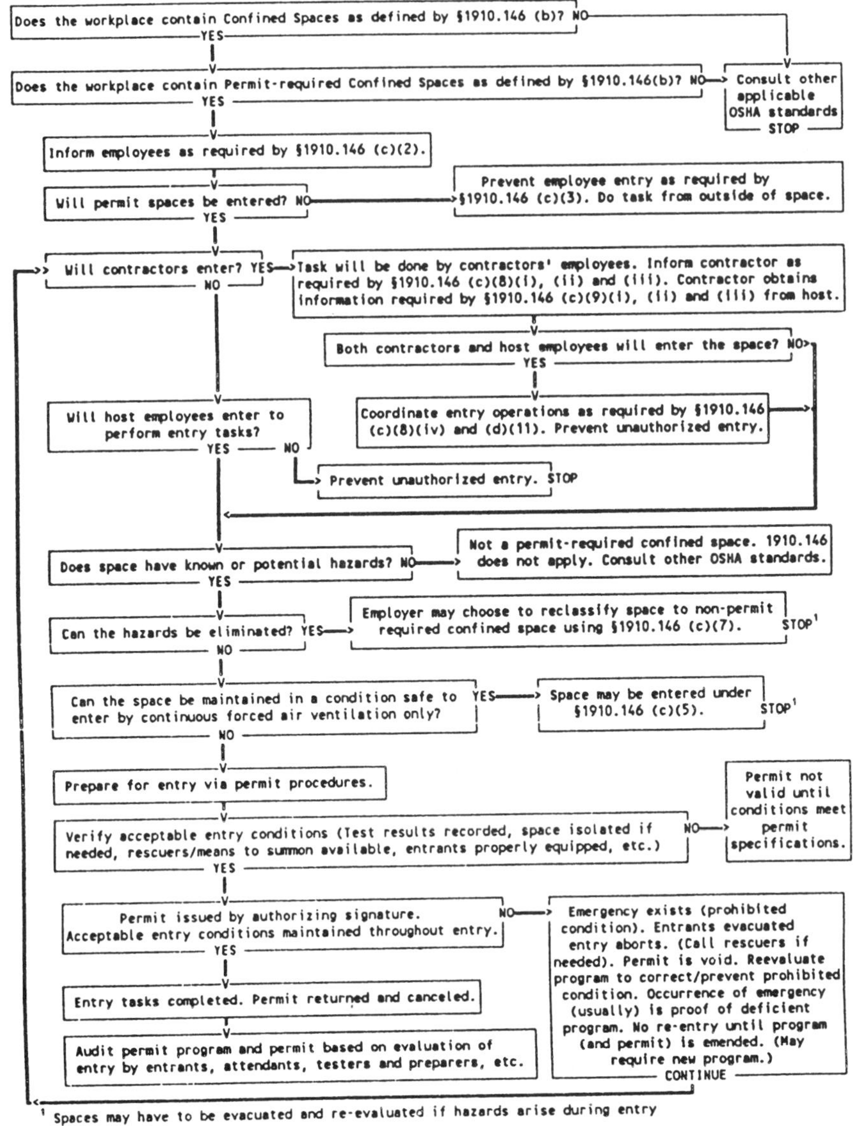

[58 FR 4549, Jan. 14, 1993; 58 FR 34846, June 29, 1993]

Figure 9.18. Permit-required confined space decision flow chart. (From 29 CFR 1910.146, Appendix A.)

1. The specific space to be entered
2. The purpose for entry
3. The date and duration of the entry permit
4. The name(s) of the entrant(s)
5. The name(s) of the attendant(s)
6. The name of the entry supervisor
7. Hazards to be encountered in the space
8. What is done to close off the space before entry, to prevent unauthorized entry
9. Acceptable entry conditions
10. Testing done on the space, signed or initialed by the testers

11. The rescue and emergency services available and how they can be called
12. Procedures for communication between entrants and attendants
13. Equipment for testing, communication, personal protection, and rescue provided
14. Other information
15. Other permits, such as a hot permit

Clearly, the preparation of a permit ensures that the procedure has been studied in all aspects before it is consummated.

The space should be entered only by authorized (trained) workers. The employer is responsible for training these entrants, and the training should be updated whenever there is a change in procedures or conditions that were not part of earlier training. The entrant should understand the hazards and necessary personal protection involved in the entry.

There must be a similarly trained attendant, who remains outside to monitor the entrant(s).[13] The attendant keeps track of who is in the space, and should be in communication with both the entrant(s) and a trained rescue team, should one be needed. The attendant monitors conditions and orders the entrant(s) to leave the space if a dangerous situation arises. The attendant also ensures that no unauthorized individuals enter the space while access is available, and notifies the entry supervisor if someone does enter. Any condition that makes it unsafe to open the space must be eliminated before the space is opened. The attendant may operate rescue equipment from the outside, such as a winch to lift an incapacitated worker from a below-grade space, but must not go into the space to attempt rescue. No duties should be assigned to an attendant that interfere with the above.

A barrier must surround the opening to prevent either workers from accidentally entering or falling through the opening or objects being dropped on employees in the space. The atmosphere must be tested for (in order) oxygen content, flammable gases or vapors, and toxic compounds in the air. If the air is safe, the employee may enter. If not, forced air ventilation must correct the problem before entry of workers.

Retrieval from a Confined Space

The worker may enter the space by a ladder or boatswain's chair, but it may be necessary to remove the worker under conditions where he or she is unable to climb a ladder or even sit in a chair. The regulations state that "Each authorized entrant shall use a chest or full body harness with a retrieval line attached at the center of the entrant's back near shoulder level or above the entrant's head. The other end of the retrieval line shall be attached to a mechanical device or fixed point outside the permit space in such a manner that rescue can begin as soon as the rescuer becomes aware that rescue is necessary. A mechanical device shall be available to retrieve personnel from vertical-type permit spaces more than five feet deep"[14] (Figure 9.19). The standards do not specify harness design, and until they do the standards for fall arrest systems apply.[15] Where the entrance to the space is small, pulling the worker out by the wrists allows easier exit through the opening by narrowing the workers width at the shoulders. The head is also protected by the arms in such a harness.

A worker may be retrieved from a confined space using a winch attached to the harness, but some guidelines are important.[16] For example, in case of a stroke or heart attack, it is important to remove the worker for care within about 4 min. This means that the burden of retrieving the worker falls on the personnel at the entrance, since most often the rescue team would not arrive quickly enough to meet this requirement. It also means that using a hand

[13] The attendant must *never* attempt to enter the space to rescue workers. There are many examples of rescuers becoming victims, leaving no one outside to summon help.
[14] 29 CFR 1910.146(k)(3)(i)–(ii).
[15] 29 CFR 1910.66, Appendix C.
[16] J. N. Ellis, "Suspenseful Suspension," *Occupational Health and Safety*, March 1994, p. 39.

Figure 9.19. A portable hoist lowers a worker into a confined space. In case of emergency, the worker can be raised again quickly. (Photo courtesy of Miller Equipment, WGM Safety Corp., Franklin, PA. With permission.)

winch that raises the worker about 12 fpm is not adequate if the worker is going more than about 50 ft down.

It is important that the attendant know if the worker in a confined space is in trouble. Direct communication systems allow the entrant and attendant to communicate and are adequate for most situations. In a large operation involving several entrants communications could seem normal while one worker, out of sight of the others, is in trouble. Firefighters frequently use a passive backup system, called the personal alarm safety system (PASS). Typically, such a device sounds a warning if the worker does not move for 20 s, then a louder alarm if there is no movement for 30 s.

TRENCHES

Working in a trench is a special case of confined space entry primarily, but not exclusively, encountered by construction workers. A trench may contain a bad atmosphere if it contains a leaking pipe or if chemicals seep through the soil from a buried tank or spill. The special hazard, however, is collapse of dirt trench walls. A worker buried in a deep trench has little chance of survival. Dirt weighs around 100 lb/ft^3, so the mass of earth compresses the victim's chest, making any kind of breathing impossible. Death by suffocation takes about 5 min. It is necessary to anticipate and prevent wall collapse, since rescue after collapse has little chance of success. Recognize that a rope tied around the worker for retrieval is useless once the worker is buried under perhaps tons of earth.

A trench wall may remain stable for a while, leading workers to assume entry is safe. However, heavy equipment near the trench or vibration, as from a nearby highway, may increase the likelihood of a cave-in. Rain may reduce the cohesiveness of the dirt, making the wall more prone to collapse, or the sun drying a stable wall may change its character enough to make collapse likely.

Protection from cave-in is provided by one of three approaches: sloping, shielding, or shoring.[17] Sloping the trench walls reduces the chance of cave-in. Shielding involves placing a reinforced box into the trench. The worker operates inside the box. Finally, shoring involves bracing wood or metal walls against the inside of the trench to hold the soil in place.

INDOOR AIR QUALITY

The chief emphasis in this chapter, and the chief concern in industrial hygiene as a whole, has been and is safety in an industrial workplace where hazardous materials and processes have placed workers at risk of dangerous exposure. Today there is increasing interest in assessing safety in other work settings, especially as employment shifts away from traditional industry toward alternatives such as office settings. OSHA estimates that 21 million workers are subjected to poor indoor air quality.

BUILDING INTAKE OF OUTDOOR AIR

Levels of contaminants in the air of an office building are kept low by three techniques: stopping emissions into the air at the source, filtering or otherwise treating recirculated air, or diluting contaminants with outdoor air. The latter has traditionally been the most significant factor in providing acceptable indoor air, and standards recommended the volume of outdoor air that should enter the building per person per minute. The original ASHRAE indoor ventilation standards (1895) called for a ventilation rate of 30 cfm/person. This was dropped to 10 cfm/person in 1936.

The 1973 oil boycott shocked Americans into considering ways to reduce energy consumption so as to reduce the depletion of resources and to become less dependent on imported petroleum. Lower fuel consumption by automobiles was mandated, and reduction of energy used to heat buildings became an important goal. To improve building efficiency, insulation was installed and leaks at doors and windows were sealed. Ventilation standards regarding the rate of building intake of outdoor air as recommended by ASHRAE were lowered to 5 cfm/person. These initiatives reduced the amount of outdoor air requiring heating or cooling that entered the building.

SICK BUILDING SYNDROME

As the flow of air through offices and public buildings was reduced, a new problem surfaced, which has been termed *sick building syndrome*.[18] Workers in these buildings experienced a range of symptoms including headaches, dizziness, fatigue, irritation of the nose and eyes, and a cough, symptoms that generally cleared when the worker left the building. These problems sometimes result from rising levels of volatile organics in the air, chemicals from cleaning products, pesticides, adhesives, and other chemicals in carpeting and furniture, copying machine chemicals, paints when decorating is done, and other sources.

[17] J. F. Rekus, "Safety in the Trenches," *Occupational Health and Safety*, Feb. 1992, p. 26.
[18] R. Laird, "Sick of the System," *Occupational Health and Safety*, Sept. 1994, p. 65.

Moisture in a building creates breeding grounds for bacteria and molds. Such moisture may result in part from humidifier condensate collecting on air conditioning condenser coils, leading to microbial growth in the ducts. Excessive moisture causes damp insulation, carpeting, walls, or ceilings, which become additional microbial breeding grounds. Distribution of these organisms or their spores in the ductwork can cause allergic response or sickness.

LEGIONNAIRE'S DISEASE

A dramatic and initially mysterious occurrence of disease spread by bacteria breeding in ductwork occurred in 1976 at an American Legion convention in Philadelphia, and was named *Legionnaire's disease*. More than 200 people displayed symptoms and 34 died. The organism, *Legionella pneumophila*, enters the body in inhaled water droplets. The disease is often diagnosed as pneumonia, and a study in Ohio of patients hospitalized with pneumonia suggests that 10,000–15,000 cases occur annually. Careful housekeeping of air conditioning and ventilating systems, giving cooling towers special attention, helps prevent growth of the organisms.

A NIOSH study of 700 problem indoor air cases covering an extended time period (since 1971) revealed that in 65% of the cases a satisfactory explanation of the cause of the problem was not reached. In those cases where the problem was explained, 31% were the result of contaminants drawn into the building. It is important to ensure that it is clean air that enters the building. Intakes should not bring in traffic exhaust or otherwise polluted air. About 9% of the cases were described as due to building materials. The majority, about 60%, were due to indoor contaminants. Of those, almost one-third were attributed to microbial contamination due to excess water.[19] Some of this moisture might have been cleared by greater flow of air through the building. To combat indoor air problems, ASHRAE increased the recommended rate to 20 cfm of outdoor air entering the building per person.[20]

As with many industrial hygiene problems, solving sick building syndrome may require some detective work to determine the exact cause.

CHEMICALS IN INDOOR AIR

Air in office settings is in a different world from the air in a foundry, and chemical contaminants in indoor air seem unimportant by contrast. However, it is also true that people depend on the equivalent of GEV indoors, but seldom consider designing the indoor system to do other than supply adequate outdoor air for dilution. Consider three airborne chemicals: hydrogen sulfide, formaldehyde, and ozone. Systems that can be installed to purify indoor air are listed in Table 9.3.

Hydrogen sulfide (H_2S) has a strong rotten-egg odor, so strong that one can detect it well below dangerous concentrations. Most likely H_2S enters rooms from the sewer line when the water trap in a little-used sewer dries out. Pouring water down the unused drain reestablishes the seal.

Formaldehyde (H_2CO) enters buildings in a number of products. Urea–formaldehyde insulations can release H_2CO, mildew and stain guards used on fabrics contain H_2CO, and H_2CO is present in combustion products. Normally levels are low enough that most building occupants are not bothered, but H_2CO is a strong sensitizer. Allergic individuals have strong reactions to even low levels.

[19] G. L. Ritter, "Indoor Air Quality — What Have We Learned?" *Industrial Hygiene News*, 18(2), 1995.
[20] ASHRAE standard 62-1989.

Table 9.3. Purification of Recirculating Indoor Air.

Method	Type of Purification	Medium or Method
Filter	Removes particulate	Porous medium traps particles
Electrostatic precipitator	Removes particulate	Ionization chamber and collection plate
Adsorption	Gases, liquids, vapors	Active surface adsorbs chemicals (charcoal, silica gel, alumina, etc.)
Chemisorption	Specific chemicals	Adsorbent surface is impregnated with reactive chemicals[a]
Catalyst	Ozone	Converts O_3 to O_2
Antimicrobial filters	Bacteria	Filter coated with antimicrobial agent

[a] Alumina impregnated with $KMnO_4$ oxidizes impurities, Na_2S reacts with formaldehyde, and metal oxides remove H_2S.

Ozone (O_3) is generated by electric motors, especially older motors, and has long been in indoor air from such sources as fans. Today some of the new office machines generate significant amounts of O_3, and these machines sometimes are placed in small rooms with poor air circulation. As a strong oxidant, O_3 can cause lung damage.

ENVIRONMENTAL TOBACCO SMOKE

Because society is increasingly concerned with secondary exposure to tobacco smoke, smoking is being banned in public buildings and workplaces. There had been a tendency to be concerned with exposure to air contaminants generated by use of or processing with chemicals in workplaces, ignoring tobacco smoke as a source of air contamination. However, in January 1993, environmental tobacco smoke was designated by the EPA as an environmental carcinogen. Smoking raises particulate levels in air and tars coat surfaces to produce an odor problem that lasts, as these tars gradually reenter the air long after smoking has ceased.

Banning smoking does not ensure that no one will ever smoke in a building. Smokers with a strong habit are likely to sneak a smoke out of sight. Many plans therefore provide a place where smoking is permitted, usually a room, which must then be well ventilated. This room should have negative pressure, so that air enters rather than leaves the room when the door is opened.

MONITORING INDOOR AIR CIRCULATION

To monitor the adequacy of air circulation it is common to measure levels of CO_2, generally the result of human respiration. Elevated levels of CO_2 are often found in problem offices. Outside air contains around 300–400 ppm. Human occupancy of a room raises the concentration as the people exhale air rich in CO_2. The author was once in a classroom with 12 other adults and a continuous CO_2 monitor. It was winter, and the room was heated with a combination of baseboard hot water radiators, involving no circulation of air, and a conventional HVAC system. With the air circulation turned off, CO_2 levels rose in less than 2 hr to between 1400 and 1500 ppm. Such a level is certainly not IDLH; the action level is 2500 ppm. However, as CO_2 levels become very high, people become less alert and attentive. The recommendation by ASHRAE is necessary to hold the CO_2 level below 1000 ppm.

The primary reason for monitoring CO_2 is as a check on the efficiency of the ventilation system. OSHA considers that values of 800 ppm or more indicate that exhaled air is not being removed at an adequate rate, and may trigger inspection of the ventilation system for correct function. However, to raise the intake of outdoor air from 20 cfm to 30 cfm per person, estimated as necessary to achieve a level of 800 ppm CO_2, would require an increase in energy consumption of 58.77 billion kilowatt hours per year (Janczewski and Caldeira, 1995).

Just as with a GEV system in a factory, enough air may move in an office to produce good air quality, but inappropriate locations for air inlet and exhaust ducts and the surrounding of desks with partitions may result in poor enough air movement in some locations that the air quality is unsatisfactory. Relocation or addition of air ducts or the provision of fans to mix air better can correct these problems.

HUMIDITY CONTROL

Indoor air should have a humidity of between 40 and 60%. During the heating season water normally must be added to the air to achieve this level. Humidifiers operate on the basis of evaporating cool water from wicks or meshes placed in the warm air ductwork, or they inject steam directly into the air. Cold water systems can be sites for growth of molds and bacteria, so steam systems are preferred. If the heating system is based on a steam boiler, that boiler is the likely source of the steam. Chemicals, usually amines, are added to boiler water to prevent acid corrosion of the boiler and pipes. Morpholine, cyclohexylamine, and diethylaminoethanol (TLVs are 20, 10, and 2 ppm, respectively) are commonly used. Maximum air concentrations of the amines resulting from this practice might be as high as 1 ppm, so there is no problem from a toxicity standpoint. However, the odor thresholds for morpholine and diethylaminoethanol are 0.01 to 0.1 ppm, and objectionable odors may be detected by some occupants. A cure is to use a steam generator dedicated to steam production for humidification which does not utilize anticorrosives.[21]

KEY POINTS

1. Good plant maintenance and housekeeping reduce airborne contaminants.
2. Hazard to workers can sometimes be reduced by stationing the operator away from the process.
3. Changing a process to use less toxic chemicals reduces hazard.
4. The removal of contaminated air is accomplished by ventilation. This is essential if the contaminants pose a health, fire, or explosion risk.
5. Air moving through ducts has a positive pressure in a blower duct and a negative pressure in an exhaust duct. It moves with a velocity resulting from the air flow and the capacity of the system.
6. A minimum velocity is needed to keep particles suspended in air.
7. A general exhaust system moves the entire air of the workplace using an exhaust fan. System capacity is measured in room changes per hour. Equations are presented to calculate the maximum limit concentration of contaminant, given the rate of generation and system capacity. Rate of clearance when generation ceases is also determined in terms of contaminant half-life.
8. A good GEV system should ensure that air flow is adequate at the site of contaminant generation and that contaminant is not carried toward workers. Maintenance is important.
9. LEV moves air at the point the contaminant is generated. Advantages include the requirement to move smaller air volumes, avoidance of spreading contaminant throughout the workplace, and minimization of the volume of air to be treated before being exhausted.
10. Local exhaust systems have hoods, ducts, a fan, and possibly an air cleaning device.
11. Hoods provide a range of degree of enclosure from none to complete. Greater enclosure increases efficiency but impedes access to the workstation.
12. Air velocity drops rapidly with distance outside the hood opening.

[21] D. J. Burton, "Humidification: What about Water Additives?" *Occupational Health and Safety*, Oct. 1995, p. 28.

13. A good hood system clears air from all parts of the contaminant source, moves it away from the workers, is able to overcome cross-drafts, and does not interfere with work at the station.
14. Ducts generate resistance to air flow, requiring more energy expenditure at the fan. Round ducts and larger area ducts have less resistance. Intake from multiple sources must be balanced.
15. Fans come in a variety of designs that vary in efficiency and noise level. Regular maintenance is important.
16. Cleaners to remove air particles operate with a certain overall efficiency, but most are more efficient for larger particles.
17. Settling chambers lower air velocity and/or provide baffles to cause larger particles to settle.
18. Cyclone collectors spin larger particles out against the system walls. Efficiency is greater in smaller diameter cyclones or cyclones with dynamic precipitators.
19. A bag house filters out particles through cloth bags.
20. Electrostatic precipitators give particles an electrostatic charge, then attract them to an oppositely charged plate.
21. Vapors and gases are usually removed either by wet scrubbers or charcoal filters.
22. Respirators, which come in designs specific for particular problems, are useful temporary measures to protect workers. These divide broadly into air purifying devices and supplied air devices.
23. Indoor air quality has become a greater concern, given the shift in employment patterns away from heavy industry and toward office work. Sick building syndrome has been the result of inadequate ventilation standards. Tobacco smoke provides special challenges.
24. Monitoring air on a regular basis ensures continued compliance.
25. Testing ventilation systems involves comparing velocities and pressures with system norms.
26. Employee health records must be kept as a check that protection is adequate.
27. Special regulations apply to worker entry into confined spaces. An unsuitable atmosphere, the possibility of engulfment by liquid or flowing solid, and entrapment in the space due to its shape are potential problems.

BIBLIOGRAPHY

29 CFR 1910.94 and 1910.252; regulations specifying ventilation for specific operations.

ACGIH, *Industrial Ventilation — A Manual of Recommended Practice,* 20th ed., Committee on Industrial Ventilation, Lansing, MI, 1988.

W. A. Burgess, M. J. Ellenbecker, and R. D. Treitman, *Ventilation for Control of the Work Environment,* John Wiley & Sons, New York, 1989.

R. P. Garrison, "Ventilation for Contaminant Control," in *The Work Environment,* Vol. 1, Lewis Publishers, Chelsea, MI, 1991.

R. J. Heinsohn, *Industrial Ventilation: Engineering Principles,* Wiley-Interscience, New York, 1991.

Industrial Ventilation Workbook, 3rd ed., IVE, Inc., Bountiful, UT, 1995.

OSHA Instruction CPL 2-2.54, February 10, 1992, "Respiratory Protection Program Manual."

J. N. Janczewski and S. J. Caldeira, "Improving Indoor Air Quality," *Occupational Health and Safety,* Oct. 1995, p. 31.

H. J. McDermott, *Handbook of Ventilation for Contaminant Control,* 2nd ed., Butterworth, Stoneham, MA, 1985.

Occupational Health and Safety, November 1995, has a large section devoted to respirators.

D. E. Risenberg, "Sick Building Syndrome Plagues Workers, Dwellers," *J. Am. Med. Assoc.,* 255, 3063, 1986.

PROBLEMS

1. You are a new industrial hygienist at the original (built in 1924) facility of Acme Foundry and Fabricating in Jackson, Michigan, where machinery beds and transmission housings for heavy industrial machines are produced. In building A we find at various stations (1) molds

being formed of sand, (2) iron being melted in a natural gas–fueled furnace, and (3) casting of iron into the molds. Exhaust blowers are mounted in the roof, and air enters everywhere because the furnaces provide an excess of heat even on cold winter days. Gases from the furnace go up a stack. There is (4) an outdoor yard where castings are cooled. In building B we find in the northeast corner (5) hydrochloric acid baths for cleaning rust from castings stored in the weather and (6) a sand blasting station, in the northwest corner (7) sites where castings are ground to provide a gasket surfaces and drilled and tapped to insert bearings and studs for assembly. In the southwest corner is (8) an assembly operation where gears, bearings, and shafts from a supplier are installed, and in the southeast corner (9) spray painting booths. Two large exhaust fans are mounted on the west wall, and air enters along the east wall where it is heated in winter. Building C (10) is a largely enclosed crating, storage, and loading dock area with diesel-powered forklifts and electric engine–driven hoists to load the finished product onto flatbed diesel trucks.

The company services and modernizes equipment it has built and sold previously. In a separate room of building C (11) there is a degreasing tank used to clean lubricants from castings from used equipment and a workshop to work on machines that are being modified. Each room has a ceiling-mounted exhaust fan and air inlets from the heating system near floor level, or through the open doors when a truck is being loaded.

There is a reasonably good rate of general ventilation throughout all buildings of the facility, but there are short periods of high activity at various stations when this is insufficient. Your boss tells you to buy disposable masks where needed for these various workstations to be used at times when air is contaminated.

A. After inspecting the operation, list the respiratory hazards at each operation (by number) in the plant.
B. In the safety equipment catalog you find the following respirator cartridges:

Catalog No.	Description
35001	O.V.
35002	A.G.
35003	Ammonia and Methylamine
35004	Non-Toxic Dust
35005	Toxic Dust/Mist
35006	Nuisance Odors
35007	Paint Spray

By phoning the supplier you learn that O.V. stands for organic vapors and A.G. stands for acid gases.

1. If you took one of each of these masks and cut it open, what would you expect to see inside?
2. What mask should you buy for each station? Does every station need masks? Do masks handle every problem?
3. When the masks arrive, the boxes are clearly labeled but the masks themselves are all white face covers with straps. Some have one-way valves for exiting exhaled air, but otherwise they look very similar. All are dispensed from the room at the plant entrance near the showers and lavatory where workers get their coveralls. What steps need to be taken?

C. By rearranging existing air inlets and exhaust fans, how could ventilation in building B be improved?
D. Profits are good and some money is going to be spent in plant modernization. Plant safety people ask for some local ventilation and are told to present a listing of sites that would benefit. What additional information would you need to prioritize the list?
E. You have enough money in your budget to buy a self-contained breathing apparatus for emergency use. Where is the best place to house it?
F. What else could be done to improve air quality as plant components and equipment are replaced?

2. A ventilation system is being designed for an industrial building with 3200 ft^2 of floor space and 12-ft-high ceilings. In this facility premanufactured components of home furnace humidifiers are assembled and packaged in corrugated cardboard boxes. There is no point source of hazardous air contamination, so GEV is considered adequate. It is planned to have a fan blowing outdoor air through a heating and cooling unit, with exhaust vents in the roof.
   A. How will pressure in the building compare to atmospheric pressure?
   B. What air flow must be generated by the fan to provide three room changes per hour?
   C. There is a change in the process such that an adhesive dissolved in a chlorinated hydrocarbon replaces some mechanical assembly. The solvent is of low toxicity, but has a strong and unpleasant odor. It is decided to run the system so that the level of the solvent does not exceed 2 mg/m^3 in the air. What would the rate of removal of the solvent be at the maximum concentration?
   D. If solvent is evaporating from the assembly process at 50 mg/min, will the present ventilation system achieve the goal for solvent concentration in the air?
   E. Using the present system and with the given rate of evaporation, what would the C_{max} be?
   F. At night and over the weekend with the adhesive containers closed, what would the $t_{1/2}$ be?
   G. What engineering change would improve the situation without use of LEV?
3. There are plans to install an additional degreasing unit in a room that already has two in operation. The present GEV system is analyzed and is found to move 62.9 m^3/min through the room. The concentration of solvent in the room is (TWA) 3.5 mg/m^3.
   A. What is the average rate of generation of solvent vapor by each of the present machines?
   B. It is proposed to increase the ventilation sufficiently to bring the level of vapor in the room down to 2.5 mg/m^3 at the time of installation of the new degreaser. Will adding a second blower identical to the first be sufficient?
4. A 10,000-m^3 room has a valve leaking tetrachloroethylene, which is adding 50 mg/min of vapor to the room air. The GEV system moves 75 m^3/min of air through the room.
   A. What will be the C_{max} of tetrachloroethylene vapor in the air?
   B. If the leak is stopped and the blower speed is increased so that 100 m^3/min is moved, what will the vapor concentration be in 1 hr?
5. Suppose to capture a particulate coming from a particular process requires a rate of air flow of 20 fpm.
   A. How close must the source be to a roughly square hood of 6 ft^2 and a velocity of air of 190 fpm?
   B. What rate of air flow would be required if the source were 3 ft away from the hood?
6. In the discussion of use of settling chambers to remove large particulate from exhaust air, the addition of baffles to the chamber was described as increasing the efficiency of the system. What price is paid for this addition, ignoring the cost of the baffle installation itself?
7. A large underground storage tank has been used to store fuel oil for furnaces in a plant. It has a single manhole entry point and a narrow above-ground vent. It is leaking and requires repair. The fuel is drained into other tanks and a welder is to be sent into the tank to weld the damaged seam.
   A. Does this task require a confined space permit? Look in 29 CFR 1910.146(b) and decide if a hot permit is needed.
   B. What testing must be done before a worker may enter? In this case special attention must be paid to what hazards this welder could face? What information is needed from MSDS or tables?
   C. List personnel required to be available at the time of entry.
   D. Suppose ventilation does not succeed in bringing the hydrocarbon level below the lower flammability level. What alternative is possible?
8. Discussing the problem of ventilating a smoking room, D. J. Burton, in *Occupational Health and Safety* (Sept. and Oct., 1994), recommended that the room have negative pressure so that air does not move from the room to the surrounding areas, and that the fresh air supply provides 20 air changes per hour or 60 cfm per smoker in the room at maximum use, whichever provides the greatest ventilation. He provides an example of a room $20 \times 30 \times 8$ ft with a maximum occupancy of 25 people. Use of which standard would provide the greater air circulation? What would the necessary air movement be?

9. You are responsible for the entry of workers into a utilities tunnel where a pipe carrying tetrachloroethylene is leaking. Trace a path through the flow chart in Figure 9.18 as the following situations apply:
   A. The space meets the definition of a permit-required confined space (29 CFR 1910.146[b]) in that it contains a hazardous atmosphere. What is the first task you should undertake?
   B. You decide to blow air through the tunnel to eliminate the hazardous atmosphere. What requirements must this ventilation meet?
   C. What part of 29 CFR 1910.146 sets the conditions under which entry is permitted after forced air ventilation is operating?
10. A small industrial building has general exhaust ventilation. The ventilation system exhausts air from the ceiling at one end of the building, while air enters through vents at the other end of the building. The chief source of air contamination is in the center of the room.
   A. What is the advantage of such a system?
   B. What is the chief disadvantage?
   C. What would be the effect of adding ceiling fans to the building?
   D. If it is found necessary to add local ventilation at the source of contamination, what would be gained and how could the cost of such an installation be recovered by changing the operation of the ventilation system?
11. In a plant building military electronics assemblies we find a workstation at which a 50-cm-wide unit, recently cleaned in solvent, is placed in a jig on a bench that has a back panel. The jig is 0.5 m from the back panel. Traces of solvent remain, and it is decided to cut a hole at the intersection of the table and back panel and connect a duct to provide local ventilation. The hood is to be at the intersection of the bench and the back panel. It must be decided whether to have a round opening or a slot of length L. Q is the volumetric air flow (m^3/min). The velocity of air (v) at distance x from the slot or radius x from the center of the circular hole is (Heinsohn, 1991)

$$v = \frac{2Q}{L\pi x} \text{ (slot)} \qquad v = \frac{Q}{\pi x^2} \text{ (circle)}$$

   A. Given constant Q, what is the effect of making the slot twice as long?
   B. If a slot is cut twice the length of the electronic assembly and Q is 20 m^3/min, what is the velocity of air at the center of the assembly?
   C. What is v at the center of the assembly when the hole is circular?
   D. What is the velocity of air at the edge of the assembly at the slotted bench?
   E. What is the velocity of air at the edge of the assembly at the bench with the circular hood?
12. In an article (J. L. Repace and A. H. Lowery, "Indoor Air Pollution, Tobacco Smoke, and Public Health," *Science,* 208, 464, 1980), the authors describe the concentration of tobacco smoke in a small room as rising steadily as the cigarette burns, then dropping due to adsorption to the walls. The drop was slow in an experiment with undisturbed air and rapid in an experiment in which fans mix the air. Explain this difference.
13. Using Equation 8 we can calculate the amount of air necessary to hold the concentration of a vapor at the legal limit. Calculate the rate necessary to vapors of methylene chloride at its TLV when
   G = 8 pt/hr
   SG = 1.336
   K = 5
   TLV = 500 ppm
   A. What is the flow rate necessary to hold this vapor at the TLV?
   B. What would the rate be if the room had twice the volume?
   C. If we were to change room volume, perhaps by doubling the room size without changing anything else, what factor in the calculation might change to reflect this change?
   D. Suppose the location of the fan with respect to the source of contamination were changed, what factor in the calculation might change to reflect this change?

14. Do the calculation of problem 13, where the contaminant is benzene:
    G = 8 pt/hr
    SG = 0.879
    K = 5
    TLV = 10 ppm
    A. What is the flow rate necessary to hold this vapor at the TLV?
    B. Compare the answers for acetone (the sample problem in the text), methylene chloride (problem 13), and benzene. The huge increase in demand for ventilation is due to what single factor in the calculation?
    C. What generalization about GEV can we draw from this?
15. An office building was built at the time when ASHRAE recommended entry into the building of 5 cfm of air per occupant. The planned occupancy was 250 people. The annual cost in that location of conditioning the air entering the building is $\$3.50/ft^3$. Reorganization of building functions raises the occupancy to 300 people, and the ASHRAE recommendation is now increased to 15 cfm/occupant. Supplemental ventilation is needed with what capacity, and what increase in operating costs will be experienced after the equipment is installed?
16. A midwestern building (elevation, 1500 ft) has a general exhaust system serving a room with an 80- × 60-ft floor and 12-ft-high ceilings. The fan, which draws 15 amp on a 220-V circuit, has a pressure drop of 3.30 in. on a water gauge. Data from the fan supplier follows: Mechanical efficiency of fan is 0.65, efficiency of the motor is 0.98, and drive loss factor on the system is 1.12. Estimate room changes of air per hour for this system.

CHAPTER 10

Fire and Explosion

Throughout our lives we hear about fire safety, particularly safety in the home. We are warned not to use space heaters near flammable materials, to maintain our furnaces carefully, and not to store oil-soaked rags in the basement. Schools have fire drills to teach children to escape a burning school building, and every public building has lighted red signs indicating fire exits. Fire has been a menace throughout the history of civilization, and major cities such as Rome, Chicago, London, and Detroit have at some time burned to the ground.

In all industries fire is a concern, and the potential for fire often is increased because particular materials are in use, so that special precautions and understandings are necessary. Actually we have a dual concern, first for the start and spread of a fire and second for the occurrence of an explosion.

FIRE: GASES AND VAPORS

In order to have a fire, two components are required: fuel and an oxidant. Thus in a gas furnace the reaction is

$$\underset{\text{fuel}}{CH_4} + \underset{\text{oxidant}}{2O_2} \rightarrow CO_2 + 2H_2O$$

However, the presence of fuel and oxidant is not sufficient to start combustion. The mixture must be raised to a high enough temperature to start the reaction, usually by a spark or a pilot light flame. Once started, combustion releases enough heat to keep the mixture temperature above that necessary to initiate the reaction.

To understand the process better, pretend you are a tiny (molecule sized) observer in the path of an approaching flame front. You are surrounded by molecules both of fuel and of oxidant that are colliding due to their kinetic energy (the energy of motion proportional to the absolute temperature of the molecules), but whose collisions do not have sufficient energy to initiate the chemical changes of combustion. As the flame front nears, energy released by the combustion radiates ahead of it, and the molecules around you begin to move faster, colliding harder and more often. The closer the flame front, the greater the increase in temperature. At some temperature (the ignition point) the collisions around you are energetic enough to initiate combustion, the combustion reaction starts, and now the molecules around you are the source of released energy. The fuel is consumed, the flame front moves by, and you are left surrounded by fast-moving combustion products.

It is possible to have a fuel and an oxidant, but in proportions such that the process cannot occur continuously. If the fuel is a small amount of gas or vapor mixed with air and we initiate combustion with a spark, the small amount of heat produced by burning might be insufficient to ignite adjacent fuel, and continuous combustion fails to happen.

Most people accept the idea of fire not starting because of inadequate fuel, but many are troubled by the concept that there can be too much fuel. However, too much fuel is really insufficient oxygen, and once again the heat produced when a spark ignites some fuel is not enough to ignite adjacent fuel. If you have trouble with this concept, think of the last time you tried to light charcoal, admittedly a solid fuel, on your grill. You had a great deal of charcoal fuel and you generated heat by burning lighter fluid, but the charcoal was stubborn until . . . you blew on it, increasing the oxygen supply. Once it was burning the hot combustion products rose rapidly upward and new air rushed in to replace them, providing enough oxygen to continue combustion.

For each flammable chemical we can determine flammability limits, concentrations of gas or vapor in air that are at the low or high extremes of ability to sustain continuous combustion. There is a lower flammability limit (combustible-lean limit mixture), a concentration in air, usually expressed as volume percent (vol%), below which the fuel–air mixture does not sustain continuous combustion. The upper flammability limit (combustible-rich limit mixture) is the highest concentration of fuel in air that burns continuously. Such values should not be treated as absolute physical constants such as melting point or density of a pure substance at 25°C, but rather as guidelines when establishing criteria for safe conditions. For example, flammability limits vary with temperature, pressure, and conditions of combustion. Most importantly, it should be remembered that a closed container whose fuel concentration is above the upper flammability limit will not ignite, but opening the container and allowing more air to enter may produce a flammable mixture.

Another variable arises from the "layering" or separation by density of gases or vapors in air. Light gases such as hydrogen or methane become more concentrated at upper levels in a container, whereas dense gases or vapors such as butane or benzene concentrate at the bottom. A dense gas has a molecular weight greater than the 28–32 kDa of nitrogen and oxygen. A container may overall contain a concentration of a gas or vapor fuel lower than the lower flammability limit, but still have the potential to burn at the top or bottom of the container.

EXPLOSION

An explosion, as opposed to a fire, occurs when the rapid expansion of gases produces high pressures. Fire and explosion may be two sides of the same coin where both are based on a strongly exothermic (heat-producing) chemical reaction, usually combustion in air. An explosion is more likely when the fuel–oxygen mixture is midway between the flammability limits.

Returning momentarily to our microscopic observation of burning, if, because of the character of the fuel–oxidant reaction, many more molecules were present after burning than before, the sudden appearance of hot new molecules rushing and colliding at high speed produces a region of very high pressure. These molecules rapidly move from the zone of high concentration (pressure) at the site of the reaction, both heating and pressing together the next layer of fuel and oxidant, causing it to ignite very quickly. The rapid progression of this pressure and heat is the explosion. Such pressures can result in structural damage and are life threatening. The difference between a fire and a combustion explosion thus lies in the rate of generation and the amount of gaseous product.

PREDICTING HAZARD

Knowing the flammability limits and other characteristics of a chemical allows us to predict the potential for a fire starting. In the absence of a table of flammability limits or when dealing with a compound that is not on the table, we may still be able to estimate or be aware of the hazard of an organic liquid. For example, on the standard diamond-shaped hazard code placed on containers of chemicals, the red upper quadrant warns of combustion hazards. Most organic liquids are flammable, the noteworthy exceptions being chlorinated or otherwise halogenated hydrocarbons.

Industry has been strongly induced to use chlorinated hydrocarbons as solvents because of their inability to burn. The key here is the replacement of carbon-to-hydrogen bonds by more stable carbon-to-halogen bonds. One chlorinated hydrocarbon, carbon tetrachloride, was commonly used as a fire extinguisher fluid before its toxic characteristics were recognized. In its place we sometimes find the chemically similar freons being used as extinguishers.

Volatility, the tendency of molecules of a liquid to enter the vapor state, is another important liquid property related to fire or explosion hazard. In the vapor state these molecules contribute to the atmospheric pressure in proportion to their numbers. Given time to equilibrate to a maximum value, this contribution to total pressure is called the vapor pressure of the liquid. At a given temperature the vapor pressure over a pure sample of the liquid is a constant. Higher temperatures increase the proportion of the liquid molecules with enough energy to leave the liquid, and consequently raise the vapor pressure. When the vapor pressure equals the atmospheric pressure above the liquid sample, the liquid boils. Consequently, a liquid with a relatively high vapor pressure has a relatively low boiling point.

If a liquid has high volatility or vapor pressure, and consequently a low boiling point, this is a warning sign for greater fire or explosion risk. If such a liquid is given an opportunity to evaporate, it enters the atmosphere more rapidly and reaches a higher concentration than one of lower volatility. The flashpoint of the liquid, the most frequently encountered measure of hazard due to flammability, is the minimum temperature that raises the vapor pressure to the lower flammability limit, thus that provides sufficient vapor for combustion on ignition. A flammable liquid therefore should never be stored in an unventilated room or cabinet in which the temperature might exceed the flashpoint. A combination of a leaking container and a spark or other means of ignition could result in a fire or explosion. Tanks assumed empty that once contained volatile combustible liquids may have enough residual inflammable material to produce an explosive mixture of vapor and air. A welder was recently killed when striking an arc to repair the tank of a gasoline hauling truck that had not been properly purged.

The ignition temperature (autoignition temperature) is the temperature to which the vapor–air mixture must be raised to initiate combustion without a spark or other artificial trigger. Values for these constants for some common solvents are listed in Table 10.1.

COMPRESSED GASES

Explosions are generally classed as any rapid expansion of gases, and are not limited to combustion. An explosion could be the result of failure of a container of gas stored under high pressure. Cylinders of compressed gas are common in manufacturing facilities, for example, being used to store gaseous reactive chemicals, fuels, gases for welding, and breathing air for respirators.

Gas cylinders should be fastened so they will not tip or roll (Figure 10.1). They should be capped when not in use. Flammable gases and oxidizers are not compatible and should be separated for storage. OSHA's general requirements (29 CFR 1910.101) refer to documents by the Compressed Gas Association (P-1-1965; S-1.1-1963; 1965 addendum; and S-1.2-1963).

Table 10.1. Flammability Data for Common Solvents.

Solvent	Flashpoint		Boiling point		Autoignition Temperature	
	°C	°F	°C	°F	°C	°F
Acetone	−16.7	2	57	134	604	1118
n-Amyl acetate	—	77	149	300	379	714
n-Amyl alcohol	57	134	138	280	327	621
Amyl chloride	3	38	106	223	260	500
Anthracene	121	250	340	644	540	1004
Benzene	−11	12	80	176	580	1076
n-Butyl acetate	39	102	127	260	421	790
n-Butyl alcohol	47	116	117	243	367	693
Butyl cellosolve	61	141	171	340	244	472
Isobutyl alcohol	22	72	107	225	441	825
Carbon disulfide	−22	−8	40	114	125	257
Cellosolve	40	104	135	275	238	460
Cellosolve acetate	51	124	156	313	379	715
Chlorobenzene	32	90	132	270	640	1184
Cyclohexane	3	37	80	176	245	473
Dibutylphthalate	157	315	366	690	403	757
1,2-Dichloroethylene	17	63	84	183	413	775
Diethyl ether	−41	−40	35	95	186	366
Ethyl acetate	−4	25	77	171	486	907
Ethyl alcohol	14	57	78	173	426	799
Ethyl chloride	−50	−58	12	54	966	—
Ethylene glycol	111	232	197	387	413	775
Ethyl formate	−19	−2	54	130	455	851
Furfural	56	133	161	322	316	600
n-Heptane	—	25	98	208	233	452
n-Hexane	—	7	69	156	247	477
Kerosene	55–73	100–165	151–301	304–574	210	410
Methyl acetate	−13	9	60	140	502	935
Methyl alcohol	0	32	64	147	475	887
Methyl *n*-butyl ketone	23	73	128	262	533	991
Methyl cellosolve	47	107	124	255	288	551
Methyl cellosolve acetate56	132	143	289	—	—	
Methyl cyclohexanone	55	130	163	325	595	1102
Methyl ethyl ketone	−7	19	80	176	516	960
Naphtha V.M.P.	−16.1–6	20–25	100–161	212–320	232	450
Naphthalene	79	174	218	424	527	979
Octane	13	56	125	257	220	428
Paraldehyde	27	81	—	—	283	541
n-Propyl acetate	14	57	102	215	842	1005
Isopropyl acetate	8	46	90	194	460	860
n-Propyl alcohol	22	72	97	207	433	812
Isopropyl alcohol	12	53	83	181	456	852
Toluene	4	40	111	232	552	1026
o-Xylene	30	85	144	291	493	920

Reprinted with permission from *Fire Protection Handbook,* 17th edition, ©1991 National Fire Protection Association, Quincy, MA 02269.

DUST FIRES AND EXPLOSIONS

When solid particles are suspended in air, and when these particles are capable of reacting with oxygen, there is a possibility of a fire or explosion. Such combustions have been most threatening in coal mines, and the ignition of coal dust has been more carefully studied than

Figure 10.1. Cylinders containing gas at high pressure present an explosion risk. They should be fastened firmly to a bench or wall to prevent the valve from being snapped by a fall. A loose tank with a snapped neck is much like a rocket.

many other dust hazards. Fine plant fiber in the air of agricultural storage bins is another common dust fire or explosion hazard.

There are ways in which the dust problem is dealt with that are very similar to the approach used with vapors and gases, and other ways in which the approach is quite different. It is possible to define lower and upper flammability limits for dust particles just as was done with vapors and gases. However, there is more uncertainty involved in making predictions about the degree of dust hazard. In a gas or vapor, the combustible chemicals are present as individual molecules with uniform access to oxygen. This lends gas and vapor hazards a degree of uniformity and predictability. In the case of dusts, the particle size and shape are important variables. Smaller particles have a higher percentage of the fuel molecules at the particle surface and thus exposed to oxygen. Similarly, flattened particles present a greater surface area than do spherical particles of the same mass. Where oxygen can contact the fuel molecules more readily, the dust will react more rapidly, increasing the tendency for an explosion to occur.

In general, dusts burn more slowly than vapors or gases because most of the fuel molecules are buried at the start of combustion and only become available as the reaction proceeds. On the other hand, the total amount of fuel present in dusts can be great, so that the total production of heat and gases can be larger than would occur in a gas or vapor burn. This means also that it does not always require a high concentration of dust in the air to create a hazardous situation.

The total fuel available also may increase as a fire or explosion begins. A process producing dust may well result in layers of particles lying on the ground or floor, and the expanding gases at the start of the combustion may add these to the air. For example, in a coal mine there may be only a small amount of dust in the air, but a good supply on the floor of the mine shaft. Stirring this up into the air may produce the conditions for a powerful explosion. In fact, the initial burn may be something other than coal, methane perhaps, but the heat and expanding gases may then generate a coal dust fire.

Fires or explosions may occur when one is deliberately collecting (therefore concentrating) dust, as in a filter or cyclone collector. Here a spark from a switch or electric motor can initiate ignition.

Finally, the violence of the burn depends on the nature of the reaction. Metal powders oxidize in air, producing metal oxides and actually reducing atmospheric volume by removing oxygen from the air. Expanding gases are the result of the heat released as the reaction occurs. A carbonaceous dust such as coal or plant fiber also uses up oxygen from the air, but it releases volumes of gaseous combustion products such as CO, CO_2, and H_2O. These add to the gas pressures produced, increasing the destructiveness of the combustion.

EXPLOSIVES

Some chemicals contain both fuel and oxidant and rapidly produce volumes of gas without need of oxygen. Such explosives are a special case, and conditions of proper handling are specific for each chemical. We will not attempt a general discussion of this subject, but strongly recommend that any use or handling be done only after a person has become well informed about the peculiarities of the particular compound.

COMBUSTION PRODUCTS

Hazard from a fire goes beyond the heat and physical destruction. Fire is a chemical reaction, so the reaction products are also a concern. Often the hot gases of the fire are swept upward and away and new air rushes in to replace them. If the fire is in an enclosed space such that there is no departure of gases from the combustion and no intake of new air, breathing that air may become hazardous. Combustion products depend on the nature of the fuel and can present a variety of health problems.

CARBON DIOXIDE

Most fuels are carbon based, and complete combustion generates CO_2. Although CO_2 is not highly toxic, as an acid anhydride it shifts blood pH to lower values, which results in a stimulation of deeper and more rapid breathing. At very high levels this effect attains serious proportions. Recognize also that the CO_2 is produced by using up the oxidizer O_2. A fire can seriously deplete air of oxygen and create the possibility of asphyxiation.

CARBON MONOXIDE

Incomplete combustion of carbon fuel results in CO production. The proportion of CO produced rises sharply as the supply of oxygen to the fire is restricted. Because CO can be oxidized further, it is a fuel. Opening the door to a room in which a fire is smoldering so that new oxygen is introduced can result in explosive combustion of accumulated CO, a serious hazard to firefighters.

Carbon monoxide is also highly toxic (Chapter 7). Levels higher than 0.1 vol% can deliver a fatal dose in a few hours, and above 1 vol% in a few minutes. This is the most common highly toxic combustion product associated with a fire.

UNBURNED PARTICULATE

Smoke from a fire contains particles of unburned carbon fuel. In the lungs irritation can be extreme enough to cause fluids to form and collect, blocking respiration. This effect is also produced by the other respiratory tract irritants discussed later in this section. Smoke inhalation is the cause of a very high percentage of the fatalities in a fire. Smoke also obscures visibility, increasing dangers to those in the vicinity of the fire.

SULFUR DIOXIDE AND HYDROGEN SULFIDE

Fuels that contain sulfur, most significantly rubber from tires, produce SO_2 on complete combustion. This is an acid anhydride, like CO_2, but is the anhydride of a stronger acid than is CO_2. As such it is an upper respiratory tract irritant, so that exposed individuals are very aware of its presence and do everything possible to avoid inhaling it.

Hydrogen sulfide is the product of incomplete sulfur combustion. A strong rotten-egg odor is its warning property, but at high concentrations it rapidly overcomes the sense of smell. In the presence of additional oxygen, H_2S burns to form SO_2.

PRODUCTS OF CHLORINE COMBUSTION

Although fully chlorinated hydrocarbons are difficult to oxidize, partially chlorinated structures are more combustible. Further, if they are mixed with vapors of a good fuel, the high temperature of burning gases breaks down chlorinated hydrocarbons to release a variety of products. A variety of compounds containing chlorine result from such combustion.

Hydrogen chloride is a potent upper respiratory tract irritant, so people try hard not to inhale it. As a strong acid it is capable of serious damage to the lungs if allowed to accumulate.

Phosgene is a much more serious decomposition product of chlorinated hydrocarbons. It is not as irritating, so it is more readily inhaled. Continued exposure to as little as 25 ppm can be lethal.

HYDROGEN CYANIDE

Combustion of a variety of substances, particularly silk and wool, several common plastics, and agricultural chemicals, can result in production of HCN. This is a very toxic substance, and even doses in the range of 100 ppm can cause an accumulated blockage of oxygen utilization in the body that is harmful to lethal in less than 1 hr.

METALS

Fires that encompass electrical equipment may volatilize some of the low melting metals used in solders. Some of these, particularly lead and antimony, represent serious health threats.

BURNS

Employees may receive burns in the event of a fire or explosion. This is a very serious type of injury and requires immediate first aid and rapid transfer to an emergency care center.[1] It is important to have first aid supplies available and personnel trained in the first aid procedures.

Burns fall into one of three classes according to their severity. First-degree burns are like a serious sunburn. Damage is limited to the epidermis; the skin appears reddened, but turns white on compression. Second-degree burns involve the dermis too. There are blisters and the skin feels moist and soft due to the destruction of the tissue and leakage of body fluids normally blocked by the skin. In third-degree burns, the skin is completely destroyed, and the body surface appears eroded. The surface has a hard, leathery feel. The percentage of the body surface burned is also an important statistic.

The first care given a burn victim is to cool the burned area, generally with water. The burns are covered with sterile dressing (remember that the barrier against infection provided by skin has been breached in more severe burns). A product recommended by the authors referenced in the footnote is a dressing specifically for burns called Water-Jel™ Sterile Burn Dressing, which combines the water cooling with the sterile dressing. It is important to look for other problems when treating a burn victim. When the face is burned, as is often the case, the person providing treatment should be alert for respiratory distress resulting from damage to the respiratory tract due to inhalation of hot gases. Then, as quickly as possible, the person should be transferred to an emergency care facility.

FIRE CONTROL

The methods used to extinguish a fire, once started, fall into two classes: deprive the fire of oxygen and cool the fire below the ignition temperature. Most fire extinguishers operate by excluding oxygen from the site of combustion. This may be done with a gas such as CO_2, commonly used in portable fire extinguishers. Such extinguishers spray a cloud of CO_2 particles, which convert to gas and blanket the area, excluding oxygen. Sizable amounts of heat are required to convert the dry ice to gas, so the site of combustion is also cooled. Other systems to exclude oxygen use nitrogen gas or freons.

The chief example of cooling below the ignition temperature is the fireman's traditional method, spraying the flames with a stream of water. Not only is water commonly available, but its high specific heat makes it an effective cooling agent. Steam produced as the water hits a hot surface also serves to displace oxygen. As an electrical conductor, water would not be used on an electrical fire, and there is a danger of scattering and spreading the fire if water is sprayed on burning liquids.

Foams are sprayed on fires to exclude air. The foam is generated in the extinguisher using water and a foaming agent.

Fires have been classified by the National Fire Prevention Association (NFPA) into four groups:

[1] T. Scully and B. Proctor, "Prompt Diagnosis, Treatment Critical in Workplace Burn Emergency Response," *Occupational Health and Safety*, March 1994, p. 80.

Class A. Fire in ordinary combustible materials such as wood, cloth, paper, and plastic.
Class B. Fire in flammable or combustible liquids.
Class C. Fire in electrical equipment carrying power.
Class D. Fire in combustible metals.

Extinguishers are marked with letters according to the class or classes of fire for which they are effective. These letters may be preceded by a number that gives a clue as to the magnitude of fire they are designed to handle — the larger the number, the bigger the fire. Extinguishers should be inspected and maintained according to a regular schedule to assure their ability to function (Figure 10.2).

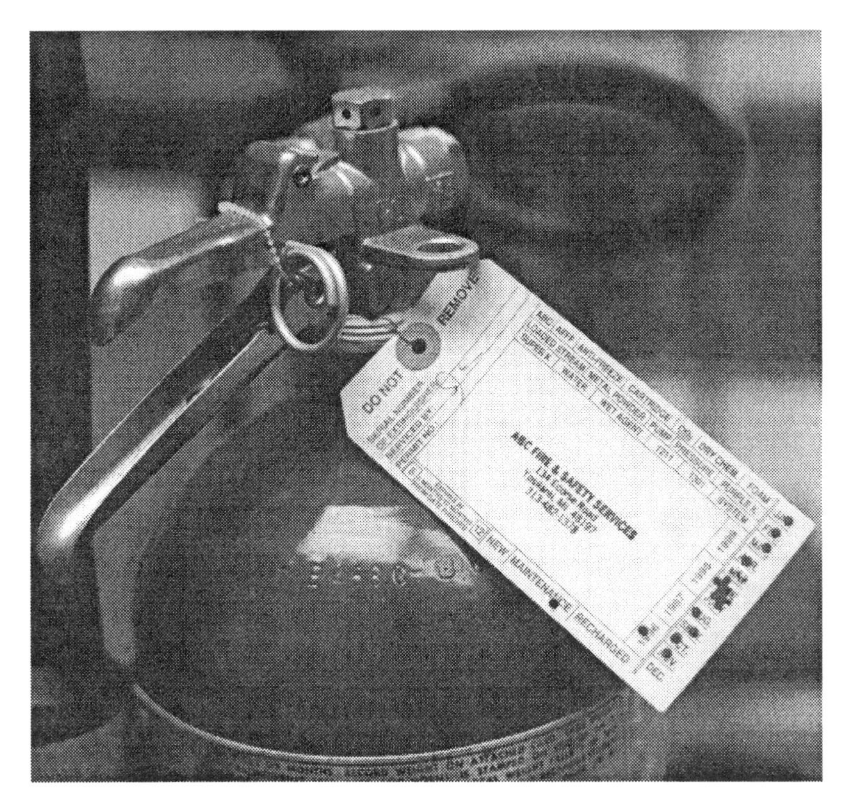

Figure 10.2. An empty fire extinguisher is useless in an emergency. The thin plastic thread on the pull pin must be broken to squeeze the trigger, so an intact thread ensures that the extinguisher has not been fired since the last inspection. Notice the inspection card is punched monthly by the fire inspector.

A porous barrier such as a screen, a flame arrestor, is often placed across air vents or access ports to containers of inflammables. If ignition occurs outside the container, the screen cools the gases, preventing combustion from starting inside the container.

Reducing the hazard of working with combustible substances requires recognition and control of sources of ignition. Electrical sparks, flames, hot wires, and static electricity are common. Glowing fragments from friction on a metallic object could be produced during grinding, bearing failure, broken parts dragging from a moving object, and a variety of other sources.

Storing solvents in cabinets that restrict the broadcast of vapors or the spread of fire, should one start, is useful (Figure 10.3A). Such cabinets may be vented as a further precaution (Figure 10.3B). Structures or buildings that store combustible liquids or gases may need to be placed far enough away from other operations that any fire or explosion occurring does

Figure 10.3. (A) Special cabinets are available for solvent storage. Vapors are confined and isolated from an ignition source. Should a fire start, the cabinet restricts the air supply, limiting combustion.

not threaten other personnel or operations. Unbreakable solvent cans for larger amounts of solvent reduce the chance of spills (Figure 10.4).

FIRE SAFETY IN BUILDINGS

Safety of building occupants in case of fire is a concern for fire marshals and building inspectors, and is a responsibility not usually assigned to industrial hygienists. OSHA is increasingly looking at this aspect of safety, and may cite employers for violations, so it deserves at least brief mention. A safe factory or office building is equipped with fire alarms, automatic sprinklers, smoke detectors, and fire hoses. In the event of a fire or power failure, emergency lighting (Figure 10.5) and illuminated exit signs are needed to assist occupants of a building to leave safely. Such lighting usually operates on rechargeable batteries equipped with chargers, and should be capable of supplying light for whatever time periods are necessary for escape from the particular facility. Standards are written for such lighting by the NFPA and the National Electrical Code (NEC). These standards require recordkeeping and monthly testing of systems.

KEY POINTS

1. A fire requires fuel and an oxidant in correct proportions, and a temperature high enough to initiate the reaction. Correct fuel mixtures range between lower and upper flammability limits.
2. Explosion is the rapid expansion of gases, generally resulting from a chemical reaction.

Figure 10.3. (Continued). (B) Such cabinets may also be vented through a hood, as shown here, or through the plant exhaust ventilation.

3. A liquid that enters the vapor state in high concentration under specified conditions is relatively volatile. A volatile liquid has a high vapor pressure under existing conditions. A liquid with a relatively high vapor pressure has a relatively low boiling point.
4. The flashpoint of a liquid is the temperature at which the vapor pressure produces a concentration of vapor equaling the lower flammability limit.
5. The ignition temperature is sufficiently hot to ignite a fuel–oxidant mixture.
6. Dust can serve as fuel for a fire or an explosion.
7. Explosives contain both fuel and an oxidant, and require no further reactants to produce volumes of gas.
8. Carbon dioxide, carbon monoxide, smoke, sulfur dioxide, hydrogen sulfide, hydrogen cyanide, and chlorinated compounds are potentially toxic combustion products. Metals can be volatilized by combustion or explosion.
9. Fires are extinguished by cooling fuel below the ignition temperature and/or by excluding air.
10. Fires are prevented by eliminating sources for ignition and by correct storage of flammable liquids.
11. Emergency systems in case of fire include fire alarms, automatic sprinklers, smoke detectors, fire hoses, and emergency lighting and exit signs.

Figure 10.4. Solvent storage cans are used to store larger but still portable amounts of solvent. The snap-shut cap ensures that the solvent is not left to evaporate into the room, and the metal or plastic construction will not break if dropped.

BIBLIOGRAPHY

P. A. Carson and C. J. Mumford, *The Safe Handling of Chemicals in Industry,* Long Scientific and Technical, Essex, England, 1988, Chapter 4.

J. Grumer, "Fire and Explosive Hazards of Combustible Gases, Vapors, and Dusts," in G. D. Clayton and F. Clayton, Eds., *Patty's Industrial Hygiene and Toxicology,* Vol. 1, Part B, 4th ed., John Wiley & Sons, New York, 1991.

W. Hammer, *Occupational Safety Management and Engineering,* 3rd ed., Prentice-Hall, Englewood Cliffs, NJ, 1985, Chapters 20 and 21.

H. R. Kavianian and C. A. Wentz, Jr., *Occupational and Environmental Safety Engineering and Management,* Van Nostrand-Reinhold, New York, 1990, Chapter 5.

National Fire Protection Association, "Fire-Hazard Properties of Flammable Liquids, Gases, and Volatile Solids," NFPA Document 325M-69.

National Fire Protection Association, "Flammable and Combustible Liquids Code," NFPA Document 30-78.

National Fire Protection Association, "Standard for Automatic Fire Detectors," NFPA Document 72E-74.

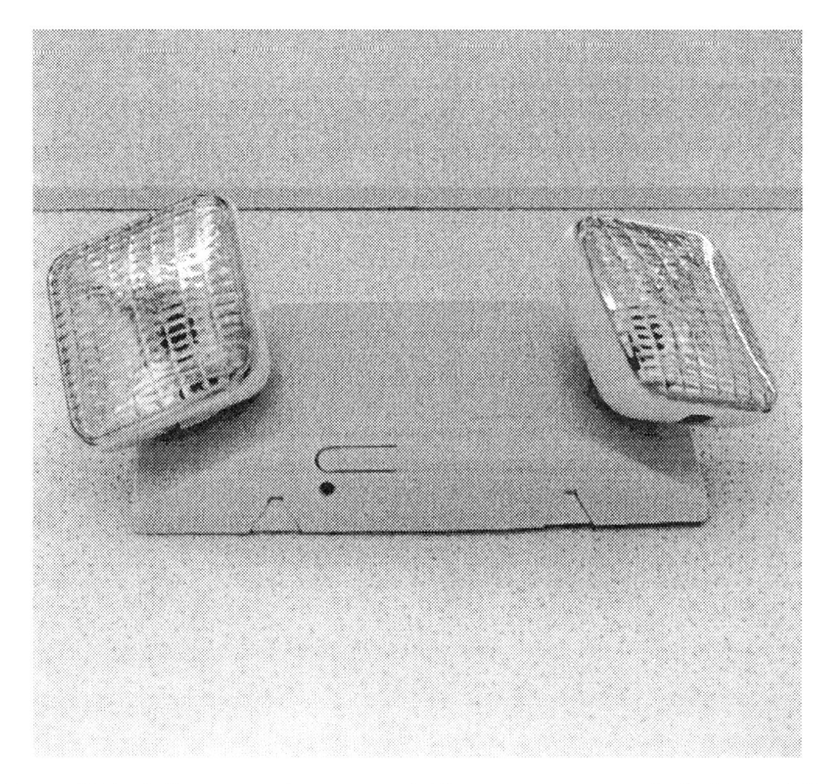

Figure 10.5. Battery-powered emergency lighting comes on automatically if power fails.

National Fire Protection Association, "Standard for the Storage and Handling of Liquefied Petroleum Gases," NFPA Document 58-76.

N. I. Sax, *Dangerous Properties of Industrial Materials,* Van Nostrand-Reinhold, New York (latest edition). (This is a comprehensive source of information about hazards, including fire and explosion hazards, of chemicals.)

PROBLEMS

1. Regulations regarding fire and explosion prevention are found in CFR 1910.106 Flammable and combustible liquids, 1910.107 Spray finishing, 1910.108 Dip tanks, 1910.109 Explosives, and 1910.110 Storing LPG. Look at this section of CFR and get a sense of the extent and amount of detail included. Notice it uses more than 100 pages of the volume.
   A. In CFR 1910.106(a)(5) we find the definition for purposes of these regulations of boiling point of a liquid. Notice that it is defined at 1 atm. Since boiling point is a physical constant, doesn't this seem to be an unnecessary restriction to place on the definition?
   B. In CFR 1910.106(a)(14) flashpoint is defined. Is the definition provided consistent with the definition in the text? Read how the flashpoint is to be determined. Why is so much detail provided? What does ASTM stand for?
   C. In CFR 1910.106(a)(18) and (19), how does flammable liquid differ from combustible liquid? Which presents the higher fire hazard? What is a Class 1B liquid? What is the value of making such distinctions?
   D. You are inspecting a "tank farm", a storage yard for above-ground tanks of flammable liquids. Of what material should the tanks be constructed? If the tanks are 21 ft in diameter and are not storing LPG, how far apart should they be?
2. A tank contains vapor or gas at an overall concentration below the lower flammable limit. However, the possibility exists that the vapor might accumulate at the top or at the bottom

of the tank due to layering in a concentration high enough to ignite. For each of the following compounds, would the hazard exist at the top, at the bottom, or not at all?

A. $CH_3(CH_2)_6CH_3$
B. CH_4
C. CCl_4
D. Ne
E. CH_3–CH_2–CH_2–SH
F. CO_2
G. CH_3-CH_3

3. Considering the possibility of a dust fire or explosion in an agricultural storage bin:
   A. Would an air suspension of small or of large plant fibers be more likely to explode, if the total mass of fiber per cubic meter were the same?
   B. How would a fire or explosion differ if the number of particles per cubic meter were the same, but in one case the average particle mass were greater?
   C. Would it lower the risk if air were constantly circulating as opposed to having a sealed bin?
   D. We can determine a lower flammability limit for dust in the same fashion as for a vapor or gas. How would it affect the limit if in two cases the concentration, composition, and size of the particles were the same, but in one they were flattened and in the other they were spherical?
4. You are engaged in the design of a new explosive. What would be the requirements for the chemical?

CHAPTER 11

Protection from Chemicals in Special Situations: Large Chemical Operations, Hazardous Waste Site Cleanup, and Emergency Response

LARGE CHEMICAL OPERATIONS — PROCESS CONTROL

We have considered a variety of hazards of chemicals from the standpoint of worker health and safety: toxicity upon entry into the body, irritation and damage to skin and eyes, damage to lungs and asphyxiation, and danger from fire or explosion. The OSHA response to these chemical hazards is directed at individual hazards, and takes the form of exposure limits to airborne chemicals (PELs), requirements for protective clothing and respiratory equipment at times of defined worker hazard, and the like.

In 1992 OSHA took quite a different approach to worker protection, one that focuses on processes using large quantities of hazardous chemicals, whatever their individual hazards may be. Called "Process Safety Management", it is found in 29 CFR 1910.119 (see Figure 11.1). Sometimes called a "holistic" approach to health and safety,[1] this regulation has the purpose of "preventing or minimizing the consequences of catastrophic releases of toxic, flammable, or explosive chemicals." It calls for industry to anticipate worst-case scenarios, to prevent them if possible, and to minimize their consequences if they do occur. This regulation differs from the usual OSHA regulatory approach by allowing industries some flexibility as to how they achieve the stated goals.[2]

PROCESSES COVERED BY THE REGULATION

The regulation first defines processes that must meet its requirements. Chemicals must present a high degree of hazard and must be present in large quantities. This avoids nuisance regulation, involving special paperwork for an operation using a small amount of a hazardous chemical, use of which would already be covered by other regulations.

Some specific chemicals covered and the quantity of each chemical defined as "large" are listed in Appendix A of the regulation (Table 11.1). In that table each chemical is identified by its CAS number to avoid any misunderstanding, and the amount involved that triggers regulation is termed the threshold quantity (TQ). In addition, flammable liquids or gases

[1] S. J. Kuritz, "A Holistic Approach to Process Safety," *Occupational Health and Safety,* Oct. 1992, p. 28.
[2] D. Barnes, "Process Safety: Dow Chemical USA Finds the Standard's Positive Side", *Occupational Health and Safety,* Oct. 1992, p. 36.

Figure 11.1. A major facility that handles large amounts of chemicals, such as the Dow Chemical Michigan Division, must meet the special regulations of 29 CFR 1910.119. (Photo courtesy of Dow Chemical Company. With permission.)

present in one location in quantities in excess of 10,000 lb require the site to comply with this regulation, unless (1) they are hydrocarbon fuels not used in a process (heating fuel, fuel for vehicles), (2) they are flammable liquids whose boiling point is not exceeded without special cooling, (3) they are stored at a fuels retail facility, (4) they are part of an oil or gas drilling or servicing operation, or (5) they are at an unmanned remote facility.

HAZARDS AT THE FACILITY

The employer must assemble a document describing the facility and including its hazards or potential hazards. This includes a description of the technology of the process. A block diagram or its equivalent outlines the workings of the process (Figure 11.2). This diagram displays the flow of chemical through the facility, indicating what happens to the chemical at various locations, for example, heating, distillation, and reaction. The document lists any chemical reactions performed.

A description of the facility provides more detail than the block diagram, including materials of construction, a piping and instrument diagram, electrical systems, ventilation systems, material and energy balances during operation (newer facilities), and safety systems such as pressure relief, alarms, and monitors.

For each hazardous chemical the document specifies toxicity and exposure limits, physical data, reactivity and corrosivity data, thermal and chemical stability data, and mixing incompatibilities. This is basically the information contained in an MSDS. The document lists the maximum amounts of chemical to be stored and/or processed (intended inventory). Finally, the document provides information about limits of operation regarding such aspects as temperature, pressure, flow, and composition of mixtures, and the consequences to health and safety of exceeding those limits.

Table 11.1. Highly Hazardous Chemicals, Toxics, and Reactives.

Chemical Name	CAS[a]	TQ[b]
Acetaldehyde	75-07-0	2500
Acrolein (2-propenal)	107-02-8	150
Acrylyl chloride	814-68-6	250
Allyl chloride	107-05-1	1000
Allylamine	107-11-9	1000
Alkylaluminums	Varies	5000
Ammonia, anhydrous	7664-41-7	10000
Ammonia solutions (>44% ammonia by weight)	7664-41-7	15000
Ammonium perchlorate	7790-98-9	7500
Ammonium permanganate	7787-36-2	7500
Arsine (also called Arsenic hydride)	7784-42-1	100
Bis(chloromethyl) ether	542-88-1	100
Boron trichloride	10294-34-5	2500
Boron trifluoride	7637-07-2	250
Bromine	7726-95-6	1500
Bromine chloride	13863-41-7	1500
Bromine pentafluoride	7789-30-2	2500
Bromine trifluoride	7787-71-5	15000
3-Bromopropyne (also called Propargyl bromide)	106-96-7	100
Butyl hydroperoxide (tertiary)	75-91-2	5000
Butyl perbenzoate (tertiary)	614-45-9	7500
Carbonyl chloride (see Phosgene)	75-44-5	100
Carbonyl fluoride	353-50-4	2500
Cellulose nitrate (concentration >12.6% nitrogen)	9004-70-0	2500
Chlorine	7782-50-5	1500
Chlorine dioxide	10049-04-4	1000
Chlorine pentrafluoride	13637-63-3	1000
Chlorine trifluoride	7790-91-2	1000
Chlorodiethylaliminum (also called Diethylaliminum Chloride)	96-10-6	5000
1-Chloro-2,4-Dinitobenzene	97-00-7	50000
Chlorormethyl methyl ether	107-30-2	500
Chloropicrin	76-06-2	500
Chloropicrin and methyl bromide mixture	None	1500
Chloropicrin and methyl chloride mixture	None	1500
Cumene Hydroperoxide	80-15-9	5000
Cyanogen	460-19-5	2500
Cyanogen chloride	506-77-4	500
Cyanuric floride	675-14-9	100
Diacetyl peroxide (concentration >70%)	110-22-5	5000
Dimethane	334-88-3	500
Dibenzoyl peroxide	94-36-0	7500
Diborane	19287-45-7	100
Dibutyl Peroxide (tertiary)	110-05-4	5000
Dichloro acetylene	7572-29-4	250
Dichlorosilane	4109-96-0	2500
Diethyizinc	557-20-0	10000
Diisopropyl peroxydicarbonate	105-64-6	7500
Dialuroyl peroxide	105-74-8	7500
Dimethyldichlorosilane	75-78-5	1000
Dimethylhydrazine, 1,1-	57-14-7	1000
Dimethylamine, anhydrous	124-40-3	2500
2,4-Dinitroaniline	97-02-9	5000
Ethyl methyl ketone peroxide (also Methyl ethyl ketone peroxide)	1338-23-4	5000
Ethyl nitrate	109-95-5	5000
Ethylamine	75-04-7	7500
Ethylene fluorohydrin	371-62-0	100
Ethylene oxide	75-21-8	5000
Ethyleneimine	151-56-4	1000
Fluorine	7782-41-4	1000

Table 11.1. Highly Hazardous Chemicals, Toxics, and Reactives. (continued)

Chemical Name	CAS[a]	TQ[b]
Formaldehyde (formalin)	50-00-0	1000
Furan	110-00-9	500
Hexafluoroacetone	684-16-2	5000
Hydrochloric acid, anhydrous	7647-01-0	5000
Hydrofluoric acid, anhydrous	7664-39-3	1000
Hydrogen bromide	10035-10-6	5000
Hydrogen chloride	7647-01-0	5000
Hydrogen cynide, anhydrous	74-90-8	1000
Hydrogen fluoride	7664-39-3	1000
Hydrogen peroxide (52% by weight or greater)	7722-84-1	7500
Hydrogen selenide	7783-07-5	150
Hydrogen sulfide	7783-06-4	1500
Hydroxylamine	7803-49-8	2500
Iron, pentacarbonyl	13463-40-6	250
Isopropylamine	75-31-0	5000
Ketene	463-51-4	100
Methacrylaidehyde	78-85-3	1000
Methacryloyl chloride	920-46-7	150
Methacryloyloxyethyl isocyanate	30674-80-7	100
Methyl acrylonritrile	126-98-7	250
Methylamine, anhydrous	74-89-5	1000
Methyl bromide	74-83-9	2500
Methyl chloroformate	79-22-1	500
Methyl ethyl ketone peroxide (also Ethyl methyl ketone peroxide)	1338-23-4	5000
Methyl fluoroacetate	453-18-9	100
Methyl fluorosulfate	421-20-5	100
Methyl hydrazine	60-34-4	100
Methyl iodide	74-88-4	7500
Methyl isocyanate	624-83-9	250
Methyl mercaptan	74-93-1	5000
Methyl vinyl ketone	79-84-4	100
Methyltrichlorosilane	75-79-6	500
Nickel carbonyl (Nickel Tetracarbonyl)	13463-39-3	150
Nitric acid (94.5% by weight or greater)	7697-37-2	500
Nitric oxide	10102-43-9	250
Nitroaniline (para-nitroaniline)	100-01-6	5000
Nitromethane	75-52-5	2500
Nitrogen Dioxide	10102-44-0	250
Nitrogen oxides (NO, NO2; N2O4;N2O3)	10102-44-0	250
Nitrogen tetroxide (also called Nitrogen peroxide)	10544-72-6	250
Nitrogen trifluoride	7783-54-2	5000
Nitrogen trioxide	10544-73-7	250
Oleum (65% to 80% by weight; also called Fuming sulfuric acid)	8014-94-7	1000
Osmium tetroxide	20816-12-0	100
Oxygen difluoride (Fluorine monoxide)	7783-41-7	100
Ozone	10028-15-6	100
Pentaborane	19624-22-7	100
Peracetic acid (concentration >60% Acetic Acid; also called Peroxyacetic acid)	79-21-0	1000
Perchloric acid (concentration >60% by weight)	7601-90-3	5000
Perchloromethyl mercaptan	594-42-3	150
Perchloryl fluoride	7616-94-6	5000
Peroxyacetic acid (concentration >60% Acetic acid; also called Peracetic acid)	79-21-0	1000
Phosgene (also called Carbonyl chloride)	75-44-5	100
Phosphine (hydrogen phosphide)	7803-51-2	100
Phosphorus oxychloride (also called Phosphoryl chloride)	10025-87-3	1000
Phosphorus Trichloride	7719-12-2	1000
Phosphoryl chloride (also called Phosphorus oxychloride)	10025-87-3	1000

Table 11.1. Highly Hazardous Chemicals, Toxics, and Reactives. (continued)

Chemical Name	CAS[a]	TQ[b]
Propargyl bromide	106-96-7	100
Propyl nitrate	627-3-4	2500
Sarin	107-44-8	100
Selenium hexafluoride	7783-79-1	1000
Stibine (Antimony hydride)	7803-52-3	500
Sulfur dioxide (liquid)	7446-09-5	1000
Sulfur pentafluoride	5714-22-7	250
Sulfur tetrafluoride	7783-60-0	250
Sulfur trioxide (also called Sulfur anhydride)	7446-11-9	1000
Sulfuric anhydride (also called Sulfur trioxide)	7446-11-9	1000
Tellurium hexafluoride	7783-80-4	250
Tetrafluoroethylene	116-14-3	5000
Tetrafluorohydrazine	10036-47-2	5000
Tetramethyl lead	75-74-1	1000
Thionyl chloride	7719-09-7	250
Trichloro (chloromethyl) silane	1558-25-4	100
Trichloro (dichloropheny) silane	27137-85-5	2500
Trichlorosilane	10025-78-2	5000
Trifluorochloroethylene	79-38-9	10000
Trimethyoxysilane	2487-90-3	1500

[a] Chemical Abstract Service number.
[b] Threshold quantity in pounds (amount necessary to be covered by this standard).
Note: These compounds are listed by OSHA as having potential for a catastrophic event at or above the threshold quantity (TQ).

From OSHA, 29 CFR 1910.119, Appendix A.

EMERGENCY PLAN AND SAFETY PROGRAM

The employer prepares a process hazard analysis using an expert team that includes at least one employee familiar with the operation. The team identifies, evaluates, and proposes controls for hazards in this analysis. A history of previous incidents at the facility is included. The document anticipates the consequences to health and safety of a system failure. The employer responds to and resolves problems uncovered in the analysis in a timely manner. Management updates the study every five years, and retains these records for the life of the process facility.

How does a team determine the potential hazards at a facility? OSHA does not specify a method. Four approaches are commonly employed, any one of which works.[3] An example is the "what if?" approach. The process is broken down into a series of steps, and for each step a series of "what if" questions are asked: What if temperature control failed and the temperature rose to the maximum possible? What if the valve locked open and flow of this chemical could not be stopped? What if the valve locked closed and the chemical ceased to be added? After the consequences have been predicted, a plan is prepared to deal with each dangerous situation, should it arise.

The employer prepares a set of clear written operating procedures for the process that cover safety considerations. These procedures cover normal operation, but also address startup and shutdown, emergency shutdown, and restart after an emergency shutdown. The document lists operating limits and consequences of exceeding them. The document also includes the hazards of the chemicals, precautions (engineering and administrative) to avoid exposure, and measures to employ if exposure does occur. Similarly, employees who maintain, inspect,

[3] R. B. Smith, "Process Hazard Analysis," *Occupational Health and Safety*, June 1995, p. 33.

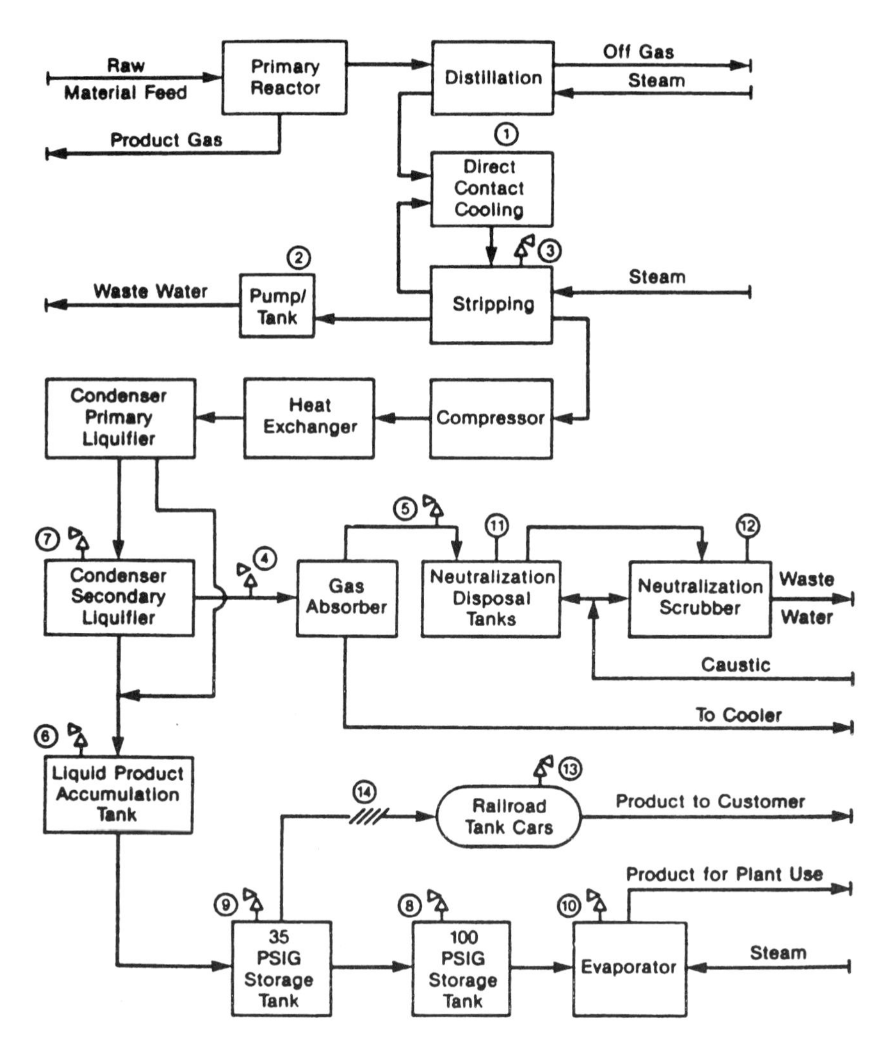

Figure 11.2. Shown here is an example of a block diagram of the operations at a major facility using chemicals. (From OSHA, 29 CFR 1910.119, Appendix B.)

and test the equipment require written procedures. These procedures are then available to all employees who work at or maintain the facility.

Employers are responsible for a written emergency action plan so that employees are prepared to handle an emergency. The plan includes escape routes. Responsibilities of specific employees for rescue and medical duties are spelled out. The plan must meet the criteria outlined in 29 CFR 1910.38(a).

OSHA places high importance on training workers. The employer provides workers assigned to the process with training about the process and with follow-up refresher training at least every 3 years. Employees involved with maintenance, inspection, and testing or with emergency response must also be trained. Records of the training include the identity of the workers trained, the date of training, and an indication that the worker understood the training.[4]

[4] Some kind of test at the end of training indicating the success of the program is a common feature of more recent OSHA regulations.

Within 48 hr the employer must investigate any incident that has occurred at the facility. Investigation must be by a competent team, and a report covering designated aspects of the incident is prepared and retained for 5 years.

This chapter does not include all details of the regulation. Instead it has presented the flavor and intent of the regulation.

HAZWOPER

Since the birth of the environmental movement in the 1960s, the U.S. has attacked problems arising from the use and disposal of hazardous chemicals in a series of legislative steps. In 1976 the Resource Conservation and Recovery Act (RCRA) established a mechanism for identifying hazardous wastes and following their history "cradle to grave". A paper trail keeps a record of waste generation, transport, and final disposal. The intent is to prevent the creation of new improper hazardous waste sites that could contaminate air and water resources.

In 1980 the Comprehensive Environmental Response, Compensation, and Liability Act (CERCLA, or Superfund) tackled the problem of existing uncontrolled and abandoned hazardous waste sites. The EPA identified and prioritized the urgency of cleanup of these sites with the National Priority Site list. Then sites were to be cleaned or contained to avoid further environmental damage. This created a new job classification: hazardous waste removal worker.

In 1986 the Superfund Amendments and Reauthorization Act (SARA) continued the funding of Superfund. SARA, under Title I, required OSHA to develop standards for employee safety and health during these hazardous waste removal operations. Following the procedures for creating new standards described in Chapter 2, OSHA published a Notice of Proposed and Public Hearings in August 1987. The final rule was published in the *Federal Register* March 6, 1989 (54 FR 9294) and appeared in the *Code of Federal Regulations* as 29 CFR 1910.120.[5] Enforcement began 1 year later, and corrections were published a month after that (55 FR 14072).

This Hazardous Waste Operations and Emergency Response regulation, usually termed HAZWOPER, was intended not only to protect hazardous waste removal workers, but also to protect workers performing RCRA hazardous waste operations at treatment, storage, and disposal (TSD) facilities and workers who respond to spills and other emergencies involving hazardous chemicals (HAZMAT teams). The regulations recognize that these are workers whose job description deliberately places them in contact with hazardous chemicals under circumstances that are often unpredictable.

Major components of the regulation focus on planning, training, employee protection, site control, and medical surveillance.

PLANNING

First, employers must develop a written safety and health program. Employers are given a degree of freedom to design a program to fit the special characteristics of the individual situation, but the program must "identify, evaluate, and control safety and health hazards, and provide for emergency response for hazardous waste operations".[6] Recognize that we are dealing here with a different situation than the more typical sort: We have a machine on

[5] The EPA and OSHA passed identical regulations (see 40 CFR Part 311). OSHA regulations are directed at private employers, whereas the EPA regulations are directed at state or local governments as employers.

[6] An interesting, brief discussion is found in C. R. Diez, "The ABC's of Emergency Response," *Occupational Health and Safety*, Feb. 1996, p. 39.

the floor with moving parts that could injure a worker, and the regulation describes guards for the moving parts. The employer here must look at the facility and anticipate what *might* go wrong, without a set of guidelines already in print for the exact situation. Specific required components include:[7]

- An organizational structure
- A comprehensive workplan
- A site-specific safety and health plan which need not repeat the employerís standard operating procedures required in paragraph (F) of this section
- The safety and health program
- The medical surveillance program
- The employer's standard operating procedures for safety and health
- Any necessary interface between general program and site-specific activities

Included within this framework is a preliminary evaluation of the characteristics of the specific site, identifying the likely hazards. There should be a site map indicating site work zones. In particularly hazardous situations employees should keep watch on one another through a "buddy system". The plan includes communications on site and the location of the nearest medical assistance.

The employer must make this plan available to any contractor or subcontractor involved in the cleanup, to the employees, and to OSHA or other agencies with authority over the site.

TRAINING

No one is allowed on the worksite until they are trained. General site workers, managers, and supervisors must have 40 hr of training off site and 3 days on site under supervision. Workers who occasionally enter and leave the site to perform their duties, such as individuals collecting samples for analysis, must have 24 hr of off-site training and 1 day on site under supervision. Eight hours of refresher training is required annually. Regulations spell out standards for competent trainers.

Training includes these components:

- Names of supervisors responsible for site health and safety
- Health and safety hazards on site
- Use of personal protective equipment
- Safest work practices
- Use of safety engineering controls and equipment on site
- Medical surveillance, including signs and symptoms of exposure
- Contents of the safety and health written plan.

SAFETY PROCEDURES ON SITE

Periodic monitoring determines kinds and levels of hazardous materials in the air. A combination of engineering controls, work practices, and personal protective equipment must ensure that worker exposure to these chemicals is below established exposure levels (PEL, etc.). Employers determine procedures and locations for decontamination of workers and equipment in advance.

Employers must anticipate possible emergency situations and prepare a plan to handle them before operations begin. Such a plan includes lines of authority and the roles of the personnel to prevent confusion should the emergency arise. Employers predetermine safe refuges, evacuation routes, and plans for emergency medical treatment.

[7] 29 CFR 1910.120(b)(1)(ii).

MEDICAL SURVEILLANCE

Baseline, periodic, and termination medical exams should be provided for certain workers in both hazardous waste and emergency response operations. When workers are exposed to a hazardous substance at or above established exposure levels, or if they wear approved respirators 30 days or more on the site, they should be given a medical exam at least once a year and at the end of employment. Any worker who is exposed to unexpected or emergency hazardous chemical release also shall be examined.

HAZMAT TEAMS

Much of what is done by a HAZMAT team is similar to the work of a waste site clearance worker. The HAZMAT team has the disadvantage of not controlling the timetable. They respond when the release of a hazardous chemical occurs, so they must be ready for such an event at all times. The planning must be done before there is an emergency, and the written plan includes basically the same components as the plan at a waste site. Also the same is the need to protect the workers at levels of chemical that may well exceed the regulated limit.

OSHA defines a number of classes of employee in a HAZMAT operation:

First Responder — Awareness. This individual is likely to discover a hazardous material release. He or she is trained to initiate an emergency response, but should not take any further action.

First Responder — Operations Level. This person responds to the release defensively to protect people, property, or the environment. No attempt should be made to stop the release, but this person may attempt to contain the release from a distance. Such a person has at least 8 hr of training.

Hazardous Materials Technician. This person responds to stop the release. A minimum of 24 hr of training is necessary.

Hazardous Materials Specialist. Such a specialist assists the hazardous materials technician and acts as site liaison with government officials. Again, a minimum of 24 hr of training is needed.

Senior Emergency Response Official. This is the person in charge of the site-specific incident command system. Functions include identifying the hazardous substance, starting the emergency operations, controlling personnel at the site, ordering use of correct personal protective equipment and other back-up equipment, terminating the operation if and when the situation calls for it, and decontaminating the site after the emergency is contained.

KEY POINTS

LARGE CHEMICAL OPERATIONS

1. OSHA has written a standard that focuses on controlling potential hazards to workers at facilities that use large quantities of hazardous chemicals.
2. The need to comply with this standard is triggered by the use of amounts of specified chemicals in excess of defined threshold quantities.
3. Employers must:
   a. Prepare documents that outline the technical process involved and lists the characteristics and hazards of the chemicals.
   b. Inspect the facility to anticipate the consequences of a system failure and identify operations in need of improvement.
   c. Provide written directions for all aspects of operation of the process.
   d. Provide a written emergency action plan.
   e. Provide training to all workers involved in the process.

HAZWOPER

4. The Hazardous Waste Operations and Emergency Response regulations protect hazardous waste cleanup and treatment workers and emergency response workers.
5. The regulations require planning, training, site control, employee protection, and medical surveillance.
6. For both hazardous waste workers and emergency response workers, classes of workers have specified training requirements.

BIBLIOGRAPHY

PROCESS CONTROL

29 CFR 1910.119 Process safety management of highly hazardous chemicals.

OSHA Instruction CPL 2-2.45A, September 28, 1992, "29 CFR 1910.119, Process Safety Management of Highly Hazardous Chemicals — Compliance Guidelines and Enforcement Procedures."

HAZWOPER

J. W. Hosty and P. Foster, *A Practical Guide to Chemical Spill Response,* Van Nostrand-Reinhold, New York, 1990.

National Fire Protection Association, *Recommended Practice for Responding to Hazardous Materials Incidents,* Standard 471, August 14, 1992.

National Fire Protection Association, *Standard for Professional Competence of Responders to Hazardous Materials Incidents,* Standard 472, August 14, 1992.

Occupational Safety and Health Guidance Manual for Hazardous Waste Site Activities, NIOSH/OSHA/USCG/EPA; October, 1985. Publication Number 85-115.

OSHA Instruction CPL 2.94, July 22, 1991, "OSHA Response to Significant Events of Potentially Catastrophic Consequences."

U.S. Department of Transportation, *Emergency Response Guidebook,* Washington, D.C., 1990.

PROBLEMS

1. Which of these facilities would be subject to process control regulations?
   A. A paper plant producing special-purpose papers chops wood fiber and extracts lignin with liquid sulfur dioxide. Used cotton rag is added and the mixture is bleached with chlorine. Vats are heated with fuel oil. A green dye is added to the mixture. Sheet paper is pressed from the mixture. Waste processing utilizes sodium hydroxide (caustic). The chlorine tank holds 890 lb, the sulfur dioxide tank holds 950 lb, the maximum caustic inventory is 450 lb, and 100 lb of green dye is the maximum inventory. The fuel oil tank holds 15,000 lb.
   B. A plant prepares solvent for spray paints. Acetone is obtained from a solvent recycle company and is distilled before use. Cleaned acetone is mixed with methyl ethyl ketone and toluene. The final mixture is packaged in 55-gal drums or **1-gal** metal cans for shipping. The maximum inventories include 9,000 lb acetone, 12,000 lb methyl ethyl ketone, and 3,500 lb toluene. The distillation apparatus is heated by natural gas piped into the plant.
   C. An agricultural chemicals facility prepares and packages bulk fertilizers. The following chemicals are stored in tanks or silos on site: urea, 9,500 lb; anhydrous ammonia, 12,000 lb; nitric acid, 1,000 lb, potassium oxide, 500 lb; and sodium phosphate, 9,500 lb.
2. Prepare a block diagram for the operation described in Problem 1.B.

3. Explain why HAZMAT team members might be firemen, but not all firemen are HAZMAT team members.
4. What would be the first step in cleaning out a hazardous waste site whose contents were unknown?
5. Which of the four EPA levels of required protection (Chapter 9) would be needed when removing rusting but intact 55-gal drums of an unknown liquid from a waste site?

Section IV
Physical and Biological Hazards in the Workplace

Chapters 12–16 deal with threats to worker health and safety that arise from the worker's physical environment. These are varied problems lacking the common underlying theme of "problems with chemicals" found in earlier sections. Many industries employ machines, particularly heavy machines, whose operation involves impact, vibration and other noise-producing functions. Hearing damage from noise exposure usually comes about gradually and without pain, and may be discounted by the worker as a real threat (Chapter 12). Radiation damage similarly occurs without the worker's awareness, and may be ignored as a threat by the worker without posted warnings, monitoring, shielding, and other precautions (Chapter 13). Prevention of such accidents as tripping, falling, and physical injury from machinery was an early safety concern (Chapter 14). More recently recognized are problems arising from working in improper positions that strain parts of the body, or from performing continuously repeated tasks (Chapter 15). and physical injury from machinery was an early safety concern (Chapter 14). More recently recognized are problems arising from working in improper positions that strain parts of the body, or from performing continuously repeated tasks (Chapter 15). and physical injury from machinery was an early safety concern (Chapter 14). More recently recognized are problems arising from working in improper positions that strain parts of the body, or from performing continuously repeated tasks (Chapter 15). Finally, stress from working in extreme temperatures is discussed (Chapter 16).

Some tasks make exposure to high or low temperatures unavoidable, but we must understand the limits to safe exposure to such environments and the signs that overexposure has occurred.

Chapter 17 covers harmful interaction with biological agents. Biological hazards are a recent concern. Chapter 17 covers harmful interaction with biological agents. Biological hazards are a recent concern. Chapter 17 covers harmful interaction with biological agents. Biological hazards are a recent concern. Such concerns focus on the possibility of infection on the job.

CHAPTER 12

Occupational Hearing Loss

The handicap caused by loss of vision is well recognized. Most people have simulated blindness by trying to perform some simple tasks with our eyes shut, and everyone has experienced finding their way through a darkened room. Few ever experience loss of hearing, and are unaware how effectively this isolates people from others. To appreciate this, search around until you find a television program in which the story line is not visually graphic (i.e., sex or violent chase scenes — good luck!), but one where a story is being told. First, attempt to follow the story line with your eyes shut, getting clues only from the sound track. Then try watching the screen with the sound off completely. Loss of hearing is a serious handicap.

In a congressional report,[1] the number of Americans estimated to suffer permanent hearing disability is between 8.7 and 11.1 million. Approximately nine million workers in the U.S. are exposed to noise levels sufficiently high to cause hearing damage, and about one million workers per year experience occupational hearing loss as a result of work conditions, resulting in worker compensation claims in excess of $800 million in the period from 1977 to 1987.[2] NIOSH has classed noise-induced hearing damage as being among the ten leading illnesses or injuries of the workplace.

Two causes are distinguished: loss due to a single event such as an explosion or a blow to the head, or loss due to continuous exposure to relatively high sound levels over a long time period. The "single-event damage" results from circumstances that are avoidable, and such worker exposure is accidental. Continuous exposure to sound, on the other hand, is an unavoidable aspect of work in many industries. When machinery with moving parts and parts in collision is run, sound is produced. This circumstance is the focus of this chapter as we deal with the ear, the nature of sound and of hearing, how sound is measured, the levels of sound that cause hearing loss, and the means for protecting hearing. OSHA noise exposure regulations arise from this understanding.

THE EAR

The ear clearly divides into three anatomical parts (Figure 12.1):

1. *The outer ear.* The ear canal spans about 1 ½ in. from the outdoors to the eardrum. The eardrum is a fibrous membrane sealing the passage at the end of the canal. This membrane vibrates in response to sound waves.

[1] "Report to the President and Congress on Noise," Administrator to the Environmental Protection Agency, 92nd Congress, Document No. 92-63, February 1982.
[2] Council for Accreditation in Occupational Hearing Conservation.

2. *The middle ear.* Behind the eardrum is an air-filled chamber. Atmospheric pressure changes cause pressure changes across the eardrum. These are equalized by air entering or leaving the chamber through the Eustachian tube, which connects the chamber with the throat. When you are in an aircraft that is descending to land you refer to the burst of air entering the middle ear chamber to counter the rising outside pressure by saying your ears "popped". Blockage of the Eustachian tube, as during a cold, allows a pressure differential to build that reduces the ability of the eardrum to vibrate, reducing hearing acuity. Three tiny bones — the hammer, the anvil, and the stirrup — bridge the chamber and transfer vibration of the eardrum to the end of the fluid-filled compartment of the inner ear. Their lever action amplifies vibrations from the eardrum.
3. *The inner ear.* Two sensory functions are served in the inner ear: equilibrium and hearing. Equilibrium is sensed in the semicircular canals, fluid-filled loops containing floating membranous structures. Hearing is served by the snail shell–shaped cochlea. The stirrup bone of the middle ear attaches to a membrane at the opening of the cochlea. The stirrup presses fluid into the cochlea, which then circulates to the center of the spiral and back out by a parallel passage. At the end another membrane in the inner ear chamber bulges outward to relieve the pressure. Waves of pressure in the fluid cause the walls of the fluid chambers to bulge outward, pressing on a parallel chamber containing hairlike nerve endings. These respond to pressure to produce nerve impulses, which are transferred to the brain. Nerve endings at the wider portion of the cochlea near the membranes separating the inner ear from the middle ear are sensitive to higher frequencies, whereas those sensitive to lower frequencies are at the center of the cochlea. Hearing loss is due to the loss of these cochlear hairlike nerve endings.

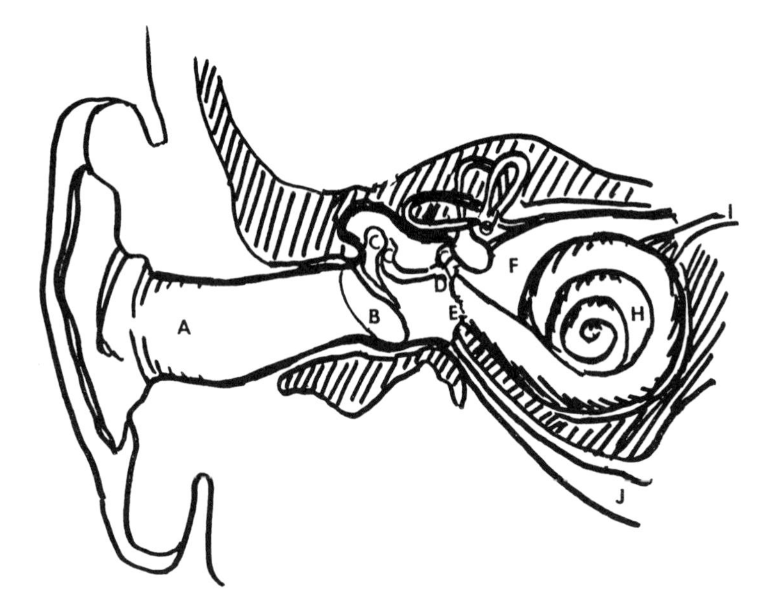

Figure 12.1. Pictured here is a cross-section of the ear. Labeled are (A) outer ear, (B) eardrum, (C) bones of the middle ear, (D) oval window, (E) round window, (F) inner ear, (G) semicircular canals, (H) cochlea, (I) auditory nerve, and (J) Eustachian tube.

THE NATURE OF SOUND

Noise is unwanted sound.

Sound is the propagation of a pattern of compressions and rarefactions, normally through air, although other matter can serve as the conducting medium. The compressions and rarefactions, which in air really are instantaneous rises and drops above and below the atmospheric pressure, move at a characteristic rate, traveling from the source to the ear. The ear responds by converting the pattern into a nerve impulse signal to the brain. The brain recognizes two properties of this pattern of moving compressions: the intensity or loudness and the tones, from high pitched to low.

Sound is visualized as a wave pattern with peaks at the air compression and valleys at the region of lower pressure that follows the compressions. The louder the sound (the greater the pressure changes), the higher are the peaks and the lower the corresponding valleys. This is termed the amplitude of the sound. Some amplitudes are smaller than the ear can detect, so the sound is not heard. Logically, the greater the amplitude of the sound, the greater is its ability to cause damage to hearing.

The brain also distinguishes sounds by the frequency with which these compressions and rarefactions stimulate the ear. The x axis in the wave model is time. Higher frequency sound has a shorter wavelength, hence more waves are propagated per unit of time. When more compressions per second reach the ear, the brain interprets the sound as being higher in pitch (higher frequency sound). Musical instruments produce tones of different frequency, and these are blended together to produce the usually pleasing sound of the instrument or instruments. Human speech, the most important sound to humans, is a blend of many tones.

FREQUENCY

The measure of frequency is the number of compressions passing a point (for example, your ear) per second. This number is expressed as hertz (Hz) and has cycles per second (s^{-1}) as units. Although everyone is a little different, typical young human hearing responds to frequencies from 20 to 20,000 Hz. Almost everyone has heard, or rather has not heard, a silent dog whistle. This produces a tone above the range of the human ear but not above that of a dog.

Conceptually, sound of a single frequency, called a pure tone, is the easiest to understand. It is easy to generate pure tones electronically or with a tuning fork, and one approximates such tones by striking piano keys. A C at the middle of the piano keyboard produces a tone of different frequency than a D or B, and they each have a characteristically different sound.

However, if one moves to the right on the keyboard to strike the next C, one hears a tone that is clearly higher in pitch but otherwise sounds the same as the lower C. The relationship between these two different tones, which are described as being an octave apart, is that the higher C has exactly twice the frequency of the lower C. One hears tones at octave intervals as similar tones.

For purposes of analysis, the range of human hearing is divided into octave bands, eight ranges in each of which the highest frequency is twice the lowest frequency. These bands are each further subdivided into three parts, termed one-third octave bands. Octave bands and one-third octave bands are identified by their center frequency, both are numbered in a standard fashion (see Table 12.1). The frequency range of noise is identified by these bands in the analysis of noise and of hearing damage.

Frequency and wavelength (λ) are interdependent measures of the sound wave, as expressed in the equation:

$$\lambda = c / f$$

where c = rate of sound propagation
f = frequency of sound (Hz)
λ = wavelength of sound wave

Table 12.1. Octave and One-Third Octave Bands.

Octave Band Number	Frequency Range (Hz)	One-Third Octave Band Number	Frequency Range (Hz)	Center Frequency (Hz)
Suboctave	22–45	14	22–28	25
		15	28–35	31.5
		16	35–45	40
1	45–89	17	45–56	50
		18	56–71	63
		19	71–89	80
2	89–178	20	89–112	100
		21	112–141	125
		22	141–178	160
3	178–355	23	178–224	200
		24	224–282	250
		25	282–355	315
4	355–708	26	355–447	400
		27	447–563	500
		28	563–708	630
5	708–1413	29	708–892	800
		30	892–1123	1000
		31	1123–1413	1250
6	1413–2819	32	1413–1779	1600
		33	1779–2239	2000
		34	2239–2819	2500
7	2819–5626	35	2819–3548	3150
		36	3548–4467	4000
		37	4467–5626	5000
8		38	5626–7080	6300
		39	7080–8913	8000
		40	8913–11234	10000

The rate of sound propagation varies with the medium and the temperature, but it can be approximated as 1129 ft/s (344 m/s) in air.[3] Under those conditions a sound wave of 13 in. or 0.344 m wavelength would display a frequency of 1000 Hz. Wavelength is most important as a measure of sound when the objective is to block the sound.

Most sound is not composed of pure tones, but rather is a complex mixture of frequencies. Human speech is complex sound, largely within the frequency range of 500 to 3000 Hz. To interpret speech one needs to hear all the frequencies in use. For example, some consonant sounds have important high-frequency components. Loss of hearing in the high-frequency range makes it difficult to distinguish words that differ only in the terminal consonant (e.g., fit, fix, fist).

AMPLITUDE

The amplitude or loudness is the sound characteristic of greatest concern in the industrial environment. Therefore, an early requirement is a method to measure sound levels and units with which to express the results. Sound is a form of power, and the sound power of a source is expressed in watts. Typical sound power levels of a variety of sources are presented in Table 12.2. However, what is actually heard or measured is fluctuation in air pressure caused by the source, which is called sound pressure and which has pressure units.

[3] The relationship of the speed of sound in air to temperature is as follows:

$$C = 49.03(460 + T_F)^{1/2} \text{ ft/s, where } T_F \text{ is Fahrenheit temperature}$$

$$C = 20.05(T_K)^{1/2} \text{ m/s, where } T_K \text{ is absolute temperature}$$

The speed of sound in other media is quite different from the speed in air. In water it is about 4 times as fast and in steel it is about 15 times as fast.

Table 12.2. Approximate Sound Power Levels of Commonplace Sources.

Source	Sound Power (watts)	L_W (dB)
Rocket	100,000,000	200
Air hammer	1	120
Riveting machine	0.1	110
Jet plane at 1000 ft	0.01	100
Shouted conversation	0.001	90
Average factory	0.00007	75
Conversation	0.00001	70
Typical office sound level	0.0000001	50
Whisper	0.000000001	30
Quietest audible sound	0.000000000001	0

Note: Sound power level for a source of W watts is calculated as follows:

$$L_W = 10\log(W/10^{-12})dB = (10\log W + 120)dB$$

One would therefore expect the loudness of sound to be expressed straightforwardly in pressure units. Instead, units are devised that measure relative sound pressure, termed sound pressure level. "Relative" indicates that there is a reference sound pressure value to which the sound pressure of interest is compared. In principle, the reference value is the quietest audible sound. Because this value is not the same for everyone, it is estimated for convenience to be 2×10^{-5} N/m^2.[4] A log scale is used to cover conveniently the broad range of values encountered, and the values are made larger by multiplying them by 10. The units of sound pressure level are based on the log of the ratio of the mean square pressure of the source to that of the quietest audible sound, and they are expressed in decibels (dB). Sound pressure level (L_P) for a sound of P sound pressure is calculated as follows:

$$\begin{aligned} L_P &= 10\log(P/P_{ref})^2 \text{ dB} \\ &= 20\log(P/P_{ref}) \text{ dB} \end{aligned}$$

The acoustic power generated by a sound source is termed the sound power level, or L_W. Once again, the reference level is the quietest audible sound, but the units now are watts and this level is assumed to be 10^{-12} watts.

$$\begin{aligned} L_W &= 10\log(W/W_{ref}) \text{ dB} \\ &= 10\log(W/10^{-12}) \text{ dB} \\ &= 10\log(W) + 120 \end{aligned}$$

Notice how large changes in sound level produce only small changes in number of decibels. If we measure a power tool at low speed and high speed, and determine it to be 0.0002 and 0.0004 watts, respectively, the sound level is twice as high at high speed. We calculate the number of decibels at each speed to be

[4] or 0.0002 dyn/cm^2 or 0.0002 microbars. The quietest audible sound varies among individuals according to their hearing sensitivity, and varies for everyone by wavelength.

$$L_W = 10\log(W) + 120$$

Low speed:

$$= 10\log(0.0002) + 120 = 10(-3.70) + 120 = 83 \text{ dB}$$

High speed:

$$= 10\log(0.0004) + 120 = 10(-3.40) + 120 = 86 \text{ dB}$$

Thus, a doubling of the noise level causes a change of only 3 dB.

SOUND LEVEL METERS

Sound level meters are portable, battery-powered devices used to measure sound pressure levels in the workplace (Figure 12.2). They include a microphone to convert sound to an electrical current, and the current is sent to an amplifier. The amplified electrical signal causes a needle to deflect on a calibrated scale, indicating the sound pressure level. Rather than having the meter cover a typical range of sound levels from 40 to 140 dB, which would make accurate reading difficult, an attenuator is added that allows the meter to be adjusted to read within an appropriate 10-dB range. Thus if the attenuator is set for the 70- to 80-dB range and the needle stops at 6.0, the sound pressure level is 76 dB. Devices are available that mount onto the microphone and produce a standard signal used to calibrate the meter before use.

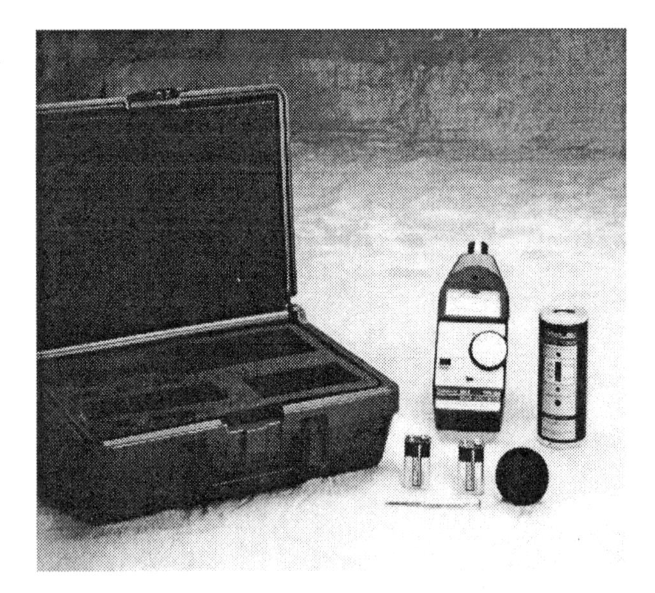

Figure 12.2. This portable sound level meter is capable of reading a broad range of sound levels. The cylinder at the top of the instrument is the microphone, and the meter reads the level of sound within the range set by the round control. To the right of the meter is the calibration device, which mounts on the microphone and generates sounds of known level. (Photo courtesy of Lab Safety Supply, Inc., Janesville, WI. With permission.)

The human ear is not uniformly sensitive to all frequencies, being most sensitive to sounds of about 4000 Hz, dropping acuity gradually as frequency becomes lower and sharply as it becomes higher. Most sound level meters include weighting networks, labeled as A, B, and C scales. The A scale varies the response of the meter with frequency to match the response of the human ear. Thus, the meter is very sensitive reading 4,000 Hz and much less sensitive to 20,000 Hz. The C scale is flat, so that the sound level meter responds equally well to frequencies across the entire range. The B scale is intermediate. Usually the A scale is used to estimate potential damage to the ear, or the C scale may be used if the signal is going to be transferred to another instrument for detailed analysis. An octave band analyzer attached to a sound level meter allows separate analysis of the sound pressure level within each octave band. Such analysis is useful in the design of a control system for a particular sound source.

Another instrumental variable is the response rate. With a fast response setting the instrument varies its readings more nearly with the instantaneous level, jumping about in environments in which the sound levels fluctuates greatly, whereas a slow response setting averages the reading and displays a relatively steady value. If you were monitoring a symphony concert, readings on the fast response setting would peak each time the symbols crashed and would not on the slow response setting.

MEASURING EXPOSURE IN THE WORKPLACE

Before taking readings with a sound level meter, be sure you are familiar with the workings of the particular instrument. Check the batteries for adequate charge. Usually, the instrument should be set to operate on slow response on the A scale. The meter should be calibrated using a device designed to produce a standard, reproducible signal. If air moves across the microphone, it creates additional noise. This can be screened out using a foam cover provided by the manufacturer.

The sound level meter is used to assess the exposure of a worker to noise at his or her workstation. Sound levels drop with distance from the source, although reflection or absorption of sound by nearby surfaces complicates the sound level/distance relationship. Readings of sound level should therefore be taken as near as possible to the location of the worker's ears. Sound levels vary, so readings should be taken at various times of the day or even on different days. If these readings are taken too close to the sound source, unrealistically high exposure would be indicated, and similarly greater distance infers improperly low exposure levels.

A dosimeter is a variation of the sound level meter that is worn by the worker and accumulates total noise exposure. It is to the noise level meter what the personal air sampling pump is to grab sampling. Spot checks with a sound level meter report the instantaneous exposure. At a workstation where noise levels vary as machines are turned on and off, or in a situation where the worker is mobile, moving into and out of noisy areas, dosimeters determine the noise burden as a TWA exposure.

HEARING DAMAGE

Brief exposure to high sound levels can cause a ringing sensation and temporary threshold shift (temporary drop in hearing acuity). After a rest period the hearing returns to normal. If exposure to excessive sound is repeated frequently, there is a permanent threshold shift (loss in hearing becomes permanent). Hearing loss caused by noise is termed noise-induced permanent threshold shift. People vary greatly in their susceptibility to hearing damage by continuous sound exposure, but at high levels everyone loses some hearing ability. Workers exposed daily to high noise levels display the greatest loss during the first years. If we set the standard for serious hearing loss at a 25-dB or more drop in acuity, over a work lifetime of 40 years 18% exposed to 90 dB and 70% exposed to 115 dB incur such loss. Approximately 75% of the workforce are exposed to sound levels above 85 dB.

TESTING HEARING DAMAGE

An audiometer is a device that generates a range of pure tones at known sound pressure levels. Hearing is tested by placing the subject in an environment that is free of interfering sound stimuli, and the hearing threshold is determined for a series of pure tones (Figure 12.3). The resulting information is displayed as a graph called an audiogram with frequencies along the x axis and the decibel level of the threshold on the y axis.

Figure 12.3. Hearing loss evaluation is done with an audiometer. The worker is isolated from external noise, ideally in a soundproof room as shown here. A pure tone is generated and delivered to the earphones, beginning at a very low level. As the loudness is increased, the worker indicates when the sound is first heard. This level is recorded on the audiogram (Figure 12.4) for each frequency tested for each ear. (Photo courtesy of Industrial Acoustics Company, Inc., Bronx, NY. With permission.)

The American Academy of Ophthalmology and Otolaryngology has established a standard audiogram presentation, which displays frequencies in octaves across the top from 125 Hz on the left to 8000 Hz on the right. The hearing threshold level in decibels is displayed down the left side from –10 at the top to 110 at the bottom. One octave on the x axis has the same length as 20 dB on the y axis. Hearing thresholds of the left ear are displayed as blue X symbols and at the right ear as red O symbols. Such a standard presentation simplifies comparing relative hearing acuity from one audiogram to the next without troublesome conversion calculations. Because the decibel scale at each frequency is referenced to the quietest sound of that frequency detectable by a person with normal hearing, a person with normal hearing generates a plot that is linear and close to the 0 decibel level (Figure 12.4A).

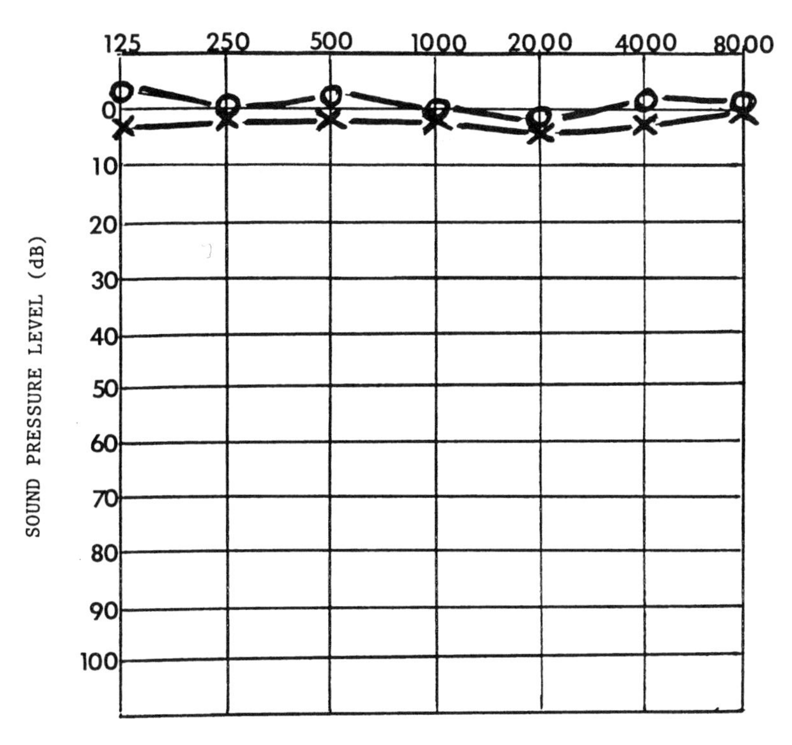

Figure 12.4A. Audiograms present the response of each ear to a series of pure tones. Zero on the y axis represents response in individuals with undamaged hearing. This plot presents the audiogram of such a person, with the response of the right ear indicated with circles (normally red) and the left ear with exes (normally blue).

Many hearing conservation programs routinely measure hearing at 500, 1000, 2000, 3000, 4000, and 6000 Hz. Frequencies from 500 to 2000 Hz are included because these cover the normal range of speech. As a rule, the first sign of hearing damage due to continuous noise exposure is a dip in acuity around 4000 Hz, a notch in the audiogram (Figure 12.4B). Including 3000 to 6000 Hz in the analysis permits detection of early damage, even though such loss may not interfere with the understanding of speech. As hearing loss progresses, this notch deepens and widens. As hearing loss extends into the range of human speech, the individual first becomes aware of hearing loss. When values for loss greater than 25 dB are obtained on testing, hearing is considered to be impaired.

Appearance of readings indicating loss of acuity on an audiogram does not establish industrial noise as the cause. Many programs add readings at 250 and 8000 Hz. Readings at 8000 Hz are important diagnostically. Observed deafness may be due to some cause other than industrial noise exposure if, rather than a notch at 4000 Hz and some recovery by 8000 Hz, the curve is flat or there is an increasing loss of acuity through higher frequencies (Figure 12.4C). A list of disease- or damage-related causes for hearing loss includes:

1. Obstruction of the ear canal. This could be due to accumulated ear wax or a foreign object in the canal.
2. *Infection.* Infection can cause swelling and obstruction of the ear canal. Infections of the middle ear, often secondary to infections elsewhere, cause temporary or permanent hearing impairment.
3. *Allergy.* Allergic response to some agents may result in ringing in the ears. Continued exposure may lead to permanent damage.

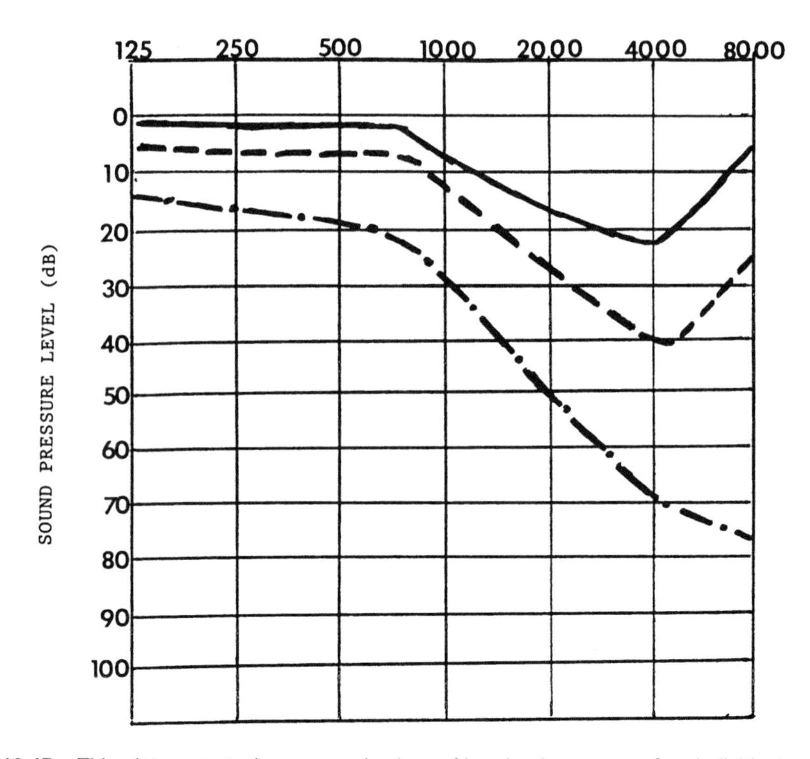

Figure 12.4B. This plot presents the progressive loss of hearing in one ear of an individual exposed to excessive noise. The top curve is the earlier audiogram, with the lower plots displaying progressive hearing loss. The "valley" at 4000 Hz is typical. In plot C the hearing loss is due to middle ear problems rather than noise exposure, producing a characteristically flat curve.

SAMPLE PROBLEM:

An audiogram is taken of a male worker at age 35 and again at age 42. The audiogram shows a drop in acuity of 12 dB at 4000 Hz. How much of this drop is estimated to be caused by aging?

Age	From Table 12.3
42	16 dB
35	11 dB
Difference	5 dB

Of the acuity shift of 12 dB, approximately 8 dB can be attributed to aging.

4. *Trauma.* An extreme sound such as an explosion can do physical damage to the eardrum or the middle ear.
5. *Brain damage.* Anything that damages the auditory portion of the brain causes deafness. This could include stroke, hemorrhage, or meningitis.

In addition to being obstructed by disease or damage, hearing acuity also normally decreases with age. Loss of sensitivity to high frequencies is particularly marked. Expected losses with age differ between men and women and are summarized in Table 12.3. In order

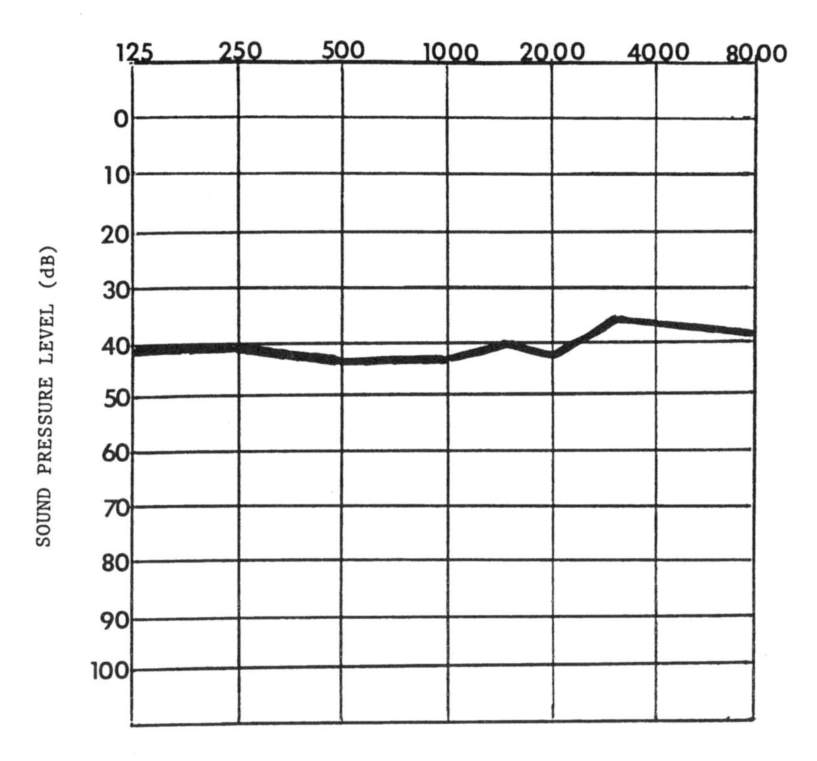

Figure 12.4C. In this plot the hearing loss is due to middle ear problems rather than noise exposure, producing a characteristically flat curve.

to correct audiograms for an individual for expected loss due to aging, subtract the value at the specific frequency at the earlier age from the value at the later age. The difference, the expected loss, should be compared to the measured loss. If the measured loss is greater, this would be attributed to hearing damage.

OSHA REGULATIONS

High levels of noise in the workplace is one of the most frequent complaints. OSHA has estimated that 2.9 million workers in production industries experience exposures of greater than 90 dB and a further 2.3 million between 85 and 90 dB for an 8-hr work day.[5] The General Industry Noise Standard [29 CFR 1910.95(a) and (b)] was written to cover general industry, exempting oil and gas drilling and service industries. There is a separate regulation for construction workers (29 CFR 1926.52 and 1926.101) that establishes the same exposure standards. The later Hearing Conservation Amendment consists of paragraphs (c) through (p).

WORKER PROTECTION STANDARDS

For purposes of testing compliance, exposures to different sound levels in the course of a work day are calculated as follows. The length and sound intensity of each exposure is logged. At each level the time exposed is totaled (C_1, C_2, . . .). Each total is divided by the

[5] 46 FR 4078.

Table 12.3. Values for Age Correction of Hearing Loss.

	Audiometric Test Frequencies (Hz)				
Years	1000	2000	3000	4000	6000
		Males			
20 or younger	5	3	4	5	8
21	5	3	4	5	8
22	5	3	4	5	8
23	5	3	4	6	9
24	5	3	5	6	9
25	5	3	5	7	10
26	5	4	5	7	10
27	5	4	6	7	11
28	6	4	6	8	11
29	6	4	6	8	12
30	6	4	6	9	12
31	6	4	7	9	13
32	6	5	7	10	14
33	6	5	7	10	14
34	6	5	8	11	15
35	7	5	8	11	15
36	7	5	9	12	16
37	7	5	9	12	17
38	7	6	9	13	17
39	7	6	10	14	18
40	7	6	10	14	19
41	7	6	10	14	20
42	8	7	11	16	20
43	8	7	12	16	21
44	8	7	12	17	22
45	8	7	13	18	23
46	8	8	13	19	24
47	8	8	14	19	24
48	9	8	14	20	25
49	9	9	15	21	26
50	9	9	16	22	27
51	9	9	16	22	28
52	9	10	17	24	29
53	9	10	18	25	30
54	10	10	18	26	31
55	10	11	19	27	32
56	10	11	20	28	34
57	10	11	21	29	35
58	10	12	22	31	36
59	11	12	22	32	37
60 or older	11	13	23	33	38
		Females			
20 or younger	7	4	3	3	6
21	7	4	4	3	6
22	7	4	4	4	6
23	7	5	4	4	7
24	7	5	4	4	7
25	8	5	4	4	7
26	8	5	5	4	8
27	8	5	5	5	8
28	8	5	5	5	8
29	8	5	5	5	9
30	8	6	5	5	9
31	8	6	6	5	9
32	9	6	6	6	10

Table 12.3. Values for Age Correction of Hearing Loss. (continued)

Years	Audiometric Test Frequencies (Hz)				
	1000	2000	3000	4000	6000
33	9	6	6	6	10
34	9	6	6	6	10
35	9	6	7	7	11
36	9	7	7	7	11
37	9	7	7	7	12
38	10	7	7	7	12
39	10	7	8	8	12
40	10	7	8	8	13
41	10	8	8	8	13
42	10	8	9	9	13
43	11	8	9	9	14
44	11	8	9	9	14
45	11	8	10	10	15
46	11	9	10	10	15
47	11	9	10	11	16
48	12	9	11	11	16
49	12	9	11	11	16
50	12	10	12	11	17
51	12	10	12	12	17
52	12	10	12	13	18
53	13	10	13	13	18
54	13	11	13	14	19
55	13	11	14	14	19
56	13	11	14	15	20
57	13	11	15	15	20
58	14	12	15	16	21
59	14	12	16	16	21
60 or older	14	12	16	17	22

From 29 CFR 1910, Appendix F.

permissible time of exposure at that level (T_1, T_2, . . . ; see Table 12.4) to produce a fraction (C_1/T_1, C_2/T_2, . . .). As long as the sum of these fractions is less than 1, the site is in compliance.

EXAMPLE

A worker is exposed for 4 hr to 90 dB, 1 hr to 100 dB, and 1 hr to 97 dB. The rest of the time exposure is below regulated levels. Is the site in compliance?

$$\text{Exposure} = C_1/T_1 + C_2/T_2 + C_3/T_3$$

$$= 4/8 + 1/2 + 1/3$$

$$= 1\,1/3$$

The total exceeds 1, so the site is not in compliance. The difficulty of this determination is avoided when an employee wears a dosimeter.

In 1972 an advisory committee to the Department of Labor was appointed to prepare a standard for occupational noise exposure, and submitted recommendations in December 1973. A standard was then published in the *Federal Register* in 1974, setting the noise exposure standards listed in Table 12.4. The standard for an 8-hr day TWA is 90 dB. NIOSH has

Table 12.4. OSHA Regulations — Permissible Noise Levels.

Sound Level (dB)	Hours of Exposure per Day
80	32
81	27.9
82	24.3
83	21.1
84	18.4
85	16
86	13.9
87	12.1
88	10.6
89	9.2
90	8.0
91	7.0
92	6.1
93	5.3
94	4.6
95	4.0
96	3.5
97	3.0
98	2.6
99	2.3
100	2.0
101	1.7
102	1.5
103	1.3
104	1.1
105	1.0
106	0.87
107	0.76
108	0.66
109	0.57
110	0.5
111	0.44
112	0.38
113	0.33
114	0.29
115	0.25
116	0.22
117	0.19
118	0.16
119	0.14
120	0.125
121	0.11
122	0.095
123	0.082
124	0.072
125	0.063
126	0.054
127	0.047
128	0.041
129	0.036
130	0.031

From 29 CFR 1910.95, Appendix A, Table G-16a.

recommended that this be dropped to 85 dB, and on review the EPA also has recommended that this be lowered to 85 dB.[6] The ACGIH also recently decided on a standard of 85 dB.

[6] *Federal Register,* October 24, 1974.

OSHA amended the Occupational Noise Exposure Standard in April 1983, adding the Hearing Conservation Amendment. The standard for exposure was not changed in this amendment. However, OSHA included all noise between 80 and 130 dB for exposure evaluation, then set an action level at a calculated exposure of 0.5 (85 dB TWA). This is similar to the action level concept applied to exposure to chemicals, in that exposure above that level triggers necessary action by the employer. In this case the action called for is the establishment of a hearing conservation program. This requires the employer to design and implement workplace noise monitoring. Monitoring should include "all continuous, intermittent, and impulsive sound levels from 80 decibels to 130 decibels",[7] and from this survey should identify employees to be included in the hearing conservation program.

Workers whose exposure exceeds the action level must be notified. Such employees must be provided, at no cost, with hearing protection appropriate to the exposure level. They must also be provided, at no cost, with an audiometric testing program, administered by qualified individuals. A baseline analysis of each exposed employee's hearing should be done within 6 months of identification of the high noise exposure levels, then repeated at least annually. If interpretation of the retest audiogram reveals a threshold shift that is (1) greater than 10 dB at 2000, 3000, or 4000 Hz in either ear (adjustment for normal loss due to aging (presbycusis) may be made[8]) and (2) not due to non-work-related factors, the employee must be notified of this in writing. The following steps must be taken:

1. An employee not using hearing protection shall be fitted with such protection, trained in its use, and required to wear it.
2. Any employee already using hearing protection shall be refitted with more effective protection.
3. The employee shall be given a clinical audiological examination or an otological examination.

Questions are raised about the recordability of hearing damage on the OSHA 200 log. First, if the hearing loss is due to an instantaneous event, such as an explosion, this is an injury. Long-term accumulated loss is an illness. A June 1991 OSHA memo states that "OSHA will issue citations to employers for failing to record work-related shifts in hearing of an average of 25 dB or more at 2000, 3000, 4000 [Hz] in either ear on the OSHA 200 log." An employer must record a 25-dB shift within 6 days of receiving notice of an employee's hearing loss.[9]

HEARING PROTECTION

PERSONAL HEARING PROTECTION

Where it is unavoidable that employees work in areas with a high noise level,[10] personal hearing protection should be worn.[11] There are two types of protection: ear plugs and ear muffs. The effectiveness of hearing protection can be estimated by determining the quietest sound audible with the protection off and in place at each of a range of frequencies (Figure 12.5). Protective devices are supplied with a noise reduction rate (NRR), a number ranging from 5 to 31, and intended to communicate the reduction in noise exposure the device provides

[7] 29 CFR 1910.95(d)(2)(i).

[8] 29 CFR 1910.95, Appendix F.

[9] This paragraph is based on J. Smith, "Hearing Losses: Which Ones are Recordable?" *Occupational Health and Safety*, May 1995, p. 65.

[10] Workers may use a rule of thumb to estimate noise exposure in the absence of noise level measurement in their workplace. If you need to talk loudly or shout to be heard by someone an arm's length away, you are probably experiencing more than 85 dB. J. D. Royster, Chap. 13 in Northern, 1996 (see Bibliography).

[11] A list of available hearing protection devices is presented by manufacturer is presented by manufacturer in *Occupational Health and Safety*, March 1996, p. 28.

in decibels. This value is determined using C scale sound level data, however, and most workplaces are monitored using the A scale. A rule of thumb correction involves subtracting 7 dB from the NRR to adjust to A scale readings.[12] Ear muffs, if properly fitted to the wearer, are generally more effective attenuators, especially at higher frequencies. No hearing protection completely blocks noise. Sounds can be transmitted through the material of the muff.

Figure 12.5. These ear muffs are worn as personal hearing protection. Notice the noise reduction rating is 20 dB. The foam padding provides a tight fit around the ear, which is important if the device is to be effective. (Photo courtesy of Cabot Safety Corporation, Southbridge, MA. With permission.)

Ear plugs may be inserted into the ear canal to block the entrance of sound (Figure 12.6). These are generally soft rubber, plastic, or wool material, or are compressible foam cylinders that are squashed and inserted into the ear, where they expand again to fill the ear canal. By filling the canal tightly, they block the direct passage of sound through the air. Some ear plugs are connected by cords, which minimize loss if they are removed momentarily for conversation. Ear plugs are small, however, and sound may be transmitted through the plug itself. They are particularly useful when exposure to high sound levels is infrequent, because they can be carried conveniently in the shirt pocket to be slipped into the ears at any time.

Ear muffs fit over the head to place a padded cover completely over the ear, which then fits tightly against the head. Because sound, particularly high-frequency sound, enters the ears through the leaks in the seals, the better the seal against the head, the more effective is the protection. Pads need to be large enough to completely cover the ear lobes, and leakage problems are created by the wearing of glasses or eye protective devices. Anything worn over the eyes should be held in place by bands around the head, rather than by earpieces.

It is necessary to conduct good training programs to convince workers of the need for protection from high noise levels. Unlike an injury due to an unguarded hazard, which is observed immediately, hearing loss takes place slowly over a long time period. Workers grow accustomed to high noise levels, are inconvenienced by the protective device, see no imme-

[12] J. Banach, "The ABC's of Noise Measurements Set the Stage for Responsive Controls," *Occupational Health and Safety*, Oct. 1994, p. 75.

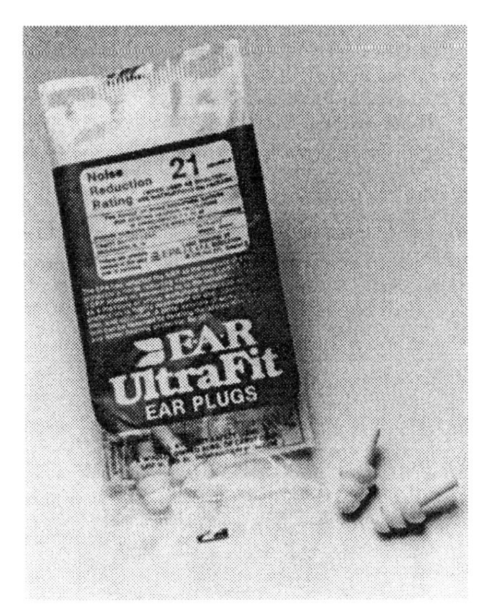

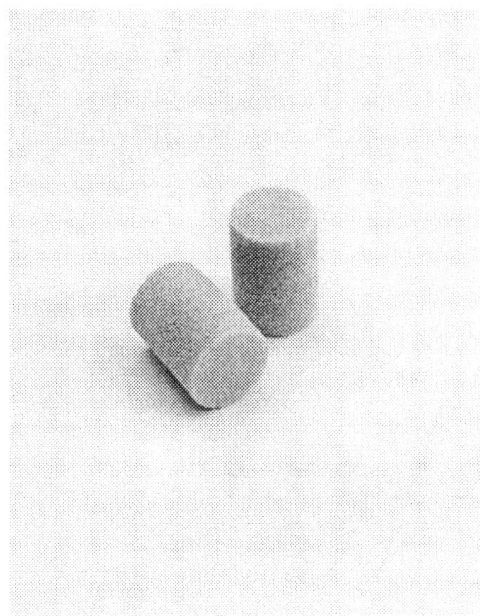

Figure 12.6. Ear plugs are a very portable form of hearing protection that can be carried in the pocket and inserted in noisy areas. In each case the plug is compressed on insertion into the ear canal, and the plug then expands to fill the canal effectively. (Photo courtesy of Cabot Safety Corporation, Southbridge, MA. With permission.)

diate benefit from suffering that inconvenience, and will remove the protection. Beyond explaining the need for protection, it may be useful to demonstrate its effectiveness. Run an audiogram of two workers at the start and finish of the day, one with and one without protection; the advantage in terms of temporary hearing loss prevention should make a strong case to the workers.

It is important that hearing protection be fitted properly and be as comfortable as possible. Proper fit is essential for the protective device to be effective. Studies show that the actual attenuation of sound provided by a protective device may be only one-third to one-half the NRR if the fit is not correct.[13] Further, a comfortable fit eases the problem of convincing the worker to wear the device.

ENGINEERING CONTROLS

It is more satisfactory to control the sound level in the workplace than to block the ears of the workers. Workers wearing ear plugs or ear muffs are out of communication with one another, and the risk of hearing damage is always present if workers are careless or recalcitrant about wearing protection. Designing engineering controls is a complex field, and such services are best provided by specialized experts. Industrial hygienists are primarily concerned with analysis of the success of such a program, so comments are restricted to a brief discussion of methods employed. References that lead the reader to more detailed presentations are provided.

Air Exhaust

Because sound is propagated vibrations of air, it is no surprise that any system deliberately moving air can generate noise. A sudden escape of compressed air is equivalent to the sudden generation of gases in an explosion. Some machines use compressed air or steam to move

[13] D. M. DeJoy, "Tactics Against Workplace Hearing Loss Motivate Employees to Wear Protection," *Occupational Health and Safety*, March 1994, p. 50.

machine parts in a work cycle. For example, the hammer of a forge may be raised or the plunger of an injection molding machine may be moved by compressed air. Leaks in these high-pressure systems add a continuous hiss of escaping air or steam. At the end of the machine cycle, the air is released, generating loud noise. Good maintenance eliminates the leaks, and the points where air is deliberately released can be fitted with mufflers.

Ventilating systems should not be forgotten. Some fan blade designs are inherently quieter, and larger but lower speed fans move the same volume of air, but emit less high-frequency sound. Sheet metal of ducts vibrates readily at high frequencies, but when coated with damping material the ducts no longer vibrate at these frequencies. Inserting flexible sections into ducts prevents transmission of fan noise throughout the facility.

Controls Related to the Machinery

Machinery that vibrates or otherwise generates sound waves is a noise source. Some benefit may result from improved machine maintenance. Worn, loose fitting, or unbalanced parts vibrate to produce noise. Improved lubrication eliminates some noise directly and postpones wear that later increases machine noise levels. Cutting oils lower sound levels of cutting operations and extend tool life.

Some machine designs are noisier than others that perform the same task. For example, hydraulic presses are quieter than mechanical presses. Analysis of the contribution of each machine to the total noise problem focuses attention on the noisiest machines, allowing considerable improvement in sound levels at the time of machine replacement. Similarly, some production methods are inherently noisier, and a change in process may result in significant noise reduction. Replacing riveting with welding is an example.

Alteration of the design of existing machines can bring about improvement. A lightweight machine part may vibrate at a high, harmful frequency. Adding weight to the part lowers the frequency of vibration. Vibration is more intense in a flexible part, so stiffening structures can be helpful.

Sound is transmitted through solids the machine contacts. Placing a machine on a resilient floor or on its own concrete pad separated from the rest of the floor by resilient spacers eliminates a major route to transmit vibration away from the machine. Changing machine fastening devices to include resistant pads may isolate and dissipate sound at its source. Placing flexible sections in pipes connected to the machine prevents this plumbing from transmitting noise to other areas.

Often alteration or replacement of machines or processes involves trade-offs of operations cost or ease of maintenance for lower noise levels. In such cases all the alternatives for noise control must be weighed and compared.

Room Design

If one suspends sound sources in the air, they radiate sound energy in all directions. Nearby persons are subjected only to sound propagated in their direction. Furthermore, the farther they are from the source, the weaker is the sound pressure they receive. In principle, sound follows the inverse square law, meaning that the fraction of sound reaching an individual is inversely proportional to the square of the distance. A person twice as far from the source should be exposed to only one-quarter the sound level.

However, if this source is placed in a room, reflection of the sound from the walls, floor, and ceiling supplements the direct sound. Walls and ceilings vary in their ability to reflect rather than absorb sound. It is very much like reading a book in a room with a single light bulb. The farther away the bulb is, the less light reaches the book (once again following the inverse square law). However, a larger fraction of the light reaches the book if the walls are painted a highly reflective white. Just as painting the room walls matte black reduces the

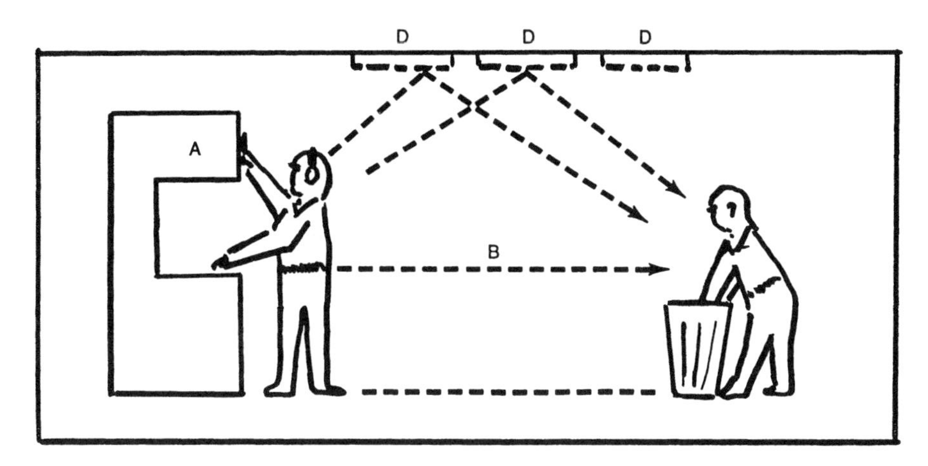

Figure 12.7. Rooms can be altered to lower sound exposure of the worker. Here a machine (A) produces a high noise level. The sound is propagated directly (B) and by reflection (C) to other workers in the room. Adding sound-absorbent panels (D) to the walls or ceiling reduces the exposure of other workers.

light reaching the book, replacing surfaces that are highly reflective to sound with acoustic panels reduces reflected sound (Figure 12.7).

Sound Barriers

Enclosures are sometimes the best choice for effective management of a noise problem. One can enclose the machine, enclose the worker, or place a barrier between them. The walls of any of these barriers transmit some amount of sound. Sound strikes the barrier, causing it to vibrate. This vibration, in turn, causes the air on the other side to vibrate. Factors in the effectiveness of sound transmission by the barrier include the mass of the barrier, its area, and its construction materials. Enclosures or walls have doors, and the tightness of the seal around that door is critical to blocking sound. Acoustic paneling inside an enclosure absorbs sound, reducing the reinforcement of sound by reflection.

Enclosing a machine usually provides a 20- to 30-dB reduction in noise (Figure 12.8). Such enclosures must take account of machine operation and employee safety as well as acoustic effectiveness. Partial enclosures that eliminate the direct path of sound from the source to the worker's ear and are covered with acoustic material can be effective, providing sound exiting from the unenclosed sides of the machine cannot simply reflect from walls or ceiling to the worker. Where a segment of the machine is the prime source of noise, just that structure might be enclosed, providing improved sound levels at lower cost and inconvenience.

Enclosing the worker allows other improvements such as air conditioning to reduce heat stress and air filtration to reduce airborne contaminants (Figure 12.9). It is also often cheaper to build a small worker enclosure than large machinery enclosures that also reduce dissipation of heat from the machine or access for maintenance.

KEY POINTS

1. The ear has three anatomical parts: (1) the ear canal, connecting the outdoors to the eardrum, (2) the middle ear, containing the bones that transmit eardrum vibrations to the inner ear, and (3) the inner ear, which includes the cochlea. The cochlea transforms mechanical vibrations to nerve impulses. The inner ear also includes the semicircular canals, which provide a person's sense of equilibrium.

Figure 12.8. Noisy operations can be enclosed to protect workers where remote operation of the machinery is possible. The effectiveness of noise control depends on good engineering practices in the enclosure design. Here, diesel engines of from 1000 to 4000 hp are tested in a Kentucky shop that reconditions locomotive engines. Unchecked noise levels exceed 120 dB. The enclosure was designed to control sound levels to no more than 75 dB in the control room. (Photo courtesy of Industrial Acoustics Company, Inc., Bronx, NY. With permission.)

2. Sound is a series of air compressions and rarefactions propagated by a vibrating source.
3. Sound is visualized to have a wave form. The frequency or wavelength of the wave determines the tone or pitch, and the amplitude measures the loudness.
4. Frequencies are measured in hertz (Hz, s^{-1}). Undamaged human hearing detects sound from 20 to 20,000 Hz.
5. Octaves are pure tones wherein the frequency of the higher pitched tone is twice that of the lower pitched tone. For analysis, the hearing range is divided into eight octave bands or 26 third-octave bands numbered from 14 to 40.
6. Human speech falls primarily in the range from 500 to 3000 Hz. Loss of hearing in this range hinders communication.
7. Sound power is measured in watts, and sound pressure, the actual sound level measured, has pressure units. The quietest audible sound is approximately 10^{-12} watts sound power or 2×10^{-5} N/m^2 sound pressure.
8. Units of sound level — decibels — are 20 times the log of the ratio of the sound pressure to 2×10^{-5} N/m^2.

Figure 12.9. Another approach to protecting workers in a noisy environment is the provision of a quiet room, such as this one in the *Chicago Tribune* pressroom. During a press run the noise level is 110 dB, but inside the quiet room it is less than 75 dB. (Photo courtesy of Industrial Acoustics Company, Inc., Bronx, NY. With permission.)

9. Sound level meters read sound pressure directly in decibels.
10. Continuous exposure to high sound levels damages hearing, particularly at the frequency of the source.
11. Hearing is tested with an audiometer, a device that generates pure tones of known loudness.
12. OSHA limits exposure to 90 dB for an 8-hr day, higher levels being permitted for shorter durations.
13. Personal hearing protection includes ear plugs and ear muffs.
14. Sound levels in the workplace are reduced by good machine maintenance, substitution of quieter machinery, and the erection of nonreflective panels and sound barriers.

BIBLIOGRAPHY

American Occupational Medicine Association, Noise and Hearing Conservation Committee, "Occupational Noise Induced Hearing Loss," *J. Occup. Med.*, 31, 996, 1989.

L. L. Beranek, *Noise and Vibration Control,* Institute of Noise Control, Washington, D.C., 1988.

R. L. Berger, W. D. Ward, J. C. Morrill, and L. H. Poyster, Eds., *Noise and Hearing Conservation Manual,* AIHA, Akron, OH, 1990.

A. L. Dancer et al., Eds., *Noise Induced Hearing Loss,* Mosby Year Book, St. Louis, MO, 1993.

C. M. Harris, Ed., *Handbook of Acoustical Measurements and Noise Control,* 3rd ed., McGraw-Hill, New York, 1991.

M. Hirschorn, *Compendium of Noise Control Engineering, Part I, Sound and Vibration,* July 1987, pp. 16–32; *Part II, Sound and Vibration,* February 1988, pp. 16–28.

B. F. Jaffe and D. W. Bell, Workplace Noise and Hearing Impairment, in B. S. Levy and D. W. Wegman, Eds., *Occupational Health,* Little, Brown and Company, Boston, 1983.

K. D. Kryter, *The Effects of Noise on Man,* 2nd ed., Academic Press, New York, 1985.

J. L. Northern, Ed., *Hearing Disorders,* 3rd ed., Allyn & Bacon, Needham Heights, MA, 1996.

J. B. Olishifski, Occupational Hearing Loss, Noise and Hearing Conservation, in Carl Zenz, Ed., *Occupational Medicine, Principles and Practical Applications,* 2nd ed., Year Book Medical Publishers, Chicago, 1988.

B. A. Plog, Ed., *Fundamentals of Industrial Hygiene,* 3rd ed., National Safety Council, Chicago, 1988.

J. D. Royster and L. H. Royster, *Hearing Conservation Programs: Practical Guidelines for Success,* Lewis Publishers, Chelsea, MI, 1990.

R. T. Sataloff and J. Sataloff, *Occupational Hearing Loss,* 2nd ed., Marcel Dekker, New York, 1993.

R. L. Stepkin and R. E. Mosely, *Noise Control: A Guide for Workers and Employers,* American Society of Safety Engineers, Des Plaines, IL, 1984.

A. Thumann and R. K. Miller, *Fundamentals of Noise Control Engineering,* Prentice-Hall, Englewood Cliffs, NJ, 1986.

PROBLEMS

1. Explain how each of these structures of the ear contributes to the hearing process:
   - A. Eustachian tube
   - B. Cochlea
   - C. Stirrup
   - D. Semicircular canals
2. Visualize the wave pattern used to represent sound.
   - A. What do the peaks represent?
   - B. What dimension is termed amplitude?
   - C. When we talk about the pitch of a sound, this relates to what dimension on the wave pattern?
   - D. Which of these dimensions relates to loudness?
   - E. What is the relationship between wavelength and frequency?
   - F. What is the relationship between wavelength and amplitude?
   - G. What are the units of frequency?
   - H. If a tone has a frequency of 1000 Hz, what tone is an octave higher? two octaves lower?
   - I. Under normal conditions, what is the wavelength (ft, m) of a 2000-Hz tone?
3. A, What is the approximate range of high-quality human hearing?
   - B. What is the approximate range of human speech?
   - C. At what frequency does damage to hearing first become evident?
   - D. Will typical early damage to hearing result in a noticeable loss in ability to interpret human speech?
   - E. Is human speech at the range of greatest acuity in human hearing?
4. A. What is sound power and sound pressure, and what are likely units of each?
   - B. What are the units of sound pressure level (L_p) and why are they not pressure units?
   - C. What is the usual reference sound pressure for sound pressure level and why was it chosen?
   - D. A sound with a sound pressure of 40 N/m^2 has what sound pressure level?
   - E. On a typical sound level meter, how would you read the sound pressure level in (D)?
5. An audiogram is a graph.
   - A. What are the x and y axes?

B. What would be the appearance of an audiogram for a person with perfect hearing?
C. With what tool is hearing tested, and how are points obtained for the audiogram?
D. What is the guideline for describing hearing loss as impairment at any tested frequency, and at what frequencies is impairment most serious?
E. Two workers employed in a moderately noisy environment are tested for hearing loss. The audiogram of F.B.T. shows a steady drop from middle toward high frequencies, whereas J.J.H. displays a sharp notch at 4000 Hz. What is a preliminary diagnosis?

6. Noise level studies are done in a metal working plant at three work stations:
   A. A worker is exposed to 89 dB for 2 hr, 87 dB for 1 hr, 93 dB for 3 hr, and 91 dB for 2 hr.
   B. A worker is exposed to 83 dB for 6 hr, 96 dB for 2 hr.
   C. A worker is exposed to 87 dB for 4 hr, 81 dB for 4 hr.
   Calculate whether each worker is in compliance and, if so, is above the action level.
7. A small concrete block factory building employs 12 workers in a single room. A worker stands in front of a small mechanical press at one end of the room, feeding sheet metal to the machine and transferring stamped parts to a bin on wheels in which they are rolled to workstations in the rest of the room. The other workers process the stampings into finished parts. The stamping machine produces noise levels above OSHA limits, and the operator must wear hearing protection. Suggest at least two engineering controls that might lower levels sufficiently for the rest of the workers.
8. Assume that audiometric readings at age 20 would produce an audiogram that looks like Figure 12.4A (probably a bad assumption considering that teenagers tend to listen to extremely loud music). Xerox three copies of this table (enlarging if you can). Using the data from Table 12.3, plot curves for males on one graph and females on the other as they would be expected to appear at ages 35, 52, and 60. On the third copy plot male and female values for comparison at age 50.
9. Three workers in a stamping plant have been tested for hearing acuity at a 5-year interval. Here is the data:

			Audiometer Readings at Indicated Frequencies (Hz)				
Worker	**Sex**	**Age**	**1000**	**2000**	**3000**	**4000**	**6000**
A	M	42	13	14	16	22	21
A		47	14	16	20	33	25
B	F	34	11	7	8	7	12
B		39	12	8	11	10	15
C	M	37	12	11	16	22	26
C		42	13	14	20	30	34

   Calculate the hearing change in each worker, and correct it for age. Comment on hearing loss in each case.
10. We are given this relationship between sound power (L_W) and sound pressure (L_P):

$$L_P = L_W - 20 \log d + 10 \log Q - 11$$

where d = the distance from the source in meters
Q = a correction factor for the reflection of sound from surfaces near the source (Thus, if the source were suspended in the air, Q = 1, but as surfaces are added nearby, Q becomes larger.)

An injection molding machine for plastics is listed as having a sound pressure of 88 dB at 10 m where Q = 1. Calculate some values for this machine in different settings.
   A. First, calculate the acoustic power (L_W) of the source.
   B. Now get some values for the effect of distance from the source. We know that the L_P at 10 m is 88 dB. What is L_P at (1) 2 m and at (2) 30 m.

C. Of course, the machine is not run suspended in space. Calculate the L_p at 10 m when the machine is (1) on a reflective floor (Q = 2) and in a corner with reflective walls (Q = 8).

11. In an article about hearing conservation (*Occupational Health and Safety,* March 1994, p. 58.), D. J. Burton lists seven steps to a good program in the workplace. Think about the material in this chapter and prepare your own such list, then compare with the list from the article, given in the Answers section at the end of this book.

CHAPTER 13

Radiation

The development and use of atomic weapons during World War II sharply increased awareness of radiation hazards and stimulated extensive research in radiobiology, the study of health effects of radiation. Research and health care uses of radioactive materials, the development of atomic power plants, and the rise of other potential pathways for radiation exposure made the acquisition of this information important and generated a field called health physics, individuals trained to enforce radiation safety guidelines.

From the standpoint of hazard to workers, radiation is generally classed in two groups: nonionizing and ionizing. Nonionizing radiation includes electromagnetic radiation (light) no more energetic than ultraviolet (UV) light. Of this class, UV light itself poses the greatest threat, and we shall limit our discussion to it. Ionizing radiation generates ion pairs as it passes through matter. Ion pairs generated in living tissue decompose or transform themselves into more stable structures by a variety of routes and produce damaging intermediates in the process. Five types of ionizing radiation pose a threat to workers: α, β, and γ radiation; X-rays; and neutrons. α and β radiation and neutrons are particles, while γ-rays and X-rays are very-high energy electromagnetic radiation.

NONIONIZING RADIATION — ULTRAVIOLET LIGHT

Light is pure energy, termed electromagnetic energy because of its character, and we most conveniently discuss various types of light by reference to wave characteristics. The energy of light relates to its wavelength — the shorter the wavelength the higher the energy. UV light with wavelengths shorter than 200 nm is of little concern here because it is rapidly absorbed by air.

Evidence of the ability of UV light to do biological damage is given by the use of UV sources to kill bacteria. In a familiar example, barbers have for years placed their tools under UV lamps when not in use. UV light has little penetrating power, so concerns for workers are limited to skin and eye damage. Eye damage was discussed in Chapter 6.

DAMAGE TO SKIN

Skin exposed to sufficient UV light burns. Skin pigmentation is a protection, so workers with dark skin withstand higher doses without symptoms. With chronic exposure there may be damage to skin structural elements, producing a premature aging of the skin. UV light with wavelengths below 325 nm is more damaging. Studies have also established that people with less skin pigment and/or with greater exposure to UV light (generally sunlight) run a higher risk of skin cancer.

WORKERS AT RISK

Obviously, outdoor workers of all sorts are at special risk from sunlight. This is particularly true in summer, especially in warmer climates, because workers take off their shirts, wear shorts, and generally "work on their tans". Lotions are available containing compounds that absorb (block) UV light. They are compounded to provide differing degrees of protection and should be used by outdoor workers in warm weather.

In addition, there are many people employed at indoor jobs where UV lamps are used for bactericidal purposes: food processors, nurses, barbers, hairdressers, and pharmaceutical company and tobacco company employees. At particular risk are arc welders, because the flash of light produced during welding is an intense UV source. The UV of the welder's arc also converts oxygen in the air to ozone, which can cause respiratory problems.

IONIZING RADIATION

RADIOACTIVE ATOMS

Stability of Atoms

Atoms have nuclei composed of protons and neutrons. These particles are similar in mass, but the proton has a positive charge, whereas the neutron is uncharged. The identity of an atom is determined by the number of protons in the nucleus. Thus, every atom with six protons is carbon and every atom with seven is nitrogen. Although every carbon atom has six protons, the number of neutrons need not always be the same. Most carbon atoms have six neutrons, but a small proportion of atoms in all natural collections of carbon have eight neutrons. These variations of carbon would be designated ^{12}C and ^{14}C, respectively, the 12 and 14 indicating the total number of nuclear particles. Such variants of an element are termed *isotopes* or *nuclides*.

Stability in an atom requires a correct ratio of neutrons to protons, approximately 1:1 in low-atomic-weight atoms, but with an increasing proportion of neutrons required for stability as the atoms become larger. Although assembly of an atom with a neutron-to-proton ratio that is far distant from the stable ratio is impossible, atoms exist which are near, but not at, a stable ratio. These atoms, termed *radioactive nuclides*, adjust the properties of the nucleus toward greater stability by emitting particles and energy. Such emissions are the ionizing radiation produced by unstable nuclei.

Degree of Instability: Half-Life

Not all unstable nuclei are identically far away from a stable arrangement. Thus, if the stable ratio is 32 neutrons to 27 protons (32/27; ^{59}Co), and a 34/27 (^{61}Co) ratio is wrong enough to be unstable, it is not surprising that a 36/27 (^{63}Co) ratio is farther from a stable ratio and should by some measure be even less stable. This statement is an oversimplification of the rules of stability, but leads to the question: What measures relative instability in radioactive nuclides?

An unstable nuclide has available a pathway involving nuclear change and emissions to become a different and eventually more stable nuclear structure. However, not all atoms in a sample of unstable nuclide embark on that pathway immediately. They remain in the unstable configuration until, at a time that cannot be predicted for an individual atom, alteration occurs.

If one observes a quantity of a radioactive nuclide, one sees an apparently continuous stream of emissions resulting from this random decaying of individual atoms in the sample. As time passes there is a predictable reduction in levels of emissions because there are decreasing numbers of unstable atoms left to serve as sources.

More unstable nuclides on average wait shorter periods of time before decaying. Relative instability is therefore measured by the rate at which radioactive nuclei decay. If there are two radioactive nuclides of differing instability, a greater percentage of the less stable nuclide decays per unit time. When comparing equal numbers of atoms the less stable nuclide thus has a higher level of radioactive emissions. As a consequence, a larger proportion of the less stable nuclide decays to become a different nuclide in a given period of time.

As an example, spent fuel removed from a nuclear power plant initially produces high levels of emissions. Radioactive nuclides resulting from the fission process vary greatly in stability. Less stable nuclides disintegrate at a very rapid rate, the primary source of those very high initial levels of emissions. At such high rates of disintegration, these unstable nuclides rapidly disappear, causing the level of radioactive emission to drop sharply. After a period of time, the emissions primarily originate from less unstable nuclides, which are decaying more gradually, and the rate of reduction in levels of emissions slows. Thus, simply storing the spent fuel in a vault at the reactor site for a short time results in a sharp reduction in the levels of emissions (and thus the hazard) of the waste. However, long-term storage is then necessary to allow the less unstable nuclides to decay and disappear.

Scientists express the difference in tendency to decay with a unit termed the *radioactive* or *physical half-life* ($T_{1/2}$) of the nuclide. Half-life is the length of time required for half the atoms in the sample to decay. If we start with 10 g of a nuclide and 10 min later we still have 5 g, the other 5 g having become decay product, the half-life of the nuclide is 10 min. Another unit to express the same concept is the decay constant (λ) of the nuclide, which is the fraction of the atoms in a sample that decay per unit time. Half-life and decay constant are related as follows:

$$T_{1/2} = 0.693 / \lambda$$

Half-life is a constant value for a given nuclide, and in various nuclides these values vary from fractions of a microsecond to millions of years. In the example above of the two unstable n/p ratios the half-life of ^{61}Co is 1.6 hr, and of ^{63}Co is 27.5 s.

Types of Radioactive Emissions

Unstable heavy atoms such as uranium or plutonium may become more stable by a reduction in size, and they accomplish this reduction by emitting a fragment of the nucleus called an α particle. An α particle is composed of two protons and two neutrons, the largest and most highly charged of the radioactive particles.

In atoms with too many neutrons for their number of protons, a neutron may spontaneously convert into a proton. An uncharged neutron must lose a negative charge to assume the positive charge of a proton. This is accomplished by emission of an electron, which has a negative charge and only a tiny mass. An electron emitted from the nucleus is termed a β particle. A β particle has only half the charge and slightly more than ten-thousandth the mass of an α particle.

In atoms with too many protons for their number of neutrons, a proton may spontaneously convert into a neutron. A proton must lose a positive charge to become neutral. This is accomplished by emission of a positron, which is a positively charged electron. Positron emitters are uncommon, and will not be considered further in this discussion.

Changes in the composition of the nucleus are likely to leave the protons and neutrons in an arrangement that is not the most stable possible. As the particles shift into a more stable

SAMPLE PROBLEM:

It is possible to calculate the level of radioactive emission for a given nuclide at some time in the future if the half-life and the present level of emission is known. The formula is as follows:

$$N_t = N_0 e^{-\lambda t}$$

where N_0 = present rate of emission
N_t = rate of emission at time t
e = base of natural logs
λ = decay constant
t = time for which N_t is calculated

The present rate of emission of a sample of ^{61}Co is 500 cpm (counts per minute). What will it be in 5 hr?

A. First calculate the decay constant:

$$T_{1/2} = 0.693 / \lambda$$

$$\lambda = 0.693 / T_{1/2}$$

$$\lambda = 0.693 / 1.6 \text{ hr}$$

$$\lambda = 0.433 \text{ hr}^{-1}$$

B. Now calculate N_t:

$$N_t = N_0 e^{-\lambda t}$$

where $-\lambda t$ = $-(0.433 \text{ hr}^{-1})(5 \text{ hr})$
= -2.17
N_t = $(500 \text{ cpm}) e^{(-2.17)}$
= 57 cpm

configuration, thus leaving a higher energy arrangement to assume one of lower energy, the energy difference is released as very high-energy light termed gamma (γ) radiation. Gamma rays are often emitted slightly after the release of a particle from the nucleus. As light, they have neither mass nor charge. They are very much higher in energy than the UV light discussed earlier.

CHARACTERISTICS OF IONIZING RADIATION

Various types of radiation have different properties, but all ionizing radiation has in common the ability to produce ion pairs when passing through matter. Chiefly they differ in their penetrating power and in the level of ion pair formation they produce.

α AND β PARTICLES

A charged particle emitted by the nucleus leaves with an amount of energy that is measured by its velocity. As it passes an atom, its charge causes it to remove an electron from the atom, creating an (electron/positive ion) ion pair. This interaction with an electron, whether from attraction by a positively charged particle or repulsion by a negatively charged particle, uses some of the particle energy, causing a reduction in the velocity of the particle. As more ion pairs are produced, the particle eventually is stopped.

The velocity of α particles differs from one nuclide to another, but is constant for any given nuclide. With their +2 charge, α particles are extremely effective at producing ion pairs. In air an α particle produces from 30,000 to 100,000 ion pairs per centimeter traveled. With the relatively high density of atoms encountered in tissue, ion pair production is very much higher, so α particles usually penetrate only 50 μm, the thickness of a few cells, before dispersing their energy completely. Alpha radiation from an external source does not even penetrate the already dead keratin cell layer of the skin, and is thus unlikely to do damage. However, if an α emitter gets into the body, the α particle produces a large number of ion pairs at the place it is released, so such an emission can do extensive local damage.

The –1 charge of the β particle results in the formation of about 200 ion pairs per centimeter in air. At a given point in tissue the passage of a β particle causes less local damage than does an α particle. However, this lower rate of dispersal of particle energy results in much longer path lengths, on the order of a few centimeters of tissue. Exposure to an external β emitter is therefore hazardous.

γ AND X-RAYS

Both γ and X-rays are photons of light, and as such are pure energy, lacking either mass or charge. The energy of such emissions is measured by the frequency or wavelength,[1] with greater energy associated with higher frequency (shorter wavelength). The greater the energy of the light, the greater its penetrating power.

X-ray tubes produce a range of X-ray energies, depending on which of the electrons of the target atoms in the tube are excited, up to some maximum value. Higher maximum energy X-rays result when higher voltage is applied to the X-ray tube. Often the lower energy, thus less penetrating or "softer", X-rays are removed by a metal screen, allowing only the high-energy rays to be used.

The frequency of γ-rays depends on the radioactive atom that is the source. It is possible to identify the nuclides present by analyzing the frequencies of the γ-rays coming from a complex source. In general γ-rays are higher in energy than are X-rays.

Because high-energy photons are without charge, their mechanism to generate ion pairs is different from that of charged particles. The photon strikes the electron layers of an atom, and energy of the light is absorbed by an electron. The electron now has too much energy to remain attracted by the nucleus and leaves, generating an (electron/positive ion) ion pair. The production of ion pairs per centimeter of air is lower for γ- or X-rays when compared to α or β particles.

Lower energy photons (less than 100 keV) impart their energy to an electron in an atom, which then leaves the atom and becomes the causative agent for formation of a number of other ion pairs. Higher energy radiation is more damaging because the photon loses only part of its energy to the electron of an atom, which is sufficient to remove it from the atom. The photon then continues in its path, retaining sufficient energy to cause the ejection of further electrons, a process called Compton scattering.

[1] Frequency is related to wavelength. The product of the two is the speed of light (3.0×10^8 m/s in a vacuum).

NEUTRONS

Neutrons are emitted with a variety of energies, and the energy of the neutron determines its penetrating power. Low-energy neutrons pass through perhaps a centimeter of tissue, whereas high-energy neutrons may pass through several centimeters.

Because a neutron is uncharged, it does not produce ion pairs directly when passing through air or tissue in the manner of α or β particles. A variety of changes may be generated by collision with other atoms. A nucleus may be broken into charged fragments or may have a proton ejected. Such charged pieces are now potent ion pair generators. As neutrons lose energy by such collisions, they become low enough in energy to be "captured" or incorporated into a nucleus, usually of a hydrogen or nitrogen atom in tissue. This nucleus is made unstable by the capture, and may decay radioactively after a short time. The ionizing effects of a neutron are termed *secondary emissions*, because they are due to the changes brought about by the neutron rather than by the neutron itself.

SHIELDING

Protection from a source of radiation is provided by shielding, the placing of matter around the source. As radiation interacts with the matter, it disperses its energy and is stopped. Lead is associated with shielding, and it is an excellent material to use for this purpose. However, its only special characteristic is that it is a very dense form of matter. Any substance can act as a shield, as long as enough matter surrounds the source to stop the emissions. Lead is often used because radiation from a source may be blocked without resorting to ungainly thicknesses of shielding.

The amount of matter necessary to shield a source depends on the penetrating power of the emissions. Relatively little shielding is needed for α particles, but γ-rays require a great deal. One method to express the penetrating power of a particular source of radiation is to measure the thickness of layer of a particular material needed to block half the emissions, termed the *half-thickness* in that substance.

EXPOSURE TO IONIZING RADIATION

We are constantly exposed to radioactive emissions because a portion of the atoms around us or even part of us are radioactive nuclides. For example, the ^{14}C previously described as being part of all carbon samples is a β emitter. In addition, radioactive nuclides are sometimes used in the workplace. As a result, in some workers natural exposure to radioactive emissions may be supplemented by exposure in the workplace. These exposures tend to be moderate. One would anticipate very high levels of exposure only in cases of nuclear war, a major power plant disaster, or a special and uncommon research situation.

Radioactive Nuclides in the Workplace

Because they act chemically exactly as do the stable nuclides, radioactive nuclides are used as tracers. By detecting their radiation we can follow compounds in which radioactive nuclides are included without disturbing the chemical changes being undergone by those compounds. This is useful in research laboratories, and generally involves only small quantities of radioactive material.

Hospitals use radioactive tracers diagnostically and in treatment. Four isotopes are commonly used: ^{99}Tc, ^{67}Ga, ^{133}Xe, and ^{131}I. All these have half-lives of 8 days or less. For example, the efficiency of concentration of known doses of radioactive iodine by the thyroid gland measures the activity of that gland. Such a study involves dosing the patient with low amounts of the isotope and risk of exposure is low, although careful handling procedures are required. Radioimmunoassay is a clinical analytical method that employs antibodies to do selective studies and depends on radioactively labeled versions of the compound being measured to arrive at the quantitative measurement. This procedure is done on specimens in an appropriately equipped laboratory and does not involve dosing patients.

Powerful γ radiation sources may be used to kill cells. This is useful as a means of killing bacteria and fungi in packaged food to preserve it. Such sources are also used in hospitals to destroy inoperable tumors. Sources of radiations such as ^{60}Co, ^{192}Ra, or ^{137}Cs are housed in heavily shielded containers that have a gate that may be opened to direct a narrow beam at the tumor.[2] Another approach is the insertion of a tiny container of a radioactive source directly into the tumor. This may be removed later, unless the isotope half-life is very short.

X-RAYS

As first observed by Roentgen in 1895, when a sample of metal is bombarded with high-voltage electricity, electrons surrounding the nucleus of the metal atoms acquire high energy. They lose this energy again in a short time in the form of high-energy light called X-rays. X-rays are similar to γ-rays, but fall in a somewhat lower range of energy. Such high-energy light can penetrate matter to a marked degree, the characteristic of X-rays that is most often utilized. Examples include medical diagnosis and the checking of luggage in airports. Although small doses bombard us from outer space and small amounts are produced by television or computer screens, our exposure to X-rays is primarily from devices deliberately designed to generate them.

Exposure to X-rays is a special concern in health care facilities. Most often diagnostic X-rays are taken in rooms designed for this purpose. Such rooms protect the technician operating the equipment by having the X-ray unit energized from a position, usually a control booth, that is shielded. The room should also be shielded so that there is no leakage of radiation to adjacent rooms. Only one examination at a time should be conducted in a room so that there is no danger that one technician is in an unprotected area while another is energizing equipment. Normally only the patient is in the room when the equipment is energized, but if it is essential for another person to be in the examination room, for example, when a child or a very weak patient is examined so that assistance in positioning is required, that person should wear protective clothing including a protective apron and gloves.

Mobile X-ray units can be useful in a hospital to deal with special situations involving the condition of the patient. Greater hazard is associated with this practice because the equipment is not in the confines of the specifically designed shielded room. Great care must be taken in positioning the equipment so that the radiologist or other patients are not in the path of the X-ray beam. The operator should wear protective clothing.

Dental X-ray units should be used with the same precautions as hospital equipment. There is actually less hazard here because the X-ray tube has lower output.

Neutrons

When a fission reaction occurs in an atomic power plant, neutrons are released. This is the only likely source of exposure to these particles, unless a person is in a research laboratory with a particle accelerator or deliberate neutron source such as ^{252}Cf.

[2] National Council on Radiation Protection and Measurements, "Dosimetry of X-Ray and Gamma Ray Therapy in the Energy Range 10 keV to 50 keV," NCRP Report No. 69, Washington, D.C., 1981.

MEASUREMENT

In order to work with ionizing radiation quantitatively, it is necessary to have devices to detect the radiation and a standard set of units used to express levels of radiation. Researchers who first detected ionizing radiation used photographic film, which is exposed by the radiation. More sophisticated devices have since been developed, but as you will see, film to estimate personal exposure to radiation still is used.

MEASURING DEVICES

Counting devices in use are based on two general principles: measurement of the degree of ionization produced in a sample of gas and the amount of light generated when certain liquids or crystals are irradiated. Personal dosimeters use additional detection techniques.

Geiger-Muller Counters

The Geiger-Muller meter (GM counter) is a very common and much used device for radiation measurement that is based on ionization of gases. Portable battery-powered GM counters can conveniently be taken into the workplace to measure exposure at the workstation, contamination of clothing, or contamination of the skin of the worker (Figure 13.1).

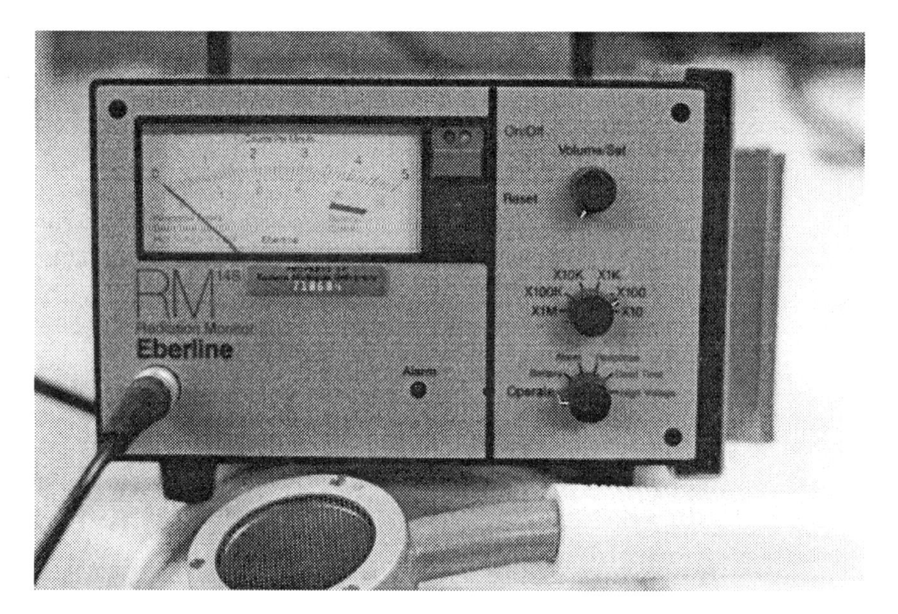

Figure 13.1. Shown here is a Geiger counter used to scan work areas for spilled radioactive nuclides.

In this instrument a tube is filled with an easily ionized gas. Electrons produced by ionizing radiation are attracted to a positively charged wire, and on striking it produce a pulse of electricity. The strength of the signal is multiplied by ionization trails as the electrons accelerate toward the wire. Such a pulse is recorded as a "count", and the instrument accumulates a total of counts (Figure 13.2).

A GM counter measures β-, γ-, and X-rays. Alpha particles cannot penetrate the window and so are ignored by the tube. By changing the design of the detector tube so that the sample is placed in a chamber with easily ionized gas flowing through, the window is eliminated (a windowless counter) and α particles can be counted. It is also possible to put sufficient

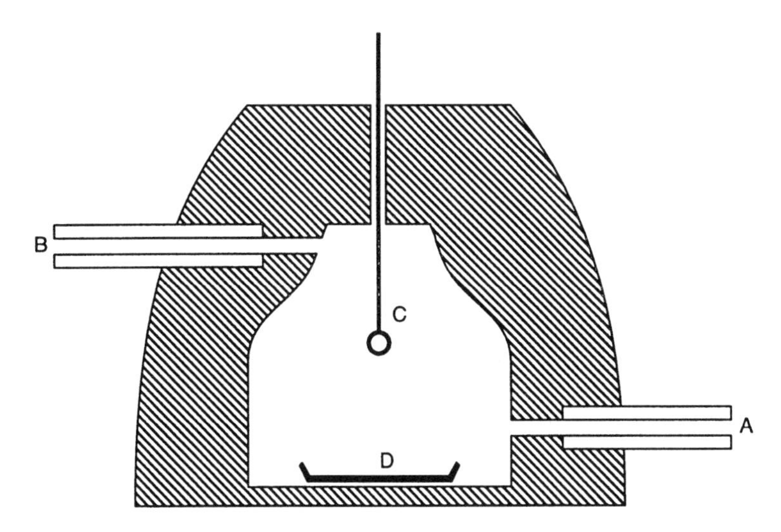

Figure 13.2. Here is one type of Geiger counter detector. Ionizing radiation passes through a readily ionized gas, causing an electron cascade to the anode. This is detected as an electric pulse in the external circuit and is recorded as a single count. Gas enters the chamber at A and exits at B. The anode is C and the sample being counted is on a thin metal holder called a planchet (D).

shielding over the window of the counter to block β particles, making the instrument sensitive only to high-energy electromagnetic radiation.

Scintillation Counters

Scintillation counters utilize the fact that certain liquids or crystals emit light when struck by radioactive emissions. The amount of light, measured using a photomultiplier, indicates the level of radioactive emissions. These are primarily research tools, used in laboratories.

Dosimeters

A dosimeter is a device worn by workers who may be exposed to radiation. One type of dosimeter, called a thermoluminescence detector (TLD), contains a chip of lithium fluoride. Radiation elevates electrons in the atoms of the chip to higher energy positions that are stable enough to remain there for an extended time. The dosimeter is read by placing it in a device that heats the chip, causing the electrons to return to their "ground state" or normal low-energy position. As they do so they emit light, which is read by the measuring device.

A film badge is a convenient and low-cost dosimeter (Figure 13.3). It contains a small piece of photographic film, which is "exposed" by radiation just as it is exposed by light. After wearing the badge for a specified time period the film is developed. The degree of exposure of the film estimates the dosage of radiation absorbed by the worker.

A pocket dosimeter to measure X-ray and γ-ray exposure looks like a pen and is worn clipped in the pocket (Figure 13.4). It contains a quartz fiber in a chamber, and the fiber carries an electrostatic charge that moves it to a zero point on a scale. Radiation generates ion pairs in the chamber, which discharge the fiber, causing it to move on the scale. The chief advantage of this device is its ability to be read by the worker at the time of exposure so as to estimate exposure as it happens, rather than having to send the device away to be read.

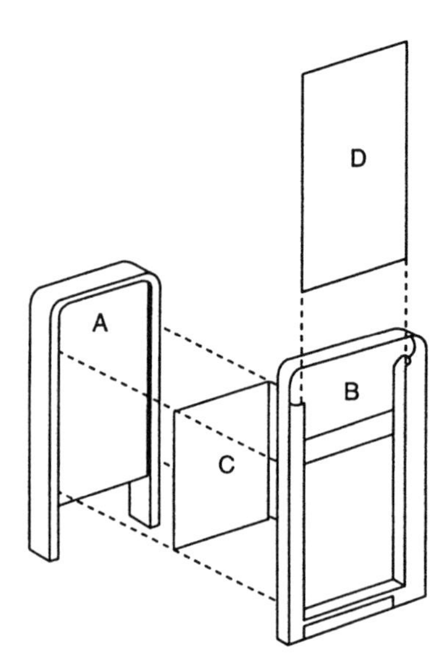

Figure 13.3. A film badge dosimeter is a very simple device. A and B are the back and front, respectively, of the housing. The film (C) is snapped in between them, and the personal identification card of the wearer (D) slides into the front.

UNITS OF RADIOACTIVITY

There are two types of measurement of interest to those who work with radioactive materials. Units of emission describe the rate of disintegration of a radioactive nuclide, and are the usual data collected in experiments involving the use of such nuclides. Industrial hygienists are more interested in units of exposure, which describe the dosage an individual has received.

Units of Emission

Essentially what we wish to know when radioactivity levels are measured is how many radioactive nuclides have disintegrated in a unit of time. The basic SI (Systeme International) unit is therefore the becquerel (Bq), which is the amount of nuclide that produces one disintegration per second (dps). The becquerel is relatively new, and older data as well as many of the newer measurements instead use the curie (Ci). One curie, originally the radiation produced by a gram of radium, is defined as 3.7×10^{10} dps, a very large measuring unit. Therefore, normally use small fractions of the basic unit are used, such as millicuries (mCi, 10^{-3} Ci) or microcuries (μCi, 10^{-6} Ci).

The number of emissions passing through the tube of a GM counter is measured in either counts per minute (cpm) or counts per second (cps). This value is not the measure of bequerels or curies for that sample. For that to be true, all emissions from the sample would need to have been beamed toward the counter tube. Emissions occur randomly in all directions, so even placing the counter right against the sample will allow detection of only about half the disintegrations, the other half having been directed away from the window. However, if counts are taken in a fashion that is geometrically reproducible with respect to the sample, valuable information about relative rates of disintegration are obtained.

Units of Exposure

The roentgen (R) is based on the energy absorbed by air from a radioactive source, and 1 R = 83 erg/g. This unit has been used principally to measure x- or γ-ray exposure, often

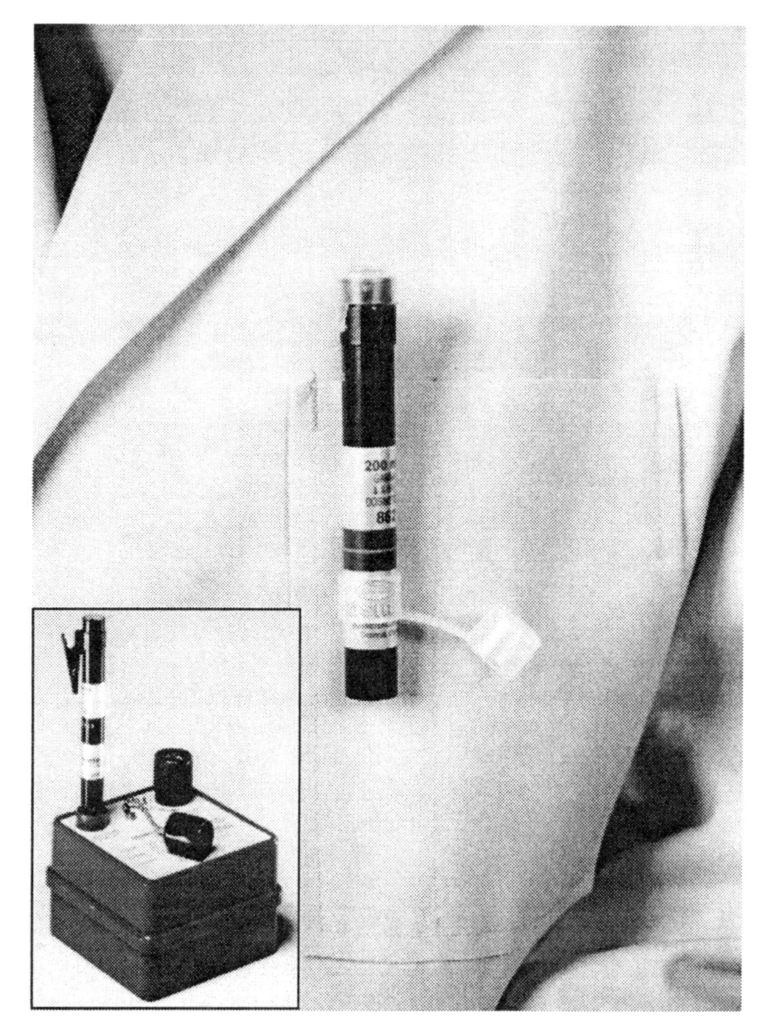

Figure 13.4. This pocket dosimeter detects exposure to X-rays and γ-rays. It is recharged in the stand shown in inset. (Photos courtesy of Lab Safety Supply, Janesville, WI. With permission.)

expressed as rate of exposure in milliroentgen per hour (mR/hr). However, the energy absorbed by more dense matter such as tissue is greater by 10–20%. Because of the difficulty of translating roentgen into tissue exposure, this traditional unit is being used less nowadays.

The rad is the most common unit in use, and is equal to an absorbed dose of 100 erg/g or 0.01 J/g of whatever material is absorbing the dose. Differences in the ability of material to absorb the energy are defined out of this unit, but actually measuring the energy absorbed by a specific sample of material is difficult. Rate of exposure is often expressed as millirads per hour. A dose of 100 rad equals one gray (Gy), an SI unit. Expressing exposure in grays is not yet commonplace in the U.S.

Different types of radiation create different levels of damage. This is indicated above, for example, in terms of the number of ion pairs produced per centimeter of dry air. This means that a rad of α particle exposure produces more damage to tissue than a rad of X-rays. The relative hazard of each type of radiation is expressed as the quality factor (QF) of the radiation. X-rays, γ-rays, and β particles have QF values of 1, whereas α particle values are about 10.

When damage is caused by taking the radioactive nuclides into the body, the chemical nature of the nuclide leads to its preferential deposition in certain parts of the body. This fact is recognized when talking about dosage absorbed by a given organ by indicating the distribution factor (DF) for that organ.

The rem is a unit that measures the dose equivalent, the likely damage to specific tissue rather than just the energy absorbed. Rems are calculated by multiplying the dosage in rads by the QF for the type of radiation, and sometimes also by the DF for the organ that is the target:

$$\text{rem} = \text{rad} \times \text{QF} \times \text{DF}$$

The SI unit for dose equivalent is the sievert (Sv), which equals 100 rem, and, like the gray, it is not in common use in the U.S.

EXPOSURE STANDARDS

For individuals in unrestricted areas the maximum permitted whole-body dose is 0.1 rem per calendar year.[3] The International Commission on Radiological Protection recommends lower dosage limits for gonads, breasts, bone marrow, and lungs.[4] Special and detailed permitted exposure levels are described for individuals in restricted areas in the *Code of Federal Regulations*.[5]

CAUTION SIGNS

In order to alert workers of a radiation hazard, OSHA prescribes a standard warning sign. The symbol is the shape of a three-bladed propeller and the colors are purple on a yellow background. Such words as, "CAUTION — RADIATION AREA" or "CAUTION — HIGH RADIATION AREA" should be included (see Figure 13.5).

Figure 13.5. The radiation hazard warning sign is lavender on a yellow field.

SPECIAL CIRCUMSTANCES IN RADIATION EXPOSURE

A variety of special circumstances influence the degree of damage caused by exposure to ionizing radiation. These include the dose rate, tissue or part of the body exposed, and the age of the person receiving the dose.

[3] See 10 CFR 20.1301(a), Standards for Protection Against Radiation.
[4] "Recommendations of the International Commission on Radiological Protection," ICRP Publication No. 26.
[5] 10 CFR 20.1301(c) and 29 CFR 1910.96(b).

Standards for exposure to radiation indicate a maximum total exposure in a given period of time. It matters whether that dose is received and distributed broadly across the indicated time (low dose rate) or entirely in a short period of exposure (high dose rate). If a small amount of damage occurs in a cell, the cell can repair the damage as it occurs. Because we are continuously exposed to radiation from our environment, damage from radiation is a natural occurrence and we have repair mechanisms. However, if damage occurs faster than the cell can repair it, the cell dies from accumulated damage. A given amount of radiation delivered at a high dose rate produces greater damage than the same amount delivered at a low dose rate.

Which cells in the body absorb the radiation is also important. We can differentiate cells as specialized and unspecialized. For example, the various types of cells in the blood have specific roles for which their special structural features adapt them; red blood cells carry oxygen, white cells engulf invading bacteria, and so on. These are specialized cells. They arise by a process of maturation, acquiring these specialized characteristics in a stepwise fashion, from very unspecialized cells in the bone marrow. Another example of this is in the epidermis of the skin, where cells at the base of the keratin layer divide to produce cells that migrate outward toward the surface of the skin and gradually become the fiber-filled cells of the keratin layer as they migrate.

Specialized cells are less susceptible to radiation damage than are the rapidly dividing unspecialized cells, sometimes called stem cells, from which they arise. Stem cells in the bone marrow, in the epithelial lining of the intestines, and in the ovaries and testes are all exceptionally sensitive to radiation.

Finally, the age of the person dosed with radiation is important. Younger individuals with a more rapid rate of cell growth are more readily damaged, with the greatest damage from a given dose level occurring in a fetus. In recognition of this, the practice of using X-rays to determine the position of an infant before delivery has been minimized, and recommended maximum exposure for pregnant women is lower than for nonpregnant women.

RADIATION FROM THE ENVIRONMENT

The limitations in occupational exposure to ionizing radiation are in addition to the background levels of exposure that everyone experiences. Largely unavoidable radiation exposure comes from the environment, health sciences exposure, and a few other sources.

Radiation strikes the earth from space, an exposure that is partially blocked by the atmosphere. Altitude above sea level is therefore a factor, exposure being about three times higher at 10,000 ft than at sea level, where one benefits from the maximum blocking by the atmosphere. Average exposures range from about 75 mrem/year in a high-altitude state like Wyoming to about 38 mrem/year in Florida. Average exposure in the U.S. is 44 mrem/year. Obviously, air travel raises this exposure. In addition we are exposed to the emissions from radioactive nuclides in our surroundings in rocks, soil, and bricks, estimated to range from 15 to 140 mrem/year according to location, and averaging about 40 mrem/year in the U.S. We have internal radioactive nuclides such as ^{40}K that add another roughly 18 mrem/year, bringing the total to around 100 mrem/year from all natural sources for the average individual.

Health care involves X-rays for many people and exposure to diagnostic radiopharmaceuticals for a few. An estimate from 1970 places the average exposure from medical sources at 73 mrem/year, but a lessened dependence on diagnosis by X-ray techniques has probably lowered this figure more recently.

A few miscellaneous sources add a little to nonoccupational radiation exposure. Television and computer screens emit soft X-rays, adding perhaps 2 mrem/year. In total, nonoccupational exposures average about 175 mrem/year per person, with individual totals varying widely according to location and experiences.

BIOLOGICAL EFFECTS OF RADIATION

Radiation disrupts chemical structures to produce free radicals and ion pairs. This is the direct effect of the radiation. These unstable products then undergo further change, generally involving interaction with the molecules around them, in order to return to being stable chemical structures. The sum of these interactions and changes are the indirect effects. Direct effects are virtually instantaneous with the passage of the emission, and the highly energetic and reactive direct products produce the indirect effects in the next tiny fraction of a second. There is no time to intervene after exposure to radiation to block these events. Treatment must involve dealing with the consequences of the exposure.

The biological effect of radiation depends on the changes in living tissue produced by these reactive species. Radiation is not selective in its chemical targets and direct damage occurs to the entire range of molecules in the path of the emission. However, the majority of the molecules in tissue are water, so if one understands what happens to water on irradiation, one knows a large percentage of the events occurring and by example understands the nature of the others. Important reactions resulting from irradiation of water are given in Table 13.1. From these typical reactions occurring in an irradiated cell, one can see that:

1. Free radicals are generated.
2. Hydrogen peroxide is generated.
3. The amount of hydrogen peroxide generated is increased by increased oxygen concentrations in the tissue.

Table 13.1. Reactions Resulting from Ion Pair Formation in Water Due to Radiation.

Reaction	
$H_2O \xrightarrow{\text{energy (34 eV)}} H_2O^+ + e^{-1}$	(Ion pair formation)
$H_2O + e^{-1} \longrightarrow H_2O^-$	(Electron capture)
$H_2O^+ \longrightarrow H^+ + OH\cdot$	(Free radical formation)
$H_2O^- \longrightarrow H\cdot + OH^-$	
$H\cdot + O_2 \longrightarrow HO_2\cdot$	(Free radical interactions)
$HO_2\cdot + H_2O \longrightarrow H_2O_2 + OH\cdot$	
$H\cdot + H\cdot \longrightarrow H_2$	(Loss of free radicals by collision)
$H\cdot + OH\cdot \longrightarrow H_2O$	
$OH\cdot + OH\cdot \longrightarrow H_2O_2$	

The first step (ion pair formation) is the direct result of the radiation, the rest of the steps result from this initial reaction. As a result of electron capture in water, there are both positively and negatively charged water molecules, which form free radicals as they react to form more stable structures. Free radicals stabilize by extracting atoms from other molecules, but the source molecule in turn is left as a free radical. This "chain reaction" of the free radicals causes miscellaneous damage to important biological molecules in the cell. This terminates by chance collision of two free radicals to form stable molecules.

Once generated, free radicals initiate a chain of destructive events. The unpaired electron is inherently unstable, and stabilizes by removing an atom from a surrounding molecule along with one of the two electrons that bonded that atom to another, as in the second step of the interaction of a hydrogen free radical with oxygen. The original free radical is now stable, but the molecule with which it collided is now a free radical. This sequence of collisions leaves the molecules of the tissue chemically altered, and to the extent that important biological species are involved, the cell functions are impaired.

Hydrogen peroxide is a strong oxidizing agent and destroys biologically functional molecules. Tissues have an enzyme that converts hydrogen peroxide to harmless products, but it can prevent damage only if it encounters the hydrogen peroxide before a sensitive functional species does.

IMMEDIATE EFFECTS OF LARGE DOSES

Regulations are designed to hold total radiation exposure (occupational and otherwise) to under one rem per year for the typical person. There are no immediate symptoms related to exposures at these levels. Here we consider the health effects of dosages that occur in a brief time period and go far beyond regulated levels, exposures that would occur only in the case of a serious accident. Symptoms appear in a short time and result from extensive direct tissue damage by the radiation. Table 13.2 summarizes severity of symptoms due to increasing levels of exposure.

Table 13.2. Effects of High Whole-Body Radiation Dosage.

Dose Level (rad)	Health Effect
1	No health effects detected in the short term
10	Developmental effects in very young embryos
10^2	White blood cell count decreases
10^3	Damage to the GI tract causes vomiting, diarrhea, and nausea; depressed blood cell production; death: 1–2 weeks
10^4	Nervous system damage; coma and death: 1–2 days
10^6	Massive cell death; death: immediate

Below 100 rad, there is little chance that exposure is fatal in the short term. However, symptoms are produced that reflect the sensitivity of less specialized cells. If the dose is confined to a small area of the body there is loss of hair as cells at the base of hair follicles die, and skin displays erythemia, as in a burn. Rapidly dividing cells that produce sperm cells are damaged, just a few rem resulting in a drop in the sperm count. Sterility problems do not require a complete disappearance of sperm production, only a reduction in numbers of sperm. Because sperm are produced over the lifetime, a man's testes may return to normal in time, but in women the lifetime supply of egg cells is at risk on exposure of the ovaries, so reproductive damage is more likely to be permanent. Cataracts result long after exposure because lens cells in the eye are not replaced when destroyed. Without these cells to maintain the lens, it gradually loses transparency.

Above 100 rad, the cells that produce the epithelial lining of the gastrointestinal tract begin to show damage. Within a few hours a loss of appetite (anorexia) is displayed, then nausea at higher doses. By 200 rad vomiting is likely and by 250 rad diarrhea is probable.

Above 500 rad, damage to the blood-cell-producing cells in the bone marrow becomes a serious, potentially lethal problem. The altered pattern of production of white blood cells is obvious in 1–2 days, and if the victim survives, this pattern may not return to normal for an extended time period. The white blood cells are the most radiation sensitive, and, because these effects are dose dependent, the drop in white blood cell production is sometimes used to estimate the severity of exposure. Cataracts may result over a longer time period from these exposure levels.

Above 1000 rad, damage to the epithelium of the GI tract becomes life threatening. Lining destruction leads to extensive internal bleeding, and death often follows in a few days. Beyond 5000 rad the destruction of the nervous system becomes the immediate problem, and at higher levels the rapid and massive cell death is immediately fatal.

LONG-RANGE EFFECTS OF MODERATE EXPOSURE

Levels of radiation described in the previous section occur only in extreme cases involving accidents, and they are not the basis of concern in setting worker exposure standards. A small proportion of the damage to cells resulting from the generation of free radicals or hydrogen peroxide in the cells is to the DNA molecules. A small percentage of this damage alters the DNA so as to convert the cell to a malignant pattern. Thus, in a dose-dependent fashion, the frequency of cancer is increased by radiation exposure. Leukemia is the earliest type of cancer to appear. Although the latent period is longer, other types of cancer result from radiation damage and in the long term are more common. Finally, premature aging for reasons sometimes difficult to pinpoint is an important long-range result of exposure to radiation. This has been seen in animal tests and in studies of exposed humans.

KEY POINTS

1. Damaging radiation is divided into nonionizing and ionizing radiation.
2. Nonionizing radiation includes UV light from the sun or artificial sources, which can damage the eyes and skin.
3. Atoms with a composition of protons to neutrons that is distant from the ideal adjust the ratio by radioactive emissions.
4. The degree of instability of a radioactive nuclide is reflected in the rate of radioactive disintegration and is measured as the half-life or time required for half the nuclides in such a sample to disintegrate.
5. Radioactive emissions include charged particles (α — two protons and two neutrons, β — an electron, and positron — a positively charged electron); high-energy light (γ- and X-rays); and neutrons.
6. Matter is ionized by the passage of these emissions. Charged particles draw or drive electrons from atoms, high-energy light excites the electrons so that they leave the atom, and neutrons drive charged particles out of the nucleus, which then draw or drive electrons from atoms.
7. Radioactive emissions can be blocked by placing matter in their pathway (shielding). Charged particles are more easily blocked than high-energy light.
8. Levels of radioactive emissions can be measured by GM counters and scintillation counters.
9. Levels of exposure can be determined by dosimeters worn by the worker.
10. Units of emission include the becquerel (Bq; = 1 dps) and the curie (Ci; = 3.7×10^{10} dps).
11. Units of energy absorbed include the roentgen (83 erg/g in air), the rad (100 erg/g in tissue), and the rem (rad × quality factor of radiation × distribution factor in tissues).
12. The exposure ceiling for workers in unprotected areas is 0.5 rem/year. In protected areas the ceiling is higher, and follows a formula.
13. It is more damaging to receive a dose of radiation in a short time period than the same dose over a long time period.
14. The average American receives 175 mrem/year from a combination of radiation from space, environmental radioisotopes, internal radioisotopes, medical treatment, and TV or computer screens.
15. The biological effect of low radiation doses depends on the formation of free radicals in body fluids. The most common final product is H_2O_2.

16. Large doses of radiation can be lethal. More than 100 rad destroys cells lining the GI tract. More than 500 rad depletes the supply of blood cells. More than 5000 rad destroys the nervous system.
17. The most serious long-range effect of radiation exposure is an increase in the occurrence of cancer.

BIBLIOGRAPHY

J. Gauvin, "Radiation Protection in Hospitals," in W. Charney and J. Schirmer, Eds., *Essentials of Modern Hospital Safety,* Lewis Publishers, Chelsea, MI, 1990.

C. H. Hobbs and R. O. McClellan, "Toxic Effects of Radiation and Radioactive Materials," in C. D. Klaassen, M. O. Amdur, and J. Doull, Eds., *Toxicology: The Basic Science of Poisons,* 3rd ed., Macmillan Publishing, New York, 1986.

R. G. Thomas, "Evaluation of Exposure to Ionizing Radiation," in L. J. Cralley, L. V. Cralley, and J. S. Bus, Eds., *Patty's Industrial Hygiene and Toxicology,* 3rd ed., Vol. III, Part B, John Wiley & Sons, New York, 1995.

A. C. Upton, "Ionizing Radiation," in B. S. Levy and D. H. Wegman, Eds., *Occupational Health: Recognizing and Preventing Work-Related Disease,* 2nd ed., Little, Brown and Company, Boston, 1988.

G. L. Voelz, "Ionizing Radiation," in C. Zenz, Ed., *Occupational Medicine,* 2nd ed., Year Book Medical Publishers, Chicago, 1988.

C. Zenz and A. L. Knight, "Ultraviolet Exposures," in C. Zenz, Ed., *Occupational Medicine,* 2nd ed., Year Book Medical Publishers, Chicago, 1988.

PROBLEMS

1. Using a periodic chart, answer the following:
   A. How many protons are in the nucleus of the elements V, Ag, and K?
   B. What particles are found in the nucleus of ^{15}N, ^{23}Na, and ^{235}U?
2. Write balanced nuclear equations of this form:

$$^{14}_{6}C \longrightarrow {}^{14}_{7}N + {}^{0}_{-1}\beta$$

   for the nuclear reactions indicated using these symbols: alpha, $^{4}_{2}\alpha$; beta, $^{0}_{-1}\beta$; neutron, $^{1}_{0}n$; proton, ^{1}p; positron, $^{0}_{+1}\beta$
   A. $^{26}_{11}Na \rightarrow$ beta
   B. $^{212}Rn \rightarrow$ alpha
   C. $^{111}Sb \rightarrow$ positron
   D. $^{16}_{8}O + {}^{1}n \rightarrow$ proton
3. Refer to the definitions of units for measuring exposure to radiation in 29 CFR 1910.96(a).
   A. What dosages are equivalent to 1 rem?
   B. What does this assume is the quality factor of β, γ-, and X-radiation, neutrons, high-energy protons, and heavy particles of high penetrating power?
4. How does the standard for maximum permitted whole-body dose for workers in unrestricted areas compare to the average dose from naturally occurring sources?
5. Find the section in CFR that lists the maximum permitted dosage for workers in restricted areas.
   A. What is the relationship between whole-body exposure permitted here and permitted for workers in an unrestricted area?
   B. Read Table G-18. The permitted dose to hands, forearms, feet, and ankles is greater than to whole body or to a series of body regions. Why, and why is each specific body region so designated?

C. Dosage can go beyond the levels of Table G-18. What is the ultimate limit?

D. J.W.J. is a 45-year-old worker with an accumulated occupational dose of 110 rem. Must she be isolated from radiation exposure this year?

E. S.D.W. is a 17-year-old worker. What are his exposure limits in a restricted area?

6. A. The half-life of ^{29}Al, a β emitter, is 6.6 min. What is its decay constant?

B. A sample of ^{24}Na, a β emitter, is counted with a GM counter and reads 2550 cpm. If the half-life is 15.03 hr, what will the emission level be in 2 hr?

CHAPTER 14

Working in Extreme Temperatures

BODY TEMPERATURE CONTROL

Humans, like all mammals and birds, have physiological mechanisms designed to maintain a constant body temperature. Such a system has advantages, allowing high physical activity in spite of decreasing ambient temperatures; but the price we pay is the need to operate within a narrow range of body temperature. When our surroundings become hotter or colder, we must have corresponding regulatory mechanisms to control our internal temperatures. Thus, in hot surroundings when these mechanisms are inadequate to prevent a rise in body temperature we experience first discomfort, then illness and finally death.

We do not have a uniform body temperature, but rather a warm core surrounded by a cooler shell. When we measure body temperature to assess our state of health, we measure core temperature by placing the thermometer in the mouth, ear, or rectum. Even these temperatures are not the same, rectal temperatures running slightly higher than oral temperatures. Everyone is aware of the standard 98.6°F or 37°C "normal" temperature.[1] In fact, normal can range from just above 36°C to just below 38°C in a resting person, and climb to above 39°C during hard physical exertion. When the core temperature drops to 35°C, we are in a state of hypothermia, and by 27°C death occurs. A core temperature as high as 40°C is described as hyperthermia, and death occurs by 42°C. Skin temperatures are much more variable, normally being around 33–34°C, but dropping to as low as 22–27°C in cold weather and rising close to core temperatures during heat stress.

HEAT GENERATION AND DISTRIBUTION IN THE HUMAN BODY

Two body processes release heat as the result of metabolic chemical reactions that overall are exothermic: the chemical alteration of nutrients and muscle contraction. Nutrients can be from the diet or from reserves such as either body fat or glycogen from liver or muscle. Metabolism operates continuously, even while we are sleeping or when effort expended is no greater than pushing the remote button to change channels. A resting adult produces about 76 kcal/hr. (~300 BTU/hr) from metabolic activity. However, when physical effort increases, rates of metabolism rise to supply the fuel for greater muscle contraction. Physical work sharply increases generation of internal heat, and heat produced is proportional to effort expended.

[1] Brush up on the conversion of °F to °C [°C = 5/9(°F – 32)]. You may be used to using °F, but references in this field use °C.

Heat is produced in the warm core, and is transferred by blood circulation to the skin to be lost, much as a car engine transfers excess heat generated in the combustion processes into surrounding water, then circulates that hot water to the radiator to transfer heat to air.

HEAT STRESS

Harm, termed *heat stress*, results from too great an increase in body temperature. Some jobs involve working at high surrounding temperatures, and so have the potential to cause such harm. Indoor examples include laundries, restaurant kitchens, bakeries, food processing facilities, power plants and other boiler operations, molten metal or glass processing facilities, steam tunnel work, and brick or ceramics firing operations. A number of outdoor jobs, including roadwork, agriculture, and construction, involve potential heat stress in hot weather.

PREVENTING BODY TEMPERATURE INCREASE

The brain, primarily in the small segment at the lower surface of the brain called the hypothalamus, regulates core body temperature. Temperature sensory input is interpreted in the hypothalamus, which directs two compensatory actions to counter a core temperature rise. First, smooth muscle lining blood vessels in the skin relaxes, causing the vessels to dilate and carry a greater volume of blood through the skin. In a sufficiently light-skinned person this is seen as a reddening of the skin or flushed appearance during exercise. Loss of heat to surrounding air increases. Second, sweat glands coat the skin with sweat, a very dilute salt solution. The skin supplies heat for evaporation of this water, cooling the skin. Evaporation of a gram of sweat eliminates about 0.58 kcal of body heat. The net loss of about a liter of water through increased sweating is tolerable, but a much greater loss produces discomfort, increased heart rate, and thirst. Water should be taken frequently during physical work in a hot environment to compensate for loss as sweat; but simply responding to thirst usually leads to inadequate water intake. Water should be taken in amounts that exceed that needed to quench thirst.

Salt also is lost as the worker sweats, approximately 3–5 g/l. Given that the worker in extreme cases may produce more than 4 l/day of sweat, and that typical daily intake of salt is 10–15 g/day, supplementation by using more salt on food or by taking salt tablets may be necessary. When the worker does not replace the lost salt, the kidneys retain less water, because the mechanism in the kidney for retaining water involves moving it back into the blood along with recaptured salt. This leads to dehydration. Diagnosis of salt deficiency involves observing abnormally low levels of chloride ion in the urine.

The effectiveness of sweating as a method of heat loss increases as:

- more skin surface is exposed to air
- the humidity of the air is lower, facilitating evaporation
- the air is moving so that a layer of water-saturated air does not build up at the surface of the skin

Both methods of losing body heat depend on the surface area of the body. Increased skin area provides more surface at which blood vessels are losing heat to the air and more surface from which sweat is evaporating. A slender person has a greater area of skin per kilogram of body mass than does a stocky or obese individual, and thus suffers less from heat stress, all else being equal.

Heart Rate

Working under conditions of heat stress leads to a higher heart rate, so heart rate is used as a measure of heat stress. One cause of this higher heart rate is the need to move additional blood to the surface of the skin as dermal vessels dilate. The muscles of a person working physically use additional nutrients and oxygen, requiring increased blood output to the muscles. The actual heart rate increase varies with the fitness of the cardiovascular system. A person in good physical condition requires a smaller increase in heart rate to respond effectively to a given increase in core temperature. A heart rate of 180 to 200 contractions per minute is the maximum rate sustainable in most adults, and that rate can be supported only for a few minutes.

OTHER FACTORS IN RESPONSE TO HEAT

Acclimatization

People acclimatize to working in a high-temperature environment. The first day working under high-temperature circumstances, a person may show signs of stress. These lessen with each succeeding day and disappear within 2 weeks. Once a worker leaves the high-temperature work environment, acclimatization is lost.

Primarily, acclimatization is seen as an adjustment in the production of sweat. A larger volume of sweat is produced in the acclimatized worker, as much as 8 l of sweat in an 8-hr shift. Another aspect of acclimatization is that the sweat contains only 1–2 g/l of salt, reducing the loss of sodium ion by this route.

Prescription Drugs

There are a number of prescription drugs that can interfere with adaptation to work in a hot environment. Diuretics and antihypertensives may reduce the volume of blood in circulation or interfere with the heart response to stress, and so could increase the hazards of heat stress. Tranquilizers and pain drugs also increase risk. A worker taking prescription drugs should consult with a physician about the implications of the particular drug to heat adaptation.

Alcohol and Social Drug Consumption

Binge-type alcohol consumption results in the production of quantities of dilute urine. The individual is then much more susceptible to further dehydration working in a hot environment, and therefore to heat stroke. As a CNS depressant, alcohol interferes with heat adaptation. Other social drugs can similarly have detrimental effects on adaptation, increasing the chances for heat stroke and death.

Age

Older workers have greater problems with work in a hot environment. Capacity to work under stress is reduced by age due to factors such as a decrease in cardiovascular efficiency. Older workers do not begin to sweat as readily as younger workers and produce a lower volume of sweat when they do, resulting in a lessened ability to lower core temperatures. As a result, the older worker has a higher core temperature doing the same task as a younger worker and requires longer to recover during a rest period.

Physical Conditioning

Two workers of the same age, but differing in physical conditioning also differ in their ability to withstand heat stress. Improvements in the circulatory system result from physical conditioning. A fit person moves blood to the skin more effectively and experiences less strain on the heart working in a hot environment.

SUMMARIZING HEAT FLOW

People studying heat gain and loss by workers express each aspect of heat flow by a symbol and place these symbols into equations that facilitate quantitation of each aspect. For a person in equilibrium, that is, whose body temperature is not changing, the sum of additions and subtractions to body heat is zero:[2]

$$M \pm R \pm C - E = 0 \qquad (14.1)$$

where M = heat production due to metabolism
R = radiative heat transfer
C = convective heat transfer
E = heat lost due to evaporation

An industrial hygienist would not use this equation for calculations in the field. However, it is conceptually useful to see the components spelled out in this fashion and to get a sense of the contribution of each.

Metabolic heat production (M) is the heat energy released by the oxidation of dietary material to produce energy storage intermediates (largely the famous ATP molecule), and the use of energy storage intermediates to drive muscle contraction and other body processes. Heat production is hard to measure directly, but can be estimated easily by measuring oxygen intake.[3] Metabolic heat production is roughly 5 kcal per liter of oxygen. Oxygen intake varies from 0.3 l/min in someone resting to 2 to 4 l/min in a hard-working fit person.

As a warm body, a person loses heat to cooler surroundings by radiative and convective heat transfer. Radiative heat transfer (R) is the transfer of heat energy through space according to the equation:

$$R = K_R \Delta T \qquad (14.2)$$

where K_R = a constant
ΔT = the temperature difference between the person and the surrounding objects

This is a two-way street, however, because hot surrounding objects transfer heat to the person. Consider standing on a blacktop parking lot on a hot summer day, then moving off the blacktop into a field of grass. The blacktop radiates much more heat to your body than does the grass.

Similarly, convective heat transfer is the transfer of heat to or from air contacting the skin. The equation is very similar:

$$C = K_C \Delta T \qquad (14.3)$$

where K_C = a constant
ΔT = the temperature difference between the person and the surrounding air

[2] Another term that could be added to the equation would represent heat conducted between the body and objects it contacts, but this is usually very small and can be ignored.
[3] Using oxygen consumption to estimate metabolic heat production is sometimes termed *indirect calorimetry*.

Once again, the direction of transfer is normally from skin to air, but in a very hot environment it would transfer in the other direction. The constant is modified here by surface area of the body, degree of contact of skin with outside air, and rate of air movement. For example, on a warm day you might switch from a long-sleeved to a short-sleeved shirt or blouse, increasing the area of skin exposed to the air.

Heat loss due to evaporation of sweat depends not on temperature differences, but on the relative humidity (water content) of the surrounding air. Dry air accepts water evaporating from the skin more rapidly, producing a higher rate of heat loss. The higher the skin temperature the more rapid the evaporation, therefore the more rapid the heat loss.

ILLNESSES DUE TO HEAT

It is hard to classify a particular heat-related illness without medical diagnosis. A rise in body temperature and heart rate are common symptoms. The safest approach to emergency treatment in what appears to be a serious situation is to assume the worst case, heat stroke, until adequate analysis by an expert has been performed. We present here classes of heat problems.

HEAT CRAMPS

A worker may experience severe muscle cramps, muscle spasms, working in a hot environment. Abdominal muscles or muscles fatigued by a task are particularly susceptible. The cause is loss of fluids and salt as a result of excessive sweating, leading to inadequate circulation in muscle tissue. Specific causes of the muscle contractions are a subject of disagreement, but rest and intake of fluids containing a small amount of salt (about 0.5%) is often recommended as treatment. Careful attention to adequate fluid intake during exertion in hot work environments helps prevent heat cramps.

HEAT EXHAUSTION

Heat exhaustion is also called heat prostration. The shift of blood flow to the skin to lower body temperature creates a demand for greater blood volume. Combined with a loss of fluids as a result of sweating, this leads to circulatory collapse. The brain receives insufficient blood and the victim experiences headaches, vertigo, nausea, and weakness, followed either by feeling faint or by actually fainting. Judgment and concentration are affected, so machine operation is dangerous. The victim should lie down, perhaps with the head low, and take slightly salty fluids.

HEAT STROKE

Heat stroke is the result of body temperature rising to very high levels, in the range of 40–41°C (105–106°F). Serious tissue damage occurs at these temperatures, especially to the liver, kidneys, and brain. Systems for temperature regulation may not be working properly as a result of high brain temperatures, so in spite of high body temperature, sweating may have ceased. The person has a headache, is fatigued, and feels dizzy. The pulse rate is rapid and the victim becomes disoriented and quickly becomes unconscious. Convulsions may occur. It is important to bring the body temperature down, but immersion in ice or cold water, as used to be recommended, is no longer considered wise. Cooling skin to that degree causes blood vessels to contract, reducing transport of core heat to the skin. Wetting the skin and moving air to increase the cooling rate by evaporation is useful. This is a very serious problem, and a high percentage of the victims that recover show signs of irreversible brain and kidney damage. The victim should receive medical care as quickly as possible.

EVALUATION OF WORKING CONDITIONS

Industries subject to heat stress problems generally fall into two categories: hot-dry industries and warm-moist industries. Hot-dry industries are those with a high temperature process that does not add water to the surrounding air. Examples include processing metals at high temperature (steel rolling mills, forges, foundries, and smelters), forming glass, and firing bricks. Evaporative cooling by a perspiring worker is effective in these settings. Warm-moist industries involve hot water and raise the water content of surrounding air as they raise the temperature. Examples are laundries, kitchens and food processing plants, and paper plants. Here evaporative cooling by perspiring workers drops in effectiveness as the humidity rises. Measurements to assess potential heat stress obviously must include measurements that indicate relative humidity.

Over the years a number of units have been devised to measure the potential of a work environment to generate heat stress. All have one failing or another. Those that include all relevant factors are cumbersome to use, whereas those that omit some factor obviously suffer for the omission. The wet bulb globe temperature index (WBGT) is the method that has been adopted in all proposed heat stress standards. It was first devised by the military, and has the important advantage that it is measured rapidly and easily. Three temperature measurements are involved: dry bulb temperature (T_A), natural wet bulb temperature (T_{NWB}), and globe temperature (T_G) (Figure 14.1).

Dry bulb temperature is simply the measurement of air temperature with an ordinary thermometer. A wet bulb thermometer has a wick that carries water to the bulb of the thermometer. The evaporation of this water cools the bulb, so wet bulb temperature is lower than dry bulb temperature. If the humidity of the air is high, evaporation is slowed, and the amount of cooling is decreased, just as would be the case with sweat evaporating from the skin. In this case the difference between wet bulb and dry bulb temperatures is smaller. A natural wet bulb thermometer is exposed to the existing air movement, and moving air expedites evaporation from the bulb, as it would from the skin.

Globe temperature is measured by inserting the bulb of the thermometer inside a thin copper sphere that has been painted nonreflecting black on the outside. The bulb picks up heat radiated from surroundings of the globe and from the sun, so globe temperature is likely to be higher than dry bulb temperature.

Each of these three temperatures is multiplied by a fraction, and the sum of these fractions is 1. The resulting value is the WBGT. Outdoors, WBGT is calculated as follows:

$$WBGT = 0.7T_{NWB} + 0.2T_G + 0.1T_A \tag{14.4}$$

The heavy emphasis given to T_{NWB} shows the importance of evaporative cooling to body temperature control. Indoors, WBGT is calculated in this fashion:

$$WBGT = 0.7T_{NWB} + 0.3T_G \tag{14.5}$$

Notice that now the T_A is omitted from the calculation.

Other measures of potential heat stress include the measurement of air velocity. This is an awkward measurement to perform in the field and was omitted as impractical in WBGT. However, the effect of air velocity is inferred by using the T_{NWB} in the calculation.

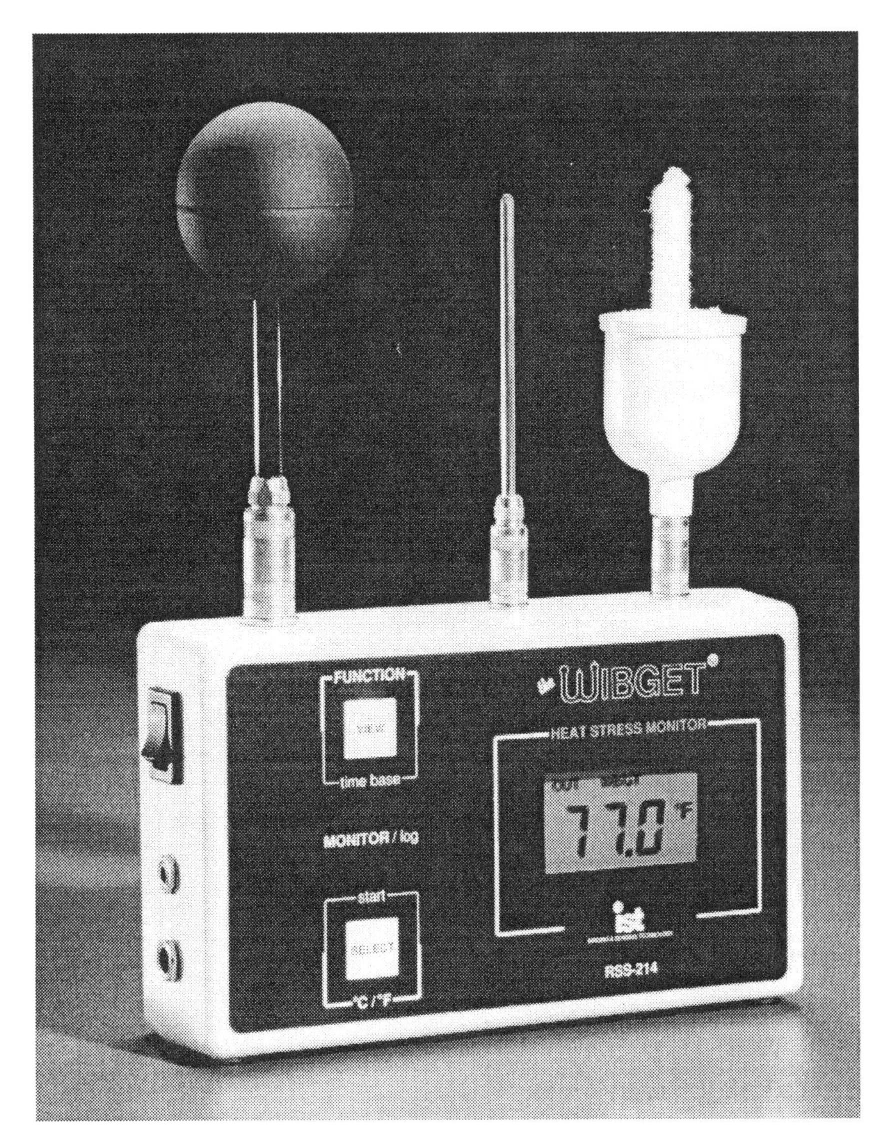

Figure 14.1. The WBGT apparatus. Here three thermometers measure the temperature. In the center the air temperature is measured directly. On the right a layer of water evaporates and cools the thermometer. On the left the temperature sensor is inside a black globe, which is warmed by radiant energy above the ambient temperature. (This is the environmental heat stress monitor of Imaging and Sensing, Cambridge, Ontario.)

EXPOSURE STANDARDS

At this time OSHA has not issued standards for protection of workers from heat stress parallel to those for exposure to chemicals or noise. There is a statement in the original OSH Act language of 1970 that provides general coverage, charging each employer to protect workers by providing: " . . . employment free from recognized hazards causing or likely to cause physical harm." Recommendations have been sent to OSHA both by NIOSH and by a Standards Advisory Committee on Heat Stress, but at the time of writing this text a standard has not been written.

ACGIH prepared a set of standards that were published in 1974 (Table 14.1). These are threshold limit values (TLVs) based on the assumption that an acclimatized worker, fully clothed, whose core temperature is at or below 38°C (100.4°F) is not experiencing heat stress. However, one cannot constantly measure core temperatures of workers, so the standards describe instead the conditions in the workplace, including the WBGT index, the severity of the work load, and the percentage of time engaged in continuous work (see Table 14.1). Standards assume workers wear clothing that does not trap air, preventing evaporation of sweat, and that conditions of the rest area are approximately the conditions of the work area.

Table 14.1. ACGIH Permissible Heat Exposure TLVs.

	Work Load (kcal/hr)		
Work Regimen	**Light (200)**	**Moderate (350)**	**Heavy (500)**
		TLV (°C/°F WBGT)	
Continuous	30.0/86	26.7/80	25.0/77
75% Work	30.6/87	28.0/82	25.9/79
50% Work	31.4/89	29.4/85	27.9/82
25% Work	32.2/90	31.1/88	30.0/86

Reprinted from "1993–1994 Threshold Limit Values for Chemical Substances and Physical Agents and Biological Exposure Indices" by permission of the ACGIH, Cincinnati, OH.

NIOSH recommended a set of standards[4] in 1972. These standards proposed a ceiling limit WBGT temperature for continuous heavy work by an acclimatized worker for more than 1hr of 26°C. Beyond that limit one of a number of work practices should be employed to ensure that core temperature does not exceed 38°C. Due to extensive objection to this proposal, a committee was established, the Standards Advisory Committee on Heat Stress, which modified the proposal, as shown in Table 14.2, and submitted the revision in 1974. In 1986 NIOSH updated their proposal,[5] defining recommended alert limits (RALs) for healthy nonacclimatized workers and recommended exposure limits (RELs) for healthy acclimatized workers. NIOSH also defined for each group a ceiling limit (CL) above which workers must be provided with and must correctly use protective clothing and equipment. The qualifier "healthy" refers to a variety of criteria relating to the worker's age, health, degree of body fat, and other physical characteristics. The proposed regulations would require heat measurements to be taken at least hourly during the hottest parts of the work day, during the hottest part of the year, and during heat waves.

Table 14.2. Standards Advisory Committee Recommendations.

Work Load	**Low Air Velocity (<1.5 m/s)**	**High Air Velocity (>1.5 m/s)**
	Threshold WBGT Values (°C/°F)	
Light (<200 kcal/hr)	30/86	32/90
Moderate (<300 kcal/hr)	28/82	31/87
Heavy (>300 kcal/hr) 26/79	29/84	

The International Organization for Standardization (ISO) published standards in 1982 that are based on WBGT measurements and the premise that worker core temperature should not exceed 38°C.[6] Their recommendations closely resemble those of the ACGIH.

[4] Criteria for a Recommended Standard . . . Occupational Exposure to Hot Environments, U.S. Department of Health, Education and Welfare, NIOSH, HSM-72-10269, 1972.
[5] Criteria for a Recommended Standard . . . Occupational Exposure to Hot Environments, Revised Criteria 1986, U.S. Department of Health, Education and Welfare, NIOSH, April 1986.
[6] Hot Environments — Estimation of Heat Stress on Working Man Based on the WBGT Index (Wet Bulb Globe Temperature), ISO 7243-1982, 1982.

ENGINEERING CONTROLS AND WORK PRACTICES

To whatever degree possible, ventilation should be used to remove sources of heat and humidity from the work environment. Providing a cooler room for worker breaks reduces total exposure and allows a chance for body temperature to normalize if it is elevated. Fans that move air over the workers speed the evaporation of sweat, providing some cooling. Radiant heat can sometimes be blocked by reflective panels or cloth, reducing intake of heat by that route.

New workers or workers off the job for a period of time must be allowed to acclimatize before doing a full day's work under hot conditions. Workers must be supplied with liquids to replenish fluids lost by sweating. Breaks to drink water should be frequent under more extreme conditions, because normal drinking does not bring enough water into the body at one time to avoid dehydration. Depending on the heat load, it may be necessary to establish a pattern of working and resting.

PROTECTIVE CLOTHING AND HEAT STRESS

Protective clothing generally adds to the problems of heat stress. Two factors are (1) degree of skin covered and (2) nature of the protective fabric. Fabrics that do not transmit vapors place a greater burden on the worker. Table 14.3 provides estimates of the extra burden as adjustments to the WBGT for various types of protective clothing.

Table 14.3. WBGT Correction Factors for Protective Clothing.

Clothing	Correction (°C)
Cotton work clothes	0
Cotton coveralls	2
Melt-blown polypropylene	4
Water barrier, vapor transmitting	6
Spun-bond polyethylene	7
Lightweight vapor barrier	8
Heavyweight vapor barrier	11

From T. E. Bernard, "Avoiding Heat Stress," *Occupational Health and Safety,* July 1995, p. 45.

Workers required to wear a completely enclosed chemically protective garment (Chapter 6) have special problems with heat. In such a garment normal cooling mechanisms such as the evaporation of sweat are restricted. Ice vests or vortex coolers are often used to cool the worker.

Ice vests are vests with pockets of frozen coolant. The cooling is effective for up to 90 min, and, because the worker is not attached to any lines or tanks, the system is completely portable. However, the vests weigh over 10 lb and are tiring to carry.

Vortex coolers are devices based on splitting compressed air into two streams, one of which is compressed by spinning it rapidly, causing it to heat. The heat is transferred to the second air stream, which is exhausted from the suit. Once leaving the spinning pattern and having transferred the heat, the first stream is now cooler than when it entered the system. This cool air is distributed to the suit interior. Vortex systems require the worker to be connected to an air line, reducing mobility, and half the delivered air is discarded.

An experimental system using high-pressure air tanks (2000–4000 psi) has been described.[7] This system is used with workers loading barrels at a hazardous waste incinerator site. In the fashion of the vortex, air gives off heat as it is compressed, then cools as it expands

[7] C. M. McClure, C. D. McClure, and M. Melton, "Breathing Air Cooling System Combats Respiratory Contamination, Heat Stress," *Occupational Health and Safety,* Aug. 1991, p. 33.

to atmospheric pressure. The heat produced by compression is dissipated at the compressor. Expansion occurs inside the suit and provides the cooling.

COLD STRESS

Most workers exposed to extremely low temperatures are outdoor workers in northern winters or at high altitudes. Outdoors, the more significant measure of exposure is the wind chill (the same as on the TV weather report). Moving air has increased ability to lower body temperature, so the wind chill index equates the effect on exposed skin of air at a particular temperature and wind velocity to colder stationary air (Table 14.4). Although many outdoor workers simply do not work when the temperature (or, better, wind chill) drops below a certain value, many safety and emergency workers, such as fire department personnel and power company linemen, go out even in extremely cold situations. A number of indoor jobs (such as butchers or freezer plant employees) expose workers to cold environments.

Table 14.4. Wind Chill Index.

Wind Velocity (mi/hr)	Thermometer Reading (°F)							
	30	20	10	0	−10	−20	−30	−40
0	30	20	10	0	−10	−20	−30	−40
5	27	16	6	−5	−15	−26	−36	−47
10	16	4	−9	−21	−33	−46	−58	−70
15	9	−5	−18	−36	−45	−58	−72	−85
20	4	−10	−25	−39	−53	−67	−82	−96
25	0	−15	−29	−44	−59	−74	−88	−104
30	−2	−18	−33	−48	−63	−79	−94	−109
35	−4	−20	−35	−49	−67	−82	−98	−113
40	−6	−21	−37	−53	−69	−85	−100	−116

Note: Numbers below thermometer readings indicate the temperature equivalent in still air of the stated conditions.

Heat transfer in a cold environment has the same components as in a hot environment, as described by Equation 14.1:

$$M \pm R \pm C - E = 0$$

Physical activity generates body heat (M). Heat is generally lost by radiation (R) and convection (C). If an individual becomes wet, as by being immersed in water, or simply because sweat is dampening the skin, then heat is also lost by evaporation (E). From this we see that an active worker is better able to cope with cold, because of the generation of metabolic heat.

BODY DEFENSES AGAINST TEMPERATURE DECREASE

When temperatures drop, the body, under control of the hypothalamus, moves to reduce heat loss. Vessels in the skin constrict. Less warm blood is carried from the body core to the skin, so less heat is lost to surroundings. Shivering, involuntary rapid muscle contractions, burns metabolic fuels and so generates additional heat. Some acclimatization to cold environments occurs, but these physiological adjustments are not as dramatic or effective as acclimatization to a hot environment.

LOWERED BODY TEMPERATURE — HYPOTHERMIA

If the body response described above is inadequate to counteract a cold environment, the body core temperature drops. The International Society of Physiologists defines the start of hypothermia as the dropping of the core temperature to 96°F (35°C). The skin becomes cold as circulation is withdrawn from the skin. If this withdrawal is extreme, skin tissue may die from lack of oxygen. As the body temperature drops further, severe shivering is evidence of a problem. At 95°F heart rate slows and central nervous system functions become less effective. The individual is drowsy, fatigued, forgetful, and uncoordinated. By 94°F speech is difficult, vision is less acute, and disorientation sets in. Consciousness is lost when the core temperature drops to between 86 and 90°F. When the core temperature drops below 86°F (30°C), weakened heart beat and accompanying slower respiration are potentially lethal.

Many workers believe that alcohol helps a person cope with low temperatures. This belief is supported by stories of mountain rescue dogs carrying brandy to those stranded in the cold. Alcohol dilates skin blood vessels, giving the person a sense of being warmer. However, the effect of greater blood flow in the skin is more rapid heat loss, speeding the drop in core temperature.

When the worker is fatigued, body responses to cold operate less effectively. Body heat is therefore lost more rapidly.

LOCALIZED TEMPERATURE DROP

Frostbite

Parts of the body not covered by clothing lose heat more rapidly and may drop in temperature to dangerous levels long before core temperature is lowered seriously. When the wind chill is below –25°F, skin must be protected. Facial skin, especially the nose and ears, is often the first location affected. The extremities, fingers and toes initially, are at the distant reaches of the circulatory system, making them more vulnerable if circulation slows due to cold. For their size, they also have larger surface area through which heat may be lost than other parts of the body. These tissues may freeze.

When tissue freezes, blood circulation ceases. Without oxygen or nutrients, cells die. The term *frostbite* is generally applied when tissue has frozen. Skin suffering from frostbite is lighter in color, all the way to a gray-white color in Caucasians. There may be pain during the freezing, but, once the tissue has frozen, it is numb. The victim may not be aware of frostbite, and may have to be told by an observer that there is a problem.

When rewarmed, cells frozen for a significant period do not recover. In mild cases, only the outer skin is affected, and it eventually regrows. As deeper tissues are involved, there is scarring.

We carry a microorganism in our tissues, *Clostridium histolyticum,* that destroys collagen, the protein of connective tissue. This organism cannot prosper in the presence of oxygen (is anaerobic), so is normally not a threat to us. Once tissue freezes and circulation stops, the oxygen supply is cut off. The organism can then multiply and begin destroying the structure of our tissues. We call this infection *gangrene*, and tissues affected may require amputation.

Trench Foot

When exposed to cold, though not necessarily subfreezing, temperatures and to dampness, the feet slowly develop a problem called trench foot. The skin first becomes red and inflamed, then later pale and swollen. In early stages there is numbness or painful prickling. Later, ulcers form and eventually, gangrene is found.

PROTECTION

Clothing

First of all, the body should be as completely covered as possible with clothing in a cold environment to minimize radiative loss of heat (R). The choice of clothing is important when working in a severely cold environment. The outer covering should block the wind, which speeds cooling by both convection (C) and evaporation (E). Layers of garments are best, particularly if air is trapped between the body and the outer garment. Air insulates the skin from the outdoors, slowing the transmission of body heat outwards. Damp clothing transmits heat outwards more rapidly, so accumulation of sweat must be minimized. Polypropylene or a hollow fiber polyester (DuPont's Thermax) next to the skin acts as a wick to carry moisture away from the surface of the skin, where its evaporation would cause further cooling. Cotton is sometimes favored, but after it absorbs moisture, it holds it next to the skin. A second layer of wool absorbs and holds the moisture, even surprisingly large quantities of moisture. Gortex®, a man-made material, is also good.[8] Insulating clothing can be compared for effectiveness by means of the "clo" unit.

The extremities, hands and feet, must be warmly clothed, because they cool faster than the trunk of the body. Insulated boots are important. Mittens are better than gloves in very cold situations because they provide a pocket of air around the fingers. A serious problem with gloves is that the better they protect the hand, the more clumsy they are. The head has a good blood supply, is capable of radiating heat effectively, and must be covered. Facial skin should be protected, and, in extreme conditions, should be checked frequently for frostbite.

Work Practices

Under cold conditions, it is important to avoid fatigue, which reduces resistance to cold. Workers should have frequent breaks, preferably in a warm environment. In a warm rest area, the windbreaker clothing should be removed to allow evaporation of accumulated moisture. Changes of dry clothes should be available if inner layers have become damp. Cold air is very dry, and breathing such air leads to gradual dehydration. Fluids should be available to the workers during breaks.

Workers should be trained in protective practices and in the symptoms of hypothermia. If a worker is shivering severely, is drowsy, or is experiencing euphoria, that worker should be removed to a warm rest area.

KEY POINTS

1. The body has a warm core temperature and a cooler skin temperature. Humans maintain core temperature between 36 and 38°C. Higher temperatures are termed hyperthermia and lower temperatures are termed hypothermia. Reaching extreme body temperatures is fatal.
2. Excess metabolic heat generated in the core is transferred to the skin to be lost.
3. Core temperature is regulated by the hypothalamus, which expands skin blood vessels to carry more warm blood to the skin and stimulates sweating to cool the skin by evaporation to dissipate excess heat.
4. Sweating leads to a loss of water and salt. Water must be replaced by frequent fluid intake, and in extreme cases salt supplements are needed.

[8] T. L. Eisma, "Controlling Cold Stress and Injury," *Occupational Health and Safety*, Dec. 1991.

5. During heat stress, heart rate increases due to greater vascular volume as skin vessels expand. Muscular activity calls for greater blood flow to muscles.
6. Workers acclimatize to a hot environment, eventually producing a greater volume of more dilute sweat.
7. Older workers control body temperature by sweating less effectively.
8. Use of certain prescription drugs, alcohol, and other social drugs reduces worker ability to handle heat stress.
9. Thinner and more fit individuals handle heat stress more easily.
10. The body heat balance is the sum of changes in metabolic heat production, radiative heat gain or loss, convective heat gain or loss, and loss by evaporation.
11. Maximum possible heat loss due to evaporation can be estimated when body dimensions, degree of exposed skin, and environmental conditions are known.
12. Heat cramps are muscle cramps caused by inadequate delivery of blood to muscles during heat stress.
13. Heat exhaustion results from circulatory inadequacy. There is a greater demand for blood flow combined with a loss of fluids.
14. Heat stroke, often fatal, results when the brain response to heat stress ceases as core temperatures rise to very high levels.
15. Wet bulb globe temperature index (WBGT), used to evaluate working conditions, is based on dry bulb, normal wet bulb, and globe temperatures. These are weighted and added according to formulae that differ for indoor and outdoor work.
16. No OSHA standards have been written to regulate workplace heat stress. ACGIH has written standards, and NIOSH and the Standards Advisory Committee on Heat Stress have proposed standards, all based on WBGT values.
17. Work in low temperatures presents two kinds of hazard: a general lowering of body temperature (hypothermia) and freezing of exposed tissue (frostbite).
18. Protective clothing should cover hands and feet particularly well. The head should be covered, and at very low temperatures the face should be protected. Layers of garments should be worn, with cotton preferred next to the skin.
19. Work practices should avoid fatigue. Frequent work breaks in a warm environment should be provided and fluids should be consumed. Coworkers should be alert for symptoms of hypothermia: drowsiness, severe shivering, and euphoria. Exposed skin should be watched for signs of frostbite.

BIBLIOGRAPHY

"Cold and Its Effect on the Worker," *Occupational Health Bulletin,* 15(11), Occupational Health Division, Department of National Health and Welfare, Ottawa, Canada.

A. J. Kielblock and P. C. Schutte, "Physical Work and Heat Stress," in C. Zenz, Ed., *Occupational Medicine: Principles and Practical Applications,* 2nd ed., Year Book Medical Publishers, Chicago, 1988.

W. J. Mills and R. S. Pozos, "Low Temperature Effects on Humans," in *Encyclopedia of Human Biology,* Vol. 4, Academic Press, San Diego, 1991.

J. E. Mutchler, "Heat Stress: Its Effects, Measurement and Control," in G. D. Clayton and F. E. Clayton, Eds., *Patty's Industrial Hygiene and Toxicology,* Vol. 1A, 4th ed., John Wiley & Sons, New York, 1991.

National Safety Council, *Pocket Guide to Cold Stress,* Chicago, 1985.

National Safety Council, *Pocket Guide to Heat Stress,* Chicago, 1985.

NIOSH, *Hot Environments,* U.S. Dept. of Health and Human Services, 1985.

OSHA, "Protecting Workers in Hot Environments," Fact Sheet 92-16.

J. D. Ramsey, F. N. Dukes-Dobos, and T. E. Bernard, "Evaluation and Control of Hot Working Environments," *International Journal of Industrial Ergonomics,* 14, 119, 1994.

E. A. Sellers, "Cold and Its Influence on the Worker," *Journal of Occupational Medicine,* March, 115, 1960.

L. H. Turl, "Clothing for Cold Conditions," *Journal of Occupational Medicine,* March, 123, 1960.

PROBLEMS

1. A small, windowless cinderblock tool shed with a corrugated metal roof nailed to beams laid on the cinderblock is constructed on the grounds of the equipment yard of a large construction company in South Carolina with the original intention that yard employees would spend only brief periods of time inside the building. As the business expanded, the building was changed to a place to house inventory and maintenance records, and workers began to spend longer times in the building. With age the roof became blackened and rusty. In summer weather, temperatures rise to uncomfortable levels in the building.
   A. Does placing an oscillating fan in the room lower room temperature? Explain in terms of the process of losing heat from the body why the situation is improved by such a fan.
   B. If instead an exhaust fan is installed in the wall, how specifically does this help? Would air inlets be of help, and if so where would they best be located?
   C. The metal roof is contributing to the heat problem. What terms in the equation for heat balance in the body are affected by the roof? What could be done to the roof to reduce the heat problem in the building?
2. Three workers operate a station in a plastics recycling facility at which an open vat is heated by gas burners. Washed shredded scrap plastic milk bottles are delivered in an overhead hopper and are dumped into the vat to be melted. The vat contents then are stirred by impellers driven from above. Excess water from the washing process boils off the vat. One side of the vat has a walkway with a pipe safety railing and devices to control the hopper, stirrer, and valve to transfer molten contents to a blow molding machine. The mix is sampled and analyzed, and pigments and plasticizers are added by dumping bags of appropriate dry solid chemicals into the vat from the walkway. The molten plastic product is then piped to a blow molding machine to be formed into squeezable red bottles to contain a detergent product. Worker 1 controls the hopper and stirrer and is stationed most of the time on the walkway. Worker 2 carries bags of chemicals up steps, opens the bags by pulling the tear tab across one end, and dumps the contents down a shute into the vat. Worker 2 must wear a face mask, safety glasses, and a plastic coverall because one of the chemicals is a skin irritant. Worker 3 removes samples from this and other vats from the walkway with a dipper and takes them to the nearby lab station where decisions are made about amounts of chemicals to add. Worker 3 wears a protective face mask at the time samples are removed. The gas burners are controlled automatically by temperature sensors in the vat. Analyze the heat stress potential of these three jobs.
3. Four male construction workers are working outdoors in hot, slightly breezy weather shoveling fine gravel into trenches. Worker 1 wears no shirt or hat, and his skin is constantly wet with sweat. Worker 2, also shirtless and hatless, is dripping sweat onto the ground. Worker 3 has on an open-weave hat and a cotton tee shirt that stays wet with sweat. Worker 4 wears a leather cowboy style hat and a blue denim jacket over a sweat-soaked tee shirt. Contrast the effectiveness of body cooling of these four workers.
4. A. In a very hot, dry indoor environment such that the air temperature is above body temperature, is it better to expose as much skin as the law allows or is clothing an advantage?
   B. In such a workplace outside air can be drawn in through a water spray, and the air is cooled by the heat removed to evaporate the water (evaporative cooling). T_A is lowered. Will T_G be improved, made worse, or be unaffected by this process? Will T_{NWB} be improved, made worse, or be unaffected by this process?
   C. Below are data taken before and after installation of an evaporative cooling system.[9] Calculate WBGT for each set of data. Do the changes in individual values agree with your predictions above?

	Before Installation	After Installation
T_A (°C)	42.8	29.4
T_G (°C)	46.1	35.0
T_{NWB} (°C)	24.2	24.4

5. In an aluminum plant one job involves skimming dross (impurities floating on the molten metal) using a ladle. The worker receives a great deal of radiant heat from the molten metal. A sheet of aluminum with a window and holes to reach the ladle was installed between the worker and the molten metal. Once the shield is installed, which of the three temperatures should show the greatest change?
Below are data from such an installation. Calculate WBGT for each set of data.

	Before Installation	**After Installation**
T_A (°C)	47.8	43.3
T_G (°C)	71.7	43.3
T_{NWB} (°C)	36.3	29.8

6. You are monitoring exposure to fluorides in an aluminum smelter by checking urine samples. Some workers are on the floor where the temperature is about 78°F, and others are on an elevated walkway near the pots of molten aluminum where the temperature is 103°F. Some workers on the walkway drink water regularly and some do not. How might these factors affect the results of your BEI testing.

[9] Data from R. S. Brief and R. G. Confer, *Med. Bull.*, 33, 229, 1973, discussed in J. E. Mutchler, "Heat Stress: Its Effects, Measurement and Control," in G. D. Clayton and F. E. Clayton, Eds., *Patty's Industrial Hygiene and Toxicology*, Vol. 1A, 4th ed., John Wiley & Sons, New York, 1991.

CHAPTER 15

Prevention of Accidents

An accident is a single event causing physical harm to the worker and is sometimes termed *overt trauma*. A slip-and-fall accident is a common example. Throughout life we learn self-protection skills: fear of heights, avoidance of obviously hot objects, leaving the path of a large moving object. This sort of acquired common sense must be involved in creating a safe workplace. This is supplemented by experience with typical work situations and records that indicate what aspects of a job lead to accidents. Such records reveal that common problems involve slip-and-fall accidents, injury from moving objects, contact with moving parts of machinery, and hazards from high-energy electrical installations.

To avoid having to rediscover common hazards in each workplace, standards are set by OSHA and presented in 29 CFR 1910. Statements range from general principles to specific requirements. Take, for example sections of Subpart D: Walking–Working Surfaces:

General Principles

1910.22 General requirements (a) *Housekeeping* (1) All places of employment, passageways, storerooms, and service rooms shall be kept clean and orderly and in a sanitary condition.

Very Specific Regulations

1910.23 Guarding floor and wall openings and holes. (e) Railing, toe boards, and cover specifications (1) A standard railing shall consist of a top rail, intermediate rail, and posts, and shall have a vertical height of 42 inches nominal from upper surface of top rail to floor, platform, runway, or ramp level. The top rail shall be smooth surfaced throughout the length of the railing. The intermediate rail shall be approximately halfway between the top rail and the floor, platform, runway, or ramp. The ends of the rails shall not overhang the terminal posts except where such overhang does not constitute a projection hazard.

SLIP-AND-FALL ACCIDENTS

In 1988 17.1% of industrial occupational work injuries and 19.1% of service occupational work injuries were falls.[1] Prevention of slip, trip, and fall accidents is the subject of Subpart D, from which the above quotations are drawn, and Subpart F (Powered Platforms, Manlifts, and Vehicle-Mounted Work Platforms). Floor openings, stairs, ladders, scaffolds and a variety of powered platforms are covered in detail in these sections.

[1] National Safety Council, *Accident Facts*, 1992.

The slipperiness of a walking surface is a concern, and the concept of "slippery" is quantified as coefficient of friction between the shoe and the surface. Coefficient of friction values range from 0 (very slippery) to greater than 1. Values of 0.7 or higher, for example, a brushed concrete surface, are very safe. The Department of Justice's Access Board Bulletin No. 4 recommends a minimum of 0.6 on level surfaces and 0.8 on ramps.

Although the concept of frictional coefficient is a means of evaluating the safety of walking surfaces, a single, standard method of measuring this constant is not in use. In the U.S. we measure a static coefficient of friction, which is determined as the horizontal force needed to start a "shoe equivalent" object into motion on the floor surface divided by the weight of the object. OSHA and Underwriters Laboratories recommend use of a leather slider. The American Society for Testing and Materials provides a test standard: ASTM C 1028-89. Many other countries determine a dynamic coefficient of friction, measured as the resistance to movement of a "shoe equivalent" object already in motion on the surface divided by the weight of the object. The static coefficient is generally higher. At the time of writing, the International Organization for Standardization (ISO) is attempting to provide universally acceptable static and dynamic test methods.[2]

A dry surface changes its coefficient of friction when covered with water, grease, or oil. With liquid on the floor surface, the shoe is somewhat separated from the surface by a film of liquid, reducing the coefficient of friction. With a lighter object on a larger contact surface the liquid film can completely separate the two, making the surface quite slippery. The term *hydroplaning* is sometimes used to describe a layer of water separating the object from the floor. This is more likely to happen with a more viscous liquid, so oil and grease present a greater problem than water, but it can happen with water under the proper conditions. The effect of floor waxes on coefficient of friction is clearly important.

Three obvious approaches providing an adequate coefficient of friction present themselves: modifying the walking surface, using anti-slip shoes, and maintaining good and appropriate housekeeping. To raise the coefficient of friction of the floor, high-friction decals can be applied in a high-hazard area. Similarly, walking surfaces coated with an anti-slip paint have a high coefficient of friction, even when wet. Such paints have particulate added that allows shoes to grip the surface effectively. The environment of the floor influences paint selection, for example, the heaviness of the traffic load, whether solvents may be spilled, or the occurrence of extreme temperatures.

Anti-slip mats are another choice where walking surfaces may have standing water, grease, or oil. These mats are usually vinyl or rubber based, may have holes to drain liquids, and may provide high-friction surfaces. They should be cleaned regularly to maintain their anti-slip characteristics. Mats may also provide a cushioned surface to ease problems associated with jobs requiring continuous standing.

The coefficient of friction depends not just on the floor surface. The other half of the picture is the worker's shoes. Shoes are manufactured for the workplace with high friction soles.[3] Rubber is commonly used in such soles, but does not grip a wet or greasy surface well. Aluminum oxide grit may be embedded into the soles to improve the grip. As the sole wears, new grit is constantly exposed.

Finally, good housekeeping to remove substances that decrease the coefficient of friction (water, grease, small round particles) reduces hazard. Cleaning the soles of shoes that do encounter grease or oil at regular intervals is a good precaution.

Studies show that an unexpected change in the surface encountered while walking or climbing raises the likelihood of an accident. This could mean a change from a surface

[2] See discussion in: G. Sotter, "Friction Underfoot," *Occupational Health and Safety*, March 1995, p. 28.

[3] The postal department defines a value less than 0.5 as unsafe and requires shoes that have a minimum coefficient of friction of 0.5 between the shoes and a tile floor.

"gripped" well by a person's shoes (high coefficient of friction) to a more slippery surface (low coefficient of friction). Floor irregularities can trip a worker. While walking, a person's heel may skim within 1/4 in. of the floor, so that even an irregularity lower than 1/2 in. can cause a person to trip. 49 CFR 31528 provides standards for floor surfaces designed to prevent such accidents. Housekeeping also helps to keep the floor free of loose objects that can cause a worker to trip.

The focus so far has been on indoor walking surfaces. Outdoor surfaces such as ladders, platforms, and walkways that become ice coated in winter are also a serious hazard. Mechanisms to remove the ice or shoes with built-in or attachable spikes or studs to grip the ice surface are needed. Footwear for outside use must also be waterproof.

MOVING PARTS AND OBJECTS

Anytime materials are moved in a plant, objects (sometimes quite heavy objects) are set into motion with risk to workers. Aspects of the design, maintenance, and operation of machinery to do this moving are presented in Subpart N (Materials Handling and Storage).

Moving parts on machinery can cause serious injury either directly to the body or by snaring clothing or hair. It was recognized early that in a safe workplace, guards are in place over moving belts, cutters, shears, saws, and other such devices to avoid worker contact. Regulations are presented in Subpart O (Machinery and Machine Guarding).

HAZARDOUS ENERGY

Factories operate at high levels of productivity because we harness energy to replace manual labor. Adapting the steam engine to run machinery started the industrial revolution, and we have become increasingly sophisticated in using energy to speed production and improve quality ever since. This same energy is capable of injury and death, however, if it is not kept under control.

Energy must be transferred to the job site to be useful, and this transfer takes several forms. Historically, the linkage between the source of energy, a water wheel or a steam engine, and the work was a direct mechanical linkage. For example, one can see old factory interiors in museums such as Greenfield Village in Dearborn, Michigan, where a steam engine turned revolving shafts that ran the length of the building. Belts turned by this shaft carried the energy to run machines at individual workstations. Energy was turned on and off by tightening and loosening tension on the belt. Although mechanical transfer of energy is still used, today we are more likely to transfer electrical energy generated at a distant power plant to run motors at the workstation. Alternatively, we may use fluids under pressure (hydraulic systems) or air under pressure (pneumatic systems), transmitting the fluid or air to the work in pipes or hoses.

ELECTRICAL ENERGY

Electrical energy is not visibly threatening, as was a belt and spinning shaft. If a machine uses electrical energy, we must take special care to guard the worker from accidental contact.

DIRECT AND ALTERNATING CURRENT SYSTEMS

Direct current electrical systems have two wires: one with a negative potential, which is the source of the electrons, and one with a positive potential, which is the place to which the electrons go. The greater the pressure for electrons to go from the negative wire to the positive wire, the greater the voltage. In alternating current systems, the potential at one end of the wire rapidly alternates between negative and positive. Electrons flow first in one direction, stop and flow in the other direction, then stop and flow again in the original direction. This is one cycle. This is the case in household wiring, where the current alternates at 60 cycles per second. Most current in workplaces is alternating current. Direct current has some special uses, however. Direct current is used to electroplate, to refine (purify) metals such as aluminum or copper, in electromagnets, and to charge batteries.

Your house has alternating current. If you disconnect power to a circuit (pulling the fuse or shutting off the circuit breaker) and open a receptacle, you can see the wiring. One wire in an alternating current system (black) is energized, and the second (white), often termed "neutral", returns to ground. The potential (voltage) between the two wires is 120 V. Where higher voltage is desired, for an electric stove or hot water heater, a three-wire system is used. Now two wires are energized and one is neutral. The difference between each wire and the neutral is 120 V, and between the two energized wires is 240 V.

OHM'S LAW

Voltage can be thought of as electrical "pressure". All else equal, higher voltages move electrons more rapidly (greater current flow or amperage) through a circuit (electrical pathway). Circuits provide resistance to current flow, and resistance is measured in ohms. The relationship between voltage, resistance, and amperage is given by Ohm's Law:

$$I = ER$$

where I = amperage, flow rate of electrons through the conductor
E = voltage, pressure driving the flow of electrons[4]
R = resistance (ohms), to the electron flow[15]

SAMPLE PROBLEM:

A circular saw is rated at 13.0 amp when attached to a 120-V source. What is the load (impedance) due to the saw motor?

$$I = ER$$

$$R = I / E$$

$$= 13 \text{ amp} / 120 \text{ V}$$

$$= 0.108\ \Omega$$

WIRING INSULATION

Sometimes, if there is sufficient resistance, no current flows. An air gap separating two electrodes that are connected to a source of electrical potential prevents any flow of current at low voltages, but if the voltage is raised sufficiently, current arcs across the gap. This is exactly the reason spark plugs in an automobile engine are connected to a high-voltage circuit.

[4] Voltage is also called electromotive force, or EMF.
[5] In a DC circuit, resistance is simply called resistance, whereas in an AC circuit, it is called impedance.

The 12 V provided by the battery are inadequate to allow arcing across the spark plug gap to ignite the fuel.

Electrical insulation, such as that wrapped around wires, is a high-resistance material designed to contain the current in the wire. Like the air gap, insulation that is sufficient at a lower voltage becomes inadequate at higher voltages.

During a workplace inspection, attention should be paid to wiring insulation to make sure it is intact and in good condition. This is most important on extension cords. Cuts or abrasion of the insulation may well mark a place where a worker could receive an electrical shock. Oil-soaked or otherwise chemically damaged insulation is also unacceptable. A wire with damaged insulation lying in water can be a dangerous combination. Inspect the insulation at plug connections on the cord, the place of greatest stress to the insulation. For example, if it is frayed or torn, this could indicate that the cord has been unplugged by pulling on the wire.

GROUNDED CIRCUITS

If the wiring in some household device becomes faulty, you do not want electrons flowing through you on the way to the ground when you touch the device. Household electrical appliances, tools, lights, and so on, are therefore grounded. The out-of-place electrons flow through this low-resistance circuit to the ground rather than through your higher resistance body.

In household (120-V) wiring there are normally three holes in the receptacle and three prongs on the plug. If you pulled a wall plug box apart , you would see a third, usually bare, wire entering the box and connecting in some fashion to the plug receptacle. Older houses have receptacles with only two holes for the plug. The difference between the two systems is the round prong, which connects by a low-impedance pathway to the ground. It is then possible to ground separately every unit that is in use. Appliances and tools that do not need to be separately grounded are sold with two-prong plugs.

Often people with older houses buy "cheater" adapters, which plug into a nongrounded two-hole receptacle and accept a three pronged plug. There is a green wire attached to the adapter that is supposed to be connected to a ground, but usually is not. Furthermore, just because a three-hole receptacle is in the wall does not ensure the ground plug is actually grounded. Sometimes homeowners, knowingly or unknowingly, buy three-prong receptacles to replace two-prong units, giving the receptacle the appearance of being safely grounded.

For all ordinary electrical usage in a factory the electrical system is exactly the same as the household system. In the same fashion, electrical equipment in a factory should be grounded. Sometimes a separate ground is in place: a heavy bare cable bolted to the machine at one end and the plant ground system at the other. Metal cages enclosing dangerous electrical installations in factories should also be grounded so that a faulty system inside the cage cannot give the cage a high potential.

The problems and abuses of electrical usage in houses all may be true in the workplace. When conducting a safety inspection, whether at a factory or an office, and especially of an older facility, the industrial hygienist should be aware of the wiring. You can pull back slightly on a plug to see if it is three prong without actually unplugging the unit. Simple portable testers allow checking to see if what is supposed to be grounded actually is. Certainly the use of cheater adapters should "wave a flag" for the inspector. 29 CFR 1910.304(f)(1) through (f)(7) list equipment grounding requirements.

CIRCUIT BREAKERS AND FUSES

In your house you have a circuit breaker or fuse box. All the circuits in the house pass through this box. The circuit breakers or fuses are designed to "open", to interrupt the flow of electricity if the flow becomes too great. Thus a 15-A (amp) circuit breaker opens if the current exceeds 15 A. If an appliance becomes faulty and diverts its electricity directly to the neutral or ground (a short circuit), the low resistance of this path would allow a large flow of electricity (high amperage), enough to open circuit breakers or blow fuses. This is important because a high flow rate of electricity against resistance generates heat, like the filament in a light bulb or the burner on an electric stove. In this case the house wiring becomes hot, a fire hazard. In factories, electric systems similarly require circuit breakers or fuses.

Breaker boxes in the workplace should be easily accessible. Standards [29 CFR 1910.303(g)] call for 3 ft of clearance in front of the box and a 3-ft aisle. The box should be unlocked and the circuits should be clearly labeled. The bus bars (copper bars carrying the electrical potential) should not be exposed.

GROUND FAULT CIRCUIT INTERRUPTERS

Fuses or circuit breakers interrupt the flow of current in the circuit if it becomes very high. However, there can be a problem such that the circuit finds another path to ground through your body. Suppose a hand tool becomes wet, so that there is a path from the energized wire to the housing or tool surface. Then suppose the worker is in contact with ground, standing in water or touching a grounded equipment housing, for example. The circuit can now complete itself through the worker. This is called a ground fault.

Ground fault circuit interrupters (GFCI) protect against ground faults. During normal use the current flowing through both wires is the same. However, when there is a ground fault, some current passes through the worker rather than through the neutral wire. The GFCI detects the amperage by means of the magnetic fields around each wire. If the ground fault is providing an alternative path to ground, the current in each wire differs by the flow through the ground fault. If this difference reaches 5 mA, the GFCI opens the circuit.

ELECTRICAL DAMAGE TO THE BODY

Your body also has electrical resistance. Low voltages, such as are generated by flashlight batteries (3–9 V), cannot drive electrons through your body and are therefore safe to handle. However, house wiring (120–240 V) can and is therefore dangerous.

In everyday language, when a current flows through your body, you "get a shock". A 10-mA shock is painful and a 100-mA shock is lethal. The borderline current that is lethal just because of effects of the current lies somewhere between. Current flowing through the body causes muscle contraction. Sufficient current in the wrong place can paralyze breathing or stop the heart. At around 15 mA muscle contraction prevents the worker from releasing the source of current. This is called the "let go" threshold. Other factors enter into electrical accidents. For example, a nonlethal shock to a worker on a ladder or platform can cause a fall.

Exposure to current can cause burns. This is especially true if the current is arcing across an air gap. The arc reaches very high temperatures.

ELECTRICAL FIRES

A breakdown in electrical systems is a common cause of fires. Electrical circuits include a "load", a resistance where work of some sort is done by the system. This could be a motor or a light bulb, for example. When a system "short circuits", the current finds a way to bypass the load. A common short circuit results from damage to the insulation of the wiring so that the wires either touch one another or press against a common conductive surface. By Ohm's law, the existing voltage produces a large current in the absence of high resistance. The high amperage heats the wires or other components of the circuit, perhaps enough to ignite combustibles. If the breakdown in the system involves an arc bridging an air gap, the high temperature of the arc is a likely ignition point. Solvent vapors in the vicinity of a short circuit may well be ignited.

REGULATIONS

Subpart S (29 CFR 1910.300–1910.399; Electrical) deals with design, safety-related work practices, safety-related maintenance, and other safety requirements for electrical equipment. For example, standards require enclosures of hazardous places in a plant electrical system to be entered only by qualified workers and require minimum working space and illumination within these enclosures.

Many of the standards refer to the system voltage. For example, many protective systems are required only when the potential is greater than 50 V.

EQUIPMENT MAINTENANCE AND ENERGY SOURCES — LOCKOUT/TAGOUT

Serious injury often results when a machine undergoing service or maintenance starts running. OSHA estimates that 10% of serious industrial accidents occur during maintenance or servicing of machinery without controlling energy going to the machine. In order to prevent this, in 1989 OSHA created a system called lockout/tagout to isolate machinery from energy sources during such operations.[6] Industry (the National Association of Manufacturers) challenged the rule in court in 1991, claiming that the injury rate varied greatly from one industry to the next, the rule was applied uniformly to all industries regardless of risk level, and the cost to those with low risk was unreasonably high. In 1994 the U.S. Court of Appeals ruled in favor of OSHA.

OSHA defines *affected* and *authorized* employees. Affected employees run or use the equipment as part of their job or work in the area where the servicing is occurring. Authorized employees are responsible for the lockout/tagout of the machine and for removal of the lock or tag after servicing is completed. The authorized person:

- Must know the source of energy that powers the equipment
- Shuts down the equipment
- Isolates the equipment from the energy source or sources. If it is not clear how to do this, the manufacturer or distributor of the equipment can be contacted. Once done, the equipment is said to be *de-energized.*
- Attaches locks to prevent the energy from being turned on again or tags to warn against turning on the energy. These locks are assigned to the authorized person and are labeled to identify that person as being responsible for that lock. Similarly, the tags have the authorized employee's name on them. No one else can legally remove that lock or tag.
- Control any stored energy, as by stopping moving parts or relieving pressure in a hydraulic or pneumatic system. The authorized employee should then do a final check that there is no possibility of a restart.

After servicing is complete, the authorized employee:

- Does a safety check (is assembly complete, are tools removed from the site, etc.)
- Informs affected employees the machine or equipment will be functional, and clears the area, if this is appropriate
- Removes the locks and/or tags

All employees must be trained about the lockout/tagout policies of the company, and authorized and affected employees receive appropriate special training.

[6] 29 CFR 1910.147.

A machine is capable of being locked out when a switch, valve handle, or blocking device is equipped to receive a lock for lockout purposes. Equipment purchased after October 31, 1989, new or refurbished machines, are expected to be capable of being locked out.

LOCKOUT/TAGOUT VIOLATIONS

Violations of lockout/tagout procedures are common violations cited by OSHA. In October 1993, an employee at a tire company was injured, and later died of those injuries, while servicing machinery. OSHA, on investigating the incident, fined the company $7.49 million, which included $70,000 for each of the 98 exposed employees, based on "willful violations of the lockout/tagout standard". OSHA claimed the company disregarded both lockout procedures and training.

From *Occupational Health and Safety,* June 1994, p. 20.

PERSONAL PROTECTIVE EQUIPMENT

Identifying a potential hazard at a workstation may lead to recommendations for protective gear to be worn by the worker. Face shields and/or safety glasses are appropriate for certain jobs, may be required, and must meet the standards in 29 CFR 1910.133.[7] Shoes or boots may be anti-slip (discussed above) and provide ankle support. However, in some situations anti-slip footwear may be needed where ankle support is not, for example, in food preparation areas or hospitals. Safety shoes have metal toe covers that protect feet from heavy objects and must meet standards in 29 CFR 1910.136 (ANSI Z41.1 of 1967). Protective equipment for electrical workers is outlined in 29 CFR 1910.335 and includes such items as nonconductive headgear, insulated clothing and tools, and fuse removal equipment.

Almost a quarter of workplace fatalities are due to head injuries, a number that would be much higher without the use of hard hats.[8] Standards for head protection (hard hats) intended to protect workers "from falling and flying objects and from limited electrical shock and burn" are presented in 29 CFR 1910.135 or 29 CFR 1926.100 (construction). This was originally the ANSI standard Z89.1 of 1969, updated in 1986: ANSI Z89.1-1986. The latter became the OSHA standard in 1994 under the new personal protective equipment standard. Type 1 hard hats have a brim completely around the hat, whereas Type 2 hard hats have just a bill over the eyes. Two aspects of construction absorb impact. The hat shell, usually molded from high-density polyethylene, flexes slightly on impact. More significantly, the hat is held away from the head by a suspension system, almost like spring mounting the hat. All classes protect against impact, but differ in degree of protection against electrical shock. Class A protects against low voltage, Class B against high voltage, and Class C provides no electrical protection.

RIGHT TO KNOW — POSTING HAZARDS

Workers should be instructed about the hazards of operations in their work area as a part of compliance with the right-to-know regulations. Signs posted to inform workers are classed

[7] Based on ANSI Z87.1-1968, updated in 1989.
[8] V. Meade, "Heads Up: What's New in Protective Gear," *Occupational Health and Safety,* July 1995, p. 33.

Figure 15.1. Hazard warnings. Right-to-know regulations require that workers are informed of hazards. When signs are used for this purpose, a color code is used to distinguish notices according to their urgency. DANGER warnings are red (the word "danger") and black (the lettering) on a white field. CAUTION signs have black letters on a yellow field.

as follows: *danger* warnings (red, black, and white), which inform about immediate danger requiring special precautions; *caution* warnings (black and yellow), which inform about potential hazards or caution against unsafe practices; and *safety instructions* (green and white) (Figure 15.1).

KEY POINTS

1. An accident is a single event causing harm, an overt trauma.2.Standards to prevent accidents are spelled out in 29 CFR 1910.
3. The frequency of slip-and-fall accidents is reduced by providing proper railings and barriers, avoiding unexpected changes of the floor surface, and practicing good housekeeping.
4. Materials handling generates risks of workers being impacted by machinery used for moving.
5. Moving parts on machinery require guards.
6. Electrical equipment requires special layout, isolation, grounding, and circuit breakers or fuses.
7. During maintenance or servicing, machines should be isolated from energy sources and the switches, valves, and other controls either locked off or tagged to indicate that they must remain off.
8. Protective attire such as safety glasses, hard hats, or protective work shoes may be required in certain jobs.
9. Hazards should be labeled with warning signs.

BIBLIOGRAPHY

American National Standards Institute, "National Electrical Safety Code," ANSI Document C2-81.
T. Ferry, *Modern Accident Investigation and Analysis,* John Wiley & Sons, New York, 1988.
W. Marletta, "Trip, Slip and Fall Prevention," Chapter 12 in D. J. Hansen, Ed., *The Work Environment: Volume 1,* Lewis Publishers, Chelsea, MI, 1991.
National Fire Protection Association, "National Electrical Code," NFPA Document 70-78.
National Safety Council, *Accident Prevention Manual for Business and Industry,* 10th ed., Itasca, IL, 1992.

PROBLEMS

1. Ladders are frequently involved in accidents, so it is not surprising that ladders are closely regulated. Even so, you may be surprised at the degree of detail of the regulation. Turn to 29 CFR 1910.21(c).
   A. What is the specific subject matter of this section?
   B. How many kinds of ladder are separately defined?
   C. What is wane?
   D. What is shake?
   E. Is section 1910.21(c) all or in part a set of regulations?
   F. Compare 1910.21(c) and (d). Why is this done?
   G. Now look at section 1910.25. What is this section?
   H. OSHA is sometimes charged with regulating too closely, creating overly detailed rules that make compliance and enforcement difficult. Suppose you were going to work as a janitor in a plant, and would be replacing fluorescent tubes while standing on a high wooden ladder. Which of these sections in 1910.25 would you eliminate in the interest of simplification?
   I. Next time you are in a hardware store or a lumber yard, look for OSHA stickers on the ladders. What does this sticker tell you?
2. You may at some time have looked up at workers washing windows on a tall building and wondered how safe such a job can be.
   A. What are the hazards?
   Turn to section 29 CFR 1910.66.
   B. How many pages are devoted to this topic?
   C. What is the kind of "rope" that can be used on such a platform and what safety margin is required regarding its load-bearing ability?
   D. Suppose a platform has two suspension wires, each rated at 5000 lb and the working load of the platform is 2000 lb. Is the system in compliance?
   E. How often must the wire ropes be inspected, and how would an inspector know if the ropes were in compliance?
   F. Look at **Figures 1, 2, and 3**. What are the systems illustrated designed to prevent?
   G. What are the regulations regarding design and operation of a platform in the wind?
3. Decide whether each of these signs should be labeled DANGER or CAUTION, or should be presented as safety instructions, and indicate the correct color code in each case.

THIS EQUIPMENT STARTS AND STOPS AUTOMATICALLY	EYE WASH FOUNTAIN	WEAR HEARING PROTECTION IN THIS AREA	OXYGEN IN USE NO SMOKING OR OPEN FLAME

4. An industrial hygienist who is inspecting facilities for safety conditions constantly encounters new types of plant operations. It is necessary to be able to locate the appropriate regulations in the CFR that apply to each new situation, and to be able to interpret these regulations in light of the site visit in progress. Suppose you are visiting a plant where ornamental wooden trim is shaped for house construction, and you have never before visited a woodworking facility.

A. Where are safeguards for use of woodworking machines found in the CFR?
B. Bandsaws are in use to cut curved shapes in 3/4-in. plywood. A bandsaw has a continuous ribbon of saw blade that is stretched over two large drive wheels that are above and below the work surface, so that the blade moves vertically downward at the work area. This machine catches your attention as a potential finger amputator. The wheels and blade are enclosed in a housing that surrounds all but one side. This side has a screen that is removable for blade replacement. The screen catches your attention because it was not one originally built by the manufacturer of the saw. It is heavy-gauge stamped sheet metal with diamond-shaped holes approximately 11/2 in. wide and 7/8 in. high, and is mounted onto the rest of the housing by bolts that were part of the original assembly. The blade passes through rollers that prevent the blade from wobbling. The gap between the work-table and lower edge of the upper wheel housing exposes the blade for cutting wood 3 in. above the table surface. (1) Where are bandsaw regulations? (2) Is this saw in compliance?

5. Looking at regulations for electrical safety, we find various voltages that are designated as dividing lines in specific rules.
   A. What are they?
   B. Why does the voltage make a difference?
6. 29 CFR 1910.304(f) discusses what should be grounded.
   A. Why is grounding necessary?
   B. Select a subsection of part (f) and consider the hazard if the regulated system were not grounded.
7. If the current flowing through a circuit is 1.5 A at 100 V, what current flows at 200 V in the same circuit?

CHAPTER 16

Cumulative Trauma

Injury that is the result of a long series of events overstressing some part of the body is termed *cumulative trauma disorder* or *repetitive motion disorder*. Work stress is the application of external force during the performance of a task. As an object is lifted, its weight is distributed across the body in a fashion dependent on posture during lifting. Work strain is the response of the body to work stress. An example is lower back pain resulting from repeated lifting of heavy loads for a long time period. Such problems are often classed as musculoskeletal injuries. Occupational biomechanics, ergonomics, or human factors engineering are various names for the field of investigation focused on the design of tasks to best fit human anatomy and physiology.[1] The major goal of occupational biomechanics is the elimination of cumulative trauma disorder generated by posture, excessive strain, or repetitive motion by matching the body as a machine to a task. This field therefore requires study of both the capabilities of the human body and engineering aspects of the task.

In 1983, a report by NIOSH indicated that musculoskeletal injuries were the cause of about a third of workers' compensation claims each year, affected 19 million individuals, and cost (in terms of these payments and lost work hours) more than any other occupational health problem.[2] A more recent report, titled "A National Strategy for Occupational Musculoskeletal Injuries", indicated that the situation had not improved in more than a decade. In fact, the category on the OSHA reporting form labeled "disorders associated with repeated trauma", which includes both musculoskeletal injuries and occupational hearing loss, rose sharply during the 1980s to represent the majority (63% in 1992) of all occupational illness.[3] A case is built that cumulative trauma injuries are underreported to a large degree. Many are reported as injuries rather than as illnesses.[4]

Cumulative trauma has become more commonplace as the nature of work has changed. Consider the task of a carpenter building a wall. Wood is cut to length, lumber is lifted in place, nails are hammered; the nature of the task continuously changes. Contrast that now with the type of job created by assembly line employment. The power wrench mounts the wheel nuts or the circuit board is inserted into the board mount repeatedly for the entire work

[1] There is confusion of terminology. Ergonomics originally studied optimization of human performance, so it was performance based rather than safety oriented. Increasingly the term is used to denote setting standards to prevent cumulative trauma. A proposed new OSHA standard is usually termed an *ergonomic standard*.

[2] P. L. Polakoff, "Without Action on Causes and Effects, Time Fails to Heal Cumulative Trauma," *Occupational Health and Safety*, Jan. 1995, p. 29.

[3] K. A. Grant, V. Putz-Anderson, and A. Cohen, "Applied Ergonomics," in L. J. Cralley, L. V. Cralley, and J. S. Bus, Eds., *Patty's Industrial Hygiene and Toxicology*, Vol. III, Part B, 3rd ed., 1995.

[4] F. E. Mirer, "Labor and Industry Should Cooperate to Reduce the Ergonomic Injury Rate," *Occupational Health and Safety*, Oct. 1992, p. 34.

day, every day. The nature of the task does not change, and the same muscles are used in the same fashion with pressures on the same joints all day and every day.

Overt traumas present a simple picture because the injury occurs immediately due to a failure of protection or error of judgment. Poisoning by chemicals, cell damage by radiation, and hearing damage by high noise levels allow quantitative prediction of the result of an unsafe condition. Cumulative traumas are harder to predict and avoid, and the industrial hygienist is more likely to recognize the problem after and because injury has occurred. On trying to perform a particular task on the job, the worker may say, "Yeah, sure I can do that", and may in fact do the task without difficulty for a long time every work day without apparent difficulty. But small excessive stresses on some part of the body caused by the task gradually accumulate, resulting in real physical damage at a future time. Experience has established that certain parts of the body are particularly susceptible to injury, including the fingers, wrists, elbows, shoulders, and back.

Work stress could be the weight of a box to be lifted, the resistance of a screwdriver to being turned, or the effort required to hold a power drill while drilling a hole. Work stress is a problem only when the strain is excessive. Stress could be pressure on joints, the production of friction at joints and other skeletal connections, challenges to muscle contraction, and tensions on tendons. Knowing the contributing factors helps one anticipate problems: jobs requiring repetitive movements, an awkward position, and the need to use force. Additional contributing factors include pressure on some part of the body such as the finger or thumb to run a tool. Spending long time periods in the same position (e.g., seated at a desk) may result in a continuous small insult to some body part that eventually leads to disorder. However, it is hard to assess the level of risk of injury due to sitting in an ill-fitting desk chair for long stretches during the work day in a fashion parallel to measuring accurately amounts of benzene in the air and assessing the degree of risk at particular concentrations.

CASE STUDY: REPETITIVE MOTION

Red Wing Shoe Company had always compensated workers on a piecework basis — pay according to the number of shoes produced per day. The employees were motivated by this system to perform repeated identical tasks at a high rate in order to maximize their income. The repetition put an excess burden on certain muscle groups, however, and cumulative trauma disorder became a problem at the plant. In 1990 they changed the system to one in which employees were paid on an hourly basis and switched jobs, ideally once every 2 hours. In addition, an effort was made to introduce flexibility into the design of tables, chairs, and machines in order for workers to adjust the workstation to body dimensions. The reward to the company has been a drop in annual workers' compensation costs from $4.4 million in 1990 to $1.3 million in 1994.

Adapted from the *BNAC Communicator*, Spring/Summer 1995.

DESIGNING THE TASK TO AVOID TRAUMA

Progress is being made in anticipating workstation layouts likely to produce cumulative trauma. Here are sample guidelines:

1. Utilize the relaxed or neutral positions for parts of the body, then design the job to keep those parts as close as possible to that position. Arms are below shoulder level, wrists are straight, elbows are close to the body, the head faces straight forward and the back is vertical with a natural curve. Being able to adjust parameters of the workstation, for example, the height of a seat or the level of the work, permits the worker to maintain these neutral positions.
2. Minimize the amount of force required to accomplish a task, restricting it to a low percentage of the maximum the worker can exert, unless the duration of exertion is very short.
3. Allow the worker to shift location with respect to the task. This begins with providing enough space for larger workers, so the worker is not locked into a single position. Then allow for the worker to approach the task in more than one fashion. This avoids the same set of muscles being used continuously. For example, being able to alternate between standing or sitting, or being able to turn and move laterally is helpful.
4. Avoid situations where the worker must support an object for long periods. For example, supporting the weight of a hand tool in one place is fatiguing and is not productive.

HUMAN DIMENSIONS

To design a task so that there is the least chance of cumulative trauma, basic information is needed about human beings. Consider the dilemma of the early automobile industry designing a car to be used by the entire range of sizes found in the general public. To lay out such dimensions as seat height, distance to foot controls, location of hand controls, and distance from leading edge of the seat to the seat back is a major challenge, one the auto industry has devoted large sums of money to research, because the penalty for not doing this correctly is the loss of a percentage of the public as potential customers. Very early, adjustable parameters were added to cars, starting with seats that slide closer to or farther from the steering wheel and mirrors that adjust angle. Now car seats can be adjusted for height, seat back angle, and tip. Steering wheels can be moved up and down. Gathering data to design the car successfully had to start with gathering data about the range of human dimensions.

Similarly, we must know the range of human dimensions to lay out a task in the workplace. People obviously vary from short to tall, with short to long arms for reaching. It is not sufficient to calculate what the average height and reach are, because you are going to encounter workers that do not fit that. This field of study is termed engineering anthropometry. At one time, only dimensions of men were considered. For example, a control reached overhead was located 77 in. or lower. Now with large numbers of women in the work force, such a dimension is reduced to 73 in. to accommodate 95% of female workers.[5]

Armstrong and Lifshitz prepared a checklist of job factors to predict upper body cumulative trauma (Table 16.1). Using this checklist in different job situations, they obtained high correlation between low scores and the occurrence of cumulative trauma. Computer comparison of the type of trauma with individual responses on the survey suggested corrective action to prevent the trauma.

MINIMIZING REPETITIVE STRESS

Sometimes repetitive movements or awkward positions are hard to eliminate, particularly in assembly-type operations. Other preventative practices may be employed. Workers can be shifted from one workstation to another to change the patterns of movement. Employees can be trained to do simple stretching exercises before or during the job that relieve stress and break harmful patterns. Sometimes a job makes demands on certain muscles that can be strengthened by exercise so as to prevent trauma.

[5] D. B. Chaffin, "Ergonomics Advances as a Science with Applications in Health, Safety," *Occupational Health and Safety*, Jan. 1992, p. 38.

Table 16.1. Checklist for Analysis of Upper Extremity Cumulative Trauma Disorder Risk Factors.

1. Physical Stress
   1.1 Can the job be done without contact of fingers or wrist with sharp edges?
   1.2 Is the tool operating without vibration?
   1.3 Are the worker's hands exposed to temperatures above 70°F?
   1.4 Can the job be done without using gloves?
2. Force
   2.1 Does the job require less than 10 lb of force?
   2.2 Can the job be done without using a finger pinch grip?
3. Posture
   3.1 Can the job be done without flexing or extension of the wrist?
   3.2 Can the tool be used without flexing or extension of the wrist?
   3.3 Can the job be done without deviating the wrist side to side (ulnar or radial deviation)?
   3.4 Can the tool be used without ulnar or radial deviation of the wrist?
   3.5 Can the worker be seated while performing the job?
   3.6 Can the job be done without "clothes wringing" motion?
4. Workstation Hardware
   4.1 Can the orientation of the work surface be adjusted?
   4.2 Can the height of the work surface be adjusted?
   4.3 Can the location of the tool be adjusted?
5. Repetitiveness
   5.1 Is the cycle time above 30 s?
6. Tool Design
   6.1 Can the thumb and finger slightly overlap around a closed grip?
   6.2 Is the span of the handle between 5 and 8 cm?
   6.3 Is the handle of the tool made from material other than metal?
   6.4 Is the weight of the tool below 10 lb?
   6.5 Is the tool suspended?

From Y. Lifshitz and T. Armstrong, "A Design Checklist for Control and Prediction of Cumulative Trauma Disorders in Hand Intensive Manual Jobs," *Proceedings of the 30th Annual Meeting of Human Factor*, pp. 837–841, 1986.

The height of a work surface or workbench is an example of fitting the workstation to the worker. Workers should not have to hold their arms above their shoulders, nor should they have to lean forward so as to bend their backs and rotate their shoulders forward. The sum of the bench height and the height of the object is the critical dimension. During assembly work, the object should be 2–4 in. below the height of the elbows, slightly lower if greater force must be exerted. Especially where more than one worker uses the bench, an adjustable-height bench is desirable.[6]

Cumulative trauma may also occur as a result of bad practices by the worker rather than bad design of the task. An example of this is the effect of poor posture practiced by workers on the job.[7] Remember your mother telling you to sit up straight and stop slouching? She was right! Sitting at work with the head and shoulders held forward results in muscle imbalances and weakening of some muscle systems. Muscles between the shoulder blades are weakened and those across the chest from the shoulders are shortened. These shortened muscles press on the nerves to the arm. The burden of the weight of the arms is shifted to muscles other than those intended to support them. The head is now off balance and must be supported continuously. All this results in fatigue and pain, particularly between the shoulder blades and in the neck. Incorrect posture becomes habitual, and because muscles that should be carrying the burdens are not used they become weakened and shortened, reinforcing the bad posture. Problems due to posture may be corrected by education of the worker, treatment for pain, and, as possible, exercise to strengthen weakened muscles. The

[6] R. Carson, "How to Select a Proper Workbench," *Occupational Health and Safety*, April 1995, p. 53.
[7] M. L. Langford, "Poor Posture Subjects a Worker's Body to Muscle Imbalance, Nerve Compression," *Occupational Health and Safety*, Sept. 1994, p. 38.

tasks performed by the worker should be analyzed, perhaps relocating some functions to eliminate operations that enforce continuation of the incorrect posture.

Computers may be used in a more sophisticated fashion in workstation design. Inputting detailed presentations of athletes competing allows coaches to pinpoint changes in style to improve performance. In the same fashion, the movements involved in performing a task at a workstation can be fed into a computer programmed to detect the potential for cumulative trauma.[8] The design of workstations and tools also benefits from thorough knowledge of human dimensions and strengths.[9]

EXAMPLES OF SPECIFIC PROBLEMS

HAND AND WRIST ANATOMY

One of the facilities that gives humans an advantage is ability to grasp and manipulate objects in a sophisticated fashion. Complex machinery is built into our hands so that we can flex and extend our fingers and also flex, extend, and move our wrists from side to side.

If all the muscles that do this were in the hand itself, it would be an unwieldy structure. Instead some muscles (one for the thumb and two per finger) are in the forearm and are connected to the bones of the hand by tough, bendable cords called tendons. Tendons attach at one end to a muscle and at the other to a bone, so that the tendon moves the bone when the muscle contracts . Take a moment to feel the tendons in one hand as you extend and flex your fingers. Appreciate the variety of movements you can accomplish with your fingers. Flexor muscles and tendons cause the fingers to close to grasp an object, whereas extensor muscles and tendons cause the hand and fingers to open. These are balanced systems, systems in which the contraction of one muscle (for example, the flexor) requires the relaxation of the opposing muscle (the extensor). You sometimes exert considerable force in grasping an object, so that flexor muscles must be strong. By contrast, you seldom "force" your fingers open strenuously, a minor effort normally being required to do this. See on your forearm that the flexor muscles in the front of the forearm are larger and much more strongly developed than the relatively small extensor muscles on the back of your forearm.

Now feel your wrist, and recognize that it is primarily a bony structure. The wrist is a relatively rigid structure made of eight carpal bones arranged in a row and connected to one another by ligaments. A closed loop is formed by connecting one end of this array with the other on the palm side by a transverse ligament. Nine tendons and the median nerve reach the hand through a passage about the diameter of a small coin between the wrist bones called the carpal tunnel. The transverse ligament does not stretch significantly, so the tunnel does not expand during heavy activity.

These tendons largely pass through coverings or sleeves called sheaths. Between the sheath and the tendon there is a fluid that serves to lubricate the tendon as it moves back and forth. When your hand and forearm are in a straight line, the tendons pass more smoothly through the sheath than when the wrist is bent. Heavy use with the wrist bent increases friction between the tendon and the sheath.

CARPAL TUNNEL SYNDROME

Consider some tasks. You are assigned to sort numbers of parts into bins and boxes in the parts room of an auto repair shop. Every time you pick up another part you are using your flexor muscles. How about harvesting peaches or potatoes, chopping vegetables for a salad bar, or typing? Normal daily activity for most people requires the use of flexor systems

[8] See Chaffin and Andersson, 1991, in Bibliography.
[9] R. M. Barnes, *Motion and Time Study, Design and Measurement*, 7th ed., John Wiley & Sons, New York, 1983.

— this is not a problem. Difficulties arise when a worker uses the same flexor systems continuously in a repetitive job. This causes excessive friction between tendon and sheath. Such continuous abrasion eventually leads to tenosynovitus, inflammation of the tendon sheaths. This inflammation is associated with pain upon using the hand.

Resultant swelling of the tendon sheath presses on the nerve, interfering with its function. This leads to loss of sensation, weakening of some muscles (primarily associated with the thumb), and eventual loss of grip and other hand functions. Early symptoms include tingling or numbness in the hands, especially at night. Eventually, nerve damage results. These problems are referred to as carpal tunnel syndrome.

Redesigning the Task

Although there may be other causes such as arthritis or a badly healed broken bone, the vast majority of carpal tunnel problems in the workplace are caused by repetitive movements, especially movements in an awkward position. Finger movements with the wrist bent, flexing the wrist up and down, pressing with the base of the palm, and using vibrating equipment all contribute.

Operating a keyboard is a commonplace cause of carpal tunnel problems. It is reported that "46% of secretaries, 36% of word processors, 13% of stenographers, and 11% of programmers" display carpal tunnel symptoms.[10] These occupations heavily employ women, and the smaller average wrist diameter of women may be a contributing factor to the frequency of carpal tunnel problems. Keyboarding should be arranged so the worker performs repetitive operations with a straight wrist. Keyboards are available that separate the right hand keys from the left, and tip them at a slight angle to one another so the the keys are approached with the wrist straight. Appropriately adjusting keyboard height and providing a wrist support helps prevent problems.

Hand tools used in repetitive jobs can often be redesigned to accommodate working with a straight wrist ("bending the tool, not the wrist"; see Sanders and McCormick, 1987). The most frequently cited example is commonplace pliers. Hold a pair of pliers, reach out, and grasp an object. Now, look at your wrist. The bend in your wrist means that the tendons are rubbing on their sheaths excessively. By bending the handles of the pliers out of line with the jaws by the same angle as your wrist assumes holding straight pliers, you can manipulate the pliers with a straight wrist.

For vertical work surfaces, a hand tool with a pistol grip keeps the wrist straight, like the typical electric drill. Work on horizontal surfaces is best done with a tool that is in line with the work and is grasped like the support pole on a bus or subway car.

Braces may be used that hold the wrist straight to prevent the start or worsening of carpal tunnel syndrome. There is evidence that exercise programs may help prevent carpal tunnel damage. Once the damage has occurred, it is difficult to reverse. Surgery, cutting the transverse ligament to enlarge the opening slightly, often relieves the symptoms, but in many cases the problem returns in a few years.

Detecting the Start of Carpal Tunnel Problems

There are some simple tests for the onset of carpal tunnel syndrome. One, the Phalen wrist flexor test, involves pressing the backs of the hands together with fingers pointed downward and the arms horizontal. If numbness or tingling appears after a minute, this is a sign the median nerve is under pressure, it is likely that the carpal tunnel is narrowed. In the

[10] S. L. Morgan and R. Pearson, "Pink Collar Workers and Carpal Tunnel Syndrome," *Occupational Health and Safety*, Oct. 1991.

median nerve percussion test, a person repeatedly taps the inside of the wrist with the fingers of the opposite hand. A tight fit for the median nerve results in nerve stimulation, evidenced by "prickling" in that hand. Neither of these are absolute indicators, and a physician should be consulted when there is any possibility of carpal tunnel problems.

CASE STUDY: THE COST OF CARPAL TUNNEL SYNDROME

Creation Windows, which produces windows for recreation vehicles, analyzed the cost of carpal tunnel syndrome at their facility. They estimated they had to produce and sell 9000 windows to pay for one carpal tunnel surgery, when lost work days, workers' compensation, and other expenses are totaled. Estimates of the cost of such surgery are about $35,000 each (*Wall Street Journal,* April 27, 1993).

Professional job site assessments were done, including videotaping workers and analyzing the motions involved in performing work operations. This resulted in major changes in operations. For example, the tools were largely sized for male hands, and the workforce is largely female. Overhead support systems were added to avoid workers having to support the weight of the tool. Tools and operations were modified to reduce wrist bending and use of excessive pressure.

Classes were run for managers on repetitive strain injuries. Instituting early intervention by physical therapy, rather than waiting until surgery was necessary, provided big savings.

Jobs on the assembly line were rotated every 2 hr to reduce the degree to which single muscle systems were constantly being used. Workers at first were unhappy with this change, because pay was on a piecework basis, and the need to learn a new task reduced productivity and therefore the paycheck. However, 12 months after this change, individual productivity rose to levels higher than before.

Creation Windows once averaged 22 carpal tunnel surgery cases each year. One year after starting the prevention program, the number of surgeries halved, and now there is only about one case per year.

Adapted from M. Strakal, "Prevention Pays Costs of Cumulative Trauma," *Occupational Health and Safety,* Dec. 1994, p. 43.

TRIGGER FINGER

Difficulties arise when you use the same flexor systems continuously in a repetitive job for entire workdays. The flexors become stronger and larger. Overused muscles tend to stay shortened and the corresponding extensors are continuously stretched and overextended.

Repetitive finger action results in tenosynovitus, and over a long period an inbalance of flexor and extensor muscles. The result can be a reduction in the range of motion of the fingers. Use of the finger becomes restricted such that the person can bend the finger, but has to pull it straight again.

Such difficulties are commonly called trigger finger. In general, controls should not be operated in a repetitive fashion by the index finger, operating a switch as one would pull the trigger of a firearm, although this seems to most people to be the most natural operation. Less trouble is experienced when thumb controls are used. Some muscles and tendons of the thumb are in the palm of the hand rather than the forearm, and with only two bones in the thumb, movements are less complex.

COMPRESSION IN FINGERS AND PALMS

Operations should minimize exerting pressure with fingers or palms. This compresses blood vessels and nerves in the hand, leading to ischemia (loss of circulation) and numbness in the hand. Tools should be designed to spread the compression load as broadly as possible to minimize pressure at a single point.

THE ELBOW: TENNIS ELBOW

There are two bones in the forearm, the ulna and the radius. These meet at the elbow with the humerus, the bone of the upper arm. When the arm is bent or straightened, the ulna and humerus operate at the elbow like a hinge. The wrist is attached to the ends of the ulna and radius. It is moved up and down by muscles in the forearm. Extending the wrist (moving it from down to a neutral position), as would be done in a tennis backhand stroke, stresses the tendon attaching the muscle performing that movement at its attachment to the humerus in the region of the elbow. Inflammation at that attachment by strong repetition of the backhand movement in tennis causes epicondylitis, nicknamed "tennis elbow". Extending the wrist is a common motion in numerous tasks. Frequent or forceful repetition can, just as with tennis players, produce this inflammation.

BACK PROBLEMS: LIFTING AND CARRYING

Lower back pain is a common worker complaint, costing U.S. industry 93 million lost work days (about 40% of the total in 1993), almost a quarter of workers' compensation costs (24% in 1993), and almost $80 billion in lost productivity and wages. In society as a whole it is the second most frequent reason people visit a physician. Perhaps 8% of all workers are affected.[11] Most important, a majority of these problems are preventable.

Visualize the vertebral column. This bony structure serves several functions. When we stand upright, it carries the weight of the body down to the pelvis, where the weight is transferred through the pelvis to the legs. For this task, the column must be strong. It surrounds and protects the spinal cord, which carries messages back and forth between the brain and the rest of the body. For this it must be hollow and have openings that allow nerves to exit from the spinal cord. Finally it must be flexible. For this purpose it is comprised of a number of individual bones shaped to allow movement and separated by flexible cartilaginous discs.

The discs, a necessary but weak feature of the vertebral column, are displaced, compressed, or even ruptured (herneated) by bearing heavy loads (or simply with age). Continuous excessive pressure speeds disc damage. A displaced or herneated disc may press against nerves as they leave the spinal cord. Such pinched nerves cause pain or numbness.

The back is capable of complex movement, including bending front to back, bending side to side, and twisting. These operations are accomplished by the action of a complex set of muscles. The situation is very much like the flexor and extensor muscles of the forearm, previously described. Overuse of a few muscles, either by repetitive work or bad posture, throws the muscle balances off.

Lower back pain can result from the handling of objects: lifting, carrying, pushing and pulling. Twisting, reaching, or bending while performing these tasks increases difficulties. Lifting is the most common cause of lower back pain. Although the worker may attribute the pain to a single event of lifting, studies show that only about 4% of back injuries occur due to a one-time effort.[12] Given the frequency of back problems in the workplace, consid-

[11] R. A. Deyp and Y.-D. Tsui-Wu, "Descriptive Epidemiology of Low-Back Pain and its Related Medical Care in the United States," *Spine*, 12, 264, 1987; D. Daniels and W. Knight, "Flesh and Bone," *Occupational Health and Safety*, Jan. 1995, p. 53; G. V. Oort et al., *Occupational Health and Safety*, Jan. 1990, p. 22.

[12] J. J. Keller and Associates, "Special Report Ergonomics," *Keller's Industrial Safety Report*, 10, Oct. 1991.

erable attention has been given to the act of lifting objects. A number of factors enter into designing a lifting operation, including the weight of the object, the ease with which it can be held, the location of the object at the start of the lift, and the capabilities of the worker.

In 1981 NIOSH developed guidelines, including an equation, to calculate maximum recommended weight limits (RWLs) for lifting and lowering tasks. This was revised in 1991.[13] Application of this formula requires a series of basic measurements related to the task. It is instructional to see an equation, considering the variables included (Table 16.2). The parameters measured are usually fed into a computer equipped with special software to handle the calculation.[14] The results of this calculation assist in redesigning the job and can be used to consider several variations of job redesign before a final pattern is adopted.

Table 16.2. NIOSH Lifting Equation.

$$RWL = LC \times HM \times VM \times DM \times AM \times FM \times CM$$

where			
where	RWL	=	the recommended weight limit
	LC	=	a load constant (51 lb)
	HM	=	horizontal multiplier = 10/H [H = horizontal distance midpoint to hands (in.)]
	VM	=	verical multiplier = (1 – 0.0075[V – 30]) [V = distance of hands from floor (in.)]
	DM	=	distance multiplier = (0.82 + 1.8/D) [D = verticle distance of lift (in.)]
	AM	=	asymmetric multiplier = (1 – 0.0032A) [A = angle of asymmetry = angle of load from neutral upright position (degrees)]
	FM	=	frequency multiplier (based on how many times per minute the lifting is performed and how long the work lasts).
	CM	=	coupling multiplier [based on ease of grasping the object, i.e., the presence of "handles"; values range from good handles (1.00) to no handles (0.90)]

From T. R. Waters, V. Putz-Anderson, A. Garg, and L. J. Fine, "Revised NIOSH Equation for the Design and Evaluation of Manual Lifting Tasks," *Ergonomics,* 36, 749, 1993.

There is lack of agreement regarding the best way to actually perform a lifting operation and about the value of training the worker. However, the industrial hygienist should look for some clear aspects of lifting when inspecting a job site:

1. It is better for the worker not to bend or twist when lifting a heavy burden. This stresses parts of the back unduly.
2. The worker should bend at the knees rather than at the waist, then let the legs, which are much stronger, do the work of lifting.
3. The burden should be held close to the body.

There is general agreement that the best plan where a task involves continuous or frequent lifting is to engineer the task to eliminate the lifting, or at least to ensure that the lift is from no lower than knuckle height (arms at the side), is straight up, is close to the body, and involves loads well within the capability of the worker. The task should be redesigned to keep the back in the vertical neutral position.

Physical conditioning by institution of a well-designed stretching and exercise program at the start of the work day, a practice many associate with athletes warming up before an event or with Japanese industry, reduces the frequency and seriousness of musculoskeletal

[13] *NIOSH Work Practices Guide — User's Manual,* 1993.

[14] T. R. Waters, V. Putz-Anderson, A. Garg, and L. J. Fine, "Revised NIOSH Equation for the Design and Evaluation of Manual Lifting Tasks," *Ergonomics,* 36, 749, 1993. The equation is available from NIOSH by fax at (513) 533-8287.

injuries.[15] Such a program should be designed by a health professional and should keep in mind the specific workforce.

SUCCESSFUL PROGRAMS

Eby Construction Company of Kansas, after winning a major contract, trained a group of supervisors on lifting, pulling, and pushing methods, and use of the equipment to be employed in the project. They then used this information to prevent back injuries on the job. The project was completed without back injuries, and back injuries dropped from 43% of job-related injuries to 24% overall.

City officials in Arlington, Texas, found that the Parks and Recreation Department had the highest rate of back injuries. Employees in that division took an educational program on back safety. Testing showed an increase in knowledge and consciousness about back problems. Two supervisors modified specific tasks to reduce risk of back injury, implemented weekly safety meetings, and started a Safety Advisory Committee to review accidents. The department's injury rate dropped from an average of more than seven injuries per year to none in the next year.

These case histories are taken from B. Gilsan, "Customized Prevention Programs Play a Vital Role in Back Protection Process," *Occupational Health and Safety*, Dec. 1993, p. 21.

An obese individual is more at risk. A high body weight is already a burden to be supported by the back. Simply straightening from bending at the waist requires the lifting of the body mass in an awkward position.

THE SEATED EMPLOYEE: DESIGN OF CHAIRS AND WORKSTATIONS

Another lower back problem, quite different in nature, involves correct support and alignment of the lower back while seated. With new job opportunities more often involving desks than power wrenches, increasing attention is paid to chair design and to location of the work surface or keyboard with respect to the body. Again visualize the vertebral column. The neutral position is not linear, rather having an S curve where the neck and central abdomen are farthest forward. When a person is seated, this (neutral) S shape is achieved by sitting upright with the buttocks pushed back. In the short term there will be less fatigue in this position and over a long period of time reduced damage to discs, muscles, and ligaments. A well-designed office chair encourages assumption of this position. Such a chair has a lumbar support in the back rest, a pad that fits against the lower back. This pad may move up and down and/or have an adjustable thickness to compensate for anatomical differences.[16] Lumbar supports are also found in auto and truck seats.

Other adjustments in the chair geometry permit the feet to touch the floor such that they are carrying the weight of the legs, rather than having the thighs rest on the chair seat and bear the weight. Circulation in the legs is difficult when thighs press against the seat as they support the weight of the legs. If the individual is operating a keyboard, the relationship of

[15] B. W. Simonson and P. Iannello, "Company's Exercise Program Mobilizes its Industrial Athletes before Work," *Occupational Health and Safety*, Sept. 1994, p. 44.

[16] M. Dagostino, "Lumbar Support Most Critical Feature to Consider During Chair Selection," *Occupational Health and Safety*, March 1994, p. 63.

the arms to the keyboard is important. This can be adjusted by moving the keyboard up or down of by moving the worker up or down (adjustable chair height). In the latter case an adjustable footrest may be needed to allow the feet to continue to support the weight of the legs with the seat raised.

The object observed by the worker, whether a video screen, a document, or a small assembly operation, should be at a height such that the worker does not have to look far down or up for long periods, a situation that leads to stiff neck and shoulders from muscle strain because the weight of the head is then supported continuously. If a variety of objects are to be viewed, it is preferred that they be close together and at about the same distance from the eyes, so that constant turning of the head and refocusing of the eyes is avoided. The latter is particularly critical for older workers, because the lenses of the eyes become more rigid, thus eyes become less responsive to refocusing, with age. This is why older people need bifocal glasses.

In summary, postures that seem to be no trouble when held for a few moments can cause strain when held for a workday, and people come in many dimensions. Well-designed and adjustable seats and workstations can overcome a multitude of these problems, improving worker comfort and productivity.

CONTINUOUS STANDING

Some jobs require the worker to stand continuously. Examples include bank tellers, grocery clerks, and some assembly line workers. In some jobs, standing is an advantage. A checkout clerk in a grocery store needs to move back and forth between the cash register and the bagging area. Frequent additional effort is required if it is necessary to stand up from a seated position at the register to do the bagging. Reaching to scan items in the laser beam is easier from a standing position. The advantage to standing, then, is the mobility it allows.

The muscle effort required to stand continuously leads to fatigue and pain in the legs and lower back. Blood tends to accumulate more in lower portions of the body, leading to the legs swelling and in the long run to varicose veins. Joint problems may be experienced in the knee, foot, and hip.

Some relief is provided by cushioned mats called anti-fatigue mats, particularly where the floor is concrete. These are reported to be most beneficial at reducing back pain. Anti-fatigue mats must be carefully selected so as not to be at extremes of softness or hardness, as either extreme fails to provide the desired benefits. The same care must be taken that the mat is not slippery, or slippery when wet, as the mat now is the floor surface at that workstation..

Particularly where the worker is moving about extensively, a mat may not be a practical remedy. Shoe inserts have been used with some success in these cases. These vary in thickness according to the shape of the foot and its need for cushioning, and are made of a viscoelastic material such as polyurethane. They may reduce the impact on the heel during walking by 30–40%. It is best to wear shoes sufficiently large that they can hold the insert and still provide a comfortable fit to the foot. This is also useful in reducing "skeletal shock", the jarring that joints and tendons receive when walking, especially on hard surfaces.

Simply ensuring that the employee wears appropriate shoes is important. Laced shoes that provide firm support and that have neither flat nor high heels are best. Leather shoes allow the foot to "breathe", so that feet do not become damp and uncomfortable. Shoes that pinch the toes or are loose at the heels lead to problems for the employee standing for long periods. Open shoes such as sandals are unsatisfactory because they fail to provide proper support, and if there is a chance of spills or dropped objects, they also provide no protection. Steel-toed work shoes may be required for safety in some work environments, and it is similarly appropriate to designate shoe standards for employees who stand for long periods.

A properly designed foot rail at the workstation, about 4-5 in. off the floor and located so as not to be a tripping hazard, can be of benefit in preventing lower back pain. Putting one foot on the rail relieves some of the pressure on vertebral discs, and the trading of one foot for the other at intervals is also beneficial.

Even better is the provision of a high stool or chair so that the employee can take brief breaks from standing. Some "stand seats" allow workers to lean back and relieve the weight on their feet momentarily, then perhaps swing out of the way at other times. Better still is a work plan that changes the task performed by the employee so that part of the work is done sitting and part standing. Ergonomically the greatest risk of health problems arises when employees do a continuous, repetitive job.

VIBRATION

Vibration and sound (Chapter 12) have much in common. Both are described as wave phenomena, and the frequency and amplitude are significant parameters. Vibration operates at much lower frequencies than sound, most vibrations of interest being in the range of 1 to 300 Hz (cycles per second). Just as sounds are generally more complex than pure tones, vibration may include a complex spectrum of frequencies.

Vibration can affect the comfort and efficiency of an employee and, in extreme cases, can damage health. Vibration is usually transmitted to the body through the floor to a standing worker, the seat to a seated worker, or to hands and arms from a hand tool. Most problems arise from the use of vibrating hand tools, such as air hammers, chain saws, and riveting machines, or by riding in vehicles such as tractors or forklift trucks on rough surfaces.

All objects have a natural (resonant) frequency such that if they are exposed to a particular frequency of vibration, they also begin to vibrate (resonate). The entire body resonates somewhere between 4 and 8 Hz, the range that therefore generates the greatest discomfort. Body organs are not rigidly assembled into a single resonating structure, but have their own individual frequencies. As a result, vibration of a particular frequency may generate discomfort or pain in a specific organ. For example, a particular worker may feel abdominal discomfort in response to a particular vibrational frequency.

Workers who use hand tools for long periods, particularly those that generate vibration of one hundred to a few hundred hertz, may suffer damage to nerves and blood vessels in their hands. These individuals then experience numbness, pain, and loss of control in their hands. The fingers appear pale. The problem has been termed "white finger disease", or Raynaud's disease. If exposure to the vibration is allowed to continue, the problem becomes irreversible (Figure 16.1).

Workers experience problems reading instruments or control systems when either they or the machine are vibrating. Adjusting to the movement is fatiguing, is more difficult as the frequency increases, and can result in inability to perform a task.

LIGHTING AND EYE PROBLEMS

Lighting at a workstation deserves more consideration than it sometimes receives. In an office setting the light levels are usually carefully planned in a new facility by architects, who have guidelines provided by the Illuminating Engineering Society for the amount of light necessary for desk work. Older facilities or places in a nonoffice setting that require office-type work, such as a supervisor's station on a factory floor, should be analyzed to determine if they meet these same standards. Adequate amounts of light facilitate easy reading of controls, displays, and the printed word. OSHA does not have a standard for illumination.

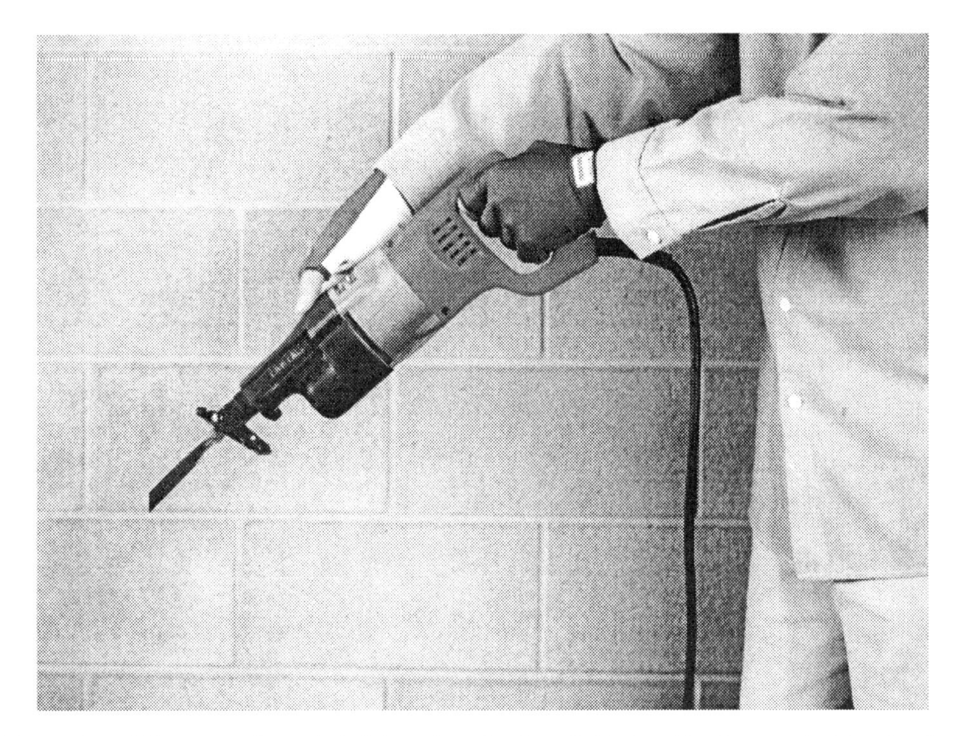

Figure 16.1. This worker is protected from the vibration of this hand-held power tool by vibration-absorbing gloves. (Photo courtesy of Lab Safety Supply, Janesville, WI. With permission.)

However, ANSI has a written standard: "Practice for Industrial Lighting", RP7-1983, which is a useful guideline. Proper illumination leads to greater productivity and lowered worker stress, and is often a factor in preventing accidents.

Equally important is the avoidance of glare. Examples include the reflection of overhead lights from a glass-covered instrument display (reflected light), a strongly illuminated white or light surface in an otherwise dark environment, or light sources directly in the field of vision (direct light). Glare is fatiguing and interferes with easy observation of the page or display of interest.

EYE PROBLEMS AND VIDEO DISPLAY TERMINALS

Fatigue from performing demanding visual work is a type of cumulative trauma. As an example, we can apply the general statements above to workers using a video display terminal (VDT). Not only is this a good specific example, but the problems are widespread because operating computers is also a fast-growing segment of the job market. Working long hours in front of a VDT produces eyestrain.

Eyestrain causes burning, redness, and irritation of the eyes. Many workers experience headaches. Vision can blur as muscles focusing the eye tire. In older workers, the lenses of the eyes are less flexible, a natural consequence of aging. Such workers' eyes may retain the near-focus adjustment necessary to read the VDT when the worker leaves the job, so that distant objects are blurred. A person whose lenses are less flexible generally wears bifocal glasses, the lower lens serving for reading or other close work and the upper lens assisting distance vision. The reading lens is generally not designed for the distance to the VDT, and the worker is frustrated by the screen not being properly in focus with either lens. Reading the screen is then a constant tiring effort. In order to bring the screen into focus, the worker may lean forward into an unnatural position. After a period of time, neck, shoulder, and back pain result from this tactic. An obvious cure is to measure the distance from the worker's eyes to the screen and have a pair of glasses prepared to correct for this distance. Neck or back pain can also result if the VDT is at the wrong height so that the head is not in the neutral position. Ideally, the screen should be 4–9 in. below eye level. Similarly, if the worker looks at the screen during the majority of the work time, the screen should be centered in front of the worker to avoid twisting stances.

Glare adds to eye fatigue. Lights behind or above the worker can reflect from the VDT, making clear vision difficult. Shields can be placed in front of the screen to prevent reflection. Lights behind the VDT that fall into the worker's view add to the effort of seeing the screen. This can be corrected by putting diffusers on the lights, shifting their location, or changing the direction the worker faces.

Finally, computer operators tend to blink less often than normal. Blinking restores moisture to the surface of the eyes, so if blinking is less frequent the film of water evaporates, leaving the eyes dry. This can be made worse if the room air is dry due to air conditioning or a failure to humidify heated outdoor air. The problem is especially bad if the worker wears contact lenses.

CONCLUSION

This discussion has looked only briefly at some major topics in occupational safety. Cumulative trauma is complex and is evolving in new directions as more is learned about the relationship of the structure of the human body and the performance of tasks. Much can be accomplished by arranging tasks so that the worker is in an optimal posture and is minimizing strain on the body as the task is performed. Redesign of controls and hand tools not only prevents injury but also increases efficiency.

Concerns are not limited to factory problems. Workers operating keyboards in front of visual display terminals are perhaps the most rapidly increasing segment of the workforce. Extensive ergonomic studies focus on this group of employees. Interested readers should consult the Bibliography for a much broader coverage of the material.

Workers becoming fatigued as a result of cumulative strain have a reduced level of output. There is also a correlation between discomfort on the job as a result of the continuous strain of awkward position at the workstation or excessive burden on certain parts of the anatomy and quality in production. Employers should therefore be motivated to eliminate ergonomic problems both to reduce workers' compensation and disability insurance costs and to raise productivity.

KEY POINTS

1. Repeated physical operations, awkward positions, heavy lifting, and long periods spent in the same position can lead to injury.
2. Awkward and/or repeated wrist bending in performance of a task can cause problems with tendons and nerves running through the carpal tunnel, termed tenosynovitus and carpal tunnel syndrome, respectively.
3. Repeated finger action results in a special tenosynovitus called trigger finger.
4. Repeated twisting movements of the forearm produce epicondylitis.
5. Pressure on fingers or palms cuts off blood circulation, with a loss of nerve activity (numbness).
6. Correct lifting and limits on the weight of objects lifted help prevent lower back pain.

7. Continuous standing leads to fatigue in the legs and back. Blood accumulates in the legs, leading to swelling and varicose veins. Some problems are relieved by soft mats or shoe inserts.
8. Vibration affects employee comfort and can cause physical harm.
9. Lighting should be adequate and glare free. Working at a video display terminal causes eye fatigue and presents special problems arising when the display is incorrectly located.

BIBLIOGRAPHY

T. J. Armstrong and Y. Lifshitz, "Evaluation and Design of Jobs for Control of Cumulative Trauma Disorders," in *Ergonomic Interventions to Prevent Musculoskeletal Injuries in Industry,* ACGIH Industrial Hygiene Series, Lewis Publishers, Chelsea, MI, 1987.

R. Carson, "Stand by Your Job," *Occupational Health and Safety,* April 1994, pp. 38–42.

D. B. Chaffin, "Biomechanics of Manual Materials Handling and Low Back Pain," in C. Zenz, Ed., *Occupational Medicine: Principles and Practical Applications,* 2nd ed., Year Book Medical Publishers, Chicago, 1988.

D. B. Chaffin and G. Andersson, *Occupational Biomechanics,* 2nd ed., John Wiley & Sons, New York, 1991.

O. B. Dickerson and W. E. Baker, "Practical Ergonomics and Work with Video Display Terminals" in C. Zenz, Occupational Medicine: Principles and Practical Applications, 2nd Ed., Year Book Medical Publishers, Inc, Chicago, 1988.

Eastman Kodak Company, *Ergonomic Design for People at Work,* Van Nostrand-Reinhold, New York, 1991.

B. S. Levy and D. H. Wegman, *Occupations Health,* Little, Brown and Company, Boston, 1983.

D. MacLeod, *The Ergonomics Edge,* Van Nostrand-Reinhold, New York, 1994.

S. Meagher, "Hand Tools: Cumulative Trauma Disorders Caused by Improper Use of Design Elements," in W. Karwowski, Ed., *Trends in Ergonomics/Human Factors III,* Elsevier/North-Holland, Amsterdam, 1986.

S. W. Meagher, "Design of Hand Tools for Control of Cumulative Trauma Disorders," in *Ergonomic Interventions to Prevent Musculoskeletal Injuries in Industry,* ACGIH Industrial Hygiene Series, Lewis Publishers, Chelsea, MI, 1987.

D. J. Osborne, *Ergonomics at Work,* 2nd ed., John Wiley & Sons, New York, 1987.

M. S. Sanders and E. J. McCormick, *Human Factors in Engineering and Design,* 6th ed., McGraw-Hill, New York, 1987. (Chapter 11 deals with hands and hand tools.)

S. H. Snook, "Comparison of Different Approaches for Prevention of Lower Back Pain," in *Ergonomic Interventions to Prevent Musculoskeletal Injuries in Industry,* ACGIH Industrial Hygiene Series, Lewis Publishers, Chelsea, MI, 1987.

W. Taylor and D. E. Wasserman, "Occupational Vibration," in C. Zenz, Ed., *Occupational Medicine: Principles and Practical Applications,* 2nd ed., Year Book Medical Publishers, Chicago, 1988.

PROBLEMS

1. Define each of these terms:
   A. overt trauma
   B. Raynaud's disease
   C. tenosynovitus
   D. carpal tunnel syndrome
   E. trigger finger
   F. epicondylitis
   G. ischemia
2. A worker is employed to custom cut 19th century style wood trim for the restoration of old houses. The job requires running a 12-lb saber saw around curved templates while the wood is in a jig at chest height. The metal handle of the saw is parallel to the wood surface during

cutting. The worker turns on the saw by squeezing a trigger with the forefinger, release of that pressure automatically stopping the saw. At the end of the cut the saw must be lifted clear of the wood surface. Referring to Table 16.1, indicate aspects of this job that are likely to produce cumulative trauma, indicate the type of worker difficulty that could result, and suggest improvements in the layout of the job to reduce strain.

CHAPTER 17

Biohazard

INFECTIOUS DISEASE

In the second half of the 20th century we have tended to become complacent about infectious diseases, a complacency based on the rather impressive success of medicine at overcoming this longstanding threat to humanity. This apathy has been shaken in recent years by events that establish clearly that infectious disease can still cause major problems. The antibiotics that have been front line weapons against bacterial diseases are suddenly found to be less effective as the bacteria develop mechanisms to resist their toxicity.

New diseases have made headlines, including AIDS, Legionnaire's disease, Reye's syndrome, and hantavirus pulmonary syndrome, and old diseases change character and plague us regularly, as does influenza. After decades of regular good health news, it is easy to convince oneself that we are in a period of losing ground to microorganisms.

TUBERCULOSIS

Tuberculosis (TB) was widespread in earlier times. In 1944 there were 126,000 reported active cases of TB in the U.S. TB was considered to be under control with the advent of effective antibiotics. Numerous hospitals built just to house TB patients were closed or converted to handle more pressing health difficulties. The appearance of antibiotic-resistant strains of TB, including some with multiple drug resistance, could put us back where we were decades ago in dealing with this infection. A Centers for Disease Control (CDC) study of 3313 TB cases in 1991 revealed that 14.2% were resistant to one or more drugs. It is worse in some locations. In a study in New York City, it was found that 33% of the cases surveyed involved organisms resistant to one drug, and 19% were resistant to both of the most frequently used drugs. In the U.S. 25,500 new cases were reported in 1990.[1]

The CDC Advisory Committee for the Elimination of Tuberculosis identifies several high-risk groups. Individuals who are HIV positive readily acquire TB, and the HIV/TB combination is lethal in 70% of these people. Because the immune system of AIDS patients is not operating correctly, they may not even give a positive skin test when they are infected. Furthermore, some of the TB drugs are dangerous to an HIV-positive individual. Other people at high risk include diabetics, those involved in immunosuppressive therapy (organ transplant recipients), alcoholics, and drug users. Those living in high-population facilities, such as nursing homes, prisons, and mental institutions, are more likely to contract the disease from another resident.

[1] C. Darius and R. L. Holdsworth, "Tuberculosis Revival Forces Health Care to Implement Strict Preventive Policies," *Occupational Health and Safety*, June 1994, p. 74.

Tuberculosis is caused by *Mycobacterium tuberculosis*, a slow growing organism. The disease is spread when an infected person talks, coughs, sneezes, or otherwise forces air out of the lungs. Air moving across lesions in the lungs picks up respirable-sized water droplets containing the organism. If these droplets are inhaled by another person and the droplets reach the alveoli, the organism can start to grow in the lungs. We associate TB primarily with the lungs, but the disease can spread to other organs.

A person can be infected but be free of symptoms of TB if the disease remains dormant. Such a person will have no organisms in their sputum samples and will have a normal chest X-ray, but they do have a positive skin test. They are not contagious, but can later develop an active infection. Millions of such individuals are in the U.S. population.

The health threat posed by the rise of new TB cases is particularly serious for health care professionals. The CDC is working with OSHA and NIOSH to create national measures to protect health care workers. For this purpose the CDC has defined health care facilities to include almost any situation where patient care is provided, including emergency medical services and dental clinics.

Control measures begin with identifying patients at health care facilities as early as possible who might have the active disease. Patients with indications of the disease are isolated, and more extensive testing for the disease is done. The patient wears a mask, and signs are posted indicating possible TB contamination. Workers entering the area must wear high-efficiency particulate respirators that have been tested for proper fit (Chapter 9, 29 CFR 1910.134). Workers at risk receive training that includes the nature of the disease and its transmission, protective measures, skin testing procedures and their meaning, and drug treatments. A periodic skin testing program should be instituted, starting at the time a worker is hired. Appearance of positive skin tests and active disease should be recorded on the employer's OSHA 200 log.

OSHA has been training inspectors, inspecting health care facilities, and issuing citations based on these guidelines and the General Duty Clause.

THE RESULTS OF THE APPEARANCE OF AIDS

Since the worldwide AIDS epidemic began, it has been the target of a major research effort aimed at prevention, treatment, and an eventual cure. The infectious agent is a retrovirus[2] called human immunodeficiency virus (HIV). HIV attacks the victim's immune system. A person infected with HIV is termed HIV positive and is generally not aware of the infection because there are no immediate symptoms. However, this person is a source of infection to others. Within a few months the victim produces antibodies to HIV, which seem to be of little protective value, but which can be detected by a screening test called the enzyme-linked immunosorbent assay (ELISA). ELISA sometimes produces a false positive test for reasons not well understood.

An HIV-positive individual progresses through four stages of the disease:

1. Within about a month of exposure the victim may display fever, rash, diarrhea, swollen glands, and symptoms of fatigue. The disease is sometimes confused with chronic fatigue syndrome in the absence of a positive ELISA test.
2. Within 6 months of exposure HIV antibodies are forming.
3. Glands enlarge, but overall health remains good. A person may remain at this level for years.
4. The virus overcomes the immune system. The victim is ill and may develop infections or cancers that have nothing directly to do with the disease, but arise because the immune system is not functioning. The victim is now said to have acquired immunodeficiency syndrome (AIDS). The victim may die of AIDS itself or of one of the "accessory" diseases.

[2] A virus whose nucleic acid core is RNA rather than DNA.

Throughout this progression, the victim is capable of infecting others.

HIV has been found in virtually every body fluid of HIV-positive individuals. Actual transmission of the disease has been shown to occur from sexual intercourse, through transfusion of contaminated blood, through use of hypodermic needles contaminated with blood containing HIV, and through contact with contaminated blood at mucous or broken skin sites.

Thus with the appearance of AIDS and its spread throughout the world, people became aware of a new type of hazardous substance, the blood of another human.[3] Medical workers began to wear latex gloves in situations where contact with patient blood was possible. The spread of AIDS by reuse of syringes among intravenous drug users focused attention on the handling and disposal of used syringes in medical settings. Athletes with a bleeding injury were pulled out of competition until the injury was treated.

In response to these concerns, in 1991 OSHA developed rules aimed at anyone contacting human blood, human blood components, products made from human blood, or specified other body fluids.[4] These rules are based on recommendations of the CDC.[5] The CDC stated universal blood and body fluid precautions: all patients should be handled as if they were HIV positive.

The OSHA rules particularly focus on the more than 4.4 million health care workers and 1.2 million public safety workers (police, fire personnel, and other emergency responders) who may deal with injuries that involve bleeding.[6] The role of emergency responders as initial providers of emergency health care is growing. As much as 80% of field emergency medical treatment is provided by firefighters.[7] Emergency responders, such as emergency medical technicians, advanced life support personnel, and paramedics, also have exposure to potentially dangerous body fluids as part of their job description. OSHA has special regulations, too, for workers in laboratories studying the bloodborne pathogens. Response to these health concerns illustrates the expansion of OSHA protective regulations beyond the traditional factory setting.

Who is not covered? This is significant, because those covered by the regulation require training and should have the option of hepatitis B vaccination available. A self-employed person without employees, such as a dentist without an assistant, is not subject to the regulations. Volunteers also are not employees, and volunteers play a significant role in health care facilities. A "good samaritan" who helps an injured coworker need not have been trained if rendering first aid is not part of the employee's job description.[8]

SOURCES OF AIDS INFORMATION

National AIDS Clearinghouse	800-458-5231
CDC: AIDS Hotline	800-342-AIDS
CDC: "Business Responds to AIDS"	800-458-5231
American Red Cross	202-434-4074

[3] In addition to detecting the infection in patients, ELISA is used for routine screening of donated blood and any products made from blood.
[4] *Federal Register*, December 6, 1991; 29 CFR 1910.1030.
[5] Centers for Disease Control, *MMWR*, 37, 377, 1988; *MMWR*, 38(6S), 1989.
[6] OSHA, Office of Regulatory Analysis, 1991.
[7] Clyde A Bragdon, Jr., testifying at OSHA's hearings on the proposed bloodborne pathogen standard, September 14, 1989, as quoted in OSHA Booklet 3130, 1992.
[8] J. F. Rekus, "Health Industry Adjusts to Specifics of Bloodborne Pathogens Exposure Rule," *Occupational Health and Safety*, Oct. 1992, p. 38.

HEPATITIS B VIRUS

Although the public is more aware of AIDS, given the high level of coverage in the media, major public health concern is also focused on hepatitis B, caused by hepatitis B virus (HBV). By some measures hepatitis B is a greater health threat than AIDS, given "that approximately 300,000 people become infected with HBV annually, that 14 people die every day in the United States from hepatitis-B related illnesses, and that hepatitis is the ninth leading cause of death worldwide."[9] By comparison there were 361,164 confirmed cases of AIDS and about a million HIV-positive Americans reported by the CDC in December 1993.

Anywhere from 6 weeks to 6 months after exposure, the victim has flu symptoms: fever, chills, joint and muscle pain, and abdominal cramps. Jaundice, the failure of liver function evidenced by bile pigments accumulating in the skin, may appear.

An important difference between HIV and HBV is the availability of a hepatitis B vaccine. Workers can protect themselves from this infection with the series of three injections. At this time HBV vaccine is being given routinely to babies in hopes that the disease eventually will be eradicated if the population lacks carriers.

OTHER DISEASES

In fact the OSHA regulations do not focus just on HIV or even HIV and HBV, but are written in general terms to include any bloodborne pathogen.[10] These include syphilis, malaria, Rocky Mountain spotted fever, and hepatitis C.[11] In 1987 there were more than 35,000 cases of syphilis reported in the U.S., and the frequency of occurrence is increasing. Syphilis is spread primarily by sexual contact, but transmission by contact with the blood of a victim has been reported. Malaria is less common, with fewer than 1000 cases reported in the U.S. in 1987. It is caused by a parasite in the blood, is usually transmitted by mosquitoes, and is far more common in tropical countries. However, here too transmission by way of the blood of a victim has occurred.

SOURCES OF INFECTION

Regulations are directed at blood in any form. Specifically mentioned are liquid or semi-liquid blood, contaminated items that would release blood if compressed, unfixed human organs or tissues, items caked with dried blood, laundry soiled with blood, contaminated "sharps" (needles, scalpels, broken glass, broken capillary tubes, etc.), pathological or micro-biological wastes containing blood, and body fluids visibly containing blood. Other body fluids are also sources of contamination. OSHA lists "semen, vaginal secretions, cerebrospinal fluid, synovial fluid, pleural fluid, peritoneal fluid, amniotic fluid, and saliva in dental procedures". Most of these fluids would only be contacted in a health care setting, and it is in such settings that the most elaborate precautions must be taken.

A worker can contract a virus infection through the eyes, mouth, any other body opening, or a cut in the skin. Having skin contact with contaminated materials, then touching any of these vulnerable areas is another route of exposure. Hand washing after handling potentially contaminated materials is therefore very important.

[9] H. Keesing, "Contractors' 'Band-Aid' Remedies May Defy Bloodborne Pathogen Risks," *Occupational Health and Safety*, Dec. 1994, p. 49.
[10] "... pathogenic microorganisms that are present in human blood and can cause disease in humans. These pathogens include, but are not limited to, hepatitis B virus (HBV) and human immunodeficiency virus (HIV)." From 29 CFR 1910.1030(b). Definition of bloodborne pathogens.
[11] T. W. Muir, "Exposure Control Plans Define Risks for Bloodborne Pathogen Infections," *Occupational Health and Safety*, April 1994, p. 75.

The likelihood of health workers becoming infected by way of exposure to HIV-infected patients through needle sticks, exposure to wounds, and exposure to mucous membranes is actually very low. Studies indicate the level of risk to be less than 1%.[12]

REGULATIONS

The rules apply to all employers, even those with only one employee. Every employer whose employees are at risk of exposure of nonintact skin, mucous membrane, mouth, eye, or parenteral exposure (through cuts, punctures, or abrasions) to blood or other infectious materials must have a written Exposure Control Plan. The OSHA standard requires that this plan include the following:

1. A list by job classification of all employees who have or may have occupational exposure and, where not all employees associated with a procedure have potential exposure, a list of all tasks or procedures in which occupational exposure occurs
2. A statement of methods of compliance
3. Agreement that the employer shall make hepatitis B vaccination available at no cost to potentially exposed employees, and provide an evaluation and follow-up to all employees who have occupational exposure
4. A provision of "biohazard" labels in orange or orange-red to be attached to all potentially contaminated containers or storage sites
5. Provision of a training program for employees
6. Agreement that individual records shall be kept for each employee with occupational exposure which meet the standards in 29 CFR 1910.20. These records must be available to OSHA and/or NIOSH on demand.

Every year the plan must be reviewed and updated, or such updating must be done when a new procedure or task that could impact exposure is undertaken by employees.

The Training Program

It is important that employees be well informed about the nature of the hazards involved, as provided for in the right-to-know legislation. The training program must be run by a knowledgeable person and must be appropriate in level and language to the audience. It must be provided by the employer at no cost to the employees during working hour, and must be given before beginning an initial work assignment. OSHA Bulletin 3127 lists the elements that must be part of the training as follows:

1. How to obtain a copy of the regulatory text and an explanation of its contents;
2. Information on the epidemiology and symptoms of bloodborne diseases;
3. Ways in which bloodborne pathogens are transmitted;
4. Explanation of the exposure control plan and how to obtain a copy;
5. Information on how to recognize tasks that may result in occupational exposure;
6. Information on the types, selection, proper use, location, removal, handling, decontamination, and disposal of personal protective equipment;
7. Information on hepatitis B vaccination such as safety, benefits, efficacy, methods of administration, and availability;
8. Information on who to contact and what to do in an emergency;
9. Information on how to report an exposure incident and on post-exposure evaluation and follow-up;
10. Information on warning labels, and signs, where applicable, and color coding; and
11. Question and answer session on any aspect of the training.

[12] E. McCray, "Occupational Risk of Acquired Immunodeficiency Syndrome among Health Care Workers," *New Engl. J. Med.*, 314, 1127, 1986.

SELECTED ASPECTS OF COMPLIANCE WITH THE RULES

Worker Protection

I do not attempt to cover all details of the regulations in this chapter. Rather particularly important aspects are highlighted to provide a sense of the important issues. Anyone becoming involved with the regulations discussed in this chapter should become thoroughly familiar with 29 CFR 1910.1030 in its most recently revised form. A recently written rule is likely to undergo frequent adjustment as application of the rule reveals deficiencies.

An elementary but important provision is a means for the employees to wash their hands [1910.1030(d)(2)(iii)]. Lacking conventional hand washing facilities, the employer should provide antiseptic hand cleaners or towels. Even with these, employees should wash with soap and water as soon as possible. Intact skin is a barrier, but transfer from the hands to the eyes, nose, or mouth is a danger. Hands should be washed immediately upon removal of protective gloves, because transfer from the gloves to the hands is a danger.

Another easy but essential requirement is that "eating, drinking, smoking, applying cosmetics or lip balm, and handling contact lenses" is banned in hazardous work areas [1910.1030(d)(2)(ix)]. Further, food or drink should never be stored anywhere that contamination is possible, for example, along with blood samples in refrigerators [1910.1030(d)(2)(x)].

Protective clothing [1910.1030(d)(3)] is necessary for workers likely to be exposed to blood or other potentially harmful body fluids. Such clothing might include "gloves, gowns, laboratory coats, face shields or masks and eye protection, and mouthpieces, resuscitation bags, pocket masks, or other ventilation devices". The worker should wear such items of protective clothing as are appropriate to the situation. The employer should make protective clothing readily available, should launder or dispose of it later without cost to the employee, and should require its use. Protective clothing should block the passage of fluids, so that the skin or personal clothing underneath does not become contaminated. If a protective garment is contaminated it should be changed and should be removed before leaving the work area. Employees should develop techniques to remove latex gloves without contacting the contaminated outside surface with the skin, removing one glove while protected by the other, then slipping bare fingers inside the other glove without touching the outside to peel it off. Used protective clothing should be placed in a designated area to be stored, laundered, decontaminated, or discarded. Disposable gloves should not be reused and should be placed in a labeled disposal container, and reusable gloves should be decontaminated and inspected for deterioration before reuse.

Of particular concern are contaminated sharps. Sharps include any objects capable of puncturing or cutting a worker, and can be classed as reusable or waste. A used scalpel is perhaps the best example of a reusable sharp. Waste sharps include such items as broken contaminated glass containers or capillary tubes and used hypodermic needles. Broken glass must not be picked up by hand, because even protective gloves do not prevent a puncture from broken glass. Instead, use mechanical means such as forceps, tongs, or a brush and shovel.

"Needle sticks" from used hypodermic needles are a very common problem for hospital personnel, particularly for nurses, laboratory workers, and housekeepers. One study estimated the frequency of occurrence ranged from 40 to 80 incidents per 100 full-time employees per year.[13] Used hypodermic needles have very high hazard, both because it is very easy to get a needle stick and because some old practices made such a stick more likely. For example, sticks commonly occurred when recapping the needle after use. The CDC now recommends that needles not be recapped. It had been routine to bend a needle once it was used to prevent accidental reuse. There is a high likelihood of a puncture wound when doing this, and the

[13] M. S. Freeley, "Occupational Needlestick Injuries," in W. Charney and J. Schirmer, Eds., *Essentials of Modern Hospital Safety*, Lewis Publishers, Chelsea, MI, 1990.

practice should stop. Similarly, removing the needle from the syringe body is a dangerous practice.

Housekeeping [1910.1030(d)(4)] is also very important. All surfaces and equipment with any possibility of contamination should be decontaminated by cleaning with disinfectant as soon as possible. This includes not only work surfaces, but also refuse containers that may be contaminated with blood. Procedures for removal, storage, and labeling of contaminated waste should be established and followed.

Hepatitis B Vaccination

An effective vaccine against hepatitis B has been available since 1982, and a new vaccine based on recombinant technology that is produced by yeast cells became available in 1987. The latter eliminates use of any human materials in preparation, and thus eliminates the risk of HIV contamination of the vaccine. This vaccination is the most important preventative action that can be taken against the spread of hepatitis B, and more than two million health care workers have been vaccinated in the U.S. since it became available. Vaccination involves a series of three shots over a 6-month period. Free vaccination against hepatitis B must be offered to at-risk employees. Of course, vaccination is of no value to someone already infected.

Disposal or Cleaning of Contaminated Materials

There is a standard symbol for waste, equipment, or storage sites to indicate the presence of biologically hazardous materials (Figure 17.1). The symbol should be fluorescent orange or orange-red, include the word "BIOHAZARD", and be posted or firmly attached as a warning at storage places or to containers of contaminated objects. Alternatively, contaminated materials can be placed in a red bag.

Figure 17.1. Symbol for biohazard warning sign.

The handling of contaminated laundry must be done with appropriate precautions. An example would be hospital bedding stained with blood from a patient whose disease status is unknown. It should be handled by workers wearing protective clothing, should be bagged immediately where it was used, and should not be sorted, rinsed, or otherwise unnecessarily handled before bagging. Wet laundry should be transported in a waterproof container. Laundering at 160°F or more for at least 25 min is recommended.

Of special importance is the handling of contaminated sharps. Reusable sharps are stored until decontamination in a puncture-resistant, leakproof container labeled as a biohazard. These containers should never be overfilled. If it is necessary to recover the object from the container for decontamination, the worker must not reach in by hand. Broken glass should not be picked up by hand. Storage of waste sharps until disposal should follow the guidelines for storage of reusable sharps, and containers should be clearly labeled as biohazard.

The AIDS epidemic has brought about vast permanent changes in the public safety and medical care professions. Given the seriousness of the consequences of carelessness, the regulations must be carefully followed.

BIBLIOGRAPHY

J. L. Gererding, C. E. Bryant, A. R. Moss, et al., "Risk of Acquired Immumodeficiency Syndrome Virus Transmission to Health Care Workers," International Conference on AIDS, Paris, France, June 1988.

J. P. Hughes, "Biological Agents," in L. J. Cralley, L. V. Cralley, and J. S. Bus, Eds., *Patty's Industrial Hygiene and Toxicology*, 3rd ed., Vol. 3, Part B, John Wiley & Sons, New York, 1995.

H. J. Sawyer, "Occupational Health Concerns in the Health Care Field," Chapter 13 in *Patty's Industrial Hygiene and Toxicology*, 4th ed., Vol. 1, Part A, John Wiley & Sons, New York, 1991.

U.S. Department of Labor, "Occupational Exposure to Bloodborne Pathogens," Booklet 3127, Occupational Health and Safety Administration, 1993.

PROBLEMS

1. Some individuals have great fear of shots, and some may have a medical reason for not being vaccinated. Employers must provide at-risk personnel with vaccinations against hepatitis B. What can an employer do in the case that an employee refuses the vaccination? (See Appendix A in the regulations.)
2. You are inspecting a clinical laboratory in a hospital that analyzes a variety of body fluids. Glass containers are used for steps in some of the analyses, and samples of the fluids are measured for analysis with glass pipets. What safety facilities would you wish to see in the room?

Section V
Industrial Examples

In this section of the text, a few important industries are described. The procedures and practices of these fields are outlined and specific hazards are described and/or methods of protecting workers. This is not meant to be an encyclopedia of industry, but rather is a sampling to introduce important industries and show the application of the principles developed earlier in the book.

CHAPTER 18

Metals I: Metals Preparation and Manufacturing

GENERAL PRINCIPLES

Throughout the history of civilization metals have been at the heart of technological advance. Early civilizations exploited copper and bronze for tools and weapons. Then the iron age replaced the copper- and bronze-based society. By the mid-20th century, the tonnage of steel a nation produced annually was the prime measure of its industrial strength. In recent years new materials have risen in importance. Aluminum replaced wood and canvas in aircraft construction. High-performance aircraft have passed through the period of aluminum construction into one of titanium structures. The annual volume of plastics produced in the U.S. exceeds that of steel, and there are even experiments being run with plastic automobile engines.

Health problems due to production and use of familiar metals generally led to regulations designed to prevent their recurrence. In this rush of progress society must learn to deal with a host of new problems. New materials, new uses for familiar materials, and new technologies for preparing and forming metals bring with them a variety of new hazards. An awareness of the types of problems that have occurred helps people to anticipate the kinds of difficulties to expect. This chapter presents ways to deal with familiar problems, in an attempt to serve as a guide to dealing with new problems as they arise.

ORE MINING, PROCESSING, SMELTING, AND REFINING

Searching for ore deposits historically was searching for riches. Early in the American west, prospectors gambled years of their lives and endured hardship and danger in the hope of finding a rich ore lode. Today, exploitation of a newly discovered ore deposit has become a more controversial issue, with developers and environmentalists often in opposition. For the most part, the richest deposits have already been utilized and new technology has been devised to allow use of "lower grade" deposits. We have even returned to the waste piles of earlier operations to find exploitable sources. On the negative side, using less concentrated sources means that a greater volume of the Earth's surface must be disturbed to obtain the same amount of raw material. Increased recycling of already extracted metals becomes more important in such a scenario, and increased amounts of scrap are being used as the raw material, a development generating its own set of hazards.

MINING

Underground Mining

Exploitation of ore finds often involves traditional mines with shafts and tunnels drilled out by miners following the seam of ore. Hazards include dust-filled air, possible collapse of shafts, use of explosives, noise, high temperatures, flooding, and accidental explosions. Explosions particularly threaten coal miners, because the substance removed is itself combustible. Silicosis plagues "hard rock miners", miners drilling in silica rock.

Underground mining dates to ancient times, but the health of miners was not a concern, because this was work done by slaves. Since the industrial revolution new methods have gradually improved the safety and quality of life of miners. Wet drilling methods (drilling under a stream of water to prevent dust in the air), wetting down broken rock and ore before transport, placing bags of water in the hole with explosives to reduce dust production, the use of protective breathing devices when necessary, and monitors to test air quality are examples of measures employed to improve mine health and safety.

Ventilation supplies oxygen to workers, since they are largely cut off from the atmosphere in a mine shaft. Ventilation also removes dusts produced by drilling or blasting, exhausts of internal combustion engines, gaseous end products of welding or blasting, and other contaminants. Combustion and blasting products include CO_2, CO, SO_2, and nitrogen oxides. Methane is a problem in coal mines, and recently radioactive radon gas was found in some mines. Where underground temperatures are high, ventilation is a means of cooling the work environment. With long, dead-end shafts and limited access to fresh air, underground mines present unusual challenges for the design of good ventilation systems.

Mining presents serious problems of noise exposure. These arise particularly from blasting and drilling, but also from some ventilating equipment and machinery used to move ore out of the mine.

Surface Mining

Where ores lie at or close to the surface, open-pit mines (surface mines, strip mines, open-cut mines, quarries, placer mines) are employed. Huge power shovels load ore directly onto trucks or railroad cars. Dust-filled air and noise are still hazards, but the job is very much safer than below-ground mining. Dust exposure is less serious in the open air where it disperses rapidly, and in surface mining relatively few workers are close to the dust sources. Wet drilling may be used to reduce worker exposure, and in large drilling rigs the worker may be in a ventilated control cab. Broken rock and ore may be wet down before loading and hauling, and once again operators of the equipment may be in ventilated cabs. Dependence on water spraying to control dust may require that protective practices change in regions with cold winters as temperatures drop. Road surfaces are always temporary in surface mines and may be serious sources of dust if not coated or wet down appropriately. Under circumstances where workers leave the ventilated cabs occasionally, training is important to inform workers about the hazards of airborne particulate and how to minimize their exposure. Silica dust is still a special threat, and air sampling is important to assure exposures are within recommended limits.

There is a tendency to be more concerned about noise in indoor situations, perhaps because of problems with the reflection of sound from indoor surfaces, but the drilling and earth-moving equipment in surface mines produces high noise levels and blasting is a special problem. The cabs from which workers control the equipment can be designed or retrofitted to control sound. Sound levels should be monitored and appropriate hearing protection provided where necessary, taking into consideration the time spent by the operator outside the cab.

Open-pit mines, particularly for iron or copper, are some of the most massive examples of humans altering the face of the Earth. These huge holes in the ground are one of the most negative aspects of this sort of endeavor. Historically, once the ore or coal had been removed, the sites were abandoned and remained as nonproductive land and scars on the landscape. Efforts to restore old mine sites and laws to prevent present day operations from becoming the scars of the future have improved this situation, but one does not need to be a dedicated environmentalist to have serious reservations about allowing new surface mining operations in areas of natural beauty or near where people live.

Mine Safety Regulations

In recent years three laws have defined the federal role in mine safety. The Federal Metal and Nonmetal Mine Safety Act[1] of 1966 required annual inspections of underground mines, empowered the inspectors to issue notices of violation and to close mines, required worker training, mandated that injuries be reported, and established procedures for generating new health and safety standards. The Federal Coal Mine Inspection Act[2] of 1969 extended coverage to surface mines. Benefits were assured to workers with black lung disease, and plans for research to identify health and safety hazards and to control them were addressed. The powers of enforcement of federal mine inspectors were increased. Before the passage of the Federal Mine Safety and Health Act[3] of 1977, responsibility for federal regulation of mining health and safety was in the Bureau of Mines in the Department of the Interior. The new law shifted it to the Department of Labor along with OSHA, setting up a parallel enforcement agency known as the Mine Safety and Health Administration (MSHA). Thereafter, NIOSH was charged with responsibilities for researching questions of mine safety and health and for recommending new standards. This legislation completely replaced the Federal Metal and Nonmetal Mine Safety Act of 1966 and extended to all miners the provisions directed at coal miners in the 1969 law. The actual health standards are found in 30 CFR, Chapter 1. As with other standards, they are a mixture of general requirements and very specific statements concerning particular serious concerns. As a result of new technology and enforcement of government regulations, great progress has been made toward improving conditions for mine workers, but mining, particularly underground mining, is still a relatively hazardous occupation.

ORE DRESSING

Typically, ores contain only a small percentage of the desired metal, so benefaction, processing to remove a proportion of the clay, dirt, or rock that does not contain the desired metal, is necessary. Ores must be broken down into a powder. This is termed *comminution*, and begins with breaking up the large fragments from the mine using large crushers. The resulting small fragments are transferred to a rod or ball mill to be ground to powder. As these operations proceed, sizing, the separation of powdered product from the larger fragments, is done. This involves the use of either screens or classifiers. In classifiers, smaller particles are separated from larger ones by their ability to be suspended in a stream of either water or air.

Then the metal-rich components must be concentrated by separation from non-mineral-bearing dirt and ground rock. Techniques for this include froth flotation, which separates ores from impurities by chemically modifying the surface of the mineral-bearing particles to reduce their attraction to water, then generating foams that selectively attach the mineral particles. The chemicals used are sometimes specific for the particular mineral. pH adjustment is

[1] Public Law 89-577 (1966).
[2] Public Law 91-173 (1969).
[3] Public Law 95-164 (1977).

important and is accomplished by use of bases such as sodium hydroxide, sodium carbonate, or lime and by acids such as sulfuric or sulfurous acid. Depressant reagents may be added to prevent the flotation of an unwanted mineral, while allowing another to be collected. These include sodium silicate, fluoride or cyanide, lime, chromates, phosphates, and polymers such as starch. Some of these are also used to clean mineral surfaces after separation. Foams are produced using organic detergent-type compounds. Obviously the harmful properties of these chemicals must be considered and worker exposure monitored. Worker training about chemical hazards is important. Finally, the separated solids are usually "dewatered" in a tank in which particles settle and are drawn off the bottom, while the clarified water layer is removed from the top. Wastes are often pumped as water suspensions to disposal sites.

Minerals may also be separated from waste by differences in density. This may employ air or water as the fluid medium. The chief dangers are from dusts, especially where the ore includes such hazardous components as toxic metal impurities or silica rock. Powdered minerals may be mixed with binders and rolled into balls, which are baked into hard pellets in a furnace. This form will be delivered to the smelter.

The mills and crushers described above are important sources of noise, as are the ventilating systems, pumps, conveyors, motors, and shakers involved. Noise levels should be measured at all worker stations and hearing protection required as needed. Ventilation may serve to control heat as well as dust when furnaces are involved.

Piles of waste from below-ground mines or from crushing operations can represent a health and environmental hazard. Rain percolating through waste piles extracts acids or traces of potentially toxic materials such as metal salts.

REFINING

Obtaining a free metal pure enough to be useful from upgraded ore is the process termed *refining*. Three refining methods are employed: smelting, electrolytic purification, and hydrometallurgy.

Smelting

In a smelter the ore is converted to free metal by a heat and reduction process. The largest volume such process, the smelting of iron, is presented in detail as an example. Iron smelting is done in a furnace over 100 ft tall called a blast furnace, which is capable of producing as much as 4000 tons of iron per day. The raw materials are iron ore (an iron oxide), crushed limestone ($CaCO_3$), and coke (a carbon material made by roasting the volatile materials out of coal in a coke furnace). Blast furnaces are usually located on waterfronts in regions close to sources of these three components so that the cost of transportation, preferably by boat, of the huge volumes required of these raw materials is minimized. Major operations are found in such cities as Chicago, Detroit, and Cleveland, utilizing ore from Minnesota and northern Michigan, limestone from quarries surrounding the lakes, and Appalachian coal, all transported on the lakes themselves by long freighters.

Coke may be produced from coal on site. Any time coal is handled in large amounts, the accumulation of dust in the air in the coke plant generates a risk of explosion. Heating coal in the absence of air drives out volatile components. Coke plants have traditionally been dirty operations, adding aromatic organics and other unpleasant or hazardous gases to the air. Modern operations are cleaner as a result of trapping the aromatics, which are salable, and using the coke oven gas, which contains hydrogen, methane, and carbon monoxide, as a fuel in the plant, even in the coke oven itself. Worker exposure to the aromatics is regulated by OSHA using a special index called the BSFTPM, or benzene-soluble fraction of total particulate matter, to describe the aromatic emissions. Burning sulfur in the coke oven produces SO_2, a significant personal and environmental hazard if exhaust gas is not scrubbed.

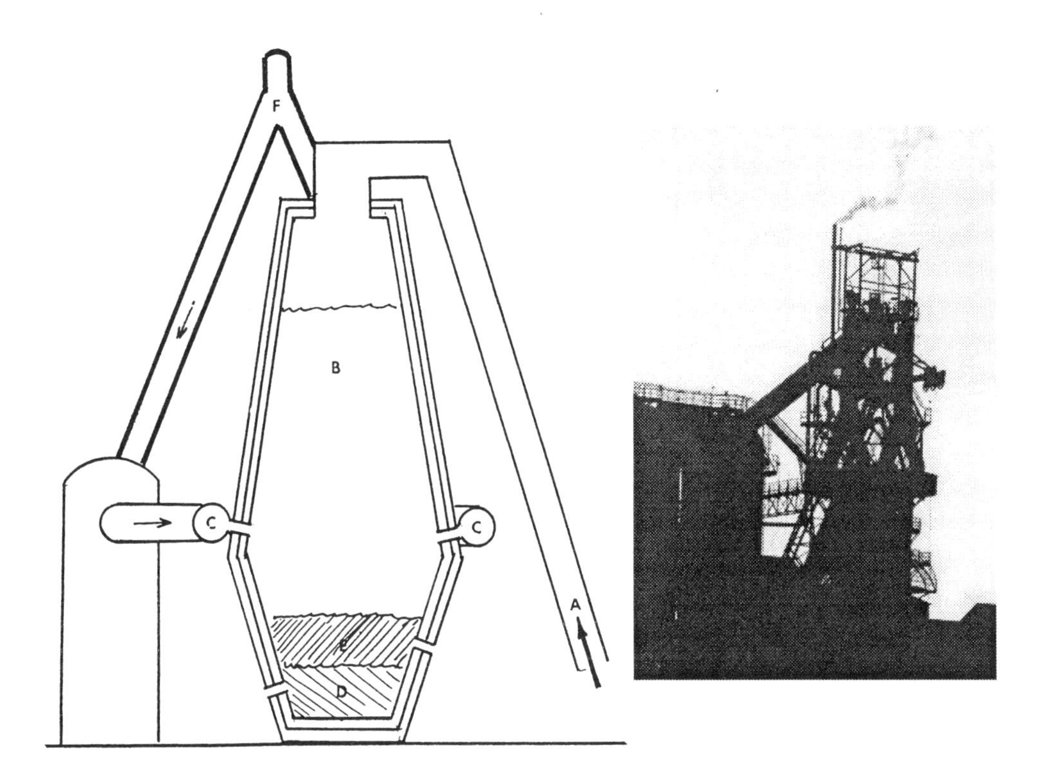

Figure 18.1. Blast furnace: iron ore, coke, and crushed limestone are moved by conveyor (A) into the furnace (B). The furnace has an inner lining of a refractory (fire brick) to withstand the high temperatures. Hot air (over 500°C), possibly enriched with oxygen, enters the furnace (C) and the coke burns, producing carbon monoxide. Carbon and carbon monoxide reduce the iron oxide to iron (D), while the limestone reacts with the impurities to produce slag (E). Hot exhaust gases are used to preheat the air (F). The furnace operates continuously, the iron being tapped every 5 hr and the slag every 2 hr at the bottom of the furnace. The photo shows a furnace in silhouette. Much of what is seen is the machinery for charging the furnace.

Turning to the blast furnace itself (Figure 18.1), limestone, coke, and ore are poured in the top. Hot air enters the lower part of the furnace, so that coke burns and further raises the temperature. Entering air may be enriched with a small amount of pure oxygen to speed the iron-making process. At these elevated temperatures two reduction reactions occur. In one the coke is the reducing agent:

$$\underset{\text{ore}}{Fe_3O_4} + \underset{\text{coke}}{C} \longrightarrow Fe + CO_2 + CO \quad \text{(not balanced)}$$

In the other, carbon monoxide is the reducing agent:

$$Fe_3O_4 + 4\,CO \longrightarrow 3\,Fe + 4\,CO_2$$

The high-density molten iron flows to the bottom of the furnace and is either poured into ingots of "pig iron" or transported in ladles directly to furnaces to be made into steel. Limestone reacts with the rock, sand, and clay impurities to form a silicate product called slag, which floats on the iron and is removed through a higher port on the furnace. Hot gases from the top of the furnaces are used to preheat air entering the bottom of the furnace, and the remaining CO in these gases is burned as a fuel in the plant.

Pig iron is rich in carbon and has some lesser impurities. Most often it is transferred still molten from the blast furnace into a basic oxygen furnace where more limestone is added and oxygen is bubbled through to oxidize the carbon. Scrap steel usually is part of the charge of the furnace.[4] Silicon, phosphorus, and manganese impurities in the iron are oxidized and react with the limestone to produce calcium silicates, phosphates, and manganates, which float as slag on the surface of the molten steel. As a result of adding the oxygen to the furnace, basic oxygen furnaces produce clouds of iron oxide particulate which must be collected. This oxide can be fed back into the blast furnace. Steel produced has a range of properties depending on the final carbon content (less carbon yields a more malleable product, more a harder product) and the possible addition of other metals to produce an endless array of alloys.

Heat stress and noise are important problems for the worker. Particulate levels and gases such as CO and SO_2 may rise to high levels in plant air.

Electrolytic Purification

Aluminum and copper are important examples of metals purified by electrolysis, and the details for aluminum are presented here. Bauxite (aluminum ore) is surface mined and must be converted to a relatively pure material for electrolytic processing. Ore is first crushed and ground, then the alumina (Al_2O_3) is "digested" by dissolving it at elevated temperatures in caustic (sodium hydroxide) to produce a solution of sodium aluminate:[5]

$$Al_2O_3 + 2\ NaOH \longrightarrow 2\ NaAlO_2 + H_2O$$

Suspended solids are removed by sand traps, settling tanks, and filters. The sodium aluminate solutions are cooled and hydrated aluminum oxide is precipitated in tanks by seeding the solution with crystals of the product and thus reversing the formation of the aluminate salt. The product is dewatered, then dried in high-temperature rotary kilns. Worker heat stress is a potential problem at all stages where elevated temperatures are employed. Throughout this process strongly alkaline solutions and suspensions, sometimes at elevated temperatures, are commonplace. All workers likely to be exposed to these solutions require protective clothing, eye protection, and training. Careful ventilation at sources of dust is important.

The electrolytic method requires first that the source of metal be melted or dissolved. Alumina has a very high melting point, but at lower temperatures it dissolves in molten cryolite ($AlF_3 \cdot NaF$). The electrolytic cells or pots are lined with carbon, which serves as the cathode, and anodes made of carbon bound with coal tar pitch dip into the molten mixture. At the cathode the aluminum ion becomes free aluminum, which collects in a pool at the bottom of the cell to be tapped off, while the oxygen produced at the anode reacts with carbon to become carbon dioxide or carbon monoxide:

$$2\ Al_2O_3 + 3\ C \xrightarrow{\text{electrical energy}} 4\ Al + 3\ CO_2\ (+\ CO)$$

Workers in the "potroom" have potential heat stress problems. The possible presence of airborne fluorides from cryolite is a special hazard of this industry. Designing ventilation systems for the pots is complicated by the need for addition of alumina and replacement of spent anodes at frequent intervals. Recent designs enclose the pots, add alumina from the

[4] The impurities in such scrap should be known ahead of time, and hazards from oils or other metals in the scrap should be considered during operations.

[5] Heat exchangers used in these processes are cleaned with sulfuric acid by workers in protective clothing.

center above, remove spent anodes from the side by opening doors, and run the stream of ventilating air across the top of the pots. General exhaust ventilation of the room is assisted by the high temperatures at which equipment operates. Air entering through floor vents is heated and rises to the room ceiling for exhaust. A creative idea in emission control is the use of alumina destined to be added to pots as the adsorbent in "dry scrubbers". Alumina dust is injected into the exhaust air stream, then is later collected in baghouses. The alumina adsorbs the fluoride without creating a filtration product that has to be disposed of carefully. There may be special filters in the ventilating system with pads of sodium carbonate to react with gaseous fluorides. Where needed, personal respirators are designed with alumina-impregnated filters that absorb the fluorides. Analysis of air samples for fluorides is backed up by testing worker urine at intervals for fluoride content.

Cells operate at only around 5 V, but at hundreds of thousands of amps. The electrical demands of this process are very high because each aluminum ion requires three electrons to become an aluminum atom, so these plants are located in regions of inexpensive hydroelectric electricity such as the Tennessee valley or the Pacific northwest.

Hydrometallurgy

Metals can be dissolved out of ores in a process called leaching. This is done with zinc, silver, gold, magnesium, manganese, and copper. In the case of copper, the metal is dissolved out of the ore using sulfuric acid, creating a solution of $CuSO_4$. The copper sulfate solution is then transferred to a bath where the copper is plated out of the solution onto a copper cathode. Other leaching agents include ammonia and sodium cyanide. When a metal is extracted as a salt, it is sometimes displaced from solution by adding a more "active" metal to the solution. The more active metal goes into solution as ions and the desired metal precipitates out as free metal.[6] As an example, copper is precipitated out of dilute leachate by adding iron to the solution:

$$CuSO_4 + Fe \longrightarrow FeSO_4 + Cu$$

Important worker hazards here involve exposure to a variety of chemicals, some toxic.

MODIFYING METAL PROPERTIES

HEAT TREATING

Hardness of a metal object depends certainly on the metal or metals involved, but for any given metal or alloy the grain or crystal lattice structure also affects the hardness. Metals can be hardened, throughout the object or at its surface, by a process called heat treating. Most commonly this is done with ferrous metals, but other metals are also hardened this way.

If steel already has medium carbon content, which itself alters hardness, it can be further hardened by heating above a temperature at which the crystal structure changes to a harder pattern. If the steel has low carbon content, heating and treatment can add carbon or nitrogen to the metal surface, resulting in surface hardening (case hardening). Hardening is commonly done in manufacturing cutting tools, gears, bearing surfaces, and similar high-wear, high-strength products.

[6] Remember the single displacement reactions you learned in general chemistry?

Case-Hardening Steel

Heating a steel object in a furnace at around 1700°F in an atmosphere with a high but controlled level of carbon monoxide results in carbon diffusion into the steel surface (carburizing). This is followed by quenching, controlled cooling.

Similarly, diffusing nitrogen into the surface hardens steel (nitriding). It is accomplished by heating the object in a furnace to around 1000°F in an atmosphere of catalytically degraded ammonia (three parts hydrogen to one part nitrogen).

Dipping the object into molten salt mixtures that include potassium or sodium cyanide or cyanate at 1600°F case hardens the object (cyaniding). Both carbon and nitrogen are added to the metal. Molten bath technology adapts to a line-type operation.

Quenching

Lowering the object's temperature after treatment is done in a controlled and uniform fashion by dipping the object into such baths as oil, water, molten salts, or brine, a process called quenching. Quenching tanks containing water include corrosion inhibitors and other additives. Wire quenching is sometimes done in molten lead baths, a process called patenting.

Hazards

Heat treating is an inherently hazardous operation. Explosions may occur when nitrate salts are used in molten salt or brine baths. These are unstable at higher temperatures and decompose to produce nitrogen oxides. When salt baths are cooled, vents should be formed in the bath, as by cooling the bath around metal rods that are withdrawn to leave a hole. Otherwise heating may cause trapped water or other liquids to convert to gases explosively, spraying about hot bath contents. Hydrogen gas in nitriding operations is explosive. Fires are possible whenever oils are heated to vapors, as in a quenching operation.

Many of the chemicals used in hardening have serious toxic properties. Carbon monoxide used in carburizing furnaces is a serious toxicant with poor warning properties. Carbon monoxide is also a product of concern from combustion processes involved in heating furnaces or baths. Lead fumes from patenting are deadly. Various processes involving cyanides must be watched closely. It is particularly important where cyanides and acids are part of a dipping sequence that thorough rinsing occur between entry into these baths. Otherwise, quantities of hydrogen cyanide gas form. All these systems should be run with engineering practices that minimize gas, vapor, or fume production, and should operate with well-designed and efficient ventilation systems.

Many high-temperature processes are involved in heat treating. The potential for heat stress should be a constant concern.

SHAPING METAL PRODUCTS

CASTING: FOUNDRIES

Molten metals are cast (poured into molds) in a foundry. Since the early days of the industrial revolution castings have been commercially made in steel and iron, bronze, and brass. More recently other materials have been handled in this fashion. These include lightweight aluminum and magnesium, high strength titanium, chromium, nickel, zinc, and a vast array of alloys, mixtures of two or more metals. Foundries employ more than 200,000 workers in the U.S. Cast iron production employs more than half the workers, so foundries are often classified into two groups: ferrous (iron) and nonferrous (all the rest). Some foundries produce small numbers of castings, as for the building of specialized machinery, whereas others rapidly produce large numbers of identical products, as would be the case in automobile parts production.

Melting metals may be done in a variety of furnaces. Arc furnaces generate an electric arc between the metal and carbon electrodes. Induction furnaces surround the metal charge with copper coils and induce a charge in the metals, causing the metals to heat. Cupola furnaces use coke as fuel in a layout much like a blast furnace. Crucible furnaces place the metal container in a refractory-lined chamber and direct a flame at it.

Casting may be done in permanent molds, which are repeatedly reused. This is particularly true for lower melting metals. However, casting is most often done in molds of shaped sand, which is high-melting silica (quartz). This is very common for casting iron, but is also used for bronze, aluminum, and other metals. Sand is placed in a holder called the flask and an exactly shaped depression is made in the sand using a pattern. Patterns are the products of skilled workers, may be made of wood, metal (iron or bronze), or plastic, and are reused many times. If the object to be formed is hollow, a core is suspended in the sand hollow formed by the pattern. The sand of a core shape must adhere more strongly than sand in flasks, so this form is usually prepared differently. Often the mold has two flasks that each have a depression in the sand and are bolted together to provide the complete shape of the final product (Figure 18.2). In this case the top flask is called the cope and the bottom is called the drag. A hole to allow the metal to enter the mold, a sprue, must either be part of the pattern or be cut into the sand. The molten metal in this channel replaces metal volume shrinkage during cooling in the mold, keeping the mold full. Gas vent holes and risers, channels through which molten metal rises to show that the mold is full, are sometimes added.

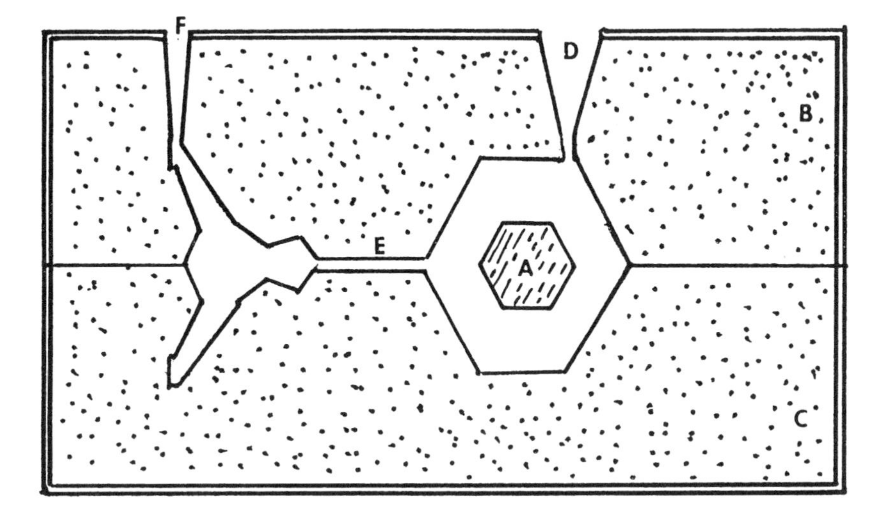

Figure 18.2. A sand mold is shown here ready for use. The shapes of objects to be cast have been pressed into sand using a pattern. A hollow casting is formed by suspending a core (A) into the depression. This mold is constructed of the two parts, the upper half called the cope (B) and the bottom half called the drag (C). Molten metal is poured down the sprue (D), fills the first space, passes through the runner (E) to fill the second space, and finally fills the riser (F). When metal appears at the top of the riser, the mold is full. After cooling and removal from the mold, metal in the sprue, runner, and riser is knocked off.

Sand is mixed with agents to hold it in the correct shape. Many foundry sands contain clay. When the sand particles are coated with clay and water is added, the particles adhere to produce a strong and accurate shape from the pattern. Chemicals forming thermoset plastics such as phenolics are also used,[7] and the shaping is sometimes done with the mixture hot,

[7] Urea-formaldehyde, phenol-formaldehyde, or furfuryl alcohol-urea-formaldehyde are plastics commonly used.

sometimes cold. Especially when the processing is done hot, or when chemicals contain catalysts and reaction follows, such chemical additives can be the source of vapors and gases. Core production requires the strong sand particle adherence produced by plastics or by use of sodium silicate (water glass). High-production lines sometimes use sand with lower binder content and vacuum-form the mold.

Several methods are used to pack sand tightly into the mold. In the most labor-intensive approach, sand is shoveled by hand and rammed by the workers around the pattern. In a jolt machine, the pattern on the bottom is covered with sand. The entire flask is then dropped and stopped suddenly, packing the sand down around the pattern. In a squeeze machine a ram presses the sand into the flask. In high-production foundries, these operations are combined in a single machine that also removes the pattern, leaving the mold ready for the molten metal, a jolt rollover squeeze machine. Such machines are sources of noise and vibration. Particularly for large castings, sand is taken from a conveyor by a device called a sand slinger and is "firehosed" into the flask using an arm with an impeller that can drive the sand at perhaps 10,000 fpm.

In sand casting systems, the sand is removed from the casting in a "shakeout" process. Some residual sand is removed by air hoses. Sand may then be collected for reuse and transported back to the mold-forming area.

Alternatives to the traditional methods described above include lost wax (investment) casting and lost foam (evaporative pattern) casting. In each case an expendable pattern is molded. Wax patterns are coated to form a temperature-resistant rigid form filled with the wax pattern. The wax is melted out of the form (shell mold), which is then used to cast the product. The shell mold is then broken off the product. In the lost foam method a pattern is made of styrofoam. This is packed in sand. On pouring the molten metal the foam vaporizes and the vapors diffuse out through the sand.

Castings need to have "appendages" removed. The sprue and riser and any runners or gates that move metal from one part of the mold to another must be knocked off. Where two halves of a mold join together there is a ridge of metal called the flashing removed by use of air-driven chipping hammers and grinders.

Hazards

Especially with some of the more recently employed metals and when the source is scrap metal, there is a problem of fumes entering the atmosphere, either of the metal itself or of impurities. For example, scrap automobiles often contain traces of lead. This is serious not only because lead is an unusually toxic metal, but because it boils at a lower temperature than iron melts and so is likely to be a serious air contaminant. Other scrap may be oily and release products of partial combustion into the air when it is heated.

Furnaces are sources of contaminants. Arc furnaces generate metal fume. Induction furnaces are relatively clean operations. Use of gas, fuel oil, or coke generates carbon monoxide, and sulfur dioxide can also be a serious contaminant when burning fuel oil. Furnaces must be ventilated, and the choices of local or general exhaust ventilation are important. General exhaust ventilation by way of ceiling-mounted blowers is usually an inefficient method of handling furnace emissions, unless the blower is directly over and close to the furnace. A local system that selectively draws air from a site of combustion is more efficient and cost effective. However, especially in a system added later to older equipment, local ventilation may need to be disconnected in order to add new metal to the furnace or to transfer the molten metal (tap the furnace). In larger scale operations, metal may be transferred from the furnace to the site of pouring in a ladle carried on an overhead monorail or by a crane. It is difficult to supply local ventilation in such a situation, although moving ducts that follow the ladle have been designed into some operations.

Respirable sand particles introduce the possibility of workers developing silicosis. In an extensive study in the 1950s more cases of silicosis were attributed to foundry work than any other manufacturing operation.[8] Silica sands may be replaced with less toxic (lower silica) sands, which has been shown to sharply reduce the incidence of silicosis. However, problems are encountered in the use of nonsilica sands, and silica sands are still heavily used.

Mixing sand with binding agents is a dusty job requiring local ventilation. Skin contact must be avoided with most chemicals involved in generating plastics as binders, and the vapors should be removed by ventilation. The sodium silicate method is less hazardous, but water glass solutions are strongly alkaline, so skin protection again is necessary.

Sand must be handled extensively, being moved from storage to the site where forms are filled, poured, and pressed into the forms and pressed in by jolt squeeze machines. After casting, sand must be blown off the casting, and used sand is moved to the next location, perhaps back to be reused. These areas should be ventilated and machines used to perform these functions in order to keep workers away from dusty zones. Moving sand about the foundry through pipes by a pneumatic conveying system is a very clean option. Sand spilled on the floor can be a significant threat. Gratings on the floors and spill pits collect this sand and prevent fans or traffic from moving fine dust into the air.

It is sometimes necessary to dry the surface of the mold with a torch before pouring in the metal. Anytime burning gas strikes a colder surface, gases burn incompletely to produce carbon monoxide. When hot metal is poured into sand molds, organic materials used as binders are destroyed, and the resulting smokes and gases, which continue to be released for some time, must be captured, preferably by local ventilation. Exposures here may include carbon monoxide, formaldehyde, phenol, ammonia, isocyanates, hexamethylene tetramine, and a variety of partial degradation products from heating the organic compounds. There is a second release of these air contaminants when the mold is opened, allowing sand that had been distant from the metal such that it still contains organic material to come in contact with the hot metal. The International Agency for Research on Cancer reports statistically higher rates of lung cancer in foundry workers, possibly from polycyclic aromatic hydrocarbons generated in these processes.[9] In high-production systems, molds on a conveyor should move directly into the hood of a local ventilation system to trap gases.

One problem for designing effective ventilation is that sources of contamination move around in the foundry. For example, the casting process involves pouring molten metal from the ladle into molds. Following this, either the molds or the ladle must move on so that the next mold may be filled. How this is accomplished depends on the size of the molds and the type of production, whether a few castings are individually produced or a large number in an assembly line fashion. Small molds may travel on a conveyor, pausing to be filled, then moving on to a cooling area and eventual removal either of the sand from the casting or of the casting from the permanent mold.

Good housekeeping is an important part of maintaining a safe work environment. Clothing can become heavily contaminated. Cleanup should be done with a thorough understanding of the nature of the dusts. Sweeping can add quantities of a fine particulate to the air and so should be replaced by vacuuming. Where very fine particulate is present, the vacuum should be equipped with a high-efficiency (HEPA) filter. Hosing an area with water can allow reactions between oxides and water to produce harmful gases. Furnace burners should be cleaned with care to prevent exposure to vanadium oxide residues from the fuel.

Air contamination is the biggest problem in a foundry, but other hazards are found. Furnaces, hot castings, and transfer of molten metal lead to potential heat stress. Eye and skin protection are required for some jobs. Striking an arc in an arc furnace is noisy. Operations

[8] V. M. Trasko, *AMA Arch. Ind. Health*, 14, 379–386, 1956.

[9] IARC Monographs, Volume 34, Part 3, International Agency for Research on Cancer, Lyon, France, 1984.

such as knocking sprues and runners from castings or chipping with an air hammer generate noise and may call for hearing protection.

FORGING

Modern industrial forging is an update of the blacksmith tradition. Blacksmiths heated metal and hammered it into horseshoes, swords, or other objects. There are a number of variations of the forging technique; this discussion does not attempt to cover all bases. Instead the focus is placed on a few typical operations. Room temperature forging is called cold forging. More often metal is first heated to make deformation into the desired shape easier (Table 18.1). Deformation involves hammering or squeezing a piece of metal to shape by forcing it into a die with one or more blows of a press, by rolling a length of material into a desired cross-section between grooved rollers, or by extruding through a die to form wire or rods of a desired cross-sectional shape.

Table 18.1. Typical Forging Temperatures (°F).

Material	Temperature
Steel	2200–2400
Stainless steels	1600–2200
Copper and alloys (brass and bronze)	1100–1600
Aluminum and alloys	600–1000
Titanium and alloys	1500–1900
Magnesium alloys	500–1000
Molybdenum and alloys	1900–2700

Most used is hammer forging. In a typical operation, metal stock is cut to the appropriate size (the blank), then is heated in a furnace. It is placed in a die, the tool that provides the correct shape for the finished forging. The die has a top and bottom half and includes places for excess metal to flow when the shaping is performed. The die itself may or may not be heated. Lubricant is added to help the metal flow during shaping and to ease removal of the part from the die later. The top half of the die is raised, then is either dropped (drop hammer forging) or is pressed down onto the lower die. The "hammer" blow then deforms the metal or a press squeezes the metal into the shape of the die. Shaping may be done in two steps. First, a breakdown hammer produces the rough shape. Then the part is reheated and forged in a second machine with the finish hammer. Afterwards the flash, the excess metal, is trimmed from the part by shearing it off in a trim press on a trimming die.

Hazards

Operating hammers or presses requires appropriate machine guarding on moving parts and safeguards so that the worker does not have a hand in danger. Controls that require two hands to activate the machine are useful.

Air contamination comes from two major sources. First, the combustion of fuel to heat the metal and possibly the dies releases carbon monoxide. Use of sulfur-containing fuel oil may add sulfur dioxide to the exhaust product. Such furnaces are normally positioned next to the forging machinery.

Second, die lubricants can burn or become suspended in air as mists. The chemical character of the lubricant determines what hazards are generated. Lubricants can be roughly divided into oil-based and water-based mixtures. Heavy oils generate mists of oil droplets or burn to produce soot and various products of incomplete combustion. Additives to the oils include molybdenum disulfide and metal soaps such as aluminum stearate and lead or zinc naphthenate. Water-based lubricants are gradually replacing the oil-based systems. Hazards here depend on the formulation of the lubricant.

Cold forging lubricants include soaps, oils, and molybdenum disulfide. Without thermal degradation of the lubricant, hazards are lowered.

Dies are sometimes heated to the same temperature as the metal, a process called isothermal forging. To avoid degradation of the dies, this must be done in an inert atmosphere, generally of nitrogen. Care must be taken that workers are not placed in an atmosphere depleted of oxygen.

Most forges have both local ventilation at forges and ovens to carry away combustion products and general exhaust ventilation to maintain acceptable air quality in the facility overall. Hard hats and work shoes protect workers when heavy objects are lifted and transported. Tinted glasses reduce the exposure of the eyes to high levels of infrared radiation.

Noise is a very serious problem in forge shops. The hammer impact obviously generates a very loud noise, but beyond that there is the release of compressed air, often used to clear the part from the die, or of a steam hoist to raise the die. Hydraulic presses are much quieter than drop hammers, and air and steam releases can be muffled, but ultimately ear protection must be supplied to the workers. Sometimes the forge is enclosed, but usually that interferes seriously with operations. Forges are simply noisy equipment.

Finally, with all the furnaces required, heat stress is a serious hazard. Furnaces run in excess of 2000°F. There is both heat radiating from furnaces, parts, and dies and heavy physical exertion resulting in high metabolic activity in the worker. Steam or air "curtains" that deflect the heat when furnace doors open is slightly effective. Reflective sheets around furnaces, especially when the door is open, block some heat from reaching workers. Sometimes heat shielding protective clothing is helpful.

STAMPING

Another common process is stamping. Sheet metal is the usual raw material. Objects are shaped in dies using presses, as in forging. Heating the metal is omitted in stamping, which avoids many of the sources of air contamination. Stamping plants generally form large numbers of parts at high speed, feeding rolls of sheet metal into a press. Noise and the possibility of accidental injury are significant hazards. As with forges, hydraulic presses are quieter than mechanical stamping machines.

MACHINING METALS: CUTTING, GRINDING, BORING

Shaping a metal object into a finished product often requires removal of metal. As an example, consider castings. The sprue or risers must be cut off a new casting using a cutting wheel (a spinning, abrasives-coated wheel) or a saw. A flat surface may be ground onto a casting using abrasives so that two structures can be joined tightly. For example, surfaces may be ground flat to connect an internal combustion engine to some sort of transmission. These two objects may be assembled by bolting them together. To accomplish this holes are drilled through one object and into the other. A hole may later be tapped, that is, threads that will anchor the bolt may be cut into the walls of the hole with a thread-cutting device.

Any process in which metal is cut creates the danger of airborne metal particulate being generated. There is also the danger of particulate arising from the abrasives used. This can be particularly serious when the abrasive contains silica, as is often the case. It is good practice to use local exhaust ventilation around these operations, arranging the system so that particulate generated is directed into the hoods. Good housekeeping also reduces the likelihood of workers inhaling particulate from these inherently dirty operations.

Cutting, grinding, boring, and other metal removing operations are noisy. Noise levels should be checked around such operations, and either appropriate engineering controls installed, as by isolating the operation, or hearing protection provided for the workers.

Cutting Oils

The operations described above generate heat due to friction between the metal and the tools in use. To extend tool life, protect the work, and lower noise levels, a stream of fluids called cutting oils or grinding fluids are directed at the tools, which reduces friction and carries the heat away. They allow the tools to be used at higher speeds, which speeds processing. More than two million workers are estimated to be in contact with cutting oils.[10]

Three classes of fluids are used. Neat oils are lubricants without a water component. These have been popular in the past, but use is declining presently. Soluble oils or oil-in-water emulsions, very common, use detergents to suspend up to 10% hydrocarbon lubricant in water. These flow very well, cool surfaces efficiently due to the water, and do not leave deposits of grease on the metal. The complexity of composition and level of performance of such mixtures has increased with time. Finally, some cutting oils contain detergent and water, depending on the detergent to lubricate the surface.

Metal debris is trapped in these fluids and washed away from the operation. In doing this, cutting oils can prevent much of the metal or abrasive particulate from entering the atmosphere directly. However, there is often an air hose directed at the cutting edge to keep that edge clear, and the compressed air can blow droplets of cutting oil and metal into the air. Splash guards can block much of the blown spray from being dispersed, but guards do not completely stop the spray, and they are sometimes removed by workers to have better access to the work in progress. Evidence that the oils are not blocked effectively is provided when the machinery and nearby surfaces become oil coated.

Skin contact with the cutting oils is commonplace in metal-working facilities, sometimes unavoidably. These jobs tend to be "dirty" jobs, and this is simply part of the "dirtiness". Workers usually do not experience an immediate problem with such contact, so attach relatively little importance to it. The usual response to getting cutting oils on the hands is to wipe the hands on a rag. The oils both on the hands and already on the rag contain tiny metal fragments that can lacerate the skin during wiping, damaging the epidermis and thus reducing the effectiveness of the skin as a barrier. Detergents in the fluids remove the lipids of the skin surface, again reducing the ability of the skin to block passage of chemicals. Concern now arises about what other chemicals the worker contacts.

Irritant dermatitis is caused by cutting oil. Causal agents in water-based systems are thought to be the detergents and the basicity of the solutions. Sensitization to the metals themselves or to formaldehyde present to prevent bacterial growth sometimes results.

The petroleum component of the cutting oils may block pores in the skin, leading to an acne condition. This is most serious with neat oils.

Contact with cutting oils, both airborne and dermal, can be reduced by well-designed splash guards on machines. Running machines at lower speeds and increasing the fluid flow rate to the tool lowers temperatures at the cutting edge, which in turn reduces mist formation.

Workers should practice good personal hygiene, washing thoroughly with soap and water to remove cutting oil residue from the skin. The practice of cleaning the oils from the skin with organic solvents such as paint thinner rather than soap and water leads to irritant dermatitis and should be discouraged. One tends to think first and only of washing the hands and arms, but the problem can extend to skin elsewhere on the body. Contact with oil-soaked clothing similarly leads to blocking of pores. Work clothes should block the cutting oils from reaching the skin to prevent this and should provide good coverage of the body. These should

[10] M. A. Lapides, "Cutting Fluids Expose Metal Workers to the Risk of Occupational Dermatitis," *Occupational Health and Safety*, April 1994, p. 82.

be changed at work at the end of the day, and the employees should be encouraged to shower before leaving work. Gloves and oversleeves that are impervious to the oils should be used wherever possible. Barrier creams are readily removed by detergents in the fluids, so they are of limited value.

Effects of inhaled droplets of cutting oils is another concern that has been less well studied.

Laser Cutting

Cutting metals with laser increasingly is being done. These are high-powered laser systems, typically hundreds to thousands of watts. Such powerful lasers represent a physical hazard, as would any powerful cutting tool. Careful training is necessary to ensure that workers appreciate the potential for harm from such light beams. The cutting process involves heating the metal rapidly to very high temperatures, melting it at the desired point. Metal vapors are an important air pollutant, so careful ventilation is necessary. ANSI has a standard for laser use.[11]

WELDING

Just under 200,000 workers are employed in joining or cutting metals using welding techniques, and several times that many do it as a part of their job description. There is a wide variation in the degree of hazard they experience. Welding done by such major employers as automobile companies is often performed by automatic machinery with very low hazard to employees. However, welders are also employed by companies too small to be regulated closely by OSHA, such as auto repair shops, and can be exposed to high levels of toxic substances. Although some jobs in industry are performed at a workstation with safety features such as guards and adequate ventilation, welding is frequently a repair or disassembly operation where welders go to the site of the job.

Welding subdivides into flame welding, using such fuels as acetylene, propane, or butane, and arc welding, where a low-voltage, high-current electric arc is generated between the work surface and the welding rod. In either method a bond is formed by melting metal, allowing two surfaces to flow together. Additional metal may be added from a welding rod that is melted in the flame or is actually one of the electrodes in the arc process. Very strong bonds can be made that are air and water tight. Welding is most successful with ferrous materials, although it can be done with a number of nonferrous metals.[12] Flame welding torches melt through metal to perform cutting. Some fabrication and repair jobs involve both joining and cutting.

Arc welding is the most common technique. Low voltages, 10–50 V, are employed, but amperages run very high, hundreds or thousands of amps. The electrode is composed of metal, similar to that being welded, and a carbonate-containing coating. On heating the coating releases carbon dioxide, which blankets the work to exclude air, which would weaken the weld. Numbers stamped on electrodes identify their composition, using a system devised by the American Welding Society (AWS).

Sheet metal is often joined by resistance welding. The two surfaces are clamped together tightly and an electrode is pressed against each side of the work. The resistance of the work to the flow of current heats the work to melting. This technique is inherently much cleaner.

Airborne Hazards to Workers

At high temperatures the metals of the working surface or welding rod vaporize. It is estimated that a small percentage of the mass of the electrode becomes vapor. Such metal

[11] ANSI, *American National Standards for the Safe Use of Lasers*, ANSI Z136.1, New York, 1986.
[12] Copper, aluminum, nickel, titanium, and magnesium.

vapors condense into respirable spherical particles called fumes, usually largely metal oxide in composition. Iron oxide fume is relatively low risk, although it can accumulate to produce a condition called siderosis. Siderosis does not involve fibrosis or other serious lung impairment. However, other metals either alloyed with iron in steel or as part of a coating on the object being welded also vaporize. Alloy metals of particular concern include chromium (stainless steel), beryllium (alloyed with copper), and manganese (welding rods). Harmful metals in coatings include lead (paints and solders), cadmium (anti-corrosion coatings), and zinc (galvanized steel). A number of fumes cause metal fume fever (Chapter 7). Studies on the rate of fume production are reviewed by Burgess (1995).

As with any combustion process, a welding torch flame produces carbon monoxide (CO). CO production increases when the flame hits a cold surface, halting the combustion process before it is complete. Concentrations of CO can exceed 300 ppm, many times the OSHA PEL of 35 ppm.

At high temperatures oxygen and nitrogen in air combine to form nitrogen dioxide (NO_2), which is irritating to the lungs and eyes. The intense UV light of arc welding catalyzes the conversion of oxygen into ozone (O_3), generating levels far above the 0.1 ppm PEL.

Clean surfaces produce better, stronger welds. Cleaning methods are described in the following section of the chapter. Chlorinated solvents used for cleaning must be completely removed to avoid their decomposition by heat into chlorine, hydrogen chloride, or phosgene. A hot flame removes grease and oxide from a surface. Alternatively, flux can be added to an oxidized surface. Fluxes contain metal sulfates, chlorides, or fluorides that dissolve oxides. Much of the flux material is volatilized in the heat of welding, producing harmful airborne materials.

In the past, asbestos was sometimes sprayed on building beams; it should be removed before welding or cutting. Some plastic coatings such as teflon produce harmful fumes when vaporized. Tanks and other containers must be empty before any welding, because toxic, flammable, or explosive contents cause serious incidents.

Skin and Eye Hazards

Arc welding produces intense UV light. This can cause skin burns essentially the same as severe sunburn. The eyes must be protected by goggles or a face mask that absorbs UV light. This reduces visibility, so masks are available that darken when struck by UV light, but are more transparent otherwise. The protective characteristics of lenses is described by a number system, and ANSI devised standards for the number required for each type of welding.[13]

The hazards to the eyes and skin due to fragments of hot metal are well recognized, so a welder wears a face mask, gloves, and other protective clothing.

Welding in Confined Spaces

Depletion of oxygen supply can occur in confined spaces. Some welding is done under a blanket of inert gas such as argon, and such release of argon could dilute the oxygen supply dangerously.

METAL PREPARATION: CLEANING METAL SURFACES

The cleaning of metal surfaces can be an important step in jobs such as applying paint or other coatings, welding, or electroplating.

[13] ANSI, *Safety in Welding and Cutting*, ANSI Z49.1-1988, New York, 1988.

SANDBLASTING

Cleaning to remove oxidation, scale, and paint and to otherwise improve appearance of metal surfaces is often done by sandblasting (abrasive blasting) the surface. This technique, which dates to the early 20th century, is the prime responsibility of perhaps 100,000 workers and the part-time job description of many others.[14] Its popularity results from its speed and low cost. Sometimes sand is replaced with pecan, walnut, coconut, or almond shells, peach pits, aluminum oxide, silicon carbide, or plastics.

Abrasive blasting is done at a number of types of workstation. Small parts may be done in an enclosed cabinet much like a glove box. The worker manipulates the hose and nozzle inside the sealed box, viewing the operation through a window. Exhaust air goes into a cyclone collector. Large numbers of small parts may be placed in a barrel and blasted while tumbling. Again the exhaust air is cleaned as it leaves the barrel. Some large work may be done in a sealed, ventilated room. The worker manipulates the hose and nozzle while wearing protective clothing and a respirator. Outdoor projects may need to be done in temporary enclosures to control the dust. High levels of contaminant are then in the worker's air, and good respiratory protection is essential.

Airborne hazards include silica or other abrasive and airborne particulate of the metal, metal oxides, or a coating such as paint that was on the surface. Special care must be taken when lead-based paints are involved. The dusts must then be collected for hazardous waste disposal. Good housekeeping is important, so that residual particulate in the work area does not later become airborne.

Abrasive blasting using sand, still the most popular abrasive because of its low cost, has a bad health history. In a study of 100 sandblasters, reported in 1974, 28 died from silicosis at an average age of 45.[15] Lower toxicity abrasives are available and have been adopted by some users. Another approach has been to suspend the abrasive in a high-velocity stream of water.

Finally, high-velocity air is usually a noise hazard, so if the work is not enclosed, workers require hearing protection. Noise levels are commonly well over the regulated levels.

DEGREASING

One of the most common ways to clean metals, and a necessary process before painting or electroplating, is degreasing. Cutting oils leave a hydrocarbon residue on newly shaped surfaces. Polishing compounds come in a wax or grease carrier, a residue of which remains on the polished part. Used parts may have a lubricant residue after long periods of service. Microelectronics work demands highly efficient cleaning and has stimulated new technologies.

Cold Degreasing

Direct degreasing with solvents (cold degreasing) may be accomplished by dipping the object in solvent, then wiping or brushing the oil or grease from the wet surface. A more sophisticated arrangement includes an agitator to vibrate the part or the solvent and/or a spray nozzle to flush the grease off the surface. Dipping in a solvent tank is straightforward, but the dirt and grease remains in the solvent, and after some time becomes a significant contaminant on the dipped part as solvent clinging to the surface evaporates. Simple degreasing operations are performed extensively across the country in repair shops and machine maintenance operations.

[14] Chapter 3, W. A. Burgess, 1995 (see Bibliography).
[15] B. Samini, H. Weill, and M. Ziskind, *Arch. Environ. Health*, 29, 61, 1974.

Cold degreasing varies greatly in hazard with the sophistication of the operation. A mechanic with a bucket of gasoline or kerosene cleaning some transmission gears is experiencing significant risk from inhalation of vapors, dermal irritation, and fire hazard. Use of a purpose-built tank with a drain rack inside the tank to prevent solvent from being carried out of the tank, exhaust ventilation at the top of the tank, and a low-toxicity, high-flashpoint solvent reduces the hazards sharply. However, as parts are removed from the solvent there is still a possibility of significant amounts of vapor contaminating the air, and handling the newly dipped part can lead to dermal exposure. Ventilation and worker protective clothing, particularly gloves and eye protection, are necessary.

Vapor Degreasing

It is estimated (Burgess, 1995) that in 1991 300,000 vapor degreasers were in use in the U.S. Vapor degreasing is done in a tank with high sides and heating coils on its floor (Figure 18.3). A layer of solvent is added to the bottom, and heating produces a high concentration of vapors above that solvent. At the top of the tank walls are cooling coils that condense the vapors back to liquid, which then runs down the sides to the liquid layer in the bottom. The part or rack of parts to be cleaned is lowered into the solvent vapors, which condense and run off into the liquid below, carrying the dirt and grease along. Cleaning may be assisted by spraying the part(s) with a stream of hot liquid. Once the parts warm to the vapor temperature, condensation stops and the parts are removed. An advantage over dipping is that the vapors are free of dirt and grease, which accumulate below.

Used properly, a degreasing tank contributes only a relatively small amount of vapor to the plant atmosphere, but misuse can cause excessive vapor loss to the room. For example, if parts being cleaned have depressions, these can carry liquid solvent out of the tank. Hanging the part with depression facing down or rotating the part before removal can minimize this problem. Elevating the part too rapidly from the tank can carry excess solvent along. Spraying the part with hot liquid must be done below the level of the cooling coils, and spray must be directed down into the tank, not splashed upward and out into the room.

Drafts across the top of a degreaser in use should be avoided, and tanks should be covered when not in use. The cooling system and the thermostat on the heater must both be in good working order to avoid vapors boiling over into the room. The unit is normally sold adjusted for use with a particular solvent. Changing to a more volatile solvent without readjusting the thermostat can lead to the vapor concentration in the tank rising to levels the cooling coils cannot handle.

An important dimension of a vapor degreaser is the freeboard. This is the distance between the highest level reached by vapors and the lowest level at which vapors can leave the tank. Freeboard should be greater for a wider tank, and present thinking calls for freeboard equal to the tank width (freeboard/tank width = 1.0). Older units were built with much less freeboard. Increasing from a ratio of 0.75 to 1.0 cuts vapor loss about 20%, and a further increase to 1.25 cuts a further 6–10% (Burgess, 1995).

Degreasing Solvents

Degreasing tanks commonly use trichloroethylene, 1,1,1-trichloroethane (methyl chloroform), perchloroethylene, methylene chloride, and trichlorotrifluoroethane as solvents. Early use of petroleum hydrocarbons was phased out because of fire hazards, in favor of much safer chlorinated hydrocarbons. Trichloroethylene, perchloroethylene, and methylene chloride are sufficiently toxic that their use is being limited, and the more recently introduced methyl chloroform is now the solvent of choice (Table 18.2). There is trend toward using more fluorocarbons that, even though they are more expensive, require less energy to boil and have a lower escape rate from the apparatus.

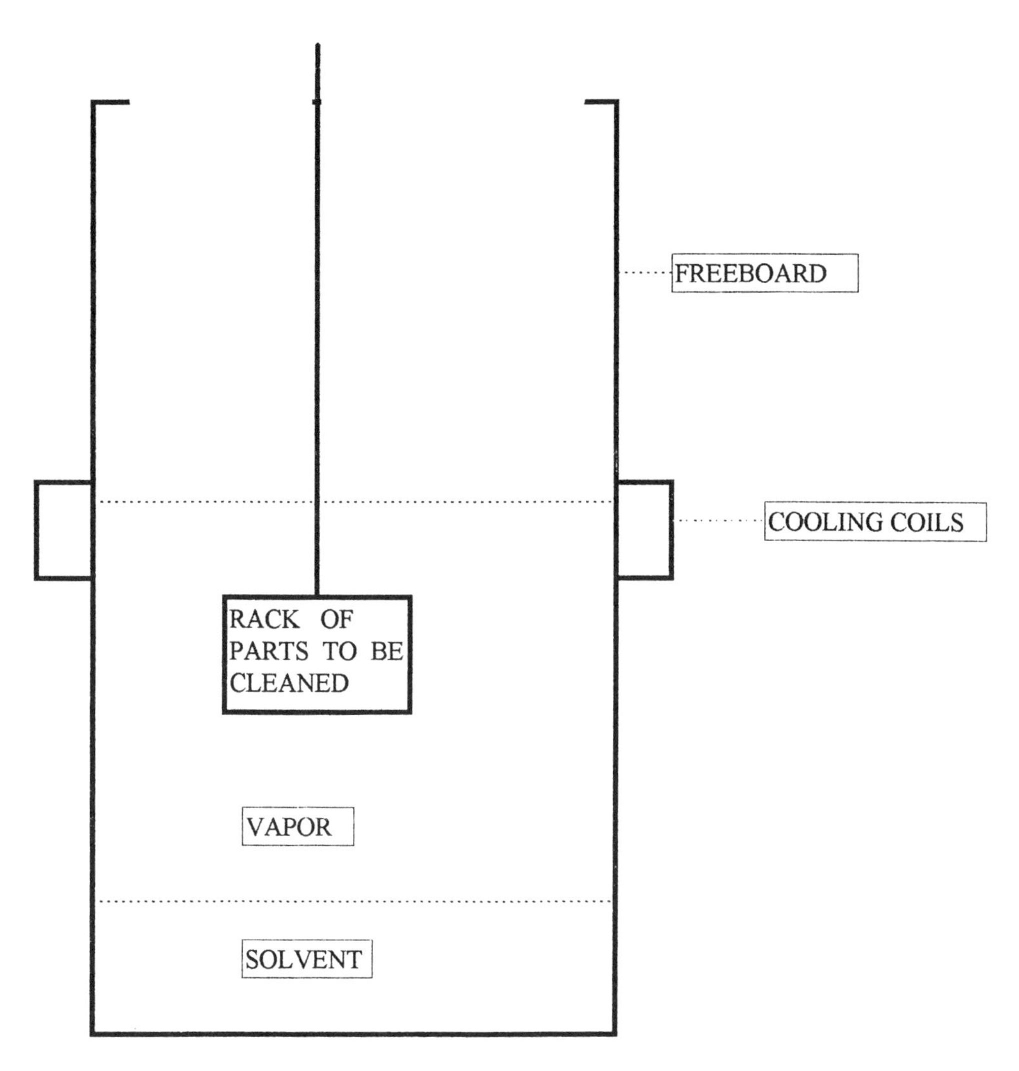

Figure 18.3. Vapor degreasing. Shown here is a simplified cross-section of a vapor degreasing tank. Liquid solvent is heated to produce hot vapors. These vapors rise to the cooling coils, where they condense and run back down into the liquid phase. Hot vapors also condense on the work, dissolve the grease, and carry it back to the liquid phase. To help scrub the parts clean, hot solvent is sometimes pumped through a wand, which is held below the vapor surface. Greater freeboard, the wall height above the cooling coils, reduces loss of solvent to the surroundings. Contaminant accumulates in the liquid solvent, which must be removed at intervals for cleaning, usually be distillation. The vapors, however, are free of contaminant.

Substitution of fluorinated hydrocarbons also is called for to replace chlorinated hydrocarbons because of ozone layer problems. The Clean Air Act of 1990 addresses this question. A specific timetable was established for a halt in production of chlorofluorocarbons in the 1987 Montreal International Agreement, with methyl chloroform being targeted specifically.[16]

If solvents are being changed, the toxic properties of the new solvent should be carefully accessed, because often one of the easiest ways to lower the level of hazard in a plant is to substitute a less toxic material.

[16] See *Science*, 259, 28, 1993.

Table 18.2. Degreasing Solvents.

Solvent	Consumption[a] (millions of pounds)	Boiling Point °F	Boiling Point °C
Methyl chloroform	258	165	74
Trichloroethylene	95	188	87
Perchloroethylene	46	250	121
Methylene chloride	30	104	40
Trichlorotrifluoroethane	10	118	48

[a] Data is for 1991. About 72% of the total use is in vapor degreasers. From EPA, *Overview of Halogenated Solvent Cleaning/Degreasing NESHAP,* for National Air Pollution Control Technology Advisory Committee, Office of Quality Planning and Standards, Washington, D.C., October 1992.

ACID AND ALKALINE CLEANING OF METALS

Acid cleaning of steel (pickling or acid descaling) is usually done with 10–25% hydrochloric or hot 5–15% sulfuric acid, both of which require careful handling. More hazardous and complex acid baths are used with stainless steel alloys. Acid pickling, often with sulfuric–nitric acid mixes, is done with more chemically reactive metals, such as magnesium, aluminum, zinc, and cadmium, but must be done properly to avoid excessive loss of metal by reaction with the acid.

Acid mists released by pickling operations are upper respiratory tract irritants, and the acids are corrosive on contact from a splash or spill. Products of reactions between acids and "scale" must also be considered, and these vary with the scale being removed.

Alkaline descaling may be done instead of or following acid pickling. Sodium hydroxide (caustic) is commonly used, but potassium hydroxide, sodium carbonate, (tri)sodium phosphate, tetrasodium pyrophosphate, and borax are also employed. In addition to the chemical action, the part may also be attached to the cathode of a direct current system in which the bath is the anode. Baths of pure molten sodium hydroxide (caustic, over 400°F) are used.

Protective clothing and eye protection are important, and employees should be trained about the hazards of contact with these chemicals, how to handle them safely, and the emergency procedures to use in case spills or splashes occur. Molten caustic particularly is extremely corrosive and very hard to remove from the skin.

SURFACE COATING OF METAL PRODUCTS

Two primary goals are reached by coating metal objects: protection from corrosion and improvement of appearance. Other goals, particularly in the new microelectronics industries, include electrical insulation or electrical conduction. Some coatings reduce flammability, reflect radiant energy, or absorb radiant energy. Many workers are employed in the application of these coatings, so the methods and health concerns need to be discussed here.

PAINTING

Painting is broadly the application of a chemical coating to surfaces, either organic or mixed inorganic/organic in chemical nature. A half million workers are estimated to be employed in these operations, about 40% in auto body repairs alone.[17] The technology of painting is changing in a revolutionary fashion in response to desires for improved finishes

[17] Burgess, 1995, Chapter 15 (see Bibliography).

and, importantly, to environmental and health concerns. Traditional paint systems involving large amounts of organic solvents are in rapid decline. Homeowners have been aware for some time of the shift from traditional oil-based paints (organic solvent systems) to systems where cleanup is done with water and the solvent smell does not have to be endured during paint drying. Some areas with serious air pollution problems banned oil-based house paints years ago.

Organic Solvent–Based Paints

Traditional paints are composed of a mixture of chemicals suspended in a reasonably volatile organic solvent. The liquid character of the mixture facilitates coating, and evaporation of the solvent is a major aspect of producing a dry solid surface. Higher-vapor-pressure solvents result in faster drying paints. A list of solvents used covers the range of organic chemicals in an undergraduate organic chemistry class, but most important are aliphatic hydrocarbons, alcohols (including glycols), ketones, aromatic hydrocarbons, glycol ethers and esters, and other esters. The amount of solvent added adjusts the viscosity of the paint, an important factor when considering the method of application to the surface and the degree of penetration desired on a porous surface.

A colored solid suspended in solvent and coated on a surface soon comes off. The heart of paint is a chemical system (binder or resin) that forms a film after coating that is like a continuous skin on the surface. This film blocks water from reaching the surface, resists abrasion, and binds the other paint components to the surface. Traditional paints use linseed oil, a highly unsaturated plant oil that (slowly) cross-links its double bonds in air to form the film. Such paints have to be left undisturbed for some time after coating to allow the film to form. Polymers familiar as plastics now dominate the list of resins, including addition polymers such as vinyls and acrylics and condensation polymers such as polyesters (alkyds), urea-formaldehydes, melamine-formaldehydes, phenol-formaldehydes, polyurethanes, and epoxys. Film formation generally involves cross-linking (curing) short-chain polymers already formed in the resin mix.

Pigments and fillers make the paint film opaque and give it color. Fillers also add "body" to the film. Pigments may be inorganic or organic compounds. Organic dyes provide a range of bright colors, but often undergo chemical alteration on exposure to UV light, causing the colors to fade.

Other chemicals impart special properties to the paint. Biocides may block molds and funguses or retard the growth of barnacles or other marine organisms on boat hulls. Stabilizers block UV light from damaging pigments or other paint components or they may make the coating more heat resistant. Some chemicals are added to speed cross-linking of the film. Mineral dusts are added to slightly roughen the smooth finish of the dried paint, cutting (flattening) its glossiness.

Water-Based Paints

Water-based paints use water as the solvent. They represent a large and growing fraction of industrial and construction paints. An obvious disadvantage of water is its relatively low volatility, but a major advantage is its lack of toxicity and environmental problems. Surfactants are added to help suspend other paint components.

Resins are generally acrylics and styrenes. Other ingredients such as pigments, biocides, and the like are similar to those in solvent-based paints. Some two-component paints, which put prepolymer materials in one component and catalysts for polymerization in the other, include highly reactive and toxic components: isocyanantes in urethanes and epichlorohydrin, bisphenol A, and amines in epoxy paints.

Powder Paints

In powder paint systems the solvent is eliminated. Use of this technology is growing rapidly and is now common in auto and appliance manufacturing. The paint components are combined into a powder, which is then deposited onto the product. The product enters a furnace at over 300°F, and the powder coating melts and flows into a continuous surface. The powder composition is not significantly different from that of solids of liquid paint.

Paint Application

Paint can be applied in a variety of ways in production, but spraying is both the most common and the most hazardous method. Direct application, as by dipping, brushing, or rolling paint on products, tends to waste less paint, but spraying produces a superior finish. In the usual equipment high-pressure air breaks liquid paint into a mist at the gun nozzle and carries it to the surface. Lower viscosity paints are successfully sprayed with lower air pressures. Lower air pressures reduce the "rebound" of paint droplets and so increase the efficiency of paint transfer to the surface from about 1/3 to about 2/3 of the paint coating the product. In airless spraying paint is forced at high pressure through fine holes that break it into droplets, and the pressure propels the paint toward the product. Electrostatic systems give paint mist particles a negative charge. The product is grounded and attracts the particles electrostatically. Efficiencies of about 90% in transfer of paint to the product result.

Powder paints also are sprayed onto products. The particles are given an electrostatic charge, resulting in high paint transfer efficiencies. Further, paint that does not adhere can be vacuumed from the floor and reused.

Hazards to Workers

Painters are at risk from inhalation of airborne particles of paint, which can be of respirable size. Solvent-based paints add the risk of exposure to solvents that evaporate during mixing, spraying, or drying of the paint. The chemicals involved in the paint mixture should be known and their toxicity characteristics understood. All else equal, a paint process that transfers a higher percentage of the sprayed paint to the product is less hazardous, and water-based paints or powder spraying eliminate the problem of organic solvent inhalation. Urethane paints include either hexamethylene diisocyante or toluene diisocyanate, which is a special concern because of the high toxicity and sensitizing potential of these compounds.

Use of respirators is often indicated. In negative-pressure systems, particulate filters are needed to remove paint droplets and should be combined with organic vapor removing cartridges when solvent-based paints are used. Supplied-air systems may be needed. Disposable positive-pressure masks are frequently used because paint coatings render the respirator unusable relatively quickly.

Because of the importance of preventing inhalation of dangerous chemicals, the problem of skin exposure is sometimes overlooked. Some chemicals may transfer across the skin into the blood. Paint solvents strip skin oils and soften the keratin layer, facilitating such transfer and encouraging contact dermatitis.

Ventilation systems in painting areas should be adequate to the task and carefully laid out so that exhaust air is not drawn past workers. Local exhaust ventilation is often called for, but general exhaust ventilation can work if carefully designed, especially if painting is done in a room or booth dedicated to that purpose and the system is well designed. Some work practices reduce the burden on the ventilation system, for example, ensuring that the distance between the spray gun and the product is correct to minimize the amount of paint not depositing on the product. In general, the most hazardous systems are found in auto body repair facilities, where the level of design expertise in laying out the spray area and ventilation may be low and sophistication in testing the air for high levels of contaminants may be lacking. Studies of worker exposure in a variety of situations, including auto body shops, are reviewed in Chapter 15 of Burgess (1995).

ELECTROPLATING

Electroplating is a method of applying a decorative, corrosion-resistant or wear-resistant finish to metal surfaces, and has been a large industry for many decades. In simple principle, the object to be plated is hung in a conducting bath as the cathode in a direct current circuit and a sample of the metal to be applied to the surface is hung as the anode. Upon application of the electrical current, metal enters the solution at the anode as ions and is converted back to metal atoms at the surface of the object that is the cathode. Sometimes more than one layer of metal is plated onto the object. For example, in chromium plating a steel object may first be coated with copper, then nickel, before the final bright chromium surface is added.

Applying a uniform coating is a challenge, especially to irregular objects, because projections coat more heavily and indentations less heavily. It is also important that the surface to be plated be very clean because impurities, especially grease and oil, block the plating action. For this reason, objects to be plated are generally run through cleaning baths that may contain acid, caustic, or organic solvents before plating. The cleaning agents must be rinsed off before the object enters the plating bath.

Plating, like aluminum purification, is a low-voltage, high-amperage operation. After the object is plated in the bath, it is rinsed in a series of tanks of water. Sometimes a second or even third metal is added, requiring more plating baths and rinse tanks. A typical installation has objects being moved automatically from one to another of a long line of tanks.

Good electroplating is a fine art. Correct orientation or movement of the object during plating affects the outcome. Bath temperatures are regulated and plating baths are prepared by following carefully worked out recipes designed to produce the best plated surface with lower electrical consumption. Assembling the chemicals is an expensive operation, so it is hoped a bath will provide long service. Furthermore, the problem of disposing of the sizable contents of a plating bath is complicated by the fact that many of the bath components are toxic and damaging to the environment: acids, alkalis, cyanide compounds, and many metal salts. Properly treating all these components for disposal is expensive. Even disposing of the water in rinse tanks presents problems.

An ingenious system has been developed to minimize the need for disposal. Because plating baths are often heated, there is significant loss of water due to evaporation. Rather than adding replacement water directly to the plating tank, water is brought back from the first rinse tank, which in turn is replaced from the rinse tank beyond. The water eventually added to replace that lost by evaporation goes into the last rinse tank. Chemicals rinsed from the objects are returned to the plating bath along with the water, rather than becoming hazardous waste for disposal, and the chemical components of the plating tank are not depleted.

Finally, plating baths may develop problems on use. Platers have become skilled in "nursing the bath back to health" by adding chemicals found to be useful for specific problems, thereby extending the life of the bath. Once in operation, a plating bath may be used continuously for long periods.

Hazards to workers focus heavily on the use of toxic or corrosive chemicals and the dangers associated with electrical machinery. Noise is not a significant problem in the plating itself, although other operations such as polishing or other metal finishing that may be associated with the plating may present difficulties. Baths are heated, but temperatures are usually moderate, so heat stress is easier to avoid.

KEY POINTS

1. Underground mining requires careful ventilation due to particulate and gases from drilling and blasting. Noise is a problem and temperature extremes may occur.
2. Airborne particulates and noise can be hazards in surface mining.
3. Mine safety regulations largely flow from three laws passed in 1966, 1969, and 1977. Standards are found in 30 CFR, Chapter 1.
4. Ores are first crushed and ground in a process to increase metal content. Froth flotation is then frequently used to separate metal compounds from dirt and rock. Hazards include airborne particulate, noise, and exposure to dangerous chemicals.
5. Smelting reduces metal oxides and sulfides in furnaces. Iron is smelted in blast furnaces using coke as the reducing agent and limestone to combine with impurities. Heat stress, noise, and airborne particulates and gases are problems.
6. Iron is converted to steel in basic oxygen furnaces that burn out carbon and use more limestone to remove impurities. Heat stress, noise, and airborne particulates are problems.
7. In electrolytic purification metals are recovered by plating ions out as free metals on cathodes. Bauxite is crushed and alumina is dissolved in NaOH. Impurities are discarded and alumina is precipitated. It is added to electrolytic cells with carbon electrodes and aluminum metal is formed. Heat stress and airborne particulates, particularly airborne fluorides, are problems.
8. In hydrometallurgy metals are leached from ores by chemicals. Hazardous chemicals are a threat.
9. Steel and other metals are hardened by raising levels of carbon or nitrogen throughout the metal or at the surface. This is accomplished at high temperatures in the presence of carbon monoxide or nitrogen from decomposed ammonia. These processes involve some toxic chemicals, and there is an explosion risk from hydrogen gases or the decomposition of nitrates. Heat stress can be a problem.
10. Metals are shaped by casting in a foundry. Molds may be permanent or shaped sand. Metal fumes, silica particulate, heat, noise, and vibration are problems.
11. In forges hot or cold metals are hammered, rolled, extruded, or stamped into shape. Serious noise problems result. Furnaces generate heat stress and problems with combustion products.
12. Stamping is like forging, but is done cold and uses sheet metal as the stock. Noise is by far the most serious problem.
13. Shaping metals can involve cutting, grinding, or boring. Airborne particulate, metal fragments or abrasives, are generated. Potentially high noise levels are reduced by use of cutting oils. Such oils can cause dermal irritation and may present problems on inhalation.
14. Welding, either by flame or arc, generates a number of airborne toxics, from combustion products to vapors from the metal or substances coated on the work. Arc welding produces intense UV light, which irritates the eyes. Extensive personal protective equipment should be employed by welders.
15. Metal surfaces are cleaned by abrasives, including sand in sandblasting. Airborne silica is a serious hazard. Solvents are used for degreasing, and vapors may contaminate the workplace. Alkalis and acids are used for pickling metals, removing oxides and scale. Inhalation and skin contact should be avoided.
16. Application of paint is done in a number of ways. Spraying paint in organic solvents generates toxic vapors and droplets including paint itself. Water-based paints are less hazardous. Dry powders are melted onto surfaces. The most important safeguard in any painting process is good ventilation.
17. In electroplating metal films are electrically deposited on surfaces. Baths include a variety of hazardous chemicals.

BIBLIOGRAPHY

American Foundryman's Society, *Health and Safety Guidelines,* Des Plaines, IL, 1993.

ANSI, *Safety Code for Design, Construction, and Ventilation of Spray Finishing Operations,* ANSI Z9.3-1985, American National Standards Institute, New York, 1985.

ANSI, *Safety Requirements for Forging Machinery,* ANSI B24.1-1989, American National Standards Institute, New York, 1989.

P. R. Atkins and K. P. Karsten, "Aluminum," in L. V. Cralley and L. J. Cralley, Eds., *In Plant Practices for Job Related Health Hazards Control,* Vol. 1, John Wiley & Sons, New York, 1989.

W. A. Burgess, *Recognition of Health Hazards in Industry,* 2nd ed., John Wiley & Sons, New York, 1995.

Center for Emissions Control, *Solvent Cleaning (Degreasing),* Washington, D.C., 1992.

M. M. Garcia, "Mining and Milling," in L. V. Cralley and L. J. Cralley, Eds., *In Plant Practices for Job Related Health Hazards Control,* Vol. 1, John Wiley & Sons, New York, 1989.

J. N. Lockington and D. L. Webster, "Steel," in L. V. Cralley and L. J. Cralley, Eds., *In Plant Practices for Job Related Health Hazards Control,* Vol. 1, John Wiley & Sons, New York, 1989.

NIOSH, *Control Technology Assessment: Metal Plating and Cleaning Operations,* Publication No. 85-102, Cincinnati, OH, 1985.

NIOSH, *An Evaluation of Engineering Control Methods for Spray Painting,* Publication No. 81-121, National Institute for Occupational Safety and Health, Cincinnati, OH, 1981.

NIOSH, *Occupational Health Control Technology for the Primary Aluminum Industry,* U.S. Department of Health and Human Services, Cincinnati, OH, June 1983.

NIOSH, *Recommendations for Control of Occupational Safety and Health Hazards — Foundries,* Publication No. 85-116, Cincinnati, OH, 1985.

Proceedings of a Conference on the Health Effects of Cutting Oils and Their Controls, University of Birmingham, England, April 28–29, 1989.

R. C. Scholz, "Sand Cast Foundry," in L. V. Cralley and L. J. Cralley, Eds., *In Plant Practices for Job Related Health Hazards Control,* Vol. 1, John Wiley & Sons, New York, 1989.

T. J. Slavin, "Welding Operations," in L. V. Cralley and L. J. Cralley, Eds., *In Plant Practices for Job Related Health Hazards Control,* Vol. 1, John Wiley & Sons, New York, 1989.

T. J. Walker, "Metal Cleaning," in L. V. Cralley and L. J. Cralley, Eds., *In Plant Practices for Job Related Health Hazards Control,* Vol. 1, John Wiley & Sons, New York, 1989.

PROBLEMS

1. Open 30 CFR, Chapter 1 (the first volume) to the table of contents on the second page. Locate safety and health standards for the following: A. metal and nonmetal surface mines; B. metal and nonmetal underground mines; C. surface coal mines; D. underground coal mines.
2. Turn to Part 56 and scan the contents.
   A. In what section do we find the limits on airborne contaminants, and what are these limits?
   B. Where are respirator standards, and what are they? How could you obtain a copy of those standards?
3. A. In each of the four parts located in question 1, locate and compare the noise limit standards.
   B. Compare these with the standards discussed in CFR Chapter 10.
   C. What are the standards for sound level meters?
4. Turn to Part 71 and scan the contents.
   A. Find the respirable dust standards. Where are they?
   B. What is the maximum permitted level of respirable dust?
   C. Where do we find the modification of this standard when quartz is present in the dust, and what is quartz?
   D. At what level of quartz is the need for lower respirable dust levels triggered?
   E. Using the formula presented, what is the maximum permitted respirable dust level when the dust is 10% quartz?
   F. Calculate the maximum concentration of respirable quartz dust permitted in the air when the percent quartz is 4%, 5%, and 6%.
   G. How often is the mine operator required to test the air at each workstation?
   H. Locate Subpart D. This deals with the procedure following an inspection that has uncovered an unacceptably high respirable dust level. What is the mine operator required to do and by when? Specifically who must respond to this submission and how is compliance ensured?

5. Some mines, particularly uranium mines, may contain the radioactive gas radon. Such exposure is dealt with in 30 CFR 57.5037.
   A. How must testing be done? Where can descriptions of these methods be obtained?
   B. What is the trigger level for compliance testing?
   C. What is the maximum permitted working level of exposure and the maximum total annual exposure of worker from this source?
6. Familiarize yourself with 29 CFR 1910.1029.
   A. The chief hazard to workers around coke ovens is exposure to coke oven emissions. What is the definition of coke oven emissions?
   B. What is the hazard of "green plush"?
   C. What problems do coke oven emissions cause?
   D. What is the permissible exposure limit for coke oven emissions?
   E. How should sampling of employee exposure be done?
   F. What is a beehive oven?
   G. What does OSHA require operators of a beehive oven to do regarding employee safety?
7. Find the Part 1910 Index near the end of the volume. Look up "welding".
   A. Where is this information located?
   B. Locate the requirements for ventilation. What is the general requirement for ventilation air flow per welder?
   C. How does this differ from ventilation regulations where a specifically listed toxic metal is involved? Use indoor welding involving cadmium as an example.
   D. What are the requirements for ventilation for welding in a confined space?
8. Locate regulations governing forging in 29 CFR using the index. What special concerns predominate?

CHAPTER 19

Metals II: Details About Specific Metals

Based on the outline of metallurgy presented in Chapter 18, discussion is extended to include uses and purification from ores of additional metals, and to deal with the specific hazards generated. When reading this chapter, recognize that although intake of a compound of a particular metal may cause a given set of toxic symptoms, it is not safe to generalize that all compounds of that metal give the same symptoms. The intake of lead salts generates symptoms quite different from those produced by tetraethyl lead. There is wide variation in the toxicity of the various forms of mercury. Initially, one must look at each metallic compound separately in dealing with hazards accompanying exposure.

One can be exposed to a metal even though that metal does not have a role in a particular manufacturing process. For example, ores generally contain a variety of metals. In fact, a number of metals are obtained as by-products of the extraction of some other material. Industrial metals are therefore seldom pure. One may be alert to the toxic consequences of exposure to the metal that is the object of a particular process, only to find that a more serious threat is posed by an accompanying metal.

An additional source of exposure to metals lies in the fact that fossil fuels contain metals. The burning of oil or coal, particularly coal, adds quantities of metal-rich particulate into the atmosphere. Combustion of such wastes as paper and plastics has a similar potential.

ALUMINUM

Once a precious curiosity, today 18,000,000 metric tons of aluminum are produced annually worldwide, 4,000,000 metric tons in the U.S. alone. Add to primary production in the U.S. 2,200,000 metric tons from recycling. Aluminum is heavily used in construction, automobiles, aircraft, and appliances. Where weight reduction is important, as in aircraft construction, the low density of aluminum is an important advantage. Good heat transfer properties make it useful for parts of internal combustion engines, compressors and radiators. Low toxicity when ingested and ability to be formed in thin flexible sheets has led to its use in food packaging, and its low porosity makes it particularly useful for wrapping frozen foods. Soft drink cans are a major use and a major source of aluminum for recycling. Because of the corrosion resistance imparted by the thin oxide coating that forms on the metal surface, aluminum finds use in items exposed to the weather, such as highway signals, lamp posts, and signs. Powdered aluminum is used in paints and fireworks. Aluminum oxide is found in abrasives and high-temperature brick.

The winning of aluminum from ore was discussed in Chapter 14. The first stage in using this primary aluminum often involves casting the molten metal. Because of the widespread use of aluminum and the high cost in energy of capturing the metal from its ore, recycling aluminum has high priority. As in steelmaking, the scrap is often added to new aluminum in furnaces as the metal is melted for casting operations. Aluminum must be dried by preheating before adding it to the furnace, because even small amounts of water can expand violently as steam in the melt. In furnaces using gas or petroleum, burners present a hazard due to the gaseous combustion products, but also generate a noise problem. Heat stress is a concern around the furnaces.

Oxides and other impurities are removed from the melt by addition of fluxes of sulfates, chlorides, and fluorides, followed by skimming the dross. Some oxides sink to the bottom, so the taphole is located high enough in the furnace to leave these behind. Hydrogen gas, derived from moisture, oil, or grease in the hot melt, dissolves in aluminum and causes metal embrittlement. Chlorine gas is bubbled through the melt through a "wand", forming chlorides of hydrogen, sodium, calcium, and magnesium. The metal salts become part of the dross and are skimmed off. Chlorine gas is corrosive to systems used to transport it, especially at connecting points, and, given the toxicity of the gas, the chlorine transport system must be maintained with great care. The furnace must be adequately ventilated.

Metals such as copper, zinc, magnesium, chromium, beryllium, nickel, silicon, and titanium are added to the melt to produce desired alloys. The molten aluminum is then cast directly into products or into ingots which later are formed into products by rolling, forging, or extruding. The dross and oxide residue at the bottom of the furnace are rich enough sources of aluminum to justify reprocessing in primary aluminum recovery systems. Dross is often wet down when it has been removed. It contains an unpredictable mixture of substances, especially if scrap was added to the melt, and must be treated carefully. Water reacts with nitrides to produce ammonia. Many of the metal salts in dross are water soluble, so runoff must be prevented from soaking into the ground such that groundwater could be contaminated.

The construction and demolition of furnaces require care. Asbestos is often used as a heat shield between fire brick and an iron vessel forming the outside container. The fire brick may contain silica.

Aluminum is forged in a manner similar to the forging of iron (Chapter 18), although temperatures are much lower. Dies are sprayed with lubricants, often lead or tin metal soaps, leading to the potential of exposure to organic or inorganic lead or tin. Cleaning parts after they are formed can expose workers to the cleaning agents. Workers may experience problems from inhalation and dermatitis. Heat stress is possible, and forges have serious noise problems.

There have been problems with fibrosis in the lungs of those working with aluminum oxide (Shaver's disease), particularly from either bauxite dust or dust produced in the manufacture of abrasives. Pulmonary fibrosis also results from exposure to metallic aluminum powders (aluminum dust lung). A sensitization has been reported in response to fumes from aluminum soldering, but this has been attributed to components of the flux, rather than to the aluminum itself. Workers in electrolysis plants have suffered from fluorosis, a problem resulting from exposure to the fluorides of the cryolite bath. This problem has been controlled by measuring the urinary fluoride content of the workers and by monitoring the plant environment (Chapter 18). Problems may arise from the release of coal tar pitch volatiles from Soderberg electrodes. Exposure to these polycyclic aromatic hydrocarbons may increase the incidence of lung cancer.

Aluminum alkyls of many different compositions are finding increasing use in industry as catalysts, for synthesis of other organometallics, and in the production of various organic compounds. These compounds ignite spontaneously, and so represent a fire hazard. The vapors convert rapidly to an aluminum oxide smoke, the inhalation of which can cause metal fume fever. Skin contact with aluminum alkyls causes burns.

ANTIMONY

Antimony is used in alloys as type, bearing surfaces, covering for electrical cable, solder, ammunition, and pewter. Compounds of antimony are found in paints, lacquers, rubbers, glass, and ceramics.

Antimony ores are largely sulfide or oxide deposits. Antimony oxide is volatilized from the ore and is collected in bag houses or precipitators. The oxide is smelted with coke, potash, or soda ash to obtain the free metal. Metallic antimony may also be obtained from sulfide ores by smelting with iron scrap, forming the iron sulfide.

Antimony spots is a skin rash much like a pox disease. Oxides and halides irritate the respiratory tract, eyes, and mouth. Antimony sulfide was associated with heart damage to workers in one case and illness with liver enlargement in others. Pneumoconiosis is reported in workers dealing with the oxide, and studies of smelter workers showed an above normal incidence of lung cancer when exposure levels were higher than standards presently allow. Problems with exposure to antimony as the ore, or during smelting, are made worse by the presence of other metals, including arsenic and lead.

Stibnine, the hydride of antimony and a very toxic gas, may be formed when antimony alloys are dissolved in acid or when lead storage cells are overcharged. Welding, cutting, or soldering are other procedures capable of generating stibnine. Early symptoms of overexposure include headache, nausea, and blood in the urine.

ARSENIC

Arsenic is certainly recognized as a classical poison by people who have never studied toxicology. Its role in the political and marital history of Europe is sizable and secure. Historically, the discovery of an analytical method to detect arsenic poisoning led to a revolution in the legal handling of deliberate poisonings.

Arsenic is used as an alloying agent with lead in type, battery plates, bearings, cable sheathing, and shot for ammunition. Arsenic compounds have been used in pesticides, leading to exposure of workers manufacturing the products and agricultural workers applying them.

It is obtained as a by-product of the processing of other ores. Arsenic trioxide is driven off during the smelting of gold, lead, or especially copper ores, and can be harvested in cooling chambers. The risk of exposure is serious in the smelting operation. The harvesting and transporting of the oxide must be done cautiously, as must repair work to furnaces, cleaning of flues, or other maintenance operations.

Arsenic causes extensive skin problems. Skin irritation is observed as reddening, possible swelling, and an itching or burning sensation. Reddening and swelling around the eyes is likely. Long-term exposure can lead to a mottling or bronzing of the skin. The nasal passages and upper respiratory tract are irritated by exposure, and the destruction of nasal tissue can lead to puncture or perforation of the nasal septum, the division between the nostrils. Less commonly, exposed workers have suffered gastrointestinal upset. Peripheral nerve damage leading to pain, loss of sensation, and weakness in the limbs has occurred as a result of exposure to arsenate sprays.

Arsine is the hydride gas of arsenic, has a garlic-like odor, and can arise in much the same fashion as does stibnine. It is likely to be produced along with stibnine, since arsenic and antimony often occur together. There is danger of arsine forming in the processing of metals in which arsenic is an impurity. For example, washing down the dusts resulting from tin refining is reported to have fatally exposed plant workers to arsine. A similar wetting of dross in a zinc furnace released dangerous levels of the gas. Arsine causes destruction of the red blood cells, resulting in anemia. The release of hemoglobin from the broken cells then can cause kidney damage.

Arsenic compounds have been implicated as carcinogens. Occupational exposure can lead to skin cancers. In addition, workers in copper smelters, where arsenic exposure can be high, or those exposed to pesticides containing arsenic have developed lung cancer to a significantly greater degree. This is more serious in cases where exposure is accompanied by the inhaling of cancer promoters, including cigarette smoke or such irritating industrial gases as sulfur dioxide.

BERYLLIUM

Beryllium finds use as an alloying agent, particularly to strengthen copper, but also with aluminum, magnesium, and steel. Beryllium oxide is used in ceramics. However, the uses of beryllium that have increased most rapidly in recent decades are as a moderator in atomic reactors and as a structural component in space vehicles.

Mining, transporting, and other handling of beryl, the ore of beryllium, does not present great hazard. However, becoming exposed to beryllium dust or fume, as when milling or otherwise working with the metal, presents a serious toxic hazard. Insoluble compounds or particulate metal deposited in the body can produce symptoms at a much later date.

Respiratory problems and skin irritations were early recognized as resulting from beryllium exposure. Chronic beryllium disease was first recognized in workers assembling fluorescent lamps, which then used beryllium phosphors. Respiratory problems can vary from acute irritation, with fluids collecting in the lungs, to a long-term damaging of the lung tissues. Symptoms of lung damage can appear years after exposure to the metal has ceased. Finally, there is evidence that excess exposure to beryllium increases the likelihood of lung cancer. Skin irritation reflects the occurrence of sensitization. Over an extended period of exposure, damage may occur to a number of internal organs.

Recently, the permitted exposure level has been sharply reduced. Other phosphors have replaced beryllium in fluorescent and X-ray tubes.

CADMIUM

About half the cadmium produced is used to apply anticorrosion coatings to metal products by electroplating. Other uses include alloying with other metals to produce bearings and solders, in nickel-cadmium batteries, and in nuclear reactors. Cadmium oxide (red) and sulfide (yellow) are used as pigments in paints.

Cadmium is obtained primarily as a by-product of the smelting of zinc, lead, and copper ores, with zinc ores being the major source. Cadmium is volatilized in the roasting process and is collected in air filters as cadmium oxide. Such by-product cadmium oxide is likely to contain other toxic impurities such as lead, arsenic, and selenium. Sulfuric acid is added to the dust to convert the oxide to calcine, cadmium sulfate. Calcine dissolves in sulfuric acid, but many contaminants precipitate and are filtered out of the solution. Hazards here include exposure to mists of sulfuric acid as well as to compounds of cadmium and its impurities. Eye protection, respirators, and protective clothing are needed.

The greatest risks of routine exposure occur in smelting operations, while calcining, and during the distillation of cadmium sponge. Dust or fume formation associated with bearing or battery manufacture presents a hazard. Workers dealing with such sources of airborne cadmium should wear respirators and coveralls that are laundered at the end of each work day (Table 19.1). Plant exhaust ventilation should be equipped with a system to trap particulate. Plants bubbling hydrogen sulfide gas for precipitating cadmium sulfide as a pigment must monitor levels of H_2S carefully.

Table 19.1. Respiratory Protection for Cadmium.

Airborne Concentration or Condition of Use[a]	Required Respirator Type[b]
10× or less	A half-mask, air-purifying respirator equipped with HEPA[c] filter.[d]
25× or less	A powered air-purifying respirator (PAPR) with a loose-fitting hood or helmet equipped with a HEPA filter or a supplied-air respirator with a loose fitting hood or helmet facepiece operated in the continuous flow mode.
50× or less	A full-facepiece, air-purifying respirator equipped with a HEPA filter, or a powered air-purifying respirator with a tight-fitting half mask equipped with a HEPA filter, or a supplied-air respirator with a tight-fitting half mask operated in the continuous flow mode.
250× or less	A powered air-purifying respirator with a tight-fitting full facepiece equipped with a HEPA filter, or a supplied-air respirator with a tight-fitting full facepiece operated in the continuos flow mode.
1000× or less	A supplied-air respirator with half mask or full facepiece operated in the pressure demand or other positive-pressure mode.
>1000× or unknown concentrations	A self-contained breathing apparatus with a full facepiece operated in the pressure demand or other positive-pressure mode, or a supplied-air respirator with a full facepiece operated in the pressure demand or other positive-pressure mode and equipped with an auxiliary escape-type self-contained breathing apparatus operated in the pressure demand mode.
Firefighting	A self-contained breathing apparatus with full facepiece operated in the pressure demand or other positive-pressure mode.

[a] Concentrations expressed as multiple of the PEL.
[b] Respirators assigned for higher environment concentrations may be used at lower exposure levels. Quantitative fit testing is required for all tight-fitting air purifying respirators where airborne concentration of cadmium exceeds 10 times the TWA PEL ($10 \times 5\ \mu g/m^3 = 50\ \mu g/m^3$). A full facepiece respirator is required when eye irritation is experienced.
[c] HEPA means high-efficiency particulate air.
[d] Fit testing, qualitative or quantitative, is required.
From 29 CFR 1910.1027, Table 2.

Some of the worst accidental exposures have occurred during high-temperature operations where the presence of cadmium in the metals being processed was not suspected. These included welding and cutting cadmium-coated pipes, heating plated rivets, and melting scrap.

Workers chronically exposed to cadmium are likely to show symptoms of kidney damage, especially protein in the urine (proteinuria). Cadmium oxide dusts have produced lung damage, including fibrosis and eventual emphysema. In other studies workers were found to have skeletal problems due to bone demineralization. The extreme example of this is the itai-itai disease, which occurred in Japan among people exposed to cadmium wastes in their water. An increased occurrence of cancer, especially prostate cancer, has been reported in workers exposed to cadmium or cadmium oxide. Cadmium has a very long half-life in the body, so that continuous exposure, even to very low levels, must be avoided.

Concern about cadmium health effects led OSHA to cut the PEL (TWA) for exposure to 5 $\mu g/m^3$ of air in 1992. Separate engineering control air limits apply to nickel–cadmium battery manufacturing, zinc–cadmium refining, cadmium pigment manufacturing, incorporating cadmium as a plastic stabilizer, and lead smelting (Table 19.2). These standards recognize difficulties in achieving the PEL in those industries through engineering controls and work practices, but in all permissible levels are lower than in the previous standard. Below the action level (2.5 $\mu g/m^3$) only training on cadmium hazards is necessary, but exposure above the action level requires exposure monitoring and medical surveillance.

Table 19.2. Cadmium: Separate Engineering Control Airborne Limits (SECALs) for Processes in Selected Industries.

Industry	Process	SECAL (μg/m³)
Nickel–cadmium battery	Plate making, plate preparation	50
	All other processes	15
Zinc–cadmium refining*	Cadmium refining, casting, melting, oxide production, sinter plant	50
Pigment manufacture	Calcine, crushing, milling, blending	50
	All other processes	15
Stabilizers*	Cadmium oxide charging, crushing, drying, blending	50
Lead smelting*	Sinter plant, blast furnace, bag house, yard area	50
Plating*	Mechanical plating	15

*Processing in these industries that are not specified in this table must achieve the PEL using engineering controls and work practices as required in f(1)(i).
From 29 CFR 1910.1027, Table 1.

CHROMIUM

Chromium ore, chromite, is no longer mined in the U.S. The ore is reduced with carbon or silicon in an electric furnace to produce an iron alloy called ferrochromium. Several techniques are then employed to obtain either the pure metal or a variety of chromium compounds.

Chromium is an alloying agent in producing stainless steels, electrical resistance wires, and cutting tools. Chromium is plated from a chromic acid bath containing sulfuric acid to form decorative and protective metal surfaces. Chromates and dichromates have a variety of applications such as tanning, dyeing, photography, and as zinc chromate in primer paint.

Hexavalent salts are irritating and destructive to tissue. Mists from electrolysis baths and plating baths cause dermatitis and damage to nasal membranes. Skin irritation and ulceration also trouble lithographers, painters using chromium-containing pigments, workers in magnesium foundries (castings are chromate treated to improve weathering), workers handling chromium-containing cements or chromium-treated timber, and welders working with stainless steels or using welding rods containing chromium. When dusts, fumes, or mists are inhaled, irritation problems extend to the lungs, and there are numerous reports of cancer in the respiratory tract from hexavalent chromium exposure. Over the years, the worst exposure problem in the U.S. has been from the chromate–chromite mixture intermediate in chromate preparation. Workers have suffered bronchial cancer on exposure to this mixture. An acute oral dose causes kidney damage. Because of the toxic nature of chromium plating bath contents, disposal must be done with regard to its serious potential for environmental damage.

COBALT

Cobalt is a by-product of mining copper and other metals. It is used largely in alloys in such applications as aircraft, turbines, and magnets. There are catalysts containing cobalt, and the oxide is used as a coloring agent in glass. A toxicologically significant application is as a binder for tungsten carbide and titanium carbide abrasives. Any abrasive is likely to become particulate in the air when used.

Cobalt is a by-product of ores of copper and nickel. Smelted ores produce matte, a mixture of the three metals. Crushed and ground matte is leached with hot acid to produce a solution of nickel and cobalt, possibly including copper. Dusts from this operation are hazardous, so workers should have eye and respiratory protection and should wear coveralls. Noise both from crushing and from leaching operations is a problem, and the leaching process may lead to some heat stress. If present in the leached solution, copper is removed, and caustic plus an oxidizing agent precipitates cobaltic hydroxide containing some nickel. Treatment with ammonia and an oxidizer completes the separation of cobalt and nickel. All these chemicals are hazardous, and handling, ventilation, and personal worker protection should be appropriate to the risk. At high pressure and temperature, hydrogen is used to reduce cobalt to the free metal.

Cobalt metal fumes and dust cause nose, throat, eye, and skin irritation. Milling and abrasive tool manufacture exposes workers to dust that produces pneumoconiosis. Workers also experience a dermatitis from abrasive manufacture or handling cobalt alloys, clays, or pottery. Cobalt in cement has been reported to cause skin irritation.

COPPER

Copper is used in a vast range of applications including wiring, plumbing, roofing, and cookware manufacturing. These depend on the high electrical and heat conductivity of copper and its good weathering properties. Many alloys of copper are produced, far more than just the familiar brass and bronze.

At one time a surprising proportion of copper mined in the world came from underground deposits of metallic copper in northern Michigan. A huge boulder of copper metal from these deposits is on display in the mall in Washington, D.C. Today, however, most copper is surface mined as low-grade sulfide ores. These are crushed and concentrated by flotation. Flotation concentrates not only the copper, but other, more toxic metals found in the ore such as arsenic and lead. Roasting may be employed before smelting, but more often today the "green charge" is added directly to the smelting furnace. In roasting, fluxes such as limestone are added to the ore and it is heated at just under 1000°F. In the smelter at above 1800°F iron and copper combine into a matte, which is drawn off into ladles to be treated in the converter. Here a flux is added and streams of air are blown through the melt using a number of tubes called tuyeres. Iron sulfide oxidizes first, and the iron forms a slag with the flux. The copper metal is freed from the sulfide and is collected in a form more than 95% pure. Further purification is accomplished by electrolysis.

Copper may also be extracted by electrowinning from appropriate ores. The ore is leached with dilute sulfuric acid as $CuSO_4$. Copper is stripped from this dilute solution into a kerosene solution by a phenolic agent, then reextracted into sulfuric acid as a concentrated solution. Copper is plated out of this solution and the acid is recycled back through the extraction process. The formation of arsine during this process is a concern, so sampling and analysis should be performed.

Mining and processing carries risks of lung damage from dusts. Noise exposure is high around crushing, grinding, and screening operations. In roasting, smelting, and converting, air may be contaminated with SO_2 and various toxic metals, so good ventilation and monitoring to assess worker exposures is necessary. Burners produce high noise levels and should be enclosed. At the converter, the tuyeres and any leakage of compressed air connected to them add to the background noise level, and clearing the tuyeres when cooled copper blocks them is a particular noise problem.

Electrolytic baths can expose workers to acid mists. Cutting, welding, and otherwise raising the metal to high temperatures increases the danger of metal fume fever and allergic dermatitis. Damage to the upper respiratory tract from metal fume and dust takes the form of irritation and atrophy of the nasal mucosa. Complaints of irritation from exposure to dusts of copper oxide and other copper salts have been recorded.

Finally, it must be noted that even at very low concentration copper salts are lethal to lower life forms. Copper salts are sometimes used to reduce snail populations in lakes, and mussel populations have been destroyed by levels of copper too low for analysis by atomic absorption. Great care must be taken concerning release of copper compounds into the environment.

INDIUM

Indium is used as a protective coating on bearings, in solders, and in a variety of high-technology roles including special mirrors, transistors, and infrared detectors. It is often found in the ores of other metals, particularly of zinc. The best source of indium is the flue dusts of zinc smelters. Exposures are possible in the course of plating and manufacturing the metal. Many hazards have not been well studied, and the literature on indium problems is sparse.

LEAD

Storage batteries are the largest single use of lead. The production of tetramethyl lead and tetraethyl lead as antiknock additives in automobile fuels was a major use of lead. However, this industry has been phased out in the U.S. due to environmental restrictions, at first because lead poisons the catalytic converters required in automobiles, then later because of concerns about levels of lead in the atmosphere. Alloyed with antimony and tin, lead is used as a sheath to weather proof electrical cables, as type metal, and in bearings. Lead is found in pigments, in solders, and in ammunition. The use of lead in shot has been vigorously and, more recently, successfully opposed by environmental groups because of long-range negative effects on wildlife. Paints once used high levels of lead, and this residue of lead on older buildings is still a hazard to workers doing maintenance or demolition.[1]

Considering levels of exposure and levels necessary to produce toxic effects, the public is at greater risk from lead than from any other metal. In the general public, the body burden of this metal is a sizable fraction of those levels at which symptoms of damage are first seen. Because of this high background, any further exposure of individuals as the result of employment should be carefully monitored and reduced to a minimum level. With the phasing out of lead additives in gasoline and other measures taken to reduce lead exposure, the levels of lead in the U.S. public have been slowly dropping. Blood lead levels and levels of intermediates in heme synthesis are common measures of the lead body burden.

The toxic effects of lead on the body (plumbism) are varied. The nervous system is most sensitive to damage. The symptoms of lead poisoning include convulsions, hallucinations, coma, weakness, and tremors. Lead palsy is seen first as wrist drop, on the right in right-handed persons. Damage to blood cell forming systems and increased fragility of the red blood cells results in anemia. Kidney damage is also commonplace in lead poisoning. The most common symptom of overexposure to lead is intestinal colic. Constipation is followed by intense abdominal pain. Once exposed to lead, the body carries a residue for a long time. It is incorporated into the bones and is only released from there very slowly. Lead causes kidney cancer and brain tumors in test animals. Recent studies in Finland confirmed a higher incidence of gliomas, a type of brain tumor, in workers with higher lead exposures.[2]

Organic lead compounds, particularly tetramethyl lead and tetraethyl lead, present a different pattern of toxicity. There is no colic, but signs of nervous system damage abound. These include insomnia, restlessness, hallucinations, and delusions.

Lead ore is galena, a sulfide ore, which is most often removed from underground mines.[3] A variety of other metals including tin, bismuth, arsenic, silver, gold, antimony, and copper

[1] Children in old houses are frequent victims of lead poisoning when they put paint chips in their mouths.
[2] *Occupational Health and Safety*, May 1996, p. 16.
[3] Carbonate and sulfate ores are also found.

are found in these ores. In fact arsenic, antimony, and bismuth are usually obtained as by-products of the processing of lead ores.

Lead ore is crushed, ground, and concentrated by flotation. Given the toxicity of lead, control of dust by wet spraying and ventilation during these operations, during transportation, and in sampling operations is essential. Noise stress is also a problem. Ore is sintered, heated with flux to burn off sulfur. The fused sinter is crushed and becomes the raw material for the blast furnace. Sinter, iron, and coke are mixed in the blast furnace and air enriched in oxygen enters the bottom of the furnace. The molten product is poured into containers where slag is skimmed from the metal product. The slag is often used as a source of zinc.

Additionally, the recycling of automobile batteries and solders provides more than a third of the lead used in the U.S. In a process called secondary smelting, old batteries are broken up and the lead recovered. Exposures may occur to lead and to sulfuric acid in this process. Further refining is similar to that performed on smelted lead ore.

Lead is melted at a controlled temperature in a drossing plant, and impurities which melt at higher temperatures, such as copper and arsenic, solidify and float to the top. Additions to the kettle separate other impurities, and the dross is removed and remelted. It then separates into three liquid layers. Matte is a copper–iron mixture and speiss is an arsenide layer. These are shipped to a copper smelter, and the third layer, lead, is returned to the kettle. Further refining removes such impurities as silver, antimony, tin, copper, tellurium, zinc, bismuth, and gold by adding various agents to the molten lead and removing the resultant slag.

Chief potential respiratory exposures of concern are to metallic dusts and to carbon monoxide from the furnaces. Rules for respiratory protection against airborne lead are shown in Table 19.3. Arsenic requires special monitoring if present. Noise from burners and heat stress are also concerns. Good housekeeping is important, and employees must be trained in the hazards of the materials being processed. Besides inhalation, ingestion of particulate is a threat. Food, beverage, and tobacco products must not be allowed in the work area, and employees should clean and change before going to the lunchroom.

Table 19.3. Respiratory Protection for Lead Aerosols.

Airborne Concentration of Lead or Condition of Use	Required Respirator[a]
Not in excess of 0.5 mg/m^3 (10× PEL)	Half-mask, air-purifying respirator equipped with high-efficiency filters[b,c]
Not in excess of 2.5 mg/m^3 (50× PEL)	Full-facepiece, air-purifying respirator with high efficiency filters[c]
Not in excess of 50 mg/m^3 (1000× PEL)	(1) Any powered, air purifying respirator with high-efficiency filters[c], or (2) half-mask supplied-air respirator operated in positive-pressure mode[b]
Not in excess of 100 mg/m^3 (2000× PEL)	Supplied-air respirators with full face piece, hood, helmet, or suit, operated in positive-pressure mode
Greater than 100 mg/m^3, unknown concentration or fire fighting	Full-facepiece, self-contained breathing apparatus operated in positive-pressure mode

[a] Respirators specified for high concentrations can be used at lower concentrations of lead.
[b] Full facepiece is required if the lead aerosols cause eye or skin irritation at the use concentrations.
[c] A high-efficiency particulate filter means 99.97% efficient against 0.3-μm size particles.
From 29 CFR 1910.1025, Table II.

One hazard for exposure occurs when spraying lead-based paints. One of the common lead pigments, white lead, is giving way to zinc oxide and titanium oxide. However, red lead in metal primers continues to find use because of its superior properties with respect to the inhibiting of corrosion. Hazard due to lead-based paints continues after the painting process is done. The removal of these paints by abrasion or burning releases harmful particulates into the air. Other sources of hazard from lead include grinding or sanding soldered surfaces and manufacturing batteries. Occasional serious exposure has resulted from cutting or welding alloys containing lead or cutting up structures painted with lead-based paint using a torch. Any process involving the manufacture or handling of organic lead compounds must be done with extreme care, because these compounds are highly toxic, particularly to the nervous system.

MANGANESE

Manganese oxide (pyrolusite) is the chief ore of manganese. Its primary use is as an alloying metal, particularly in steel. Manganese dioxide is used in the production of dry cell batteries.

Exposure occurs through dusts in mining, transportation, crushing, and sieving ores. There are dust and fumes near reduction furnaces. Chronic exposure can produce a pneumonia-like problem with possible fibrosis in the lungs. Manganism is a disease of the central nervous system resulting from absorption of manganese through the digestive tract from the swallowing of dusts moved up from the lungs over a period of time. Symptoms range from headaches, cramps, apathy, weakness, and insomnia to psychosis with delusions and hallucinations.

MERCURY

Mercury is obtained from the sulfide ore cinnabar. It is used in types of electrical apparatus and in caustic soda plants as part of the electrode. The ability of mercury to form solutions or amalgams with other metals is used to extract gold and silver from ores. As a catalyst in urethane foam production, it remains in the foam product. Mercury has been added to paints to prevent mildew and in marine applications to inhibit growth on boat hulls. Mercury compounds have been used to discourage mold production in paper and pulp and on seed grain.

From the toxicologist's standpoint, mercury is found in three forms that have different health effects. First, hazards from mercury include inhalation of metal vapors. These can occur in work areas containing liquid mercury and in an unsuspected manner by evaporation of spilled mercury that has soaked into crevices in floors. Special risks occur when mercury vapor is driven off an amalgam by heating. This can occur during the application of gold to objects by dipping them in a gold amalgam, then heating to remove the mercury. Metallic mercury has a high attraction to the central nervous system and causes a broad range of symptoms from headache, tremors, weakness, insomnia, or drowsiness to emotional and psychotic disturbance with extreme excitability and irritability. Early symptoms of mercury poisoning include salivation and tenderness of the gums. Second, mercury salts often cause skin problems. This problem is particularly troublesome in handling mercury fulminate, which is used as a detonator in explosives. Third, alkyl mercury compounds present special problems of somewhat different character than those of the inorganic mercury compounds. Alkyl mercury compounds are used to kill molds, especially in stored seeds. They are easily absorbed and cause serious damage to the nervous system. Motor control is affected and narrowing of the visual field to the point of complete blindness occurs. The damage is long lasting or permanent.

NICKEL

About half of all nickel produced is used in steel making. Most of the rest is used in other alloys or is plated onto other metals. Lesser uses include: as a component of nickel–cadmium batteries, as catalysts, and as a coloring agent in glass.

Canada is the major source of nickel. Mining and purification was partly covered in the discussion of cobalt. A variety of flotation and magnetic separation techniques are used to purify the ore. Nickel sulfide is reduced using coke. Purification of nickel can be done electrolytically or by conversion to nickel carbonyl using carbon monoxide, followed by deposition onto nickel pellets.

Hazardous exposure to fumes and dusts has occurred during high-temperature processing of nickel sulfide ores into nickel oxide, resulting in nasal and lung cancer. Not all ore sources seem to present this threat. Nickel carbonyl, which is sometimes an intermediate in the refining of nickel, is a very toxic gas that causes a hemorrhagic pneumonia.

Nickel platers suffer from a dermatitis caused by skin contact with nickel salts. There is also a chronic eczema that is not necessarily produced at the point of contact. High temperatures around the plating baths contribute to the risk. Some individuals are more susceptible to becoming sensitized to nickel than others, and once sensitized they respond even to contact with nickel alloys.

TIN

Major uses for tin include tinplate, solder, and copper alloys. The latter include brass, bronze, babbitt (bearing metal), and pewter. Organotins are used as catalysts, polymer stabilizers, antimicrobials, and in marine antifouling paints.

Not much tin is mined in the U.S., and about one-quarter of the supply is obtained by recycling. Most ores are sulfides, which are smelted and refined. Long contact with tin oxide dust or fumes from smelting operations causes stannosis, a lung deposition problem that produces little fibrosis and little in the way of serious symptoms.

Organotins are likely to be irritating to skin and eyes. The trimethyl, triethyl, and tributyl tins are particularly toxic. The EPA has proposed restrictions on the use of tributyl tin in antifouling paints, because it is harmful at levels of less than 1 ppb to nontarget fish and shellfish. Presently, 624,000 gal of such paint are sold annually.

TITANIUM

Pigments containing titanium oxide are used in paints, papers, plastics, rubber, and ceramics. The metal is used in aircraft and missiles, where a high strength-to-weight ratio makes it a useful structural material. Titanium cooling coils are used in power plants.

Titanium is a very abundant metal, the chief ores of which are oxides. Isolation is by conversion of the oxide to a chloride, then reduction of the chloride to the metal. Pure metal is obtained using a consumable titanium electrode in an electric arc furnace.

Titanium chloride causes skin and eye burns, and serious lung damage if inhaled. The oxide seems to be relatively innocuous when inhaled. Reports of problems may relate to other compounds mixed with titanium oxide.

TUNGSTEN

Most of the tungsten produced is used in very hard, wear-resistant alloys. These are used largely in cutting tools. Tungsten carbide is an important abrasive. The chief ores of tungsten are tungstate salts.

Lung problems (hard metal disease) experienced by workers with exposure to dusts from tungsten carbide abrasives cemented in place using cobalt. These have already been mentioned under the cobalt heading. Reports of pulmonary problems are numerous. Removal from exposure is usually a sufficient remedy when symptoms such as coughing and wheezing first appear. However, the correlation between length of exposure and severity of lung damage is poor, and in more advanced stages the disease is progressive and can lead to death. In some workers, early symptoms indicate sensitization. This may be a factor in the progressive form of the disease. Exposure to the dust also produces skin irritation, inferring sensitization of the skin.

URANIUM

Uranium is primarily used to produce fuel for nuclear reactors and to produce weapons. Much of the mining of uranium ores in the U.S. occurs in western states. Early speculators, hopeful of quick riches, established a large number of small uranium mining operations. As the industry matured these often poorly ventilated small mines gave way to a relatively small number of much safer operations. The uranium is leached from ore using acidic ferric sulfate. In the process, the ferric ion is reduced to ferrous ion. Its reoxidation, allowing its reuse, is accomplished using a microorganism. The uranium is electrolytically reduced and is precipitated as green cake (uranous fluoride) by the addition of hydrofluoric acid.

Mining hazards include exposure to such radioactive species as uranium itself, radium, or radon gas. Inhalation of these radioactive species increases the risk of lung cancer. Chemical exposures are less important and include exposures to other metals found with the uranium such as vanadium, lead, thorium, manganese, and arsenic. The silica rock of western mines adds the risk of silicosis. Because the ore is processed by wet techniques the risks are relatively low. Uranium hexafluoride, a gas produced for isotopic separation in the gas centrifuge, is hazardous if a leak should occur in the apparatus handling this compound.

VANADIUM

Alloys are the chief use of vanadium. The ores are widely distributed and the metal is usually extracted by leaching. Other metals are usually produced along with vanadium. In one process, uranium is precipitated from the leach solution first, followed by vanadium. In another, vanadium is leached from iron ore or from the slag produced in steel making. Metallic vanadium is produced by carbon reduction of the vanadium oxide or by a calcium metal reduction process. The oxide is also reduced using silicon or aluminum.

The greatest hazards of mining vanadium are exposures to uranium and radon in the case of carnotite ores and to the silica in the rocks of the mines. Much of the extraction of vanadium is a wet process, minimizing the respiratory exposure risks. Vanadium oxide can produce a respiratory sensitization with inflammation, irritation, and sometimes pneumonia. Sensitization of the skin produces redness, itching, and eruptions.

A problem not related to the production or use of vanadium occurs in the case of oils with high vanadium content. Workers cleaning burners, or dealing with the oil ash of residual oils, display an inflammatory respiratory problem.

ZINC

Most zinc is used in automobiles. Large amounts are used in building products and appliances, and there are numerous lesser uses. Frequently, the use of zinc is to coat iron or steel in a corrosion-preventing process called galvanizing. Zinc compounds are used in paint and in rubber.

Zinc ore is largely mined underground. The ores are enriched, sulfides are roasted to convert them to oxides, and the oxides are smelted with coke or coal, or leached for electrolytic deposition.

Zinc oxide fume causes fume fever. Reports also have appeared of a gastrointestinal disturbance from zinc oxide exposure. Other metals associated with zinc, such as arsenic, cadmium, lead, and manganese, add significantly to risks. For example, dissolving the zinc in acid or alkali in a leaching operation can lead to the release of arsine gas.

KEY POINTS

1. Extraction processes thati nvolve heating in furnaces create the hazards of heat stress, exposure to combustion products such as CO, and noise from the burners.
2. Ores are often sources for more than one metal. In a given process exposure to an unwanted metal can be a hazard.
3. Airborne aluminum, aluminum oxides, and fluoride compounds are special problems in aluminum purification and processing. Aluminum alkyls are highly reactive.
4. Antimony causes a rash, and its oxides and chlorides are irritating if inhaled. Stibnine is a very toxic gaseous hydride of antimony.
5. Arsenic trioxide, produced in the smelting of other metals, is hazardous. Arsenic compounds cause peripheral nervous system (PNS) damage and GI tract problems and are skin and eye irritants. Some compounds are carcinogens. Arsine, the highly toxic gaseous hydride of arsenic, may be produced accidentally in the purification of metals containing arsenic as an impurity.
6. Beryllium causes skin and lung irritation. It may accumulate to cause chronic problems.
7. Cadmium is a common commercial metal that is very damaging to kidneys. Its oxide causes emphysema, and in extreme cases bone demineralization is reported. It may be carcinogenic, and has a very long half-life in the body.
8. Chromium salts are very irritating to skin and lungs, and some are carcinogenic.
9. Cobalt metal fumes cause irritation to nose, eyes, and skin.
10. Copper metal fumes and dusts damage the upper respiratory tract.
11. Indium is not well studied.
12. Lead is broadly dispersed in the environment, and most people already carry a relatively high body burden. It causes nervous system damage (CNS and PNS), kidney damage, and colic. Lead alkyl compounds are potent CNS poisons.
13. Manganese exposure causes CNS and lung damage.
14. Mercury is highly toxic in some forms, causing serious CNS damage.
15. Nickel is a carcinogen. Salts cause skin irritation and can be sensitizers.
16. Tin can cause lung fibrosis, and organotins are very irritating.
17. Titanium is low in toxicity, but titanium tetrachloride is very irritating.
18. Tungsten particulate causes lung problems and may be a skin sensitizer.
19. Uranium hazards focus strongly on its radioactivity and that of compounds found with it.
20. Vanadium oxide can cause lung sensitization and skin irritation.
21. Zinc oxide causes metal fume fever.

REFERENCES

D. J. Burton and J. P. Sieverson, "Primary and Secondary Smelting — Lead, Zinc and Cadmium," in L. V. Cralley and L. J. Cralley, Eds., *In Plant Practices for Job Related Health Hazards Control,* Vol. 1, John Wiley & Sons, New York, 1989.

C. E. Dungey, "Copper," in L. V. Cralley and L. J. Cralley, Eds., *In Plant Practices for Job Related Health Hazards Control,* Vol. 1, John Wiley & Sons, New York, 1989.

L. Friberg, C. G. Elinder, et al., *Cadmium and Health: A Toxicological and Epidemiological Appraisal, Vol. II, Effects and Response,* CRC Press, Boca Raton, FL, 1986.

R. A. Goyer, "Toxic Effects of Metals," in C. D. Klaassen, M. O. Amdur, and J. Doull, Eds., *Toxicology — The Basic Science of Poisons,* 3rd ed., Macmillan, New York, 1986.

NIOSH, *Occupational Hazard Assessment Criteria for Controlling Occupational Exposure to Cobalt,* DHEW 82-107, October 1981.

OSHA, *Prudent Practices for Controlling Lead Exposure in the Secondary Lead Smelting Industry — A Guide for Employers and Employees,* U. S. Dept. of Labor, Washington, D.C., 1981.

B. R. Roy and S. A. Thielke, "Cobalt-Nickel Refining," in L. V. Cralley and L. J. Cralley, Eds., *In Plant Practices for Job Related Health Hazards Control,* Vol. 1, John Wiley & Sons, New York, 1989.

U. S. Department of Health and Human Services, *Occupational Respiratory Diseases,* DHEW 86-102, September 1986.

T. J. Walker, "Aluminum Metalworking," in L. V. Cralley and L. J. Cralley, Eds., *In Plant Practices for Job Related Health Hazards Control,* Vol. 1, John Wiley & Sons, New York, 1989.

O. Wong, D. F. Liart, and R. W. Morgan, *Critical Evaluation of Epidemiological Studies of Nickel-Exposed Workers,* Environmental Health Associates, 1983.

PROBLEMS

1. Outline the conversion of scrap soft drink cans into a casting for an air conditioner compressor, indicating hazards to the workers.
2. Summarize where workers are likely to be exposed to arsenic, including exposure not related to arsenic production.
3. In what industries is chromium exposure likely in the U.S.?
4. When cleaning equipment used in the leaching of cobalt, what are the hazards and how may workers be protected?
5. Locate the standards for working with lead using the index of 29 CFR 1900–1910.
   A. What is the PEL for lead?
   B. What must be the accuracy of the monitoring device?
   C. Who must have blood lead levels measured, and how often?
   D. What must be the accuracy of the lead blood level measuring device?
   E. How frequently must employees get a medical exam, and what must be included?
6. What metals are suspected or proven carcinogens?
7. Locate the specific regulations for cadmium exposure, and determine how cadmium is classed as a carcinogen causing lung or prostate cancer.
8. Locate the medical surveillance and monitoring requirements for workers exposed to lead.

CHAPTER 20

Polymers

Polymers are composed of extremely high-molecular-weight molecules, which are assembled by joining together chemically large numbers of relatively small molecules, monomers, in a process called polymerization. Many components of living systems are polymers: proteins, polysaccharides, and nucleic acids. A very early human industry involved the utilization of some of these natural polymers, including wool (a protein) and cotton (a polysaccharide), to produce clothing. Only relatively recently has chemical technology reached the stage of generating a variety of synthetic polymers. In 1869 celluloid, a nitrate derivative of the plant polysaccharide cellulose, was the first plastic invented, and in 1909 bakelite, a phenol formaldehyde polymer and the first completely synthetic plastic, was patented. Once launched, the growth of production and use of synthetic polymers has been impressive. The production of plastics, elastomers, and fibers tripled in the decade following 1963.[1] Large industries have grown around artificial polymers or commercial utilization of natural polymers. This chapter focuses attention on some of these industries.

PLASTICS

Plastics production is a giant industry by any measure. In 1994, more than 72 billion pounds were produced in the U.S., with an annual growth rate in production of about 4%. More than 100 commercial polymers and numerous copolymers are produced, but a relatively small number, shown in Table 20.1, represent most of the market. According to Chem Systems,[2] packaging companies use the largest share of plastics (28%), followed by the construction industry (22%). In fact, plastics have moved into every aspect of daily life, replacing wood, metal, glass, oil-based paint, traditional adhesives, paper, and natural rubber. Substitution of plastics for metals in automobiles has improved styling, eliminated some corrosion problems, improved crash safety, and reduced weight to improve fuel economy. Uses such as films represent new areas of materials consumption. The disposal of all this plastic material is of increasing concern, the EPA estimating that by the end of the century, plastics will comprise 9.8% of municipal wastes. Increasing effort is directed toward methods for plastics recycling.

[1] *Chem. Eng. News*, May 6, 1974.
[2] *Chem. Eng. News*, August 24, 1987.

Table 20.1. Leading U.S. Plastics Products in 1995.

Plastics	Production in Billions of Pounds	Growth (1985–1995)
Thermoplastics	60.8	4%/year
Polyethylene		
Low-density	12.9	
High-density	11.2	
Polyvinylchloride and copolymers	12.3	
Polypropylene[a]	10.9	
Polystyrene and copolymers	8.7	
Polyester	3.9	
Nylon	1.0	
Thermoset Resins	7.5	3%/year
Phenolics	3.2	
Urea resins	1.8	
Polyesters	1.6	
Epoxies	0.6	
Melamine resins	0.3	

[a] Includes Canadian production.
Data modified from *Chem. Eng. News,* June 24, 1996.

THE CHEMISTRY OF PLASTICS

Plastics are structural polymers and generally are synthesized using petroleum starting materials. There are, however, some important plastics based on modification of cellulose, a natural polymer, and silicones are based on silicon, rather than carbon, structures. In order to understand the hazards associated with plastics manufacturing operations, it is useful to understand some of the basic chemistry involved.

Linear and Thermoset Polymers

Two classes of plastics result from the polymerization process. Thermoplastics are linear polymers. This means the monomers are joined linearly, like beads on a string. The final solid plastic could be likened to a bowl of spaghetti, a mass composed of long strands packed together (Figure 20.1). The name *thermoplastic* arises from the fact that on heating (*thermo*) the material softens (*becomes plastic*). This is because the individual strands of hot polymer are capable of sliding past one another. Such a plastic can be heated, then molded or extruded into desired shapes. On cooling, the mass becomes rigid once again. Such a process can be repeated, so scrap can be remelted and formed again.

The other class of polymers, the thermoset plastics, connect the monomers into a three-dimensional grid (see Figure 20.1B). It is possible that once formed, any given part of such a plastic object could be connected to any other part through a continuous succession of chemical bonds, so in a sense a thermoset plastic object is like a single molecule. Thermoset plastics are very rigid. Any attempt to change the shape of the finished object requires massive breakage of chemical bonds, thus destruction of structural integrity. On heating, thermoset plastics retain their shape until the temperature is high enough to destroy chemical bonds. In the presence of air this results in charring as the carbon structures oxidize.

Linear polymers can be altered to assume thermoset properties by cross-linking or curing the plastic. The linear strands are bridged by a chemical agent to assemble the strands into larger units. By performing such cross-linking to varying degrees, properties result that fall anywhere in the range from a meltable and flexible thermoplastic to a rigid and hard thermoset. A familiar example is latex, which is found as the weak, extremely elastic uncross-linked linear polymer of rubber cement, as the still elastic but more rigid product in rubber bands, and as the much more rigid and abrasion-resistant material found in an automobile tire. The

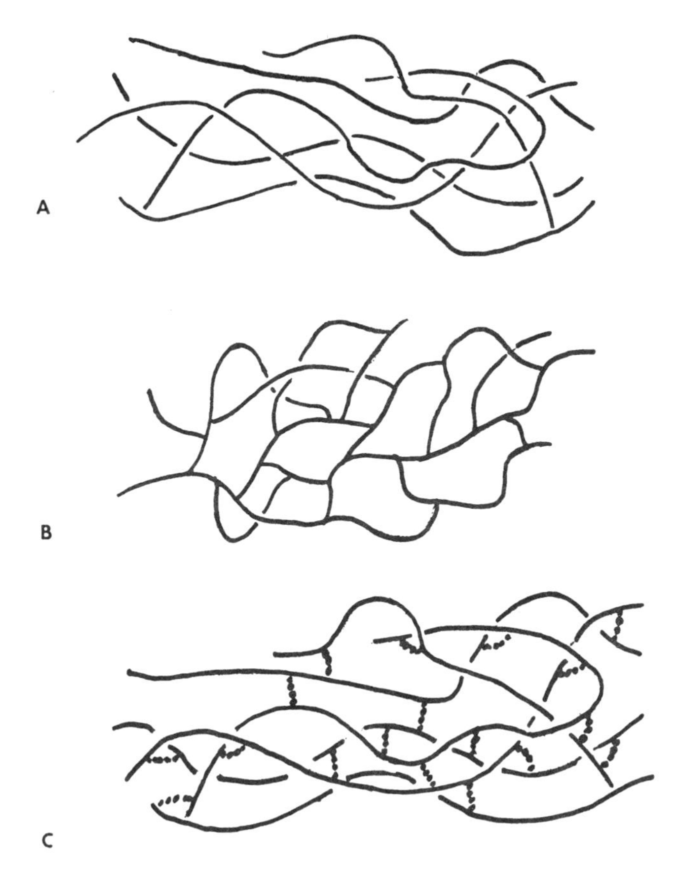

Figure 20.1. Here we see three physical forms of plastic. (A) A typical thermoplastic in which long strands are mixed, but are independent of one another. On heating these strands become more mobile, and the shape of the polymer mass changes as they slide past one another. (B) A thermoset plastic. Strands are interconnected into a three-dimensional array, and so have no ability to slide past one another. Such structures remain rigid until the temperature is high enough to break chemical bonds. (C) The same thermoplastic structure as (A), except that the plastic has been chemically cross-linked (cured). Physically this product more closely resembles (B) than (A).

difference lies in the degree of curing. This is an example of fine tuning the chemistry of polymers to obtain exactly the compromise one wishes in final properties.

Addition and Condensation Polymers

The chemical reactions used to join monomer units fall into two classes. The common feature of monomers of addition polymers is the presence of a carbon-to-carbon double bond. A number of addition monomers are shown in Table 20.2. On polymerization, the double bond becomes a single bond. The resulting addition polymers have a continuous row of carbon atoms joined by single bonds, and differ only in the specific groups sticking out from this carbon backbone. Addition polymers are linear polymers.

Condensation polymers are the other major class of polymers. In order to form a condensation polymer, it is necessary to have two chemical groups that react to join to one another. For example, carboxylic acids react with alcohols to form esters and with amines to form amides. Furthermore, it is necessary to have at least two such groups on each monomer. The formation of nylon-6,6 involves the use of two monomers, one with two carboxyl groups and the other with two amine groups. If one carboxyl group reacts with one amine group, two monomers are now joined by an amide linkage. However, at one end of the combination

Table 20.2. Commercial Addition Polymers.

Polymer	Commercial Name	Monomer
Polyethylene	Polythene	Ethene, ethylene
Polypropylene		Propene, propylene
Polystyrene	Styrofoam	Styrene
Polyvinylchloride	Vinyl	Vinyl chloride
Polyacrylonitrile	Orlon	Acrylonitrile
Polymethylmethacrylate	Lucite™, acrylic plastic	Methyl mathacrylate
Polytetrafluoroethylene	Teflon®	Tetrafluoroethylene

there is an unreacted carboxyl group available to react with another amine, and at the other end is an unreacted amine available to react with another carboxyl group. If one of the monomers were to have three reactive groups, the polymerization reactions would form the thermoset type of structure. Table 20.3 provides examples of condensation polymers.

Table 20.3. Some Commercial Condensation Polymers.

Polymer	Monomer
Nylon-6	Caprolactam
Nylon-6,6	1,6-Diaminohexane, adipic acid
Polycarbonate	Bisphenol a, phosgene
Polybutylene terephthalate	1,4-Butanediol, dimethylterephthalate
Polytetramethylene terephthalate	
Polyethylene terephthalate	Ethylene glucol, dimethylterephthalate
Whooly aromatic copolyester	*p,p'*-Dihydroxybiphenyl, terephthalic acid
Aromatic polyester	Bisphenol a, terephthalic acid
Polyurethane	Bisphenol a, toluene diisocyanate, polyglycol
Urea-formaldehyde	Urea, formaldehyde
Melamine-formaldehyde	Melamine, formaldehyde
Phenol-formaldehyde	Various phenols, formaldehyde

THE RELATIONSHIP OF CHEMICAL STRUCTURE TO PHYSICAL PROPERTIES

Major differences exist between thermoplastic and thermoset polymers or cross-linked thermoplastics. However, considering just uncross-linked thermoplastics, such properties as strength, elasticity, and hardness differ according to the chemical structure of polymers in interesting ways. Most significant factors are the ability of two strands to attract one another, the flexibility of the polymer backbone, and the regularity of the repeating pattern along the polymer strand. Such characteristics differ with the monomer selected. For example, in nylon-6,6 the C=O on one strand can hydrogen bond with the N-H of another strand, an unusually strong strand-to-strand attraction helps explain nylon's strength.

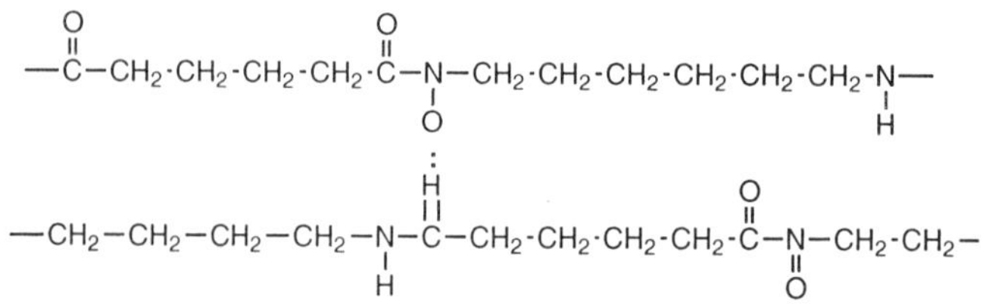

$$-\overset{O}{\overset{\|}{C}}-CH_2\cdot CH_2\cdot CH_2\cdot CH_2\cdot \overset{O}{\overset{\|}{C}}-\underset{\underset{O}{|}}{N}-CH_2\cdot CH_2\cdot CH_2\cdot CH_2\cdot CH_2\cdot CH_2\cdot \underset{\underset{H}{|}}{N}-$$

$$\vdots$$

$$-CH_2-CH_2-CH_2-CH_2-\underset{\underset{H}{|}}{N}-\overset{\overset{H}{\|}}{C}-CH_2\cdot CH_2\cdot CH_2\cdot CH_2\cdot \overset{O}{\overset{\|}{C}}-\underset{\underset{O}{\|}}{N}-CH_2\cdot CH_2-$$

Nylon-6,6

Linear polymers show varying degrees of ability to take on an orderly arrangement in the solid state. Such orderliness depends on structural variables and is favored by high regularity of the polymer structure, strong attractive forces between chemical structures on the strands, and high rigidity of the strands. A plastic that is very disorderly is termed *amorphous*. Amorphous plastics are more flexible, but have lower strength. When a plastic has orderly regions, called *crystallites*, it becomes more rigid and stronger (Figure 20.2). Even for a given plastic the degree of crystallinity varies. The manner of processing the plastic influences degree of crystallinity. When plastic is cooled slowly after molding, the strands have more time to respond to their mutual attractions and to arrange themselves into orderly patterns. This could mean a higher proportion of crystalline versus amorphous regions, or it could mean larger crystallites.

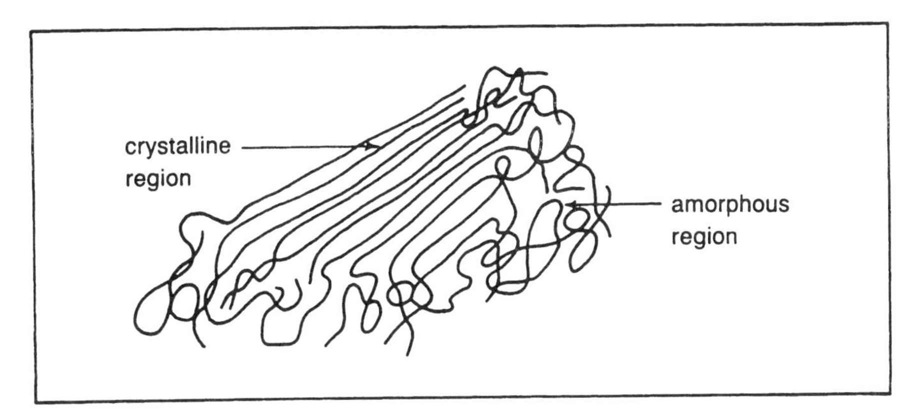

Figure 20.2. Crystallinity. When a heated polymer cools, the polymer strands may arrange themselves into orderly arrays called crystalline regions or crystallites. These impart a strength and rigidity to the plastic that it would not have in the totally amorphous state. To keep plastics flexible, compounds called plasticizers are added to physically block alignment of the strands.

Polymer properties are also fine tuned by mixing more than one monomer together to produce a copolymer. The final product should display contributions from each of the monomers. Examples of common commercial copolymers include Saran and ABS plastics. Saran is a copolymer of vinyl chloride and vinylidene chloride, whereas ABS has acrylonitrile, styrene, and butadiene in its formula. A longer list of copolymers commonly used in industry is provided in Table 20.4.

Table 20.4. Commercial Copolymers.

Polymer	Monomer
ABS	Acrylonitrile, butadiene, styrene
OAS	Olefin, acrylonitrile, styrene
ACS	Acrylonitrile, chlorinated polyethylene, styrene
ASA	Acrylic, styrene, acrylonitrile
LLDPE (linear low-density polyethylene)	Ethylene, hexene, butene
VLDPE (very low-density polyethylene)	Ethylene, other olefins
Ionomer	Ethylene, methacrylic acid salts
EMA	Ethylene, methyl acetate
EEA	Ethylene, ethyl acetate
EVA	Ethylene, vinyl acetate
SAN	Styrene, acrylonitrile
SB	Styrene, butadiene
VDA copolymers	Vinylidene chloride + vinyl chloride, acrylates, or acrylonitrile

ADDITIONS TO PLASTICS

Catalysts must be added to monomers to stimulate polymerization, and remain in the final product. A number of other compounds are added to plastics to modify properties. In all about 2500 chemicals are used for this purpose, representing a three-billion-pound market annually (Burgess, 1995, Chapter 21).

Catalysts

Conversion of monomers to polymers progresses very slowly unassisted. Polymerization reactions are often run at elevated temperatures and catalysts are added. Catalysts stimulate reaction rates markedly, even at low concentration. Catalysts are highly reactive chemicals that are added to the reaction vessel, so they remain in the finished polymer.

Plasticizers

It is possible to prevent crystallites from forming simply by adding compounds called plasticizers to the molten plastic. Plasticizers interfere physically with the alignment of the strands to form crystallites, ensuring a more flexible product. Thus, plasticizers facilitate use of vinyl plastics for clothing or upholstery. When vinyls were first used for auto interiors, the plasticizers would "cook out" of the plastics in hot weather, coating the insides of windows and leaving the vinyl shrunken and brittle. Today, plasticizers are more stable and permanent and are also generally less toxic.

Pigments

Most plastics contain pigments distributed throughout the bulk plastic. Two major classes of chemicals comprise such pigments. Metallic inorganics include compounds of lead, cadmium, titanium, iron, and molybdenum. These are very resistant to fading and are stable. An estimated 80% of U.S. cadmium is used in pigments for plastics. Organic dyes constitute the other group. These have problems of fading under UV exposure and breaking down during high-temperature plastic processing. Labels and decorations are painted onto plastic products, and a large variety of chemicals are used for this purpose.

Fillers

Plastics sometimes contain fillers, which are intended to modify the properties of the product. Fiberglass is an example of a filler used to make the finished thermoplastic product stronger and more rigid. In thermoset plastics, fillers may simply be inert material such as clay that "dilutes" the more expensive plastic components. This can be done successfully because of the great strength and rigidity of thermoset plastics.

Flame Retardants

A variety of flame retardants are added, particularly to construction plastics. Hydrocarbon plastics generally burn like a fuel, an undesirable property in a building or house fire. Flame retardants slow the combustion of the plastic. A negative effect of retardants is the production of smoke by the smoldering plastic, a smoke that obscures a clear view of exits and may contain toxic components. Chlorinated and brominated organics (polybrominated diphenyloxide is popular), organic phosphate esters, boron compounds, and antimony oxide all serve as flame retardants.

Blowing Agents

Compounds called blowing agents that either vaporize or decompose into gases are added to plastics to produce foams. Chlorofluorocarbons are heavily used, but are being phased out to reduce damage to the atmospheric ozone layer. Some azo compounds that release nitrogen on decomposing are used (e.g., azodicarbonamides).

Other Additives

If the compound must be durable outdoors, stabilizers are included that block UV light and so protect the material from the effects of sunlight. Lubricants such as metallic soaps or waxes assist molten plastics flowing into a mold.

WHERE ARE POLYMERS SYNTHESIZED AND USED?

Plastics (resins) are generally synthesized in a large chemical plant, often located near a refinery, since monomers are often derived from petroleum sources (Figure 20.3). Chemical companies produce the polymer for sale to manufacturers. Thermoplastics are usually completely synthesized, mixed with additives, and shipped as bags of pellets (Figure 20.4). These pellets are dumped into the hoppers of machines in the manufacturing plant, which melt them and form them into a product. This product may then be machined or otherwise reshaped and may be decorated. In the case of thermoplastics, very little chemistry is performed in the manufacturing plant (Figure 20.5).

Figure 20.3. Here we see the type of plant used to produce polymers. Chemicals are largely contained inside the reactors and pipes out of contact with workers during normal operation.

Thermoset plastics cannot be handled in this fashion. Once the full polymer structure is formed, the shape is permanent. The final polymerization is therefore performed in a mold by the manufacturer. The chemical company may prepare and ship low-molecular-weight prepolymers. These are mixed with catalysts (curing agents) and cross-linking structures, which are then fed into the mold. Alternatively, monomeric structures and a catalyst are mixed.

HAZARDS OF POLYMER SYNTHESIS

The synthesis of polymers is a chemical process carried on in plants, normally large plants, with towers, tanks, and reactors connected by endless pipes with valves and gages.

Figure 20.4. Here are shown pellets of plastic as produced by a chemical company to be used by manufacturers. The pellets are poured into the hopper of a plastics forming device at the time of use. (Photo courtesy of Dow Chemical Company. With permission.)

Figure 20.5. Little chemistry goes on in a plastics manufacturing plant, which looks like any of a number of facilities forming products, for example, a metal stamping plant.

As is typical of such operations, chemicals are in closed systems, so workers seldom contact them under normal circumstances. It is important to focus on times when such contact does occur, to be sure all employees are informed of the hazards of the compounds, of necessary personal protection if contact does occur, and of emergency spill procedures (Chapter 11).

Chemical exposure is most likely during unloading of chemicals at the plant site. Leaks in pipes, valves, and gages are always possible. There is special concern any time systems are opened, as during plant maintenance. Routine cleaning of tanks and other vessels involves worker entry into confined spaces and direct contact with chemicals, even if the tank is purged in an appropriate fashion beforehand.

TYPES OF PLASTICS PROCESSING

It is useful to understand what operations occur in a manufacturing facility in order to fully appreciate the generation of hazards. Manufacturing operations range from huge factories such those operated by auto companies to produce the plastic components of cars to very small plants contracting production of a few components or of short runs of a few parts. Often the machinery used is quite standard, and only the changing of a mold or die is required to convert a machine from production of one part to another in a small operation.

Mixing

As has been stated, a finished product is likely to contain a variety of additives to the pure polymer, such as pigments, plasticizers, fillers, and stabilizers. Dry mixing is accomplished with mechanical stirrers, much as a cake batter is made in the kitchen. A better mix is obtained if the plastic is melted, and this may be the total method employed, or it may be preceded by dry mixing. The plastic may be formed immediately into product once melted and mixed, or it may be formed into strips or sheets that are then cut into pellets to feed into the hopper of a molding or other forming machine later.

Casting and Injection Molding

When molding is done without pressure, it is termed *casting*. Casting could involve pouring the plastic into a mold. There are also continuous casting processes where a sheet or film is produced by pouring the plastic onto a belt moving through an oven. Casting can be done with polymers or with polymer–monomer mixes that complete polymerization in a hot mold.

The most common way to form plastic into products is injection molding (Figure 20.6). Plastic pellets are melted, then the melt is forced into a mold. The mold is cooled and the product removed, often by use of a high-pressure air jet. The shapes of the sprues, runners, and gates are trimmed off and recycled as scrap, but otherwise the surface is well finished and usually needs no further work. Injection molding of thermoplastics can form large objects and can be run at a high production rate.

Thermoset plastics can be injection molded, but the process is more difficult. There is a need to adjust the heating process to soften the reactants before molding, but the product must not harden in the heating chamber.

Blow Molding

Blow molding is used to produce hollow objects such as bottles and containers. In extrusion blow molding, a hot plastic tube is formed and the end is pinched shut, usually by the mold (Figure 20.7). The molds are water cooled, and the plastic tube is inflated to fill the mold cavity, then solidified against the cool surface of the mold. Scrap is cut off and recycled. In injection blow molding a quantity of plastic is injected on a rod and forced into a mold where the container is blown. It differs from extrusion blow molding in that the cavity is the shape of the final product, so there is no scrap.

a

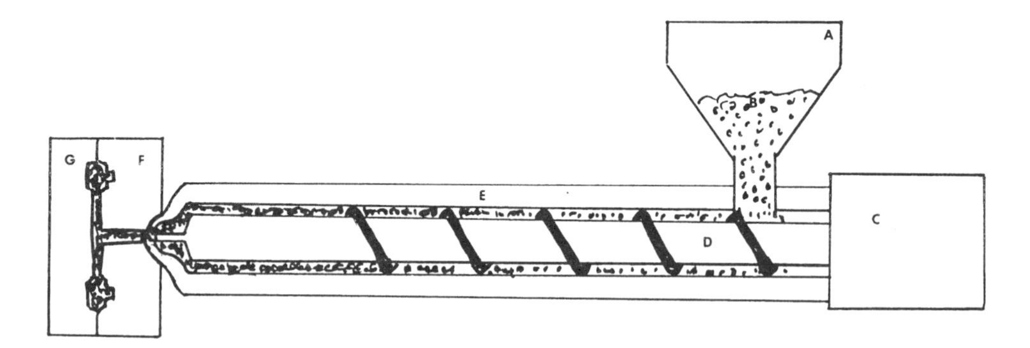

b

Figure 20.6. Injection molding. (a) An injection molding machine. At the right is the hopper containing the pellets of plastic to be formed into product. These drop down, are melted, and move to the left toward the mold. (b) The hopper (A) holds the plastic pellets (B) being fed into the machine. Under the housing (C) is the machinery to rotate the threaded plunger (D), which moves the plastic toward the mold and plunges forward to fill the mold. The housing fits tightly against the stationary platen of the mold (F). Once filled, the mold is cooled to solidify the plastic product. The moving platen (G) of the mold shifts to the left for removal of the product.

Vacuum and Pressure Forming

In vacuum forming, a thin sheet of plastic is heated and laid over a mold, either "male" or "female". The space between the sheet and the mold is then evacuated, forcing the sheet to conform to the shape of the mold. The mold is cooled to set the plastic in the desired shape, and the part is removed (Figure 20.8). Pressure forming is a very similar process, the only difference being that pressure is applied outside the sheet rather than a vacuum being formed between the sheet and the mold. These processes adapt well to high production rates. Vacuum molding is also used to package objects, especially fragile objects, by placing a plastic sheet over the object and evacuating the space between the object and the sheet.

Figure 20.6c. A close-up of the housing connection to the die.

Extrusion and Calendaring

A number of products are prepared by extrusion, including pipe, tubing, sheets, films, and fibers. The molten plastic is forced continuously through a die to generate the desired cross-section. Some films are prepared by blowing a bubble, then drawing the bubble away from the die continuously. When the desired final surface is obtained by squeezing a sheet between heated rollers, producing whatever cross-section and surface is desired, the process is called calendaring. Coating textiles or other substrates with plastic can be done on a calendaring machine.

Molding Thermoset Plastics

Usually the starting material for thermoset molding is a mixture of low-molecular-weight polymers, filler, and cross-linking agents. This molding is more challenging, because the material first melts and flows throughout the mold cavity, then hardens into its permanent shape under the influence of heat. Melting must be done to allow the material to flow into the mold, but hardening must not occur until the material is in place. Often it is also necessary to open the mold before hardening to allow gases to escape. Clearly, timing and temperature must be carefully adjusted. The plastic may be added to the mold, then the mold is closed and the material melts and flows (compression molding), or it may be melted first, then forced into the mold (transfer molding). Thermoset plastic scrap normally cannot be recycled.

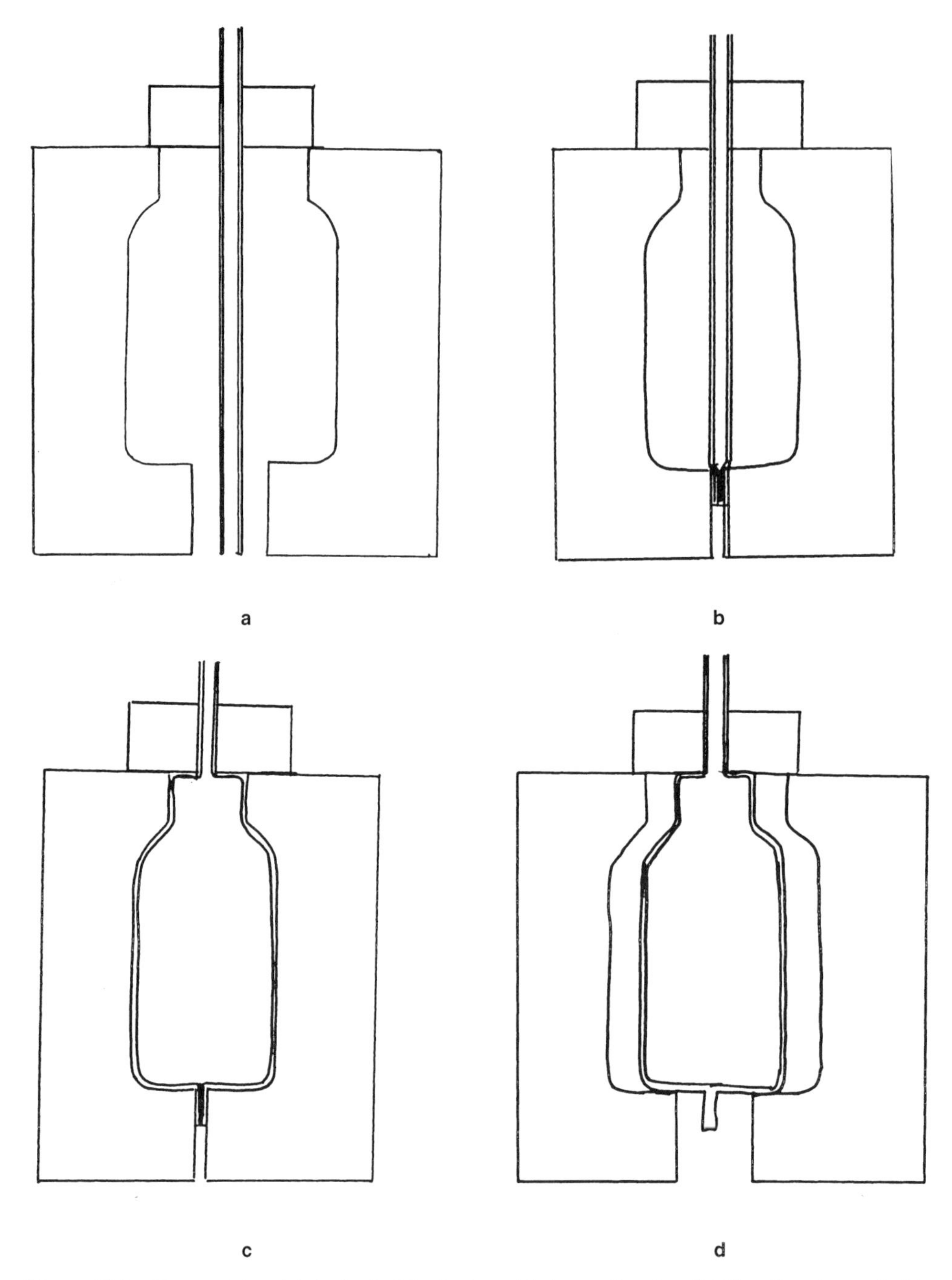

Figure 20.7. Blow molding. (a) A plastic tube is placed between the dies. (b) The dies close, pinching shut the bottom of the tube. (c) The tube is inflated to fill the opening in the dies. (d) The dies open and the finished product is released.

Foam Processing

Foams, plastics with gas bubbles throughout the finished product, can be produced using either thermoplastics or thermoset resins. Foams are strong for their mass and have excellent thermal, electrical, and acoustical insulation properties. Bubbles are produced either because the polymerization reaction releases a gas or a foaming agent (chemical blowing agent) is added to the plastic that generates a gas as the plastic is heated. For example, nitrogen may

Figure 20.8. Vacuum forming. A sheet of plastic is laid across the upper surface of the machine. The space under the sheet is evacuated, drawing the sheet down into the mold and shaping it. The mold is cooled and the finished product is removed, assisted by air pressure.

be added to the polymer melt under pressure. When the pressure is released, the gas leaves solution as a multitude of tiny bubbles.

Liquids such as freons, chlorinated–fluorinated hydrocarbons, have been very popular blowing agents because of their low toxicity, but these will need to be replaced as the ban on such compounds is imposed to reduce their damaging effect on the atmospheric ozone layer. A variety of organic compounds that decompose to gases when heated are used. These include azodicarbonamide, 1,1′-azo-bisformamide, *p*-toluene-sulfonyl semicarbazide, and *p*-toluene-sulfonyl hydrazide. These compounds are often strong irritants and present a hazard to workers if they become airborne during mixing. Azo compounds are incompatible with acids or the ketone peroxides, which may be involved in curing thermoset plastics. Contact with traces in the plastic initiates the desired nitrogen formation, but direct mixing produces a highly exothermic reaction and combustible gases.

Foams can be injection molded, extruded, or coated onto plastic sheets. The latter can later have embossed patterns pressed into them with hot patterned rollers or dies.

An example is the padded decorative door panel material used in cars.

HAZARDS IN PLASTICS MANUFACTURING

In the manufacturing plant, thermoplastic polymers come to the plant already synthesized. The least hazardous material encountered by workers is the polymer itself, because it cannot enter the body. Huge polymer molecules are not volatile, so entry as vapors into the lung is ruled out. Contact with the skin, lining of the gastrointestinal tract, or walls of the respiratory passages similarly does not lead to entry into the body, because such large molecules cannot cross the membranes blocking such access. The thermoplastic beads used in the manufacturing plant generally present little hazard.

It is possible to have unreacted monomers in a resin sample escape as the polymer is heated to be processed, since monomers are volatile compounds. In the case of highly hazardous monomers such as vinyl chloride, plastics are "stripped" after synthesis to remove most unreacted monomer. Additives, such as dyes, blockers, and plasticizers, may also be released during manufacturing operations, and their hazard must be considered. Additives increasingly are shipped as pellets rather than powders, eliminating the possibility of the dusts of these sometimes reactive chemicals entering the air during mixing operations. In some cases a series of additives are premixed and pelletized. Reducing the steps in a process reduces opportunities for exposure.

Generalizations about the relative safety of thermoplastic materials do not extend to thermoset plastics. Much more handling of potentially hazardous chemicals is involved in thermoset plastic manufacturing. Generally, in forming such a plastic, one of the starting materials is a polymer, and the curing operation is one of cross-linking the material into a three-dimensional rigid array. In that case, the curing agent is a small and reactive molecule. There are also many examples where the reactants for forming thermoset plastics are all low-molecular-weight, potentially hazardous substances. Finally, additives, including dyes, screening agents, and fillers, are often added at the time of manufacture.

After forming the product, whether thermoplastic or thermoset plastic, any operation involving sawing, grinding, milling, or sanding the product creates dusts. Excess thermoplastic that is trimmed off a molded product is often ground up, possibly mixed with new pellets, and is recycled back into processing. Once again, there is opportunity for dust to enter the air.

Any step in processing the plastic product or disposing of scrap that involves heat can lead to decomposition of the material. Such breakdown of plastics produces a variety of particularly offensive compounds, including such gases as HCN, HCl, and phosgene. Even the relatively noncombustible teflon polymers release a fume that causes an influenza-like condition with high fever, headache, and a cough. Once again the problem of the worker who uses cigarettes arises. Particulate in the air from polymer grinding can coat the tobacco, and this particulate will decompose when the cigarette is burned.

Molding requires that the dies be opened to remove the product, then closed against the machine again for the next cycle, which is a noisy process. This is laid over the general noise level of the hydraulic pumps, valves, and lines. Air jets used to remove the parts from the mold and other aspects of operations and the noise produced is quite loud. The cutting off of scrap can be noisy. Noise levels are a major problem in plastics manufacturing.

COMMENTS ON SPECIFIC COMPOUNDS

With more than 100 polymers and many more copolymers in commercial use it is not within the scope of this book to provide a comprehensive list of the hazardous substances that could possibly be encountered. However, some general comments are certainly appropriate, and discussion of some of the more commonly encountered serious hazards should be useful. These are organized by the function of the compound in producing the final polymer.

Monomers

A few monomers for which special problems exist are discussed below (see also Table 20.5).

Table 20.5. Toxicity of Selected Monomers.

Monomer	Toxicity (Airborne) (LC_{50})		Exposure Limits (PEL)	
	ppm	mg/m³	ppm	mg/m³
Acrylamide	—	—	0.3	—
Acrylonitrile	425	—	2	—
Butadiene	—	259,000	1,000	2,200
Chloroprene	—	2,300–11,800	25	90
Diglycidyl ether	30	—	0.5	2.8
Epichlorohydrin	250	—	5	19
Ethylene	950,000	—	—	—
Formaldehyde	—	92–400	1 (10 ppm peak)	—
Isopropyl glycidyl ether	1,100–1,500	—	50	240
Maleic anhydride	—	—	0.25	1
Methyl acrylate	1,350	12,800	10	35
Methylene bisphenyl isocyanate	—	178	0.02	0.2
Phenol	—	74–177	5	19
Phosgene	—	1,000–10,000	0.1	0.4
Phthalic anhydride	—	—	2	12
Styrene	—	9,500–24,000	100, 200 (ceiling)	—
Toluene diisocyanate	10	—	0.02	0.14
Vinyl chloride	180,000	—	cancer suspect agent	
Vinylidene chloride	6,350	—	1	4

Vinyl Chloride

Polymerization of vinyl chloride is accomplished in a variety of ways using peroxide catalysts. Vinyl chloride does not have good warning properties such as odor or irritant action. Historically it was considered to be a safe compound with a TLV of 500 ppm. Workers are reported to have leaned over open vats and inhaled deeply to experience its narcotic effects. Its use as an anesthetic had been studied. In 1961 it was suggested by Dow Chemical that the TLV should be lowered, based on preliminary studies. A series of problems surfaced in the period that followed. Workers who cleaned vinyl chloride polymerization vats suffered degeneration of bones (acroosteolysis) in their hands. In 1972 evidence of liver damage was observed at levels greater than 300 ppm. Then in 1974, studies showed that workers displayed a higher than expected level of a very rare cancer, angiosarcoma of the liver. The OSHA standard for exposure to vinyl chloride became 1 ppm with a 5-ppm ceiling as of January 1981. Stripping techniques are now used to remove unreacted monomer from the polymer product.

Acrylonitrile

This is the monomer in the production of orlon and is one of the components of the copolymers ABS (acrylonitrile–butadiene–styrene) and SAN (styrene–acrylonitrile). Acrylonitrile can be absorbed through the lungs or skin. It penetrates rubber gloves, making these inadequate as a protective measure. In the body it serves as a source of cyanide ion, producing asphyxiation. It has been shown to be a cause of cancer in both the lung and the bowel. The OSHA standards (January 1981) were 2 ppm with a 10-ppm ceiling.

Styrene

This compound is employed not only to prepare polystyrene and styrofoam, but as one of the copolymers in ABS, styrene–butadiene, and SAN plastics. The vapors are irritating, in fact irritating enough that dangerous exposure is not likely to be tolerated, and there is the usual skin irritation due to removal of skin oils. Styrene affects the central nervous system, producing depression. A problem called "styrene sickness" involves nausea, vomiting, dizziness, and fatigue. OSHA standards limit exposure to styrene to 100 ppm with an acceptable ceiling of 200 ppm and an acceptable maximum peak of 600 ppm for no more than 5 min in any 3-hr work period. Styrene is shipped mixed with polymerization inhibitors such as butylcatechol and hydroquinone to prevent spontaneous premature polymerization. Both of these are sensitizers.

Epoxy Resins Monomers

Toxicologically, this is a very troublesome group of compounds. Epichlorohydrin is an irritant to the skin, causing burning, itching, and redness, with pain and blistering appearing after contact. It can be absorbed through the skin and is irritating to the eyes, causing damage at higher concentrations. Epichlorohydrin is severely irritating to the lungs, producing pneumonitis (fluid in the lungs) hours after exposure. In the body, liver and kidney damage result, and sterility is caused. Finally, it is a sensitizer, leading to later allergic response upon contact with even small quantities of epichlorohydrin. The PEL is 5 ppm. Compounds used with epichlorohydrin to produce epoxy resins include bisphenol A, glycidyl ethers, and aliphatic polyamines such as *p*-phenylenediamine, diethylenetriamine, and triethylenetetramine. All these are irritants and sensitizers. The polyamines are particularly hazardous, causing severe irritation, chemical burns, reddened and itching skin, blistering, facial swelling, and asthma. They can cause bronchospasms and coughing for days after exposure. Sawing or machining finished plastic products can release amines and unreacted epichlorohydrin.

Phenolic and Amino Resin Monomers

The phenolic thermoset plastics are produced by the reaction of a variety of phenols with formaldehyde or furfural. The phenols include phenol itself, cresol, xylenol, *p-t*-butylphenol, and resorcinol. All these compounds, the phenols and the aldehydes, are potent irritants. Above 3 ppm, formaldehyde is irritating to the eyes and upper respiratory tract. Furthermore, formaldehyde is a sensitizer. Other findings about formaldehyde, including the possibility that it is a human carcinogen, are under review, and standards for its safe use were revised down to a level of 1 ppm in 1987. Amino resins are similar to phenolics, substituting urea or melamine for the phenols. Here too, formaldehyde is a serious problem. Hexamethyltetramine, used in forming those polymers, decomposes to give formaldehyde and ammonia, both of which are irritating.

Diisocyanates

Polyurethane and polyisocyanurates are produced by reacting diisocyanates with polyesters or polyethers having terminal alcohol groups on the chain. The diisocyanates include TDI (toluene diisocyanate) and MDI (methylene diphenyl diisocyanate), and these compounds present the serious threat. The isocyanate chemical group was responsible for the massive loss of life in the chemical release at Bhopal, India. The difference between the methyl isocyanate that escaped at Bhopal and these compounds is basically one of volatility. Isocyanates react rapidly with water, and so are irritants on moist surfaces such as the eyes and respiratory tract. In addition they are potent sensitizers. The PEL for MDI is 0.02 ppm and for TDI is 0.05 ppm.

Polyester and Alkyd Monomers

These polymers may utilize acid anhydrides, such as phthalic anhydride or maleic anhydride, along with polyalcohol structures as monomers. The acid anhydrides react vigorously with small amounts of water, producing heat and concentrated acid. They cause burns and irritation on contact with eyes, nasal passages, throat, and moist skin. Unlike many of the compounds in this chapter, they are not volatile liquids. However, they can enter the workplace air as dusts produced during transfer and handling. Polyethylene terephthalate is made by reacting dimethyl terephthalate with ethylene glycol to polymerize at higher temperatures by ester exchange. Antimony oxide and zinc acetate are commonly used in this process, and dusts result as the bags are opened and dumped.

Acrylic Monomers

Acrylic acid and a wide variety of its derivatives, especially esters, are used as monomers. These are skin irritants, and the vapors produce nervous system effects such as headaches, irritability, numbness in the extremities, slurred speech, and fatigue. The nervous system symptoms for acrylamide are particularly severe, and the PEL for this solid is 0.3 mg/m^3 in the air.

Polycarbonate Monomers

The preparation of polycarbonate plastics involves the use of bisphenol A, dioxane, and phosgene. Phosgene gas was described in Chapter 7 as a lower respiratory tract irritant. As such, it must be handled with extreme care. The odor and immediate level of irritation experienced on contact with this compound are insufficient warning to workers to ensure they will not voluntarily allow dangerous doses into the lungs. Dioxane has been classed as a carcinogen.

Nylon Monomers

Nylons may be prepared using diacids or diacid chlorides that are reacted with diamines. The diacids are not serious problems, but diacid chlorides are similar to acid anhydrides in properties and cause severe skin or eye damage. The diamines are irritants and sensitizers and so also must be carefully handled. The amino acid caprolactam, used to make nylon-6, causes nose and throat irritation, irritability, and nervousness.

Vinylidene Chloride

Most commonly, this monomer is copolymerized with vinyl chloride. It is irritating to the skin and eyes and may include a phenolic inhibitor, which increases the irritant potential (see "Styrene"). When inhaled, it produces a state of drunkenness.

Vinyl Acetate

This monomer is used to prepare polyvinyl alcohol. Ester (acetate) groups are removed after polymerization. Vapors of the monomer are upper respiratory tract irritants and are narcotic.

Ethylene

Ethylene is obtained from a refinery cracker and is polymerized into polyethylene using organic peroxide catalysts. Ethylene has low toxicity, but it is a high fire and explosion hazard.

Viscose Rayon Components

One starts with the polymer (cellulose) and modifies it. The most hazardous compound in the process is carbon disulfide, which is extremely volatile and has a very low flashpoint. It affects the central nervous system, producing anything from dizziness and fatigue to psychic disturbances. It is also a respiratory irritant.

Cellulose Acetate Components

Cellulose (wood pulp or cotton) is mixed with glacial acetic acid, sulfuric acid, and acetic anhydride to produce this polymer. Removal of some ester linkages by acid-catalyzed hydrolysis is then done to adjust polymer properties. Milling the cellulose source is a noisy process, and dust from the cellulose source and acid vapors are obvious problems. Washing with water generates high humidity conditions, adding to the potential for heat stress in hot weather. Drying and handling the final product can also present dust problems.

Catalysts (Curing Agents)

A sampling of commonly used catalysts is shown in Table 20.6. Reactions to form addition polymers very often proceed by a free radical mechanism. Such a mechanism requires a source of free radicals, structures with unpaired electrons, to initiate the chemical reaction. Most often organic peroxides are used. Approximately 50 different such compounds are available, primarily peresters and percarbonates. Benzoyl peroxide and methylethylketone peroxide are representative and frequently encountered examples. Peroxides share similar properties, all being strong oxidizing agents capable of reacting violently with appropriate substances. Skin irritation and burns are likely on contact with peroxides, and severe eye damage is possible. Benzoyl peroxide is a sensitizer, chronic exposure leading to an allergic rash. It has a PEL of 5 mg/m^3, and an IDLH level of 1000 mg/m^3. From the standpoint of hazard due to direct contact and because of a potentially violent reaction, waste or spilled peroxides should be diluted with water, and any rags used in the cleanup of spills similarly should be put in water. The rate of chemical decomposition of peroxides increases with temperature, and contact with combustibles such as paper or wood can lead to fires.

Table 20.6. Commonly Used Catalyst Systems.

Monomer	Product	Catalysts
Ethylene	High-density polyethylene	Aluminum alkyls plus titanium tetrachloride
	Low-density polyethylene	Chromium oxides on alumina or silica *tert*-butyl peroctoate; other peresters
Propylene	Polypropylene	Aluminum and magnesium alkyls plus titanium halides
Styrene	Polystyrene	Aluminum alkyls plus titanium trichloride
Vinyl chloride	Polyvinyl chloride	Percarbonates
Various monomers	Elastomers	Aluminum alkyls plus vanadium trichloride or oxychloride

Reprinted with permission from *Chemical and Engineering News*, (Feb. 17, 1986). Copyright 1986, American Chemical Society.

The aluminum alkyl catalysts used with polyolefins are physically very hazardous. They react violently with water and burn spontaneously in air. In solution they cause serious burns, and their fumes are damaging to the lungs.

Organic metal salts or tertiary amines are used to cure polyurethane or polyisocyanurates. Both catalysts penetrate the skin and may be sensitizers.

Accelerators

Although not catalysts themselves, accelerators make catalysts more effective. For example, cobalt naphthanate stimulates the decomposition of peroxides into the very reactive free radicals. Some structures are shown in Table 20.7.

Table 20.7. Some Accelerators.

Hexamethylenetetramine	Tetramethylthiuram disulfide (thiram)
Thiocarbanilide	Mercaptobenzothiazole
Benzothiazolyl disulfide	Thioureas
Tetraethylthiuram bisulfide (disulfiram)	Xanthates
Diphenylguanidine phthalate	

In general these compounds are irritants, airborne samples causing irritation of eyes, nose, and throat and direct contact with the skin causing a rash. They are frequently sensitizers. Thiram and disulfiram share with antabuse the ability to block the metabolism of ethanol and aldehydes. Nausea and vomiting follow intake of ethanol after contact with these compounds. The PEL for the solid thiram is 5 mg/m^3. Dimethylaniline is used as an accelerator in polyester synthesis. It absorbs readily through the skin, and once in the body it is a central nervous system depressant. Dimethylaniline also causes kidney and liver damage. It has a PEL of 5 ppm and an IDLH level of 100 ppm.

Stabilizers

Stabilizers or screening agents are added to plastics to slow degradation of the plastic caused by heat or light. Metal soaps are commonly used, and many have serious toxicity problems. Lead, barium, cadmium, calcium, magnesium, and tin compounds are used. Lead and cadmium are serious hazards to health (Chapter 19). Once in the body they incorporate into bone structure and remain there for long periods, releasing slowly into the blood. Lead causes damage to internal organs such as liver and kidney and causes lasting damage to the central nervous system. Cadmium remains in the body even longer than lead. It damages the kidney and at high levels has caused skeletal problems. Organotin compounds are strong irritants and cause chemical burns. Liver and urinary tract damage result from contact. After contact, there is evidence of central nervous system damage, including headaches, loss of visual acuity, and motor control problems. The PEL for organic tin (as tin) is 0.1 mg/m^3 and the IDLH level is 200 mg/m^3. By contrast, calcium, magnesium, and barium soaps have low toxicity.

Plasticizers

The major plasticizer in recent years has been dioctylphthalate. The listing of dioctylphthalate by the EPA as a potential carcinogen has led to its gradual replacement, often with other phthalate esters. Phthalate esters are generally of low toxicity and seldom cause problems. For example, dibutyl phthalate is relatively nonirritating and nonvolatile. In animal testing, high doses are necessary to produce toxic effects. Esters of adipic acid and of citric acid are also used, and are of low toxicity.

Early plasticizers were sometimes quite toxic. Tricresyl phosphate attacks the peripheral nervous system, and some of the chlorinated hydrocarbons used cause liver damage.

Pigments

There is a movement to stop the use of pigments containing toxic metals such as cadmium, lead, and chromium. Later burning of waste plastics releases the metals into the air as particulate. Organic dyes include phthalo blues and greens, nigrosines, anthroquinones, and a number of others. Each should be checked for toxic characteristics when used. Industries buying precolored plastics should be aware that producers of plastic beads and other stock for molding into products are changing pigments to eliminate metallic pigments, and buyers should ensure new stock does not have hazardous properties.

Fillers

Used most frequently with thermoset plastics, fillers are generally stable and inert solids. Clays, silicates, and glass fibers are typical of the materials used. Hazard is largely present when particulates from these enter the air, either while being added to a reaction mixture or when a finished product is milled or sanded. Silicates can cause fibrosis of the lungs on chronic contact. Irritation from glass fibers is a special problem with fillers. Restrictions on mineral dusts depend on the composition (crystalline silica content) but are typically 20 mppcf (millions of particles per cubic foot of air).

Other Additives

Many other compounds can be added to plastics, depending on properties desired in the product. These include fire retardants (in construction plastics), antioxidants, and agents to help lubricate during extrusion or ease release from a mold. Some flame retardants, for example, tris-(2,3-dibromopropyl) phosphate, are suspected carcinogens. The monobenzylether of hydroquinone is used as an antioxidant and is a sensitizer. Chloronaphthalene is used as a mold release compound and can cause chloracne. Most pigments are inert and present little hazard. However, they can generate a dust problem during transfer to the plastic preparation facility.

SUMMARY

These discussions serve only to point up classes of compounds and several specific problem compounds. Toxicity information supplied with chemicals used in a plant or available from sources such as an MSDS should be consulted for all chemicals used in an industrial facility. The possibilities for trouble should be anticipated so as to prevent a serious accident or injury.

ELASTOMERS

Elastomers are polymers that possess a special property. If distorted by stretching, they return to their original shape when the distorting force is removed, .

Natural rubber (latex) is a polymer of the diene isoprene. The importance of elastomers to mechanized warfare was an important early stimulus to research in Europe to develop synthetic rubbers to eliminate dependence on the shipping of natural latex from distant, tropical sources. Efforts focused on variations of 1,4-addition polymers of butadiene or a butadiene derivative, since the surviving double bond in such a polymerization was early recognized as central to elastic properties in the product. The elastomers developed had advantages in properties and price that made the continuation of this effort there and elsewhere commercially viable.

THE USE OF ELASTOMERS

Worldwide, the use of elastomers is estimated to be 14.9 million metric tons in 1992, and about 65% of the elastomer used is synthetic rubber.[3] North America consumes about 28% of the synthetic rubber and 19% of the natural rubber. U.S. production of synthetic rubber in 1994 was more than two billion pounds.[4] Tires alone account for three quarters of the natural rubber and more than half the synthetic rubber use. Of the several synthetic rubbers produced (Table 20.8), styrene–butadiene rubber (SBR) has the greatest share of the market at about 35%. Some properties of SBR are not as good as those of natural rubber or some other synthetics, but it has a relatively low price, and is therefore very competitive. Second to SBR is polybutadiene (BR), with 17% of the market. Other important elastomers include ethylene propylene copolymer (EP) and terpolymer (EPDM), polyisoprene (IR), isoprene isobutylene copolymer (butyl rubber, IIR), polychloroprene (neoprene, CR), and acrylonitrile–butadiene copolymer (nitrile, NBR). Some low-volume, usually expensive elastomers are produced for special purposes. These include silicone rubbers, chlorosulfonated polyethylenes, acrylics, copolyesters, epichlorohydrins, polysulfides, and urethanes.

Table 20.8. Production of Selected Synthetic Rubbers in 1995. (U.S. and Canada, thousands of metric tons)

Elastomer	Total Production	Used for Tire Production (%)
SBR	880	70
Polybutadiene	530	75
Ethylene propylene	270	7
Nitrile rubber	84	
Neoprene	70	

Data modified from *Chem. Eng. News*, June 24, 1996.

RUBBER PROCESSING

In the U.S. about 200,000 rubber workers are employed at about 1500 locations (Burgess, 1995, Chapter 18). The processing of rubber is similar to that of plastics. Hazards at various steps of the process are also similar.

Compounding and Mixing

The first elastomers processing step is compounding a mix of elastomer (polymer), plasticizer, curing agent, catalyst, antioxidant, and other components. Sometimes agents, especially toxic agents, are shipped in preweighed plastic containers that are added, plastic and all, without opening. Materials delivered in drums may have to be hand weighed and added, an opportunity for worker exposure to the chemicals. Large operations such as a tire plant may automate parts of the mixing, commonly done when adding carbon black.

Mixing is usually done in two stages. The first stage omits curing agents and catalysts and involves thoroughly breaking up the elastomer stock and mixing it with the other agents.

[3] *Chem. Eng. News*, May 10, 1993.
[4] *Chem. Eng. News*, June 26, 1995, p. 42.

This may be stored until needed. Then, shortly before processing, mixing is done again, this time adding agents required for curing. The mixer may be cooled to prevent premature curing.

After mixing, the stock is milled, which further blends the components and delivers the mix as a sheet that is coated with a chemical to make the surface less sticky (antitackifying). Compounded rubber is stored as folded sheet awaiting the second mixing or is transported to be processed after the second mixing.

Extrusion

Extrusion is a common rubber forming process and is similar to plastics extrusion. The product is a strip of material with a particular cross-sectional shape. Temperature control is important to adjust the fluidity of the rubber mix without triggering premature curing. In tire production, different rubber stock for sidewalls and tread are fed together into the extruder to form a layered strip in a single operation.

Calendaring

In calendaring, fabrics are coated with rubber. Rubber sheets are formed, the fabric is coated, and the rubber is forced into the fabric strands by frictioning with a hot roller moving faster than the coated fabric sheet. Tackifying oils may be added to soften the rubber surface facing the fabric. Both sides of the fabric are coated in tire production.

Forming and Curing

The uncured rubber intermediate prepared by extruding and/or calendaring is now molded or assembled into the shape of the final product. Compression molding places the stock between heated dies to which pressure is applied. Excess stock leaves through channels. In transfer molding, the stock is rammed into the mold, and in injection molding it is heated to become fluid and is injected into the mold.

Curing (vulcanization) generates a product with the desired properties. The mix contains a cross-linking material and an accelerator such that under heat and pressure, applied for the correct time interval, the polymers cross-link to the correct degree. When the product is molded, curing occurs in the heated mold. Articles otherwise assembled are usually cured using high-pressure steam (autoclaving) or by dry heat in an oven.

Tires

Given the importance of tires in the overall elastomer picture, a separate word about tire construction is worthwhile (Figure 20.9). Tires are built around a body of layers of rubber-coated fabric reinforced with steel, glass, or polymer strands (which nowadays are generally run at right angles to the tread), termed the plies of the tire. The sidewall is an extruded rubber layer outside the plies, and the tread is the extruded strip that is the part of the tire contacting the road. At the inner edge of the plies is the bead, rubber-covered steel wires that are at the point where the tire seals against the rim. Finally, a calendered sheet called the liner forms the inner layer of the tire and serves to retain air. The uncured assembled tire[5] is sprayed with mold release and placed in a molding machine. A bladder inserted into the tire is expanded, forcing it against the mold that includes the tread and sidewall pattern. The mold is heated and the tire is cured.

[5] An uncured product is termed "green".

Figure 20.9. Tire production is a big business.

HAZARDS

Hazards include exposure to noise from the mechanical devices used during compounding, extruding, calendaring, and molding. Engineering controls and careful maintenance of the machinery are important. The use of compressed air and high-pressure steam add to noise levels. Airborne particulate may be generated at several points in the transfer of mix components, weighing, addition to the mix, and mixing. Carbon black is especially troublesome because of its very small particle size. Well-designed equipment, good maintenance, ventilation, and careful housekeeping minimize this problem. As the stock is heated to be extruded or calendered, volatile components of the mix may enter the plant air. This potential is even greater during molding and curing of the final product, the greatest hazard occurring when the molds are opened and "curing fume" is released. Cements and tackifiers (agents to make rubber more sticky) may include volatile solvents, which are released after application. Benzene used to be used in such operations, but has been replaced with petroleum hydrocarbons. In addition, because some operations are run at elevated temperatures, there is a potential for heat stress.

Chemical Hazards with Specific Elastomers

The operations involving chemical processing in a manufacturing plant preparing elastomers are generally the same as in a plant producing thermoplastics. Many of the chemicals are the same, but some are unique to elastomer production (see Table 20.8).

Natural Rubber

Latexes are used in a number of processes. As a natural polymer, workers are not exposed to the monomer form or the polymerization process, and the polymer itself presents little hazard. When dipping gloves, footwear, and the like, dilute acids such as acetic or formic acid are present. The chief problem lies in diluting these acids from the concentrated form, which can involve the possibility of acid burns, heat production, and inhalation of excessive amounts of acid vapors. Hydrazines may be used in foam manufacture and are toxic materials. Ammonia fumes may be present, but usually are at low levels. The manufacture of latex adhesives involves the use of quantities of organic solvents, often chlorinated hydrocarbons. Natural rubber is included in tires, and organic solvent fumes must be removed when compounding tire stock.

Polyisoprene

Methods have been devised to stereospecifically polymerize[6] the latex monomer isoprene to generate a product with properties like those of natural rubber. Isoprene itself is not well studied and a TLV has not been established. It is an anesthetic, as are most unsaturated hydrocarbons. Synthetic polyisoprene does not exactly duplicate the properties of natural rubber. In fact, tire manufacturers continue to use natural latex because of the properties imparted by some of the natural rubber impurities.

Styrene–Butadiene Rubber and Polybutadiene

A low-molecular-weight hydrocarbon is usually used as solvent for solution polymerization of *cis*-polybutadiene, and the polymerization catalyst includes diethylaluminum chloride and cobalt octanoate. SBR is polymerized as a soap emulsion of a mix that is about three-quarters butadiene. Peroxide catalysts (potassium persulfate, alicyclic hydroperoxide) are used.

Butadiene is weakly irritating and narcotic. For the most part, unsaturated aliphatic compounds are not serious problems. However, butadiene may be weakly carcinogenic. Current exposure levels may be causing up to 23 excess cancer deaths per year among the approximately 5900 workers exposed. ACGIH has recommended a 10-ppm standard, contrasting sharply with the present PEL of 1000 ppm. Styrene was discussed earlier in this chapter.

Nitrile Rubber

Acrylonitrile and butadiene are copolymers in nitrile rubber. Polymerization is essentially the same as for SBR.

Acrylonitrile is clearly the more dangerous of the two compounds. It absorbs through the skin, lungs, or GI tract and produces cyanide poisoning. An overdose is treated exactly as one would treat cyanide poisoning. Care must be taken in handling acrylonitrile, especially because rubber protective wear does not exclude it. It may also be carcinogenic, as indicated by epidemiological studies of workers.

Neoprene

Chloroprene, the monomer of neoprene, is irritating to the respiratory tract and to the skin and is a nervous system depressant. It causes damage to the kidney and to the liver of test animals, and exposed workers have experienced temporary loss of hair.

[6] It is necessary to form an all-*cis* polymer to have elastic properties.

Butyl Rubber

Polyisobutylene is prepared largely from isobutene, but another monomer such as isoprene, butadiene, or chloroprene is added to modify the properties. The polymerization is solution polymerization in methyl alcohol, with aluminum chloride as the catalyst. Isobutene, like many other unsaturated aliphatic hydrocarbons, has anesthetic properties and low toxicity.

Silicones

Chlorosilanes are very irritating and react at moist surfaces such as mucus. The alkoxysilanes are less reactive and damaging.

Chlorosulfonated Polyethylenes

In the production of these elastomers, polyethylene is reacted with sulfur dioxide and chlorine gases. Both these gases are irritating, producing acid in the lungs. Of the two, chlorine is less irritating. As a result one willingly inhales more chlorine and thus it is allowed to penetrate more deeply into the lungs. There, chlorine reacts slowly with moisture to produce acid.

Polysulfide Rubbers

Polysulfide elastomers are formed from polysulfides and ethylene dichloride (ethylene chloride). The polysulfides are low-risk materials. Ethylene dichloride is very irritating to eyes and lungs, causing coughing, watering eyes, and salivation. It has an anesthetic effect and it damages the kidney and the liver.

Polyurethane

Diisocyanates are the main concern in the production of polyurethanes (see "Plastics"). Exposure is regulated at extremely low levels for these compounds because they are potent sensitizers.

Other Toxicity Problems

As with plastics, a number of compounds are involved in processing in addition to the monomers themselves.

Catalysts

A variety of peroxides are involved in catalysis of polymerization. Metallic catalysts are used with olefins, including diethylaluminum chloride or vanadium oxychloride (EP and EPDM) and aluminum chloride (butyl rubber). These compounds are highly reactive and are able to cause serious chemical burns. Handling bags of aluminum chloride creates the risk of exposure to dust if the bags leak. Moisture reacts with $AlCl_3$ to release HCl, creating a serious risk to eyes and lungs.

Cross-linking Agents

Vulcanization (cross-linking) of elastomers involves sulfur in the cases of natural rubber and a number of the synthetics. The sulfur itself presents little risk. Hydrogen sulfide or sulfur oxides that may result from such processing present some hazard and should be controlled by ventilation. Neoprene is cured using zinc or magnesium oxides, which present few problems. EP rubbers utilize peroxides for cross-linking. All these cross-linking agents are solids, and transfer operations should be designed to avoid dusts entering the plant air.

Accelerators

Accelerators are also employed in vulcanization. These compounds, listed in Table 20.7, are very effective sensitizers, so worker contact should be minimized. There are special problems with thiram and disulfiram, which block the metabolism of alcohol or aldehydes. They produce severe reactions in workers using beverage alcohol or being exposed to paraldehyde. Occasionally metallic accelerators such as lead oxide are employed. Contact with any lead compounds should be minimized not only because of the immediate toxic effects of the metal, but also because it is retained for periods of time in the body so that regular exposure can result in an elevated body burden of lead. The use of amines, nitrosamines, and other nitroso compounds as accelerators can lead to potentially carcinogenic *N*-nitroso compounds in plant air.

Antioxidants

Rubbers, particularly tires, often contain antioxidants to prevent ozone damage resulting from static charge build-up. These antioxidants are generally aromatic amines or hydroquinone derivatives and can be irritants or sensitizers. In addition, hydroquinone monobenzylether causes loss of skin pigments.

Fillers

A number of fillers are added to rubbers. Carbon black is a powder composed of very small particles (25–100 μm in diameter). Such small particles can readily penetrate deep into the lung, causing damage that leads to a loss of lung capacity. At levels two to three times the present TLV of 3.5 mg/m^3, workers have displayed problems, both with a loss of lung capacity and, in a few cases, with the appearance of fibrosis. At levels below the TLV the problems were sharply reduced, although they were not completely eliminated. Carbon black is known to contain carcinogenic polycyclic aromatic hydrocarbons, but it has been concluded that these are so strongly adsorbed to the surface of the carbon that they pose no threat to health. Studies show no increase in cancer, either in test animals or in workers with a history of high exposure to carbon black. With respect to this unusual ability of carbon black to adsorb substances, it has been reported to release adsorbed materials such as carbon monoxide when stored in a poorly ventilated location.

Silica and asbestos are also used as fillers. If they enter the atmosphere as dusts, they cause lung fibrosis. Asbestos also is a carcinogen. Most other fillers and pigments are low in hazard. Such compounds as aluminum or calcium silicate, calcium carbonate, clays, and magnesium, titanium, or zinc oxides present a nuisance as dusts, but do not pose major health threats.

Lubricants

Talc is often used as a lubricant to prevent rubber from sticking to itself or to facilitate its extrusion. Chemically, talc is hydrated magnesium silicate. Long-term inhalation of talc, even at the recommended TLV level of 1.0 mg/m^3, produces increases in bronchitis, cough, and other respiratory problems due to obstruction or irritation of the respiratory tract. At higher levels, fibrosis can occur. This is especially serious if the talc contains traces of asbestos-type fibers, as some sources do.

Blowing Agents

A number of blowing agents have irritating or toxic properties. Azocarbonamide produces respiratory sensitization when inhaled as a dust. Benzenesulfonyl hydrazide and *p,p'*-oxy-bis(benzenesulfonyl hydrazide) are skin or eye irritants. Azo bis-isobutyronitrile breaks down above 26°C to give tetramethylsuccinonitrile, which has been responsible for headache, nausea, and convulsions in workers making PVC foam. The TLV is 0.5 ppm. In general, care should be taken to avoid inhaling any of these products as dusts.

TEXTILES

Textiles are prepared from a number of natural and synthetic polymeric materials. The industry produces cloth for clothing, bedding, curtains and drapery, towels, wall coverings, and upholstery. A large and closely related industry is carpet manufacturing. U.S. production of synthetic fibers in 1995 was almost 10 billion pounds. In spite of the rise in use of synthetic fibers in the period following World War II, there is still a large market for natural fibers such as cotton and wool. This section focuses on natural fibers and the processing for textile use of polymers described earlier for use as plastics and elastomers. Textiles are a major end user of such polymers as polyesters, nylons, and polyolefins (polyethylene and polypropylene). Production of major synthetic fibers is summarized in Table 20.9.

Table 20.9. U.S. Production of Synthetic Fibers in 1995. (billions of pounds)

Fiber	Production
Polyester	3.9
Nylon	2.7
Polyolefin	2.4
Cellulosics (acetate and rayon)	0.5
Acrylic	0.4

Data modified from *Chem. Eng. News,* June 24, 1996.

COTTON

Harvested cotton contains seeds, leaf and stem parts, and other impurities. Some sorting of fiber by length is needed, an elaborate process that includes blowing, brushing, and twisting. Processing of raw cotton takes place between taking cotton from the bale and spinning it into yarn. A picker partially cleans the cotton fiber and forms it into a flat "web". This is then carded to open tufts of fiber, to remove the rest of the trash, and to remove tangles, a dusty operation. The flat web is formed into a round sliver. Slivers are combined and pulled to make the fibers more nearly parallel. Spinning twists the sliver into tight dense yarn or thread. Machinery is cleaned of dust and lint by blowers to reduce the possibility of fire, another dusty operation.

After cotton is twisted into thread, it is woven or knitted into cloth. The fabric is then cleaned (scoured) using caustic and detergents. If the cloth is to be dyed, it is treated with chlorine or peroxide bleach. Mercerization involves soaking the cloth in about 20% caustic, after which the appearance, strength, and ability to take dye are improved. Dye is applied in vats such that the length of exposure controls the depth of color obtained, then excess dye is washed off. Patterns may be printed onto the cloth using either rollers or screens and water-insoluble pigments.

Finally, a series of finishing steps may be used. A cross-linking agent may be added to make the cotton "wash and wear". Flame retardants may be added to the cloth. The surface of the cloth may be brushed or sueded (to form a soft nap of loose fibers) or glazed (rubbed with a fast moving roller).

Hazards of Cotton Processing

A serious lung problem called byssinosis, or brown lung disease, has historically been associated with cotton processing. Early stages of the treatment of cotton, starting with the opening of the cotton bale, generate high levels of cotton dust, particularly the step called carding. An endotoxin associated with bacteria is suspected to be the toxic agent present in the smaller dust particles ($<15\ \mu m$) and is damaging to the upper respiratory tract. Irreversible damage accumulates with time and dosage of exposure. The PEL for respirable cotton dust is 200 $\mu g/m^3$. Good air filtration and ventilation are important.

Another respiratory problem, gin mill fever, occurs when workers are first exposed to cotton processing or return to the job after a long absence. The symptoms resemble influenza or metal fume fever.

The "projectile" in weaving machines is oiled. This generates a mist of oil in the atmosphere that can reach measurable levels.

Noise is a serious problem in cotton processing, the result of high-speed machinery. New machinery designs and abatement methods are gradually reducing noise levels in the early stages, but noise associated with weaving is still a serious problem.

Chemical exposures to dyes and printing inks are possible. Benzidine dyes, which cause bladder tumors, and β-naphthylamine particularly concern safety workers. Formaldehyde resins used in cross-linking may have carcinogenic potential.

WOOL

Wool is the fleece of sheep. It is washed in soap (scoured), rinsed, and possibly bleached with 35% H_2O_2. Carding, combing with fine wire teeth, removes bits of vegetation and arranges the fibers in parallel array. Wool is spun into yarn and may be dyed at this point. Cloth is woven or knitted and may be dyed later.

Problems with worker health are not common in wool processing. Handling raw wool carries some risk of infection and there may be exposure to insect-killing sheep dips. Dyes formerly included some with serious toxic potential, but these have fallen out of use.

SYNTHETIC FIBERS

Cellulosics: Rayon

Wood pulp is the raw material for rayons. Cellulose from wood is soaked in lye, shredded, and dissolved in carbon disulfide. This produces a cellulose derivative, cellulose xanthate, that is soluble in water and forms a syrupy water solution called viscose. Viscose passes through holes in a spinnerette, a device like a shower head, into a solution of sulfuric acid. Here the carbon disulfide complex is broken down, restoring the original water-insoluble cellulose structure. The result is a continuous fiber.

The greatest hazard in this process is exposure to carbon disulfide. This toxicant affects the nervous system, producing psychosis or tremors similar to Parkinson's disease.

Other Synthetics

Plastic chips are heated and forced through an extruder (spinnerette) to form a fine continuous filament, which is immediately cooled and lubricated. The strands are heated and stretched, which pulls a portion of the polymers into parallel array in the direction of the strand and so increases the strength of the filament. Some of the strands are wound on bobbins; others may be "crimped" (given a wave pattern), heated to set the wave, and cut into lengths of a few inches. This product becomes yarn. The product may be blended with another fiber such as cotton or wool. Dyeing, weaving, and knitting operations are as described above.

As in much of the textile industry, noise is a serious concern, given the use of high-speed machinery. Heat exposure at the spinnerettes, especially during service, and where the strands are heated to be drawn or crimped can be a problem. In order to reduce static electricity generated by the rapid movement of dry filament through the machinery, plants are sometimes maintained at high levels of humidity, adding to heat stress. Chemical exposures to release agents used to smooth flow through the spinnerette and to monomers and other plastic components at the melting process may occur. Lubricants applied to the fibers can cause contact dermatitis.

ADHESIVES

Industry is increasingly using adhesives in assembly operations. To plant engineers with a long history of joining objects mechanically, the use of adhesives may come as a novel idea, and they are surprised to learn it can serve their purposes. Increased use is also a response to the availability of more effective agents to bond objects. Finally, it may be a response to the need to automate to save production costs, and such bonding methods lend themselves well to this purpose.

Principles of Adhesive Bonding

It is useful to lay a groundwork of understanding about the nature of attractive forces in order to understand why some substances are easily bonded and others are not. Basically all attractions in chemistry are electrostatic, that is, the result of the attraction between opposite charges. However, within that framework are subclasses of forces. Four will be discussed here, in order of decreasing strength:

1. *Covalent bonds.* Full chemical bonds between molecules are very strong attractions. These can form only when the adhesive and the surface have chemical groups that are capable of reacting with one another. For example, if the surface included alcohol or amine groups, an epoxy resin could attach itself covalently to that surface. Such a circumstance is unusual.
2. *Hydrogen bonds.* This type of bonding is weaker than full covalent bonding, but if a large number of these bonds can form the total force is great. In general they occur when an oxygen, nitrogen, or fluorine atom can come close to an oxygen or nitrogen that has a hydrogen attached to it. Both atoms attract the hydrogen toward them so that in effect the hydrogen is partially bonded to each. Polyesters, polyamides, polyacrylates, polyacrylamides, and epoxy resins are examples of plastics capable of forming hydrogen bonds with an appropriate other surface.
3. *Polar bonds.* Many times atoms that are chemically bonded to one another do not share equally the electrons that hold them together. In such a case, the electrons are found closer to the atom with the greatest attraction for them. The bond has a negative charge at that end and a positive charge at the other end. Such bonds can line up so that the positive end of one is next to the negative end of another, producing attraction. Structures in which carbon is attached to oxygen, chlorine, or fluorine display polar bonds. The list includes those forming hydrogen bonds, as well as structures with aldehyde or ketone groups.
4. *van der Waals forces.* Because of dislocations or vibrations of their electron clouds, all atoms display temporary polarity and attract one another weakly if they are very close together. Although each attraction is very weak, a large number add up to a measurable force.

All but the van der Waals forces depend on the presence of specific functional groups. Sometimes in plastics, small amounts of an appropriate monomer are added to a polymer formulation just for this purpose. Acrylic acid, with its carboxylic acid structure, or acrylonitrile, with its cyano group, are examples of monomers used in this fashion. Some plastics by their chemical nature do not adhere well. Polyolefins and teflon are inherently very poor candidates to be used as or with adhesives.

Solvents used in adhesives must be able to "wet" the surfaces to be joined. Solvents such as water, alcohols, or ketones (acetone) wet surfaces containing polar bonds. We can see whether or not this is happening by placing a drop of solvent on the surface. If the drop interacts with the surface it spreads out, if not it remains as a bead.

The surface may be wettable based on its chemistry, but have a coating of a greasy (nonpolar) impurity that blocks intimate contact between the solvent and the surface. Therefore, testing a solvent for its ability to wet should only be done on a very clean surface.

A number of factors are important in the joining of two surfaces. The chemistry of the surfaces to be joined is the first consideration. Some sort of interaction between the surface and the adhesive is necessary to permit attachment. This is often an electrostatic attraction. The physical nature of the surfaces is also considered. A smooth surface is an advantage when the adhesive does not flow well into irregularities. On the other hand, an etched or otherwise roughened surface has a greater surface area, and, when the adhesive binds to the irregularities, this produces a greater bonding force. Finally, the cleanliness of the surface is important, because impurities may greatly weaken the bonding by preventing intimate interaction of the adhesive with the surface. All this infers the need for processes in addition to the actual use of the adhesive, including etching, roughening, or solvent cleaning, that may introduce their own specific hazards.

General Hazards

From the standpoint of industrial toxicology, introducing the use of adhesives can bring unanticipated hazards into the workplace. This is especially true if a plant has been running in a safe fashion using mechanical joining of the parts and switches to adhesives. The adhesive may not be looked upon with alarm because there is a relatively small amount in use initially or simply because workers are less alert to hazards in an ongoing, trouble-free operation. Anytime there is a change in process, the hazards must be reevaluated.

Evaluation of the hazards introduced by adhesive bonding may be more difficult than for some other new processes that are introduced. This is because the adhesive may have a complex formulation, so that several new chemicals are entering the plant at once. Furthermore, it may not be clear what these chemicals are in such a premixed medium. The need for local exhaust ventilation may be introduced in a plant where, up to that time, general circulation of air was adequate to protect workers. The manner of use of the adhesive may affect the risk. If the bonding process is being run at room temperature, and it is found to work more effectively at an elevated temperature, workers must be alert to the probability that the levels of volatile components will increase sharply in the plant air.

Polymerization Adhesives

Adhesives may be classified by their mechanism of bonding. In many cases the components necessary to form a polymer are mixed at the site where joining is to take place. The surfaces are pressed together onto this mixture and the material is polymerized or cross linked in place. Obvious advantages to this process include the opportunity for intimate contact and close fit between the polymer and the surface and, often, the lack of a need to remove solvent from the joint.

Perhaps the best known such adhesives are the epoxy resins. The polymerization reaction is initiated by mixing two preparations, the epoxy monomer and a mixture of polyamine or polyamide, with the catalyst. Unreacted epoxy or amine groups may react with groups on the surfaces being joined to provide an exceptionally strong attachment.That failing, the groups at least provide sites for electrostatic attraction between the adhesive and the surface. Cure times can be adjusted by the choice of catalyst or by the temperature selected. Epoxy phenolics are used where strength at higher temperatures is important. Such strength is also characteristic of polyimines and silicones.

Polyurethanes are useful for difficult bonding applications. These might occur where the surfaces to be joined have poor qualities of adhesion, but do have some polar structures. Joining such plastics as polyesters, polyamides, polycarbonates, and polyurethanes themselves are examples of good polyurethane applications. Polyesters are used as adhesives where toughness is less important and a less expensive material is desired.

Acrylate-type polymers are often used as adhesives. In order to obtain a viscous, sticky material to apply to the surfaces, a monomer–polymer mixture is sometimes employed. Either a peroxide catalyst or a peroxide and an activator may be used to initiate the polymerization reaction. In the latter case, the mixture with the catalyst may be applied to one surface while the mixture with accelerator is applied to the other. The reaction begins when the surfaces are pressed together. Cyanoacrylates can be used in much the same way, because they do not polymerize until air is absent and use moisture from the surface as a catalyst.

Hazards of Polymerization Adhesives

The dangers here relate to the fact that instead of the polymers themselves being employed, the considerably more toxic monomers, catalysts, and accelerators are being handled as the adhesive is applied. Special note should be taken of the properties of epoxy monomers, amines, formaldehyde, and other volatile or toxic monomers. Catalysts, such as peroxides, are dangerous both because they cause irritation and burns on contact and because they react violently on contact with certain other chemicals. Accelerators are notorious irritants. If unreacted adhesive contacts the skin and the reaction begins, there is a possibility of both chemical damage and burns from the heat produced by the polymerization reaction.

Polymers as Adhesives

Solutions of polymers are used as adhesives. They are generally applied either in a volatile solvent or as a water suspension, and in each case the liquid must evaporate to complete the bonding. Preformed polymers are less likely to form a strong joint, because the closeness of fit between the polymer and the surface is not likely to be as good and attractive interactions between the polymer and chemical groups at the surface are far less likely as a result of this poorer fit.

Substances applied in volatile solvents include nitrocellulose (the familiar model airplane glue), elastomers such as natural rubber (rubber cement), and acrylics. Elastomers and other polymers may be suspended as droplets in water, for example, in adhesives used in construction. Natural rubber, casein, and polyvinyl acetate are used in this fashion. Sometimes the properties of such an adhesive are improved if the finished joint is heated to fuse the polymer together and to drive out the last of the solvent.

Hazards of Solution Adhesives

Overall, this approach is much safer than the polymerization techniques. The polymers themselves are generally not dangerous. Care must be taken concerning the properties of the solvent used, and any necessary ventilation must be provided.

Welding

Thermoplastics may be used as adhesives by melting them, then joining the two surfaces while the polymer is still soft. Polyamides and polyesters have been used in this fashion. Thermoplastics also can be joined to one another without an adhesive by heating the surfaces to be joined until they melt, then pressing them together.

Hazards of Molten Plastic Adhesives

Any time plastics are heated to the melting point there is the danger of volatile chemicals entering the atmosphere. Primarily, these would be unreacted monomer, but other volatile chemicals may be trapped in the solid plastic. If above-minimum temperatures are employed to do the melting there is increased possibilities for degradation of the plastic, plasticizers, or other chemicals in the plastic, accompanied by a release of volatile liquids or gases. The manner in which the welding is done influences the possibilities for contamination of the workplace atmosphere. For example, if a sheet of plastic is placed between two objects to be joined and those objects are heated, the hot plastic has minimum exposure to the atmosphere. On the other hand if the plastic is melted on an open surface, the possibility for the air becoming contaminated is much greater. Adequate ventilation is the most important safeguard at the workstation.

Solvent Bonding

When two plastic surfaces are to be joined together, it is sometimes possible to soften the plastic with a solvent, press the surfaces together, and wait for the solvent to evaporate. Obviously this is possible only if the polymer dissolves in a volatile solvent. The joint so produced has properties that depend on the intimacy of mixing of the polymers at the surfaces, the elimination of gaps, and the success of the solvent removal. At best, the polymer strands retain a random orientation at the interface, so that the joint remains the weakest part of the combined structures.

Having a solvent of the correct volatility is important. If it is too volatile, it may evaporate before the joining is accomplished. However, if it is too low in volatility, an excessive length of time is required to remove it from the assembly. Commonly employed solvents include toluene, xylene, methyl ethyl ketone, methyl alcohol, and methylene chloride.

Hazards of Solvent Bonding

Clearly, the hazards of solvent bonding arise from the release of solvent vapors into the workplace. The solvents used are commonplace chemicals with well-known toxicity characteristics. Ventilation is essential.

Coupling Agents

Coupling agents are linear organic chains with a silane group on one end and some chemical functional group such as a double bond, an epoxy group, or an amine group at the other end. The silane group adheres well to glass or metals. When these substances are used to coat fillers such as glass fibers, a functional group handle is added to the filler with which a polymeric adhesive can react. These agents are chemically quite reactive and so are generally toxic and irritating. The manufacturer's recommendations should be consulted regarding handling and personal protection when coupling agents are used.

Surface Preparation and Its Hazards

A clean surface is essential to tight bonding, but the cleaning procedures often involve the use of quite hazardous chemicals. Removal of oily residues from the surface may be done either with a detergent or with a solvent. Solvents are most frequently used and introduce the problems associated with solvent vapors in the air. Skin irritation due to the removal of skin oils or the passage of solvent through the skin into the body are added risks.

Surfaces are often roughened or cleaned by abrasion before bonding. Measures must be taken to prevent the inhalation of particles broadcast into the air by such operations. Metallic surfaces may also be treated to remove oxide coatings that would reduce adherence. Although abrasive cleaning can remove oxides, etching the surface with chromic acid does a more satisfactory job. Such cleaning results in bonding to adhesives that is as much as eight to ten times as strong as solvent cleaning alone. The etching solution contains sulfuric acid, a very strong acid capable of producing serious chemical burns, and a dichromate salt, which is a very strong oxidizing agent. The combination is deadly and must be treated with extreme care. Not only is there the potential for chemical burns, but particulate dichromate or droplets from the bath entering the atmosphere are very irritating to the nasal cavities and on long exposure can cause bronchitis. Chromium has been listed as a carcinogen, with respiratory tract tumors occurring in workers who inhale chromium compounds. The use of chromic acid baths should be accompanied by extensive safeguards.

KEY POINTS

1. Plastics are solid polymers that divide into two groups: linear polymers or thermoplastics and cross-linked (three-dimensional) or thermoset plastics. Thermoplastics can take on thermoset properties by cross-linking (curing) the plastic.
2. Addition polymers are formed by joining unsaturated monomers by addition reactions, eliminating the double bonds. Condensation polymers are formed by reaction of functional groups to join monomers together. Copolymers have more than one type of monomer.
3. Polymer properties are affected by the degree of crystallite formation in the solid. Plasticizers are inert additives that prevent crystallite formation.
4. Other plastics additives include pigments, stabilizers to block UV light, and fillers to extend the plastic and/or modify properties.
5. Monomers are mixed and catalysts added to stimulate polymerization. This is done to thermoplastics before shipping and to thermoset plastics at the time of molding.
6. Plastics may be shaped by injection molding, casting, blow molding, vacuum and pressure forming, extrusion, and calendaring. Plastics may also be foamed and shaped.
7. Hazards in plastics manufacturing arise from the chemicals and from particulate produced by grinding, sawing, or otherwise shaping.
8. Elastomers are linear polymers that return to their original shape when distorted. They are generally addition polymers and may be copolymers. A major use is in the manufacturing of tires.
9. Manufacture of elastomers is similar to that of plastics, and many of the same hazards are involved.
10. Textile polymers include natural polymers (cotton, wool) and synthetic polymers (rayons, nylons, orlons). Cotton processing involves a special hazard from airborne cotton fibers, which produce brown lung disease. Synthetic fibers have hazards similar to those of the corresponding plastics.
11. Adhesives join surfaces by generating attractive forces at the interface of the pieces. Polymerization adhesives form polymers from monomers at the interface. Solutions of polymers lose solvent at the interface and join surfaces by their mutual attraction to the polymer. Polymers may be melted in the interface to join the surfaces. Solvents may dissolve the

surfaces, usually plastic surfaces, at the interface to join the surfaces. Finally, compounds that react with functional groups at the interface surface can join the faces covalently. Hazards include the chemicals used and the formation of particles as the surfaces are prepared for joining.

BIBLIOGRAPHY

AIHA, *Toxicological, Industrial Hygiene and Medical Control of Polyurethanes, Polyisocyanurates, and Related Materials*, 2nd ed., Akron, OH, 1983.

W. A. Burgess, *Recognition of Health Hazards in Industry*, 2nd ed., John Wiley & Sons, New York, 1995.

L. F. Dieringer, "Rubber," in L. V. Cralley and L. J. Cralley, Eds., *In Plant Practices for Job Related Health Hazards Control*, Vol. 1, John Wiley & Sons, New York, 1989.

International Agency for Research on Cancer, *Monographs on the Evaluation of Carcinogenic Risk of Chemicals to Humans: The Rubber Industry*, Vol. 26, World Health Organization, 1982.

International Labour Office, *Encyclopedia of Occupational Health and Safety*, Geneva, Switzerland.

J. Jarvisalo, P. Pfaffli, and H. I. Vainio, Eds., *Industrial Hazards of Plastics and Synthetic Elastomers*, Alan R. Liss, New York, 1984.

E. B. Katzenmeyer, Jr., "Butadiene Rubber," in L. V. Cralley and L. J. Cralley, Eds., *In Plant Practices for Job Related Health Hazards Control*, Vol. 1, John Wiley & Sons, New York, 1989.

A. J. Kinloch, *Adhesion and Adhesives*, Blackie and Son (Chapman and Hall), New York, 1987.

J. R. Lynch, "Butyl Rubber," in L. V. Cralley and L. J. Cralley, Eds., *In Plant Practices for Job Related Health Hazards Control*, Vol. 1, John Wiley & Sons, New York, 1989.

M. Morton, Ed., *Rubber Technology*, 2nd ed. Van Nostrand-Reinhold, New York, 1985.

J. E. Mutchler, "Plastics," in L. V. Cralley and L. J. Cralley, Eds., *In Plant Practices for Job Related Health Hazards Control*, Vol. 1, John Wiley & Sons, New York, 1989.

J. D. Neefus, Ed., *Cotton Dust Exposures*, Vols. I and II, American Industrial Hygiene Association, Akron, OH, 1984 and 1987.

NIOSH, *Control Technology in the Plastics and Resins Industry*, Publication No. 81-107, DHHS, Washington, D.C., 1981.

A. R. Nutt, *Toxic Hazards of Rubber Chemicals*, Elsevier Applied Science Publishers, New York, 1984.

Rubber and Plastics Research Association, *Clearing the Air — A Guide to Controlling Dust and Fume Hazards in the Rubber Industry*, Shrewsbury, England, 1982.

V. O. Sheftel, *Handbook of Toxic Properties of Monomers and Additives*, CRC Press, Boca Raton, FL, 1995.

I. Skeist, *Handbook of Adhesives*, Van Nostrand-Reinhold, New York, 1977.

U. S. Department of HEW, *Health and Safety Guide for Plastic Fabricators*, NIOSH, Cincinnati, OH, 1975.

P. F. Woolrich, "Polyurethanes and Polyisocyanurates," in L. V. Cralley and L. J. Cralley, Eds., *In Plant Practices for Job Related Health Hazards Control*, Vol. 1, John Wiley & Sons, New York, 1989.

PROBLEMS

1. Recycling plastics is a relatively new industry. Imagine inspecting a plant that is recycling polypropylene milk cartons and converting the scrap into fishing tackle boxes. The cartons are compressed into blocks and chipped into small pieces. These are washed with strong detergent in hot water, rinsed, and dried in a stream of hot air. The chips are then transferred to a grinder where they are mixed with pigments and transferred to the hopper of an injection molding machine where they are converted to the product. What hazards would you look for in this operation?
2. Thermoset plastics are rigid because of a massive progression of covalent bonds with fixed angles. Consider what generates the degree of rigidity found in a thermoplastic.
   A. Why would nylon be more rigid than polyethylene, if the only differences were the structure of the polymer?
   B. Why would polypropylene that had been cooled slowly be more rigid than a sample that had been cooled quickly?
   C. How does cross-linking affect rigidity?
   D. How does a plasticizer affect rigidity?
3. You are the safety officer in a plant that assembles display shelving for retail stores. The company is considering switching from fasteners (screws and bolts) to adhesives to lower production costs. The plant engineer has decided that a polymerization adhesive would be stronger, producing a better product. You are asked to prepare a comparison of the advantages and disadvantages of using polymers in solution versus polymerization adhesives from the standpoint of the cost of modifying the plant to accommodate the new process safely.
4. You have been hired as an industrial hygienist at a textile mill that converts raw cotton to woven cloth.
   A. Use the index in 29 CFR 1910 to locate the standards for cotton dust.
   B. What is the definition of respirable cotton dust?
   C. What is the standard method for its measurement?
   D. What is the PEL for cotton dust?
   E. How often must monitoring be done?
   F. Read in Appendix A the details of the sampling procedure.
5. Locate vinyl chloride in the index of 29 CFR 1910.
   A. What is the PEL, ceiling, and action level of vinyl chloride?
   B. Contrast controls on exposure to the polymer forms of acrylonitrile (29 CFR 1910.1045) and polyvinylchloride.
6. Select and read an article from the following list and indicate (1) what workplace problems are discussed and (2) what analytical methods are employed: *Am. Ind. Hyg. Assoc. J.*, 44, 521, 1983; 38, 205, 1977; 41, 204, 1980; 41, 212, 1980; 38, 394, 1977 .

Glossary

A

AAIH — American Academy of Industrial Hygiene; a professional organization whose members all are certified industrial hygienists

ABIH — American Board of Industrial Hygiene; the accrediting organization that administers the certified industrial hygienist and industrial hygienist in training exams.

absorption coefficient (α) — the ratio between sound energy absorbed by a surface and the sound energy striking that surface

accelerator — a compound that improves the effectiveness of a catalyst

acceptable ceiling concentration — a ceiling level of exposure that is permitted only for a specified time interval

acclimatization (to heat) — adjustment of the body to working at high temperatures

accuracy — in chemical analysis accuracy indicates how close the measured value is to the true value

ACGIH — American Conference of Governmental Industrial Hygienists; this professional organization enrolls individuals who work for governmental or educational agencies

acid descaling (pickling) — cleaning a metal surface with acid

acne — a swelling due to the blockage of flow of oil out of a hair follicle

acoustics — the science of sound

acroosteolysis — the degeneration of bones in the hands as the result of exposure to vinyl chloride

action level — a level of an airborne toxicant equal to half the PEL; when the workplace is below this level, regulations are simplified

acute exposure — exposure to a harmful agent for a brief time

acute lethal dose — the LD_{50} of a compound when tested for a very short period, usually as a single dose

acute testing or dosing — testing of the toxic effects of a compound for a short period of time, usually a single dose

addition polymer — a polymer formed by an addition reaction from monomers with carbon-to-carbon double bonds

administrative controls — reduction of employee daily exposure to a harmful agent, usually by job rotation or imposition of rest periods

adsorbent tubes — tubes used for collection of gas or vapor for analysis that adsorb gas or vapor on the surface of a solid

aerosols — airborne liquid droplets

AHERA — Asbestos Hazard Emergency Response Act (1986)

AIHA — American Industrial Hygiene Association

AIHA-WEEL — Workplace Environmental Exposure Level; safe exposure level recommendations by the American Industrial Hygiene Association

air flow — the volume of air per unit time moving through a duct

airline respirator — a respirator that supplies air from a tank through a hose to a mask

alkaline descaling — cleaning a metal surface with caustic

allergic contact dermatitis — skin inflammation due to allergic response to a chemical upon skin contact
allergic response — the activation of the immune system by contact with a chemical to produce swelling, irritation, asthma, fluid retention, or inflammation
alloy — a deliberate mixture of metals prepared to obtain desired properties
alopecia — loss of hair due to toxic response to a chemical
alpha radiation (α radiation) — radioactive particle emitted from the atomic nucleus composed of two protons and two neutrons
aluminum dust lung — lung fibrosis from inhaling aluminum powder
alveoli — the air sacs at the ends of the finest branches of the lungs in which gases exchange between air and the blood
ambient temperature — the air temperature at the worker
ampere — the unit of rate of flow of electrical current
amplitude (sound) — when sound is visualized as a wave, amplitude is the peak height and is proportional to loudness
anemia — inadequate capacity to transport oxygen in the blood
anemic hypoxia — reduced oxygen supply due to impaired oxygen transport by the blood
anemometer — a device employing cups on a spinning wheel to measure air velocity
anoxic hypoxia — reduced oxygen supply due to interference with respiration
ANSI — American National Standards Institute
antibody (immunoglobulin) — one of a large population of proteins intended to defend against viral or bacterial infection that may also initiate allergic response
antidote — an agent that counteracts the effects of a poison
anti-fatigue mats — floor mats that are somewhat padded for reducing the stress of continuous standing
antigen — a chemical to which an antibody binds that may stimulate an allergic response
antimony spots — a skin rash caused by exposure to antimony
aqueous — a water solution
aromatic solvent — an organic compound including six-membered benzene rings in its structure
asbestosis — a lung disease caused by inhaling asbestos fibers
ASHRAE — American Society of Heating, Refrigerating, and Air-Conditioning Engineers, 1791 Tullie Circle NE, Atlanta, GA, 30329
asphyxiant — a gas that causes a reduction of the oxygen supply to tissues
ASSE — American Society of Safety Engineers, a professional society
ASTM — American Society for Testing and Materials; a resource for sampling and testing methods, for safety problems with chemical, biological, and physical agents, and for safety guidelines
atm — abbreviation for atmosphere, a unit of pressure
atomic absorption spectrometry — an analytical method for metals in which light passing through a hot gas stream is selectively absorbed by specific metal atoms or ions
audiometer — a device that generates pure tones at known sound pressure levels to test hearing
autoignition temperature (ignition temperature) — the lowest temperature that causes combustion of a gas or vapor without a spark
AWMA — Air and Waste Management Association; an environmental pollution professional society [P.O. Box 2861, Pittsburgh, PA 15230; (412) 232-3444]

B

bag house — a chamber with filters at the exhaust point of a plant ventilation system in which particulate is removed from the air
barrier cream — a skin cream containing oils or silicones that helps prevent contact of the skin with chemicals
basal cell carcinoma — a type of skin cancer
basic oxygen furnace — a furnace for converting iron to steel in which oxygen is passed through molten iron and resulting oxidized impurities react with limestone
BCSP (Board for Certified Safety Professionals) — the accreditation organization administering the certified safety professional and OHST exams

becquerel (Bq) — a unit for measuring radioactivity equal to 1 dps (disintegration per second)
behavioral toxin — a toxic agent that damages or otherwise affects the nervous system
BEI — biological expsoure index; a measure of exposure to a harmful biological agent
benefication — removal of clay, dirt, or rock from ore after its removal from the mine
beta radiation (β radiation) — radioactive particle composed of an electron emitted from the atomic nucleus
bias (determinate error) — error in an analysis such that the measured value is always higher or lower than the true value
biohazard — used to indicate a location or materials potentially contaminated with a harmful biological agent
biopsy — the removal of a sample of living tissue for examination
black lung disease (coal miners pneumoconiosis) — a disease of the lungs caused by inhaling coal dust
blast furnace — an iron smelter using coal and limestone to convert ore to iron
blood–brain barrier — refers to the relative difficulty of transporting chemicals from the blood into the central nervous system
bloodborne pathogen — an infectious microorganism that can be transmitted by contact with the blood of an infected individual
blow molding — formation of a hollow object such as a bottle by inflating plastic inside a mold
blowing agent — a reagent added to plastic to generate gas, creating a foam
body burden — the amount of a chemical retained in the body after exposure
boiling point (BP) — the temperature at which a liquid converts to a gas with boiling at a given pressure (usually one atmosphere)
BOHS — British Occupational Hygiene Society; a professional organization [1 St. Andrew's Place, Regent's Park, London NW1 4LB, England (01-486-4860)]
BOM — U.S. Department of the Interior Bureau of Mines
Bq (becquerel) — a unit for measuring radioactivity equal to 1 dps (disintegration per second)
breakthrough (sample collection) — the amount of sample that saturates the collecting device
bronchial tree — the branched structure of air passages in the lung
bronchioles — smaller branches of the respiratory tract
brown lung disease (byssinosis) — a lung disease caused by inhaling cotton fiber
BSFTPM — the OSHA index for worker exposure to the benzene soluble fraction of total particulate in a coke oven
bubbler (impinger) — a gas- or vapor-collecting device that bubbles the air sample through a container of solvent
BuMines — U. S.Department of the Interior Bureau of Mines
byssinosis (brown lung disease) — a lung disease caused by inhaling cotton fiber

C

CAA — Clean Air Act (40 CFR 5080); this law sets levels for atmospheric contamination and industrial discharge
calendaring — coating a fabric with an elastomer
capillary (anatomy) — the smallest diameter blood vessels through the walls of which nutrients, oxygen, and wastes are transported to and from the tissues
capture velocity — the air velocity needed to move molecules or particles in the atmosphere into a duct
carcinogen — a compound exposure to which increases the likelihood of cancer
carpal tunnel — a passage through the bones of the wrist for the tendons and median nerve
carpal tunnel syndrome — inflammation of the median nerve as it passes through the carpal tunnel resulting in a loss of sensation and grip
carrier gas — the moving phase in gas chromatography
CAS — Chemical Abstracts Service; a division of the American Chemical Society that abstracts the chemical research literature
CAS number — an identification number assigned to a chemical for identification by the Chemical Abstracts Service
catalyst — a chemical that accelerates the rate of a chemical reaction

cataract — the clouding of the lens of the eye
caustic — NaOH
CDC — Centers for Disease Control znc Prevention; a government organization assigned to monitor and control the spread of disease
ceiling value — the maximum level of exposure to a chemical permitted in the workplace, with the duration of this exposure being unspecified
centrifugal (cyclone) collector — air contaminated with particles is swirled in a cone-shaped chamber, spinning particles out against the chamber walls
CERCLA — Comprehensive Environmental Response, Compensation, and Liability Act, which established Superfund for toxic waste site cleanup and requires notification of the Coast Guard National Response Center of a toxic release
cfm — in air flow, cubic feet per minute
CFR — *Code of Federal Regulations*; a government publication listing the federal regulations
CGA — Compressed Gas Association; a private group concerned with problems with compressed gases; it is the source of some OSHA standards
chemical asphyxiants — compounds that block the transportation or utilization of oxygen
chemical pneumonitis — lung inflammation and fluid accumulation caused by a chemical irritant
CHEMTREC — Chemical Transportation Emergency Center; a 24-hr source of chemical information (800-424-9300) for transportation emergencies, established by the Chemical Manufacturers Association
chloracne — the formation of acne-like skin eruptions as the result of exposure to certain chlorinated hydrocarbons
chromatography — a separation method in which the compounds under study are partitioned between a stationary and a moving phase
chromatography column — the holder of the stationary phase in chromatography
chromosome — one of the several DNA double helices found in the nucleus of a cell
chronic testing or dosing — testing of the toxic effects of a chemical for an extended period of time, usually the lifetime of the test animal
Ci (curie) — a measure of radioactivity equal to 3.7×10^{10} dps (disintegrations per second)
CIA — Certified Industrial Hygienist
CIH — certified industrial hygienist; a person who has passed the exam of the American Board of Industrial Hygiene
cilia — hair-like structures under the mucous layer of the respiratory tract that move mucus toward the nose
cirrhosis — scarring of tissue due to a toxic agent or agents
classifier — a screen used to sort particles of crushed ore by size
clearence — the removal of substance from the blood by the kidneys
Cleveland Open Cup — a test method to determine flashpoints
CMA — Chemical Manufacturers Association; a private group of chemical industry representatives
CNS — central nervous system, including the brain and spinal cord
coal miners pneumoconiosis (black lung disease) — a disease of the lungs caused by inhaling coal dust
cochlea — the nerve-ending-lined, fluid-filled spiral in the inner ear where nerve impulses of hearing originate
coke — coal roasted to remove volatile components
colic — constipation followed by intense abdominal pain (as in lead poisoning)
collection efficiency — the percentage of gas or vapor that a sampling device collects
combustible liquid — a liquid with a flashpoint between 100 and 200°F; a term used by the National Fire Protection Association, the Department of Transportation, and other groups
combustion-lean limit mixture (lower flammability limit) — the lowest concentration of a gas or vapor able to sustain continuous combustion
combustion-rich limit mixture (upper flammability limit) — the highest concentration of a gas or vapor able to sustain continuous combustion
comminution — breaking ore into a powder

complementary DNA — the two strands of DNA in the double helix are complementary in that each base on one strand always pairs with a specific but different base on the other strand
compressed gas — a contained gas or gas mixture whose pressure exceeds 40 psi at 70°F or 104 psi at 130°F, or a liquid having a vapor pressure exceeding 40 psi at 100°F
condensation polymer — a polymer formed from monomers each of which has two functional groups, and these groups are capable of reacting to join the monomers together
confined space — a space that has limited access ports and inadequate to nonexistant ventilation that threatens entrapment, suffocation, or engulfment of workers that enter
conjugation — the attachment of chemical structures to foreign substances in the body to facilitate their removal by the kidney
conjunctivitis — irritation of the eye by exposure to UV light
contact dermatitis — dermal response to a chemical upon direct application of that chemical to the skin
continuous sampling (integrated sampling) — sample collectors that gather the sample continuously for a period of time
control group — a set of test animals in a study that are not dosed, but are otherwise treated the same as dosed animals
convective heat transfer — transfer of heat to or from the body by contact with hotter or colder matter
cope — the top flask in a two-piece mold
copolymer — a polymer composed of more than one kind of monomer
core temperature — the internal body temperature, taken at the mouth, ear, or rectum
cornea — the transparent front surface of the eye
corrosive — a chemical that destroys tissue on contact
covalent bond — a system for strongly joining atoms together based on a sharing of electrons in which the distances between and angles formed by the atoms are relatively fixed
CPSC — Consumer Product Safety Commission; the agency regulating hazardous materials in consumer goods
cross-linking (curing) — joining strands of polymers together to create a three-dimensional polymer
crystallite — an orderly region within the mass of a polymer
CSP — certified safety professional (see *BCSP*)
cumulative trauma — physical injury due to repeated insults to the body over a period of time
curie (Ci) — a measure of radioactivity, one Ci equaling 3.7×10^{10} dps (disintegrations per second)
curing — (plastics) cross-linking or catalyzing the formation of a thermoset plastic; (elastomers) cross-linking an elastomer, vulcanization
cyanosis — the condition of an organism experiencing a shortage of oxygen
cyclone (centrifugal) collector — air contaminated with particles is swirled in a cone shaped chamber, spinning particles out against the chamber walls

D

dB — see *decibels*
decibels (dB) — units of sound pressure level
decay constant (radioactivity) — the fraction of atoms that decay per unit time
deflecting vane velometer — a device used in the analysis of a ventilating system that measures the air velocity
dermal — pertaining to the skin
dermis — the second layer of the skin which includes blood vessels, nerves, and collagen fibers
desorption efficiency — the percentage of a collected sample recovered from the collection medium
determinate error (bias) — error in an analysis such that the measured value is always higher or lower than the true value
DF (radioactivity) — distribution factor; the relative dosage absorbed by a specific organ or tissue
dike — a barrier erected to contain a spilled liquid

dilution ventilation (GEV, general exhaust ventilation) — movement by a ventilation system of the entire mass of air through a plant, diluting and eventually removing impurities
diploid — cells that contain two sets of DNA, one from each parent
displaced disc (in the spinal column) — when a disc shifts from its correct location, creating pressure on nearby nerves
distribution factor (DF, radioactivity) — the relative dosage absorbed by a specific organ or tissue
DNA — deoxyribonucleic acid; the very large molecules in the nucleus of the cell carrying the genetic information
DNA double helix — the spiral assembly of a pair of related DNA molecules, which is the form in which cell DNA occurs
DOL — the Department of Labor
dose rate — dose per unit time
dosimeter (radioactivity) — a portable device worn by workers handling or near sources of radioactive materials that measures accumulated exposure
DOT — the Department of Transportation
dpm (radioactivity) — disintegrations per minute
dps (radioactivity) — disintegrations per second
drag — bottom flask in a two-piece mold
dust — airborne particulate generated by grinding, crushing, blasting, or agricultural processes
dynamic precipitator — a motor-driven cyclone collector for removing airborne particulate
dysplasia — abnormal development
dyspnea — shortness of breath

E

eardrum — the membrane separating the outer and middle ear that vibrates in response to sound waves
edema — the retention of fluid in an organ or in the body as a whole
EHS — extremely hazardous substance; one of about 400 substances designated by the EPA under SARA Title III as sufficiently dangerous that local planning committees must be informed if it is in a facility
electrolytic purification (metallurgy) — pure metal is obtained at the cathode from ore or from an impure metal anode
electron capture detector — a gas chromatography detector in which interruption of a constant stream of electrons from a β source by electronegative atoms in the compounds being analyzed is the method of detection
electrostatic precipitators — a system to remove particulate from air utilizing charged plates
electrowinning (metallurgy) — obtaining a metal from ore by leaching followed by plating the metal onto a cathode
emergency response — removal of a spilled or leaking hazardous chemical
emergency standard — the exposure standard established by OSHA without following standard procedures upon discovery of a serious threat by a chemical in common use
emphysema — the loss of lung capacity due to tissue damage
enclosure (in ventilation) — a construction at the site of contamination generation that is swept by air entering a hood
engineering anthropometry — matching the dimensions of devices to the range of human dimensions
engineering controls — reduction of the hazard due to an agent by reducing its levels
enzyme — a catalyst for chemical reactions produced by the cells
EP — ethylene propylene copolymer
EPA — the Environmental Protection Agency
EPCRA — Emergency Planning and Community Right to Know Act. SARA Title III
EPDM — ethylene propylene terpolymer
epicondylitis (tennis elbow) — inflammation of the tendon attaching at the elbow and connecting to muscle in the forearm, used to extend the arm

epidemiology — the study of the relative characteristics of large exposed and nonexposed populations for the purpose of detecting toxic effects in the exposed population
epidermis — the outer layer of the skin, the outer portion of which is interlocked dead cells
epithelium — the tissue covering outer or inner surfaces of the body
ergonomics (human factors engineering) — the study of locating tasks, laying out work schedules, and designing tools to prevent cumulative trauma and/or optimizing human performance
evaporation rate — the relative rate at which a substance evaporates compared to *n*-butyl acetate, which is assigned a value of 1; the rate is termed fast if greater than 3 and slow if less than 0.8
evaporative heat loss — the loss of body heat due to evaporation of sweat
explosive — a substance that can be triggered to react so that destructive pressure and heat are very suddenly produced
extensor muscles — muscles that are used to open or extend a movable joint
extrusion — forming bars or tubes by forcing plastic or metal through a die

F

f/cc — fibers per cubic centimeter of air
face velocity — the average air velocity through the opening of a ventilation hood
FDA — the Food and Drug Administration
Federal Register — the published record of the activities of congress
fetus — a child from week 7 of gestation until birth
fibrosis — the formation of scar tissue, especially in the lungs
FID (flame ionization detector) — a detector in gas chromatography in which ions produced by burning the compound under study then pass between charged plates allowing current to flow in the external circuit
FIFRA — Federal Insecticide, Fungicide, and Rodenticide Act; requires labeling of health hazards on poisons sold to the public
fillers — compounds added to plastics to influence the properties or allow use of less plastic
filter — a porous medium used to remove particulate
fire diamond — a commonly used warning label used on chemical containers
first aid — emergency care given to an injured person prior to the provision of regular medical care
fit test — tests to check the correct fit of a respirator to the face
flame arrestor — a screen placed across a duct or opening to a container to cool approaching burning gases below the ignition temperature of container contents
flame ionization detector — see FID
flammability limits — the concentration of a gas or liquid at the low or high extremes of ability to sustain continuous combustion
flammable aerosol — produces an 18-in. flame at full valve opening on testing (method of 16 CFR 1500.45) or a flashback to the valve (any valve opening)
flammable gas — forms a flammable mixture with air at 13% (volume) concentration or less, or a range of flammable mixtures wider than 12% (volume) with any lower limit
flammable liquid — a liquid with a flashpoint below 100°F
flammable solid — ignites and burns with a self-sustained flame faster than 0.1 in./s along its major axis when tested by the method in 16 CFR 1500.44
flashpoint — the minimum temperature to raise the vapor pressure of a flammable liquid to the lower flammability limit; test methods include Tagliabue Closed Tester (ANSI Z11.24 1979), Pensky-Martens Closed Tester (ANSI Z11.7 1979), and Setaflash Closed Tester (ASTM D 3278-78)
flask — a container in which a sand mold is shaped
flexor muscles — muscles used to close a movable joint
fluorosis — a disease resulting from exposure to airborne fluoride, experienced primarily by aluminum workers
follicle — the passage for a shaft of hair passing through and exiting the skin
forging — shaping metal by hammering

foundry — a plant where metals are cast into products
fpm — feet per minute; measure of rate of air flow through a duct
FR — *Federal Register*
frequency (sound) — when sound is visualized as a wave, frequency is the number of waves passing a point per unit time and determines the tone
friction coefficient — a measure of the friction between a shoe and a floor surface
frictioning — forcing elastomer into a fabric with a hot roller moving faster than the sheet of fabric
froth flotation — use of foams to float out minerals from powdered ore
FTIR — fourier transform infrared spectrophotometry; a rapid method of infrared spectrophotometry
full protective clothing (a moon suit) — a completely enclosing garment that includes a supplied air respirator
fume — small diameter airborne particulate produced by high-temperature metal processing; particles usually react in air to become oxides

G

gamma radiation (γ radiation) — very-high-frequency electromagnetic radiation emitted by the atomic nucleus
gas chromatography — a chromatographic separation method in which the moving phase is a gas
gastrointestinal tract — the digestive system including everything from the mouth through the stomach and intestines
GC — gas chromatography
gene — the message for assembling a single protein coded as a base sequence on DNA
general duty clause — the statement in the OSH Act [Sect 5 (a)(1)] that states an employer must provide workers with a hazard-free work environment
general exhaust ventilation (GEV, dilution ventilation) — movement by a ventilation system of the entire mass of air through a plant, diluting and eventually removing impurities
generic name — any name given to a chemical besides its correct chemical name
Geiger-Muller counter — a device for measuring radioactivity based on the creation of ion cascades in a gas by ionizing radiation
GEV — general exhaust ventilation, dilution ventilation
GI tract — the gastrointestinal tract
gin mill fever — an illness like metal fume fever experienced by new or returning workers in a cotton processing facility
glomerulus — the structure in the kidney at which the fluid portion of the blood is filtered into the tubules
grab sample — sample collecting in which a sample is collected in one pass
gray (Gy) — an SI unit of radioactive exposure equal to 100 rads
ground or grounding — connection of an electrical circuit or device literally to the ground

H

half-life (toxicology) — the time required for half of a dose of a chemical to leave the body
half-life (radioactivity) — the time required for half the radioactive atoms of a given isotope to disintegrate in a sample
half-thickness (radioactivity) — the thickness of a shielding material required to block half of a specific kind of radioactive emission
hazard (of a chemical) — the likelihood under conditions of use of harm being caused by a chemical
Hazard Communication Standard — an OSHA standard established in 1983 requiring all employers to inform employees of the hazard of chemicals in the workplace and of the steps necessary to avoid harm (29 CFR 1910.1200)
HAZMAT teams — teams trained to respond to chemical spills under HAZWOPER
HAZWOPER — hazardous waste operations and emergency response; SARA Title I; 29 CFR 1910.120

HBV — hepatitis B virus; a serious viral pathogen attacking the liver
heat cramps — muscle cramps due to loss of fluid and salt while working in a hot environment
heat exhaustion (heat prostration) — fatigue, weakness, and/or fainting due to reduced blood supply to the brain as a result of working in a hot environment
heat stress — increase in core body temperature resulting from working in a high-temperature environment
heat stroke — a very high core body temperature leading to tissue damage resulting from working in a high temperature environment
hematopoietic system — the system forming blood cells
HEPA — high-efficiency particulate filter
hepatotoxin — a toxin whose target organ is the liver
herneated disc (in spinal column) — the discs are like bags with viscous filler; these can rupture, resulting in pressure on nearby nerves
histotoxic hypoxia — a shortage of oxygen due to inability of tissues to use delivered oxygen
hertz (Hz) — the units of frequency (s^{-1})
HHS — Department of Health and Human Services (formerly the Department of Health, Education, and Welfare); the federal bureau that includes the Public Health Service and NIOSH
high-efficiency cyclone collector — a cyclone collector with a narrow cone, swirling the air faster and improving the efficiency of separation
highly toxic — a chemical that has an LD_{50} of 50 mg/kg or less, an LC_{50} of 200 ppm (2 mg/l; mist, dust, or fume) on dosing 1 hr, or a dermal LD_{50} of 200 mg/kg in 24 hr
HMIG — Hazardous Materials Identification Guide
hood (ventilation) — the site of air uptake in a local exhaust ventilation system
HPLC — high-performance liquid chromatography; a chromatography method in which the liquid moving phase is pumped through a column containing the stationary phase
human factors engineering (ergonomics) — the study of designing objects such as tools used by workers to prevent cumulative trauma
humerus (anatomy) — the bone of the upper arm between the shoulder and the elbow
HVAC — heating, ventilating, and air conditioning system
hydrogen bond — a moderately strong attractive force between a pair of atoms (oxygen, nitrogen, or fluorine), both of which are attracted to a hydrogen between them
hyperplasia — a growth of new cells in tissue, as in tumor formation
hyperthermia — body core temperature above normal
hypokinetic hypoxia — a shortage of oxygen due to interference with blood flow
hypothermia — body temperature below normal (below 35°C)
hypoxia — a shortage of oxygen in an organism
Hz — see *hertz*

I

IAFF — International Association of Fire Fighters
IAQ — indoor air quality
IARC — International Agency for Research on Cancer
ID_{50} — irritating dose to 50% of the test sampling; a measure used in studying agents causing irritant contact dermatosis
IDLH — immediately dangerous to life and health; the maximum level of exposure from which a person could exit in 30 min without escape-impairing symptoms or irreversible health effects
IES — Illuminating Engineering Society, 345 E. 47th St., New York, NY 10017; (212) 705-7926
ignition temperature (autoignition temperature) — the lowest temperature that causes combustion of a gas or vapor without a spark
IHIT — Industrial Hygienist in Training; the intermediate stage on the path to the CIA exam
immunoglobulin — see *antibody*
impervious — blocks the passage of chemicals
impinger (bubbler) — a gas- or vapor-collecting device that bubbles the air sample through a container of solvent
inches of water (ventilation) — pressure as measured on a water manometer

incompatible substances — substances that react on contact
indeterminate error (random error) — error in analysis where the measured values scatter around the true value randomly
infectious waste — waste potentially containing infectious agents that is stored in a red bag
inflammation — reddening and swelling due to chemical contact
infrared spectrophotometer (IR spectrophotometer) — an analytical instrument that measures the absorbance of specific frequencies of infrared light by a compound under study
ingestion — taken into the body by mouth
inner ear — the part of the ear including the cochlea that converts vibration to nerve impulses
injection molding — forming plastic by forcing molten plastic into a mold
integrated sampling (continuous sampling) — sample collectors that gather the sample continuously for a period of time
ionizing radiation — radioactive emissions that drive electrons off atoms as they pass
IR spectrophotometer (infrared spectrophotometer) — an analytical instrument that measures the absorbance of specific frequencies of infrared light by a compound under study
irritant — a compound that causes inflammation and edema
irritant contact dermatitis — irritation due to skin contact with a chemical
ischemia — loss of circulation and numbness due to blockage (as by compression) of blood vessels
isotopes — atoms of the same element with differing numbers of neutrons
itai-itai disease — bone embrittlement from exposure to cadmium

J

jolt machine — a machine for packing sand in molds by dropping and suddenly stopping the flask
jolt rollover squeeze machine — a machine that forms sand molds in a single operation at high speed

K

keratin layer — the outer layer of dead cells in the epidermis
kHz — 1000 Hz
kcal — energy needed to raise 1 kg of water 1°C

L

lacrimator — causes the eyes to tear
layering (gases) — separation of gases in a contained environment by density
LC_{50} — the concentration lethal to 50% of the animals in a test group, usually of gas or particulate in air
LC_{LO} — the lowest concentration of a compound proving to be lethal, usually of gas or particulate in air
LD_{50} — the dose lethal to 50% of the animals in a test group
LD_{LO} — the lowest dose of a compound proving to be lethal
leaching (metallurgy) — metal ions are dissolved from ore using acid, alkali, or a complexing agent
LEL — lower explosive limit; the lowest concentration of a vapor forming an explosive mixture in air
LEPC — local emergency planning committee
LFL — lower flammability limit; the lowest concentration of a vapor forming a flammable mixture in air
Lfm — air velocity in linear feet per minute
ligaments — connections between bones
liquid trap (impinger, bubbler) — a gas- or vapor-collecting device that bubbles the air sample through a container of solvent
local exhaust ventilation — air is drawn into a hood at the site of generation of contamination
lower respiratory tract irritant (whole lung irritant) — a gas whose irritating properties develop slowly, so that a person willingly allows it to enter the entire lung

lower back pain — muscle pain due to excessive lifting, carrying, pushing, or pulling, particularly lifting
lower flammability limit (combustion-lean limit mixture) — the lowest concentration of a gas or vapor able to sustain continuous combustion
L_P (sound pressure level) — relative sound pressure, compared to a standard
lumbar region (of spine) — the lower vertebrae of the spine
lumbar support — a pad in the seat back that helps maintain the neutral S curve of the spine

M

MAC — maximum allowable concentration; a recommendation for the highest safe level of exposure to a chemical
malignant cell — a cancer cell
malignant mesothelioma — a bronchial carcinoma (cancer) caused by inhaling asbestos
manometer — a fluid-filled (water or mercury) U tube used to measure gas pressure
Material Safety Data Sheets — pamphlets required by the Hazard Communication Standard that list the identity of chemicals, hazard information, and information about the manufacturer
mCi; μCi — respectively, milliCurie and microCurie
melanin — the pigment in skin
melanoma — skin cancer of the pigment-producing cells
melting point — temperature of conversion of a solid to a liquid
membrane — a sheet-like lipid structure serving as a nonpolar barrier around cells and cell substructures
metabolism — the system of chemical reactions by which chemicals are altered in the body
metal fume fever — a disease like influenza caused by inhaling certain chemicals, usually metals
mg/kg — milligrams of compound per kilogram of animal weight; the usual units for expressing the levels of dosing of test animals
mg/m^3 — milligrams per cubic meter of air; a unit used to express concentrations of particulates in the air
micron — a micrometer or 10^{-6} m
middle ear — the region of the ear containing tiny bones that transfer vibration from the eardrum to the inner ear
mist — finely divided liquid droplets stably suspended in air
mm_{Hg} — millimeters of mercury; a gas pressure unit (also termed torr) based on the height of the mercury column the gas pressure supports; 760 mm_{Hg} is one atmosphere
mold — a form used for casting or molding metals or plastics
mppcf — million particles per cubic foot of air
MSDS — Material Safety Data Sheets; informational sheets about chemicals required to be in the workplace by the right-to-know law
MSHA — Mine Safety and Health Administration
mucus — a moist sticky layer of protein/carbohydrate composition lining internal organs, including the respiratory tract
mutagen — a compound that causes genetic (DNA) damage
mutation — an alteration of the genetic message on DNA

N

narcosis — unconsciousness or drowsiness caused by a drug
nasal cavity — the portion of the respiratory tract between the nostrils and the pharynx
NCI — National Cancer Institute
necrosis — death, particularly cell death
NED — normal equivalent deviations (standard deviations); the change in dose that produces one standard deviation change in the likelihood of an effect of a toxin on an organism
negative pressure (in ventilation) — in a ventilation system with an exhaust blower, the static pressure in the vent is lower than atmospheric pressure, thus has negative pressure
NEL — no effect level; same as NOEL
neonates — newborn animals
neoplasia — formation of a tumor or tumors

nephron — the functional tubular unit of the kidney
nephrotoxin — a toxin whose target organ is the kidney
neurotoxin — a toxin whose target organ is the nervous system
neutron — a component of the atomic nucleus sometimes emitted as radioactivity
NFPA — National Fire Protection Association
NIOSH — National Institute of Occupational Safety and Health; a research arm of the Federal Bureau of Health and Human Services studying the health hazards found in the workplace
NOAEL — no observed adverse effect level; same as NOEL
NOEL — no observed effect level; a measure of the highest level of dosing that produces no symptoms
nonionizing radiation — electromagnetic radiation no more energetic than UV light
nonsparking tools — tools manufactured from aluminum-bronze or beryllium-copper emit sparks of sufficiently low temperature that they will not ignite most flammable liquids
NPIRS — National Pesticide Information Retrieval System; source of information about pesticides, housed at Purdue University
NRC — National Response Center; a center that must be notified of significant chemical or oil spills; 1-800-424-8802
NSC — National Safety Council
NTP — National Toxicology Program

O

occupational asthma — an allergic (asthma) response to an allergen in the workplace
occupational dermatosis — dermatosis caused by exposure on the job
occupational disease — a disease resulting from performance of work duties
occupational physician — a physician who specializes in occupational illness
octave bands — a dividing of the range of audible sound into octaves to discuss hearing problems
octave interval — a change in tone such that the frequency exactly doubles
odor threshold — the lowest concentration of a chemical in air that can be detected by odor
oncogene DNA — segments of DNA involved in the conversion of a normal cell to a cancerous cell
open-pit mine (surface mine, strip mine) — a mine in which minerals are removed from the surface of the ground
organogenesis — during the growth of the fetus this is the stage at which the organs first form
OSHA — Occupational Safety and Health Administration; a branch of the U.S. Department of Labor established to enforce the safety standards set by the government for the workplace
OSH Act — public law 91-596 that describes the federal involvement in worker health and safety
OSHRC — Occupational Health and Sefety Review Commission; the board that reviews appealed OSHA decisions
oxidizing agent — an agent that supports the oxidation of other chemicals; oxygen is the most common such agent, but other chemicals also may perform that reaction; oxidizing agents are incompatible with combustibles for storage purposes
outer ear — the ear canal connecting the outside with the eardrum

P

PAH — polycyclic aromatic hydrocarbons
PAT — the Proficiency Analytical Testing program for assessing the accuracy of analytical methods. This program is run by NIOSH
pathology — the study of the effects of disease
pattern — a shape used to prepare the depression in a sand mold
PEL — permissible exposure limit; a recommendation by OSHA for the highest safe level of exposure to a chemical
penetration time (protective clothing) — length of time until a liquid passes through a fabric used for protective clothing
permanent threshold shift — a permanent drop in hearing acuity
permeation (protective clothing) — length of time required for a liquid to diffuse through a protective garment

permeation rate (protective clothing) — rate at which a liquid diffuses through a protective garment in μg/cm²/min
phagocytic cells — cells in the body that engulf small objects, particularly foreign material
photodermatosis — skin problems caused by a chemical that is activated by light
pickling (acid descaling) — cleaning a metal surface with acid
Pitot tube — a water manometer modified to measure both static and velocity pressure in a vent
placenta — the structure in the uterus that interfaces fetal blood with maternal blood to transfer nutrients, oxygen, and wastes between mother and fetus
plasticizer — compound added to a thermoplastic polymer to prevent the formation of crystallites
plumbism — toxic effects of lead
pneumoconiosis — damage to the lungs resulting from inhaling solid particles
pneumonia — a disease in which fluid collects in the lungs
pneumonitis — fluid in the lungs
PNS — peripheral nervous system; the nerves outside the CNS that control muscles and glands and receive sensory signals
polar bond — when two atoms that are joined covalently do not share the electrons between them equally, the atom with the greater attraction has a small negative charge and the other has a small positive charge
pore — an opening in the skin arising from the sweat gland and sebaceous gland
positive pressure (in ventilation) — in a ventilation system with a blower forcing air into the duct, the static pressure in the duct is higher than atmospheric pressure, thus has positive pressure
potentiation (synergy) — the result when simultaneous exposures have a greater effect than the sum of the individual exposures
ppb — parts per billion, as grams per one billion grams or micrograms per kilogram
PPE — personal protective equipment
ppm — parts per million, as grams per one million grams or milligrams per kilogram; a common concentration unit for dilute samples of dissolved substances or airborne substances
precancerous cell — a cell that has been transformed such that it is predisposed to be further coverted to a malignant cell
precision — the reproducibility of a measurement in analytical chemistry
prepolymers — short polymer chains that are cross-linked into a thermoset plastic
presbycusis — loss of hearing that occurs with age
pressure (in ventilation) — the effect of air molecules colliding with a surface
pressure forming — shaping plastic by forcing a sheet against a mold with air pressure
probit — the change in dose that produces one standard deviation change in the likelihood of an effect by a toxin on an organism
procarcinogen (secondary carcinogen) — a compound that is converted to a carcinogen by cell metabolism
proposed rule — a proposal to establish or change an exposure standard published by OSHA in the *Federal Register*
psi — pounds per square inch, a pressure unit; 14.7 psi is one atmosphere
pulmonary — anything related to the lungs
pulmonary edema — fluid accumulated in the lungs
pulmonary fibrosis — fibrosis in the lungs
pure tone — a tone of a single frequency
push–pull ventilation system — air is blown across a contaminated site toward a hood with negative pressure that picks up the now contaminated air
pyrophoric — a chemical that spontaneously burns in air at 13°F or below

Q

QF (radioactivity) — quality factor; the relative hazard of a type of radioactive emission
qualitative fit testing — checking the fit of a respirator by exposing the wearer to a readily detectable vapor or particulate
quantitative fit testing — checking the fit of a respirator by measuring concentrations of a contaminant both inside and outside the face mask

R

rad — radioactive emission energy absorbed by a target ~100 erg/G
radiative heat transfer — transfer of heat to or from the body through space
radioactive isotopes — isotopes of an element that spontaneous disintegrate, releasing a particle or energy from the nucleus
radius (anatomy) — the bone of the forearm that rotates in the elbow to allow arm and hand rotation
RAL — recommended alert limits in WBGT (wet bulb global temperature) for healthy nonacclimatized workers
random error (indeterminate error) — error in analysis where the measured values scatter around the true value randomly
range (statistics) — the difference between the highest and lowest values in a data set
Raynard's disease (white finger disease) — damage to nerves and blood vessels in hands due to the vibration of hand tools
RCRA — Resource Conservation and Recovery Act (1976); federal plan for regulation of handling of hazardous waste
receiving hood — a hood placed in the path of propelled impurities
red bag waste — hazardous infectious waste
refining (metallurgy) — converting ore to metal
REL — recommended WBGT alert limits for healthy acclimatized workers
rem (radioactivity) — dose equivalent for energy of radioactive emission absorbed for a specific organ or tissue [= rad(QF)(DF)]
renotoxin — a systemic toxin whose target organ is the kidney
replacement air (ventilation) — air brought into a building or room to replace air removed by an exhaust fan
replication — the synthesis of new DNA using existing DNA as a sequence template
reproductive toxin — a toxin affecting fertility
resistance (in ventilation) — the relative difficulty of moving air through ducts
resonance frequency — for each object there is a resonance frequency such that exposure to that frequency results in the object vibrating
respirable particles or dust — particles generally under 5 μm in diameter capable of reaching the alveoli when inhaled
respirator — a device worn over the nose and mouth designed to provide air free of contaminants
retrospective study — a study to establish a relationship between an agent in the workplace and a health effect
riser — a channel through which metal rises to show the mold is full
risk assessment — evaluation of the risk to health of exposure to an agent
roentgen (R) — a unit to measure energy absorbed by air from a radioactive source equaling ~83 erg/G
rotating vane anemometer — a device used in the analysis of a ventilating system that measures the air velocity
RQ — reportable quantity; the minimum amount of a chemical that upon release must be reported to the EPA
RTECS — registry of toxic effects of substances; a large list of toxic effects published by NIOSH
runners — channels to move metal within a mold

S

sand slinger — a high-pressure hose spraying sand into a mold
SARA — Superfund Amendments and Reauthorization Act
SCBA — self-contained breathing apparatus
scintillation counter — radioactivity counter in which light is produced in a (scintillation) fluid by the emission, then light is measured to estimate the level of radioactive emission
sebaceous glands — glands in the skin at the base of the hair follicle that produce skin oil
secondary carcinogen (procarcinogen) — a compound that is converted to a carcinogen by cell metabolism

secondary emissions — radioactive emissions resulting from changes brought about by the original or primary radiation
sensitizer — a compound that generates an immune response and triggers allergies
SERC — State Emergency Response Commission; agency in each state responsible for implementing SARA Title III
settling chamber — air in a duct enters a large chamber, causing the air velocity to drop below that sufficient to keep larger particles suspended
sharps — objects that can penetrate the skin, which wen contaminated by blood can spread bloodborne pathogens
Shaver's disease — fibrosis in the lungs from exposure to aluminum oxide
shielding (radioactivity) — interposed matter that blocks radioactive emissions
SI units — Systeme Internationale; an internationally established set of units of measurement used in most of the world
SIC — Standard Industrial Classification; a list coding industrial work into a series of numbered categories
siderosis — accumulation of iron compounds in the lung
sievert (Sv) — 100 rem
silicosis — lung disease caused by inhaled silica particles accumulating in the lung
silicotic nodules — characteristic pockets of silica that appear in X-rays of the lungs of individuals with silicosis
simple asphyxiant — a gas that dilutes the oxygen in the air
sizing — separation of crushed ore by size
slag — waste in metal purification, especially the silicate waste formed in a blast furnace
smelting — conversion of ore to free metal by heat and reduction
smoke — airborne particulate arising from combustion of organic material
sound level meter — a device to measure sound pressure levels
sound pressure — the fluctuation in air pressure caused by sound waves
sound pressure level (L_P) — relative sound pressure, compared to a standard
sprue — a hole to allow molten metal to enter the mold
squamous cell carcinoma — a type of skin cancer
squeeze machine — a machine that shapes sand in a mold by ramming the pattern into the sand
stabilizer — a compound added to plastic to protect it from the destructive effects of sunlight
stable isotopes — isotopes of an element that do not disintegrate radioactively
stack (ventilation) — the exhaust chimney of a ventilating system on the building roof
stain tube — a device for estimating gas or vapor concentration in the air by the length of a colored reaction product produced in a tube through which the air sample has passed
standard deviation (statistics) — a statistical presentation of the variability around the mean of a set of data
static pressure — air pressure in a duct connected to a fan
STEL — short-term exposure limit; the threshold limit value for a limited period of time, typically for 15 min
STP — 1 atmosphere pressure and 25°C
stratum corneum — the keratin layer of the epidermis
strip mine (surface mine, open-pit mine) — a mine in which minerals are removed from a hole in the surface of the ground
styrene sickness — nausea, vomiting, and fatigue resulting from styrene exposure
subacute testing or dosing — daily dosing of a test animal for a moderate period of time, usually 90 days
subcutaneous — under the surface of the skin
Superfund Ammendments and Reauthorization Act — see *SARA*
surface mine (open-pit mine, strip mine) — a mine in which minerals are removed from a hole in the surface of the ground
surrogate — one species substituting for another in testing
Sv (sievert) — 100 rem
sweat glands — glands in the skin at the base of the pore that produce sweat
systemic toxin — a poison causing immediate damage throughout the body, usually cell death

T

T_A — dry bulb (ordinary thermometer) temperature
T_G — globe temperature, which measures the effect of convective heat transfer
T_{NWB} — natural wet bulb temperature; evaluates cooling by evaporation using a wet bulb thermometer exposed to natural air movement
tackyfier — a compound used to make an elastomer more sticky
target organ — the specific organ that a toxin most heavily damages on exposure; hepatotoxins damage the liver
target population — the group of individuals exposed to a particular agent
temporary threshold shift — a temporary drop in hearing acuity due to exposure to loud noise
tendons — cords that run between bones and muscles facilitating the movement of the bone by the muscle
tennis elbow (epicondylitis) — inflammation of the tendon attaching at the elbow and connecting to muscle in the forearm, used to extend the arm
tenosynovitus — inflammation of tendon sheaths due to abrasion resulting from excessive use
teratogen — a compound that causes birth defects when the developing fetus is exposed to it
thermoluminescence detector (TLD) — a personal dosage detector that stores energy on exposure to radiation, then emits light proportionately when discharged
thermoplastic — a linear polymer that can soften and melt on heating
thermoset plastic — a three-dimensional polymer that does not soften on heating
threshold toxicity — the dose that is at the interface of the lowest producing a toxic effect and the highest not producing such an effect
third-octave bands — a dividing of the range of audible sound into intervals, each one-third of an octave, to discuss hearing problems
tinnitus — ringing in the ears
TLD (thermoluminescence detector) — a personal dosage detector that stores energy on exposure to radiation, then emits light proportionately when discharged
TLV — threshold limit value; a recommendation by the ACGIH for the highest safe level of exposure to a chemical
TLV-C — ceiling threshold limit value; an exposure level that should not be exceeded even for a very short time
torr — a unit of pressure equal to a mmHg; 760 torr is one atmosphere
total pressure (ventilation) — the sum of static pressure and velocity pressure
toxic chemical — a chemical is termed "toxic" if it has an LD_{50} between 50 and 500 mg/kg, an LC_{50} (1 hr) between 200 and 2000 ppm, or between 2 and 20 mg/l of dust, mist, or fume, or a dermal LD_{50} (24-hr exposure) between 200 and 1000 mg/kg
tracers — in research, the fate of particular atoms may be determined or followed by use of radioactive isotopes of those atoms
trachea — the upper passage of the respiratory tract between the nasal cavity and the splitting out of the bronchial tubes
transport velocity — the air velocity needed to keep particulate contamination suspended in a duct
trigger finger — tenosynovitus from repetitive finger action on the trigger switch of a power tool
TSCA — Toxic Substances Control Act; gives the EPA authority to act to prevent injury from chemicals
tumor — a mass of cells displaying an alteration of expected pattern and unusually rapid growth
TWA — time-weighted average; refers to the practice of sampling exposure levels at several times and averaging the exposure over the day; used in monitoring compliance with threshold limit values or permissible exposure limits

U

UEL — upper explosive limit
UFL — upper flammability limit
ulna (anatomy) — the rigid bone of the forearm forming one side of the elbow
upper explosive limit — the highest concentration of a gas or vapor able to generate an explosion

upper flammability limit (combustion-rich limit mixture) — the highest concentration of a gas or vapor able to sustain continuous combustion

upper respiratory tract irritant — a gas that is so irritating that one allows it to be inhaled only into the upper part of the lung

USDA — the U.S. Department of Agriculture

UV spectrophotometer — an analytical instrument that measures the absorbance of specific frequencies of ultraviolet light by a compound under study

V

vacuum forming — forming a plastic sheet by evacuating the space between the sheet and the mold

van der Waals forces — the electron clouds around atoms vibrate temporarily giving the atoms a positively and negatively charged end; two atoms very close together are weakly attracted to one another by these charges

vapor — a normally liquid chemical in the gas phase due to its evaporation

vapor degreasing — dipping an object in solvent vapors to remove grease

vapor pressure — the partial pressure due to the vapor of a liquid at a given temperature when it is saturating the air above the liquid

VDT — a common abbreviation for video display terminal

velocity pressure — pressure due to air movement in a duct

vermiculite — a common absorbent material used to control spills

viscosity — the property of a liquid that measures its resitance to flow; a high viscosity substance is syrupy and flows very slowly

visible spectrophotometer — an analytical instrument that measures the absorbance of specific frequencies of visible light by a compound under study

VOC — volatile organic chemical

volatile — the property of a liquid with a low boiling point such that it evaporates readily

voltage (electrical) — the electrical potential driving a current through a circuit

vulcanization — a process by which elastomers are cross-linked to produce harder, more abrasion-resistant products

W

warning properties of a gas or vapor — the smell or irritation experienced on exposure to a specific concentration of a gas or vapor

water manometer — a curved tube containing water used to measure the difference between static pressure in a duct and atmospheric pressure

wavelength (sound) — when sound is considered as a wave, wavelength is the distance from one peak height to the next

WBGT (wet bulb globe temperature index) — the most used method to evaluate the heat stress potential of the workplace

WEEL — Workplace Environmental Exposure Level; guidelines for exposure developed by the AIHA

wet scrubbers — water sprays used to remove gases and vapors from the air

white finger disease (Raynard's disease) — damage to nerves and blood vessels in hands due to the vibration of hand tools

Williams-Steiger Occupational Safety and Health Act — the act of 1970 that created OSHA

X

X-rays — high-energy electromagnetic radiation generated by bombarding a metal target with a high-voltage electric arc

Z

Z list (Z tables) — the OSHA PELs found as an appendix to 29 CFR 1910.1000

Answers

CHAPTER 2

1. Some states already had a high level of worker protection required in their state laws and thus changed to only a small degree, whereas others needed to change a great deal. Because providing protection for workers is expensive to manufacturers, a state could use lax rules to attract new industry by advertising itself as "low cost" or "friendly to industry".
2. (a) 1904; (b) 1903. Nothing; (c) 1900 is "reserved" for future use.
3. 29 CFR 1910.215 contains abrasive wheel regulations, which are very detailed and specific.
4. A. OSHA may publish an early warning to invite reaction to a proposed new standard, and must publish the new standard as a proposed rule, then wait at least 30 days before instituting the standard. Public hearings provide opportunities for input.
5. The employer notifies the OSHA area director in writing of intent to contest the decision within 15 working days of notification of the penalty. The area director then immediately transmits it to the Review Commission.
10. General content of the agreement is found in 29 CFR 1901.3. The process is spelled out in Subpart C of 29 CFR 1902 (parts 1902.10–1902.23). The process by which the Department of Labor decides about approval or rejection is covered in Subpart D (parts 1902.30–1902.53).
11. 29 CFR 1901.5.

CHAPTER 3

1. A. (1) Renal histopathology—microscopic study of kidney tissue to find abnormal or diseased areas. (2) Proximal renal tubules—the beginning part of the tubes that are the functional structures in urine formation in the kidney. (3) Necrotic areas of tubule ule ule epithelium—dead cells' surface layers on the inside surface of the kidney tubule. This is a systemic toxin, specifically a renotoxin.
   B. Massive hepatic necrosis—extensive regions of dead liver cells. This is a systemic toxin, specifically an hepatotoxin.
   C. (1) Aplastic anemia—a reduction in red blood cell count or concentration in the blood resulting from damage to the red blood cell-generating tissues in bone marrow. (2) Leukemia- proliferation of abnormal white blood cells: a cancer of the tissue producing white blood cells. This is a carcinogen.
   D. (1) Hematologic effects—effects on blood or blood-forming tissues. (2) Peripheral neuropathies—problems of the peripheral nervous system (PNS), of the nerves running from exit from the spinal cord to various organs or tissues. This is a systemic toxin, specifically a neurotoxin.
   E. (1) Maternal toxicity—poisoning of the mother. (2) Cardiovascular—having to do with the heart or blood vessels. (3) Skeletal malformations—misshapen or missing bones. This is possibly a teratogen, depending on time of dosing.

F. Degeneration of the testicular germinal epithelium—damage to the cell layer in the testicles that generates sperm cells. This is a systemic toxin, specifically a reproductive toxin.

G. (1) Hyperplasia—increase in the numbers of some cell type. (2) Squamous cell carcinoma—skin cancer involving the keratin-forming cells of the outer skin layer. This is a carcinogen.

2. A. Having a three-symbol code and four different symbols means we can have 4^3, or the 64 different code words shown in the table. These represent 20 different amino acids.

   B. Leu-Gln-Thr-Pro-Arg-Arg-Ile-Glu-STOP. The next Gly is beyond the stop signal.

   C. Pro is replaced by Leu in the sequence.

   D. The code word still indicates Arg. This is a silent mutation because there is no change in the protein sequence even though the DNA sequence has been altered.

   E. The sequence now reads: Leu-STOP. The protein has been shortened by the elimination of the rest of the amino acids.

   F. The sequence now reads: Leu-Gln-Thr-Pro-Arg-Arg-Ile-Glu-Trp-Gly- . . . . The STOP signal is gone, and if there is DNA to the right of the problem sequence, the protein will be extended according to the continuing DNA sequence until a STOP is encountered.

   G. The sequence now reads: Leu-Lys-His-Pro-Asp-Gly-Ser-Asn-Glu- . . . . .In other words, the sequence is completely changed.

   H. The "frame", the group of three bases read to determine each amino acid, is no longer the same, so that what was the second base in the code word before is now the first base. All code words are changed. It is very unlikely that a protein resulting from a frame shift mutation will be functional unless the shift occurs very near the end of the sequence. A few such mutated hemoglobins have been found that still function.

   I.

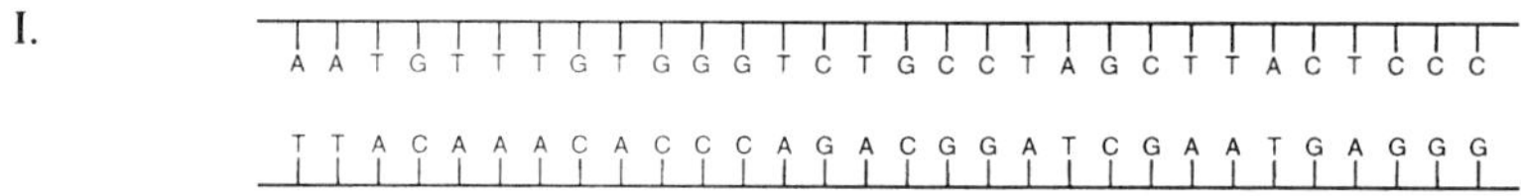

   J. The 5th base is a T. The complementary strand has an A at that point, and the repair enzyme would replace the damaged base with the complement of A, which is T.

3. Teratogens would be the most important chemicals uniquely dangerous to a woman, or more correctly to the fetus of a pregnant woman. Men might be banned from jobs where a worker must handle chemicals that damage sperm production.

4. Compounds may have a teratogenic toxic effect on the fetus only during the period of organ development — after conception. If a teratogen were also a mutagen, however, male exposure before conception could cause malformations by damaging sperm DNA.

CHAPTER 4

1. The child should not be given the same dose. Dose should be measured as mg/kg, and since the weight of the child is about one-third adult weight, the dose should be about 100 mg/day. This should not be given as one 100-mg pill, because this would be a very high dose at one time. Depending on how the drug is packaged, it might be 2×50 mg or 4×25 mg, and so forth.

2. It would still be 75 ppm. Toxicants taken by inhalation do not depend on body mass, because the lungs are proportionately smaller.

3. The likely routes of intake are inhalation of the vapors and intake through the skin from handling wet parts. D is more toxic by the dermal route, and A is more toxic by inhalation. Furthermore, A has a lower boiling point, indicating that vapors of A will be present in the air over the solvent at higher concentration. In choosing between B and C, we see that C is more toxic by the oral route. However, oral intake is unlikely. More important, the 90-day oral study reveals that B accumulates in the body and B has a lower boiling point than C. C is the best choice.

4. A. TEPP to males; phorate to females

   B. Malathion to males; ronnel to females

C. Chlorthion, dicapthon (oral only), fenthion, guthion, malathion, methyl parathion (dermal only), methyl trithion, npd, phosdrin, (dermal only), phosphomidon (oral only), trichlorofon.
D. No. Phorate is twice as toxic to females, whereas for trichlorofon the difference is slightly more than 10%.
E. Ronnel and schradan are more than twice as toxic to males, whereas carbophenothion, co-ral, demeton, di-syston, EPN, ethion, parathion, and phorate are more than twice as toxic to females. This family is more toxic to females. .
F. Oral
G. Diazinon is more toxic to females, and it is more toxic orally. Schradan is more toxic orally to males. Phosdrin is more toxic dermally to females. If a compound is more toxic dermally, it crosses the skin barrier more readily than the mucosal barrier of the GI tract.
H. Yes.

5. A.

$$1\ \mathrm{m}^3 \times \left[\frac{100\ \mathrm{cm}}{\mathrm{m}}\right]^3 \times \frac{1\,\mathrm{l}}{1000\ \mathrm{cm}^3} = 1000\,\mathrm{l}$$

B.

$$V_1 / T_1 = V_2 / T_2;\ 22.4\,\mathrm{l} / 273\ \mathrm{K} = V_2 / 298\ \mathrm{K};\ V_2 = 24.5\,\mathrm{l}$$

C.

$$1\ \mathrm{ppm} = \frac{1\,\mathrm{l}}{10^6\,\mathrm{l}} \times \frac{10^3\,\mathrm{l}}{\mathrm{m}^3} \times \frac{1\ \mathrm{mol}}{22.4\,\mathrm{l}} \times \frac{100\ \mathrm{g}}{\mathrm{mol}} \times \frac{1000\ \mathrm{mg}}{\mathrm{g}} = 4.46\ \mathrm{mg/m^3}$$

D.

$$2\ \mathrm{ppm} = \frac{1\,\mathrm{l}}{10^6\,\mathrm{l}} \times \frac{10^3\,\mathrm{l}}{\mathrm{m}^3} \times \frac{1\ \mathrm{mol}}{24.5\,\mathrm{l}} \times \frac{320\ \mathrm{g}}{\mathrm{mol}} \times \frac{1000\ \mathrm{mg}}{\mathrm{g}} = 26.1\ \mathrm{mg/m^3}$$

$$4\ \mathrm{ppm} = \frac{1\,\mathrm{l}}{10^6\,\mathrm{l}} \times \frac{10^3\,\mathrm{l}}{\mathrm{m}^3} \times \frac{1\ \mathrm{mol}}{24.5\,\mathrm{l}} \times \frac{150\ \mathrm{g}}{\mathrm{mol}} \times \frac{1000\ \mathrm{mg}}{\mathrm{g}} = 24.5\ \mathrm{mg/m^3}$$

6. A. NOEL/100
B. Make the test level 100 times the concentration anticipated in workplace air.
C. Testing only one animal carries the risk that this is a highly resistant individual.
D. We do not know what the true safe level is, and we do not know if that individual animal is highly sensitive to the chemical.
7. A. Dosing in the drinking water is easy to do, but we do not know what volume of water the animal actually drank and what was spilled.
B. 2.5 mg/kg
C. 1.25 mg/kg
D. LD_{50} is approximately 4.0 – 4.4 mg/kg.
E. Extrapolating the line of the probit plot to interception with the x axis is the easy way to determine a threshold value, here close to 1.25 mg/kg.
8. A + B, additive; A + C, antagonistic; B + C, synergistic
9. Antagonistic
10. 14/100 + 12/25 + 25/200 = 0.795. The value is not greater than 1, so this is not an emergency. It is high enough that on a bad day it could go over 1, so putting improved ventilation on a priority list would be wise.

CHAPTER 5

1. A 70-kg worker would need to transfer 13.7 G of the compound onto the apple to ingest the lethal dose, a highly unlikely event. However, if the compound accumulated in the body effectively, the worker could add small amounts to the body burden each day and perhaps

reach a level that was hazardous in time. A 90-day LD_{50} would give an indication of the likelihood of this happening. The LD_{50} measures only systemic toxicity, and other toxic characteristics are also of concern.

2. Anything that decreases the ability of skin to block the chemical raises the hazard of handling this solvent. This includes cuts and abrasion of the skin and previous prolonged skin exposure to hot, soapy water or to organic solvents (possibly including this solvent), which loosen the skin keratin layer. One must always consider that the compound may be entering the body by another route at the same time, perhaps by inhalation of its vapors, adding to the body burden.
3. A. e > c > a > b > d
   B. Yes.
   C. a, 0.052 ml/min; c, 0.066 ml/min
   D. No.
   E. Compounds in the series may be metabolized to more polar structures at different rates. Some may bind to proteins and others do not.

CHAPTER 6

1. Through the skin of the back of the hand, where it is thinnest.
2. Cuts and abrasions; loosening of the keratin layer by solvents or water
3. A. They will respond strongly to even slight contact, and so should be shifted to a job away from the chemical.
   B. Becoming sensitized depends on having the chemical bind to a protein in the skin, which then stimulates an immune response. That complete series of events occurred only in those who were sensitized.
   C. The initial reaction with a skin protein must be most likely to happen with that particular chemical.
4. A. Workers 1, 2, 4, 5, 6, and 7 require hand protection. Cotton gloves with leather palms are adequate for workers 1,2, and probably 7 (depending on temperatures).
   B. Workers 3 and 8 handle the signs after the sharp edges are removed and at a time when the signs have no chemical contamination. Gloves would be optional and might interfere with job performance.
   C. (1)You require more than one type of glove. (2) and (3) A variety of gloves are needed. All do not have to be discarded each day, and buying longer lasting higher quality gloves for some stations may save money while providing better protection. (4) This is true. These gloves should be inspected for leaks and deterioration each day. (5) Probably not. Because these gloves become contaminated by paint, it is hard to decontaminate them, and attempted use for several days would involve a build-up of paint. Disposable (one-use) gloves may be the best option at these stations.
   D. The hazard to worker 4 is contact with the acid. Reference must be made to a table listing the glove material suitable for use with aqueous acid to choose the correct gloves.
   E. Worker 4 may be exposed to acid splashes and should have at least an acid-resistant apron, safety glasses, and a face shield.
   F. From the index of 29 CFR 1910 we learn that eye and face protection is found in section 133. In 29 CFR 1910.133 we find general regulations and a reference to ANSI Z87.1-1989 for detailed regulation.
5. Gloves should be cleaned free of the chemicals being handled after each day's use. Gloves should be inspected for leaks by inflating them. This could be done in a simple fashion by folding the glove shut at the cuff and rolling back toward the fingers. If the surface of the gloves becomes tacky or shiny, or if the gloves become stiff, this is a sign of impending failure.
6. (a) vascular plexuses –; avascular — has no blood vessels; plexux of lymphatics –; regional lymph nodes –; thoracic duct; sensory nerves — nerves that detect and carry sensory information, such as sense of touch; sensorimotor nerves –.
   (b) biotransformation — alteration of the chemistry of a compound of biological catalysts (enzymes); xenobiotics — foreign substances in the body.

(c) erythemia –; edema — accululation of body fluids; vesiculation –; spongiosis –.
(d) erythemia –; actinic elastosis — premature skin aging; proliferative changes — cells multiply at a high rate; supression of T lymphocytes — prevention of fromation of additional immune system cells; actinic keratosis –; squamous cell cancers — tumors of the epidermal cells; basal cell cancers –; melanomas — tumors of pigment-producing skin cells.

7. Totally encapsulating suits are discussed in 29 CFR 1910.120(g)(4). Decontamination is discussed under 29 CFR 1910.120(k). No specifics are given.
8. The employee should see a doctor first, preferably an occupational medicine specialist, to determine if the rash is job related and if it is contact irritation or allergic in origin. Do a survey of the possible exposures of the employee at his or her workstation. Consider airborne contaminants from grinding plastic (release of chemicals from plastic or airborne dust) or spray painting that might reach the employee's workstation. If the problem is allergic, patch testing of the potential chemical exposures by the physician may pinpoint the problem.

CHAPTER 7

1. A. Starting with the nose and extending all the way to the entrance to the alveolus, the passages are lined with wet, sticky mucus. If a particle impacts the mucus it is caught like a fly on flypaper. The pathway to the alveolus includes numerous turns, at any one of which the particle can "spin out" onto the sticky surface. Further, the particle remains suspended by the movement of the air, and the farther into the lung it travels, the slower the air moves. Finally, in the alveolus if the particle touches down on the wet surface, it may be engulfed by phagocytic cells.
   B. Particles that weigh more are more likely to be deposited at turns in airways or to settle out of slower moving air. Particles weigh more if they are larger in diameter for a given composition or have a higher density for a given diameter.
   C. The key to being stopped by the mucous lining of the air passages is the ability of the chemical to interact with water, because mucus is highly hydrated. Methane cannot hydrogen bond with water and is not polar so is not soluble in water. If it does strike the lining, it simply migrates out again and continues moving with the air.
   D. HCl does dissolve in water and dissolves into the mucus on first contact.
2. This is an acidic irritant that reacts with water to form H_3O^+ and $CF_3{\cdot}CH_2{\cdot}O^-$. The H_3O^+ lowers pH in the passage lining, which stimulates (1) coughing and sneezing, which expel air from the lungs, and (2) production of additional mucus, which is carried out with the coughing.
3. There are now fewer vessels carrying the blood leaving the heart, so each vessel must carry more blood, producing back pressure on the heart.
4. A. (2) The forklift trucks are a source of CO. (3) Heated metals may be a source of metal oxide fume and gas flames are a source of CO. (5) Metal particulate could get in air from drilling (low risk). (6) Metal particulate and abrasive particulate could be getting in air from grinding (high risk). (7) Metal particulate could be stirred by the compressed air. (8) Chlorinated solvent may be entering plant air. (11) Metal particulate and abrasive could be entering plant air. (12) paint aerosol and (12 and 13) paint solvent may enter plant air.
   B. The chemicals carried to the rinse water would have to be replaced in the plating tanks, now they have been replaced from the rinse tanks. If the cyanide solutions in the alkaline copper plating solution were carried into the acidic nickel plating tank, HCN gas would be produced.
5. A. A known volume of air is pumped through a filter and the number of fibers of particular dimensions are counted in sample fields under a microscope.
   B. This is the upper size for respirable particles.
   C. Crocidolite is the carcinogenic type of asbestos.
6. Most often agree; ACGIH is more conservative.

CHAPTER 8

1. A. 0.94%
   B. Now roughly 95% of the particulate is small particles. The small particles are the respirable particles, and almost none of them have been removed by this ventilation improvement. If hazard depends on particles reaching the alveolus, the health problem is unchanged.
   C. Filter weight changes would not have changed significantly, but to the degree that small particles were captured, the hazard would have been reduced.
2. A. Two things must be measured before someone enters the tank: (1) the tetrachloroethylene, which would be a vapor, must be at a safe level and (2) there must be enough oxygen for breathing. Tetrachloroethylene would be tested using a grab sample, and gas chromatography would probably be used for the analysis. Meters are available to measure oxygen directly.
   B. Cadmium would be present as a particulate. An integrated sample would be obtained using a personal air sampling pump equipped with a filter. A grab sample would be obtained at the time of welding to ensure that the ceiling value is not exceeded at that time. Analysis would be by atomic absorption spectrometry.
   C. As long as the standards for the solvent specified TWA values for an 8-hr day, you would do integrated sampling, probably with a personal air sampling pump and an adsorbent tube. If there was a ceiling value specified, a grab sample should be taken at the touch-up site during spray painting. Analysis would be by gas chromatography.
3. A. (1) Determinate error producing values that are too low. (2) Indeterminate error. (3) Determinate error producing values that are too low.
   B. (1) Determinate error producing values that are too low. (2) Values at low concentrations are reliable, but at the high end there is determinate error producing values that are too low. (3) Indeterminate error. (4) No effect. The lowered flow rate affects the passage of both standards and samples through the column. (5) Determinate error producing values that are too high.
4. A. 0.00984
   B. 0.00222
   C. 0.00276
5. 147 ppm
6. Asbestos: polycarbonate filters
   Fe_2O_3 fume: cellulose acetate filters
   Benzene vapor: charcoal adsorption tubes
   Total mass of particulates: Polyvinylchloride filters

CHAPTER 9

1. A. (1) Silica particulate (silicosis)
   (2) Metal and metal oxide particulate in the air from handling hot metal. Unless alloys used are more exotic than iron for cast iron, metal fume fever should not be a problem. Furnaces produce CO.
   (3) Metal and metal oxide particulate
   (4) None
   (5) Anhydrous HCl gas and aqueous HCl aerosols may enter the air from the bath.
   (6) Silica particulate, a greater problem quantitatively than in (1)
   (7) Metal and abrasive particulate
   (8) None
   (9) Solvent vapors and paint aerosols
   (10) CO from internal combustion engines
   (11) Solvent vapors from degreasing operation, higher during use and highest when the unit is serviced

B. 1. 35004, 35005, and 35007 would contain filter elements. An element to adsorb vapors such as activated charcoal should be part of 35001, 35006, and 35007. 35002 would have an alkaline reactive medium and 35003 would have an acid reactive medium.
2. For stations 1, 3, 6, 7, and 10, 35004 or 35005 depending on size of particulate; for stations 4 and 8, none; for station 5, 35002, station 9, 35007, and station 11, 35001.
3. You need to (1) color code all masks before they go to the dispensing room, (2) instruct workers about the importance of using the right mask, and (3) place signs at the dispensing room and in each workstation indicating the correct color to use in that station.
C. Shift exhaust fans on the east wall and the air inlets to the west wall so that HCl fumes and solvent vapor are not drawn past the rest of the workers before leaving the room.
D. Areas 5, 6, 7, 9, and 11 could benefit from local ventilation. Air quality studies should be done to determine the sites of greatest need.
E. In the degreasing room. A spill there would quickly saturate the air with vapors.
F. Replace the forklifts with electric units.

2. A. Because air is moved into the building the building will have a positive pressure, which is higher than atmospheric.
B. 1920 cfm
C. 108 mg/min
D. Yes.
E. 0.92 mg/m^3
F. 13.9 min
G. Locate the adhesive operation and an exhaust vent as close together as possible.

3. A. 110 mg/min each
B. Almost; 69 m^3/min more ventilation capacity is needed.

4. A. $C_{max} = G/C = 0.67$ mg/m^3
B. $C_2 = C_1 \exp(-Q[t_2 - t_1]/V) = 0.36$ mg/m^3

5. A. 2.3 ft
B. 320 fpm

6. Baffles add to the resistance of the system, requiring more energy to move the air.

7. A. There is the possibility of a dangerous atmosphere, so a permit is needed, and welding is to be done, so a hot permit is needed.
B. The air should be tested for oxygen level, levels of flammable vapors, and levels of toxic vapors or gases. The literature should be consulted to learn the toxicity of the fuel vapors and the LFL and LEL of the fuel stored in the tank.
C. An entry supervisor must sign the permit. An attendant must be outside the tank, and a rescue team must be alerted and available.
D. The tank could be flooded with an inert gas to prevent ignition of the remaining fuel. The welder will then require a breathing apparatus to supply oxygen.

8. The room has 4800 ft^3. Twenty air changes per hour would require (20 h^{-1})(4800 ft^3)/60 min/h = 1600 cfm. 60 cfm × 25 persons requires 1500 cfm. The design volume for air movement would therefore be 1600 cfm.

9. A. Inform the employees by posting a sign that says, "DANGER — Permit-Required Confined Space — DO NOT ENTER".
B. 29 CFR 1910.146(c)(ii)(E): (1) An employee may not enter the space until the forced air ventilation has eliminated any hazardous atmosphere; (2) The forced air ventilation shall be so directed as to ventilate the immediate areas where an employee is or will be present within the space and shall continue until all employees have left the space; (3) The air supply for the forced air ventilation shall be from a clean source and may not increase the hazards in the space.
C. Section(c)(5).

10. A. Such a system is easy to install and design. It is necessary to calculate how much air must flow through the room to keep the level of the contaminant at a safe level, then install fans of appropriate capacity.
B. The chief disadvantage is that air from the source of contamination is drawn past all the workers between the source and the exhaust fans.

C. Ceiling fans disperse the air. With the fans on, the contaminant would not move toward the exhaust fan in a narrow concentrated plume, but would be diluted more fully in the room air, producing something closer to the well-mixed model that is the ideal for GEV. This would reduce the maximum exposure possible for a worker in the wrong place, but it would also extend exposure to more workers. Ceiling fans do not reduce the total room burden of the contaminant, only redistribute it.

D. Once local ventilation is installed, the contaminant is largely removed from the workplace at the site of generation. It may now be possible to run the exhaust fans at a much lower speed, because they are not responsible for removal of the contaminant, saving energy by substituting a small local blower for large exhaust blowers. This would also reduce the amount of heated air removed from the building in winter, lowering heating costs.

11. A. Air velocity at any given distance from the opening will be half as great at the same Q.
    B. v = 12.8 m/min
    C. v = 12.8 m/min
    D. v = 12.8 m/min
    E. v = 10.2 m/min
12. In order to adsorb onto the walls the smoke must strike the wall. Fan mixing caused much greater impact of smoke on the walls.
13. A. About 3700 cfm
    B. Room volume is not a factor in this calculation.
    C. Everything but G and K are fixed by the nature of the compound. K measures the efficiency of removal of the contaminant by the system, and could be affected by room volume.
    D. Again, K could change, decreasing if the new location were more effective.
14. A. About 300,000 cfm
    B. G and K are unchanged, and changes in SG and MW are small. TLV is the important difference.
    C. GEV is most successful when contaminants are of low toxicity.
15. Increased capacity is needed to provide 3250 cfm of ventilation, with an annual increased operating cost of $11,375.
16. 5.3 room changes per hour

CHAPTER 10

1. A. Look at the description of boiling point in the text: the temperature at which the vapor pressure equals the atmospheric pressure. Vapor pressure is a constant at a given temperature, but if the atmospheric pressure varies, the vapor pressure necessarily varies with it. Thus the temperature (boiling point) needed to obtain the correct vapor pressure changes.
   B. The definitions say the same thing. It is possible to measure such a value in a variety of ways, each of which may not produce the same experimental value for a given substance. Defining the assay method ensures the recording of a single value, avoiding confusion in the literature. ASTM stands for American Society for Testing Materials.
   C. A combustible liquid has a flashpoint at or above 100°F (37.8°C), whereas a flammable liquid has a flashpoint below 100°F. The lower the flashpoint, the greater the volatility of the liquid and consequently the greater the fire hazard. A Class 1B liquid is a flammable liquid that has a flashpoint below 73°F (22.8°C) and a boiling point at or above 100°F [CFR 1910.106 (a)(19)(ii)]. This allows regulations to be adjusted to the level of hazard. As an example see CFR 1910.106(b)(2)(iv)(f) and (g), normal venting for ABOVE GROUND tanks.
   D. From CFR 1910.106(b)(1)(i): "Tanks shall be built of steel, except as provided in (*b*) through (*e*) of this subdivision." CFR 1910.106(b)(2)(ii)(a): "The distance between two flammable or combustible liquid storage tanks shall not be less than 3 feet"; (b) "Except as provided in subdivision (*c*) of this subdivision the distance between any two adjacent tanks shall not be less than one-sixth the sum of their diameters." 2 X 21 ft/6 = 7 ft.

2. A. bottom, B. top, C. vapor is not flammable, D. gas is not flammable, E. bottom, F. gas is ot flammable, G. gas is about the same molecular weight as O_2 and N_2, so would not layer.
3. A. The small fibers expose a higher proportion of the combustible molecules to the air, and so would burn more rapidly.
   B. The larger particles would burn more slowly to complete consumption and would produce more heat, assuming the oxygen supply were sufficient.
   C. Slow oxidation of the plant material would consume oxygen and reduce the risk of combustion. Circulation of air replaces the oxygen. However, if the sealed bin supported anaerobic breakdown of the plant source with methane production, the methane could rise to the top to reach flammable or explosive concentrations.
   D. Flattened particles expose a higher percentage of their molecules to the air, and would thus present a greatër fuel-effective concentration at a given total concentration. They would have a lower limit.
4. The chemical must supply fuel, an oxidant, and reaction that produces large volumes of gases rapidly.

CHAPTER 11

1. A. All quantities of regulated substances are below the TQ. Fuel oil exceeds the minimum, but is used just for heating. Since the "heating" includes heating the process vats, a ruling by OSHA as to whether this qualifies the site for regulation would be wise.
   B. The inventory of flammable liquids not for heating purposes exceeds the 10,000-lb limit.
   C. Both nitric acid and ammonia exceed the TQ.
2.

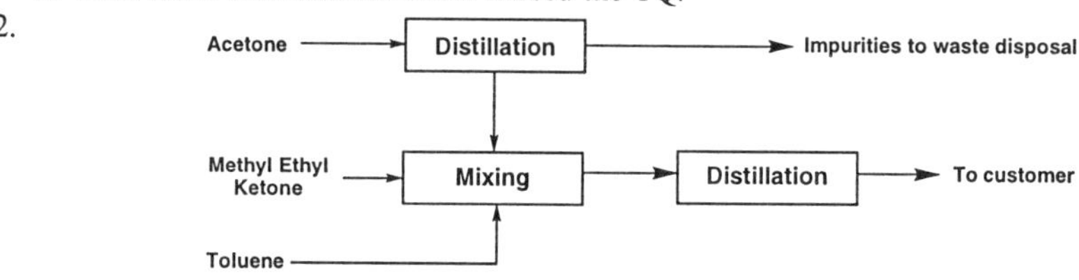

3. Special training is required to belong to a HAZMAT team.
4. The employer must develop a written safety and health program.
5. Level A

CHAPTER 12

1. A. The Eustachian tube equalizes the pressure on the eardrum by allowing air to enter or leave the middle ear.
   B. The cochlea is a fluid-filled structure in which vibration transmitted from the eardrum is converted into nerve signals.
   C. The stirrup is the last of the three bones connecting the eardrum to the cochlea.
   D. The semicircular canals have nothing to do with hearing, but rather are associated with balance.
2. A. The peaks are compressions of air in the atmosphere.
   B. The height of the wave
   C. The pitch is related to the wavelength, or peak-to-peak distance.
   D. Loudness is related to amplitude.
   E. Frequency = (speed of sound)/wavelength
   F. These are independent variables.
   G. Frequency units are Hz, (s^{-1}).
   H. 2000 Hz; 250 Hz
   I. 6.5 in.; 0.172 m

3. A. 20–20,000 Hz
   B. 500–3000 Hz
   C. 4000 Hz
   D. No, because the frequency is above the range of human speech.
   E. No. Greatest acuity is at 4000 Hz.
4. A. Sound power is the expenditure of energy generating the sound, with watts as the units. Sound pressure is the pressure of the compressions in the air (amplitude of the sound wave) and has pressure units (N/m^2).
   B. Sound pressure level is 20 times the log of the ratio of two pressures, so the units cancel. It is expressed in decibels (dB).
   C. 2×10^{-5} N/m^2 is chosen as the quietest sound detected by a person with perfect hearing.
   D. $L_P = 20 \log (40\ N/m^2/2.0 \times 10^{-5}\ N/m^2)$ dB = 126 dB
   E. You would set the range at 120–130 dB. The scale would read 6.
5. A. The x axis is frequency by octaves, usually from 250 to 8000 Hz. The y axis is the lowest sound pressure level in decibles detected by the person being tested.
   B. The graph would have a straight line drawn through 0 dB.
   C. Hearing is tested with an audiometer, and the person tested indicates the lowest sound level detected at each frequency.
   D. A loss of 25 dB at any frequency is considered impairment, and impairment is most serious in the range of human speech.
   E. F.B.T. may be experiencing normal high-frequency hearing loss as age increases, but J.J.H. may have occupationally induced hearing loss.
6. A. 1.145 not in compliance
   B. 0.855, in compliance but above action level
   C. 0.643, in compliance but above action level
7. (1) Replace the press with an automatic hydraulic press accepting steel from a roll and feeding parts into the bin, eliminating the full-time worker at the site. (2) Place acoustic tile around the machine and on the ceiling. (3) Place the press on a resilient pad. (4) Enclosing the press would be difficult, but a heavy barrier directly between the press and the workers might be possible.
9.

	Change at frequency (Hz)				
Worker	1000	2000	3000	4000	6000
A	1	2	4	11	4
A (corrected)	1	1	1	8	0
B	1	1	2	3	3
B (corrected)	0	0	0	1	1
C	1	3	4	8	8
C (corrected)	0	2	2	4	4

Worker A displays a pattern typical of noise damage, with a notch forming at 4000 Hz. The pattern of worker B would indicate normal hearing change. Worker C has a loss of hearing at higher frequencies, which might indicate excessive hearing loss due to a disease or damage situation, rather than noise damage.

10. A. L_W = 119 dB
    B. (1) L_P = 102 dB, (2) L_P = 78.5 dB
    C. (1) L_P = 91 dB, (2) L_P = 97 dB
11. These are the seven points in the article:
    (1). Noise measure and analysis
    (2). Install engineering controls where possible
    (3). Administrative controls and personal protection where necessary
    (4). Audiometric testing
    (5). Employee training
    (6). Record keeping
    (7). Evaluation

CHAPTER 13

1. A. V = 23p; Ag = 47p; K = 19p
   B. ^{15}N = 7p, 8n; ^{23}Na = 11p, 12n; ^{235}U = 92p, 143n
2. A. $^{26}_{11}Na$ ------> $^{26}_{12}Mg + ^{0}_{-1}\beta$
   B. $^{212}_{86}Rn$ ----->$^{208}_{84}Po + ^{4}_{2}\alpha$
   C. $^{111}_{51}Sb$ ----->$^{111}_{50}Sn + ^{0}_{1}\beta$
   D. $^{16}_{8}O + ^{1}_{0}n$ -----> $^{16}N + ^{1}_{1}p$
3. A. One rem equals: 1 R of X- or γ-radiation; 1 rad due to x-, γ-, or β radiation; 0.1 rad due to neutrons or high energy protons; 0.05 rad of particles heavier than protons and with sufficient energy to reach the lens of the eye.
   B. The quality factor of x-, γ-, or β radiation is 1; of neutrons or high-energy protons is 10; and of particles heavier than protons and with sufficient energy to reach the lens of the eye is 20.
4. The ratio is 100/58 rem/calendar year, or 1.7 times as much.
5. A. 5 rem/calendar year vs. 0.1 rem/calendar year = 50 times greater
   B. Hands, forearms, feet, and ankles are not highly sensitive to radiation, but the named tissues are radiation sensitive. Dosing active blood-forming organs (e.g., bone marrow of long bones) can cause loss of blood cells. Dosing the lens of the eye can cause cataracts. Dosing the gonads can reduce fertility.
   C. From 29 CFR 1910.96(b)(2)(i–ii): Whole body exposure can be as high as 3 rem/calendar quarter if the accumulated occupational exposure is less than $5(N - 18)$ rems, where N is the current age.
   D. No, she need not be: 5(45 – 18) = 135 rem
   E. From 29 CFR 1910.96(b)(3): 10% of values on Table G-18
6. A. 0.105 m^{-1}
   B. 2325 cpm

CHAPTER 14

1. A. Moving the air increases the rate of evaporation from the skin, providing cooling.
   B. Air in the building becomes hotter than outside air because of the sun heating the blackened roof. An exhaust fan removes this heated air, which must then be replaced by the entry somewhere of outside air. Air inlets would facilitate the entry of outside air, speeding the process of removing the heated air. By locating these inlets on the opposite wall of the room, the entire room is swept by entering outside air, minimizing pockets of hot air remaining in the room.
   C. *R* and *C*. The roof, heated by the sun, is radiating heat to objects in the room (*R*). It is also heating the air, and in South Carolina on a hot summer day, air is very likely to reach temperatures above body temperature (*C*). Painting the roof with a reflective paint, perhaps aluminum paint, would reduce effectiveness of the sun at raising the temperature of the roof.
2. Worker 1 is continuously in the high temperature zone, so this person should be watched for heat stress. Worker 2 spends only part of the time in the hot zone. However, this worker is climbing steps carrying heavy bags, and is wearing a coverall garment that prevents cooling of the body by circulating air. Add to this the time spent in the high temperature zone. This person should be watched closely for signs of heat stress. Worker 3 has only brief encounters with the high temperature zone, and thus is probably not threatened by heat stress.
3. Worker 1 is losing body heat due to the constant evaporation of sweat. However, by being unprotected from the sun, this worker is taking in radiative heat energy. Worker 2 is a similar case to worker 1; however, because some of the sweat is not evaporating from this worker's body, water is being lost without corresponding cooling being achieved. Worker 3 has blocked much of the radiation energy from the sun, and is still getting a large degree of benefit from the evaporation of sweat. Worker 4 has blocked out the sun, but has also blocked air movement across the body. Worker 4 may be at greatest risk.

4. A. The air is at a higher temperature than the body transfers heat to the body (*C*). However, heat loss due to evaporation of sweat depends not on temperature differences, but on the relative humidity (water content) of the surrounding air. Dry air accepts water evaporating from the skin more rapidly, producing a higher rate of heat loss. The higher the skin temperature the more rapid the evaporation, and again the more rapid the heat loss. Therefore, exposure of skin in this situation is an advantage.
   B. T_G, when compared to T_A, measures the contribution of radiative heating to body heat burden. Since the contribution of heat radiation is not changed by the spray cooling of the air, when T_A is lowered, T_G is also. T_{NWB} starts at T_A, and is then lowered by the cooling effect of evaporation. T_A is lowered by the water spray, but the humidity is raised, reducing the cooling effect of evaporation. Because of these counterbalancing effects, it is hard to predict the effect on T_{NWB}.
   C. WBGT = $0.7T_{NWB}$ + 0.2TG + 0.1TA
   Before installation:
   WBGT = 0.7(24.2) + 0.2(46.1) + 0.1(42.8) = 30.4
   After installation:
   WBGT = 0.7(24.4) + 0.2(35.0) + 0.1(29.4) = 27.0
5. T_G should show the greatest change.
   WBGT = 0.7TNWB + 0.2TG + 0.1TA
   Before installation:
   WBGT = 0.7(36.3) + 0.2(71.7) + 0.1(47.8) = 44.5
   After installation:
   WBGT = 0.7(29.8) + 0.2(43.3) + 0.1(43.3) = 33.9
6. The total volume of urine is unknown when you simply test the urine for the compound of interest excreted by the worker in the work day. Workers on the floor will produce a normal volume of urine. Those on the catwalk drinking water may produce a normal urine volume or they may produce more or less depending on the volume of water consumed. Those not drinking water are dehydrated and are producing a low urine volume. The latter group will appear to have a higher fluoride exposure, because they have more concentrated urine.

CHAPTER 15

1. A. Portable wooden ladders
   B. 1910.25(c)(2-10); 9 kinds
   C. See 1910.25(c)(18): Wane is bark or the lack of wood for any cause on the corner of a piece of wood.
   D. See 1910.25(c)(16): A shake is a separation along the grain, most of which occurs between the rings of annual growth.
   E. No, it is all definitions.
   F. (c) describes wooden and (d) describes metal ladders. (1) is repeated so that each section is complete and independent of the other.
   G. This section is regulations regarding wooden ladders.
   I. The ladder conforms to OSHA standards in 1910.25.
2. A. Examples: suspension system failing, as from a broken rope or failed mount on the roof; worker falling from the platform; wind blowing platform about; power system for moving platform failing
   B. The July 1, 1992, edition has 34 pages.
   C. From 1910.66(f)(7)(i): Each specific installation shall use suspension wire ropes or combination cable and connections meeting the specification recommended by the manufacturer of the hoisting machine used. From 1910.66(f)(7)(iii): Suspension wire rope grade shall be at least improved plow steel or equivalent. From 1910.66(f)(7)(ii): Each suspension rope shall have a "design factor" (ratio of rated strength of suspension wire to rated working load of at least 10.
   D. From 1910.66(f)(7)(ii):

$$F = S(N) / W$$

$$= (5000\ \text{lb})\ (2) / (2000\ \text{lb}) = 5$$

< 10 means system is not in compliance

E. From 1910.66(f)(5)(iii): A thorough inspection of suspension wire ropes in service shall be made once a month.
From 1910.66(f)(7)(vi): A corrosion-resistant tag shall be attached to one of the wire rope fastenings.
From 1910.66(f)(7)(vii): A new tag shall be installed at each rope renewal.
From 1910.66(f)(7)(viii): The original tag shall be stamped with the date of resocketing.

F. These control lateral movement of the platform in the wind.

G. From 1910.66(f)(1)(iii–iv): Equipment that is exposed to wind when not in service shall be designed to withstand forces generated by winds of at least 100 mph . . . . Equipment that is exposed to wind when in service shall be designed to withstand forces generated by winds of at least 50 mph.
From 1910.66 (i)(2)(v–vi): The platform shall not be operated in winds in excess of 25 mph . . . . On exterior installations, an anemometer shall be mounted on the platform.

3.

CAUTION THIS EQUIPMENT STARTS AND STOPS AUTOMATICALLY	EYE WASH FOUNTAIN	**CAUTION** WEAR HEARING PROTECTION IN THIS AREA	**DANGER** OXYGEN IN USE NO SMOKING OR OPEN FLAME
yellow + black	green + white	yellow + black	red, black,+ white

4. A. From the index you find that "Woodworking Machinery" is section 29 CFR 1910.213.
B. (1) Bandsaws are covered in 29 CFR 1910.213(i). (2) The regulations for the guard, sect. (1), indicate that the wheel cover should be enclosed by solid material, wire mesh, or perforated metal. Metal should be 20 gauge (0.037-in thick) and openings should be no larger than 3/8 in. There is supposed to be a sliding guide with a guard for the blade that raises and lowers with the guide. This is missing.

5. A. 50 V, 300 V, 600 V, 1000 V
B. Greater voltage has greater ability to cause bodily harm and to bypass insulators and safety systems.

6.A. A circuit failure such as a failure of wire insulation could place a potential on ungrounded equipment.

7. 3 A.

CHAPTER 16

1. A. An injury to a worker as a result of an accident, such as a fall.
B. White finger syndrome: A vibration injury to the hands involving nerve and blood vessel damage.
C. An inflammation of the tendons.
D. A tenosynovitus and nerve damage caused by abrasion in the carpal tunnel (passage through the wrist) due to working with the wrist bent.
E. A tenosynovitus due to continuous use of the forefinger to run a tool.
F. A tenosynovitus due to continuous twisting to the forearm, as when a screw is tightened.
G. Damage due to a loss of blood circulation in a tissue, which could be caused by continuous pressure of the tissue.

2. A saber saw produces vibration. Gloves probably should be used to prevent slivers in the hands from handling the wood. The amount of force required to push the saw along the template should be checked. The curved templates require constant turning of the saw, and

with the saw at chest height the wrist will be bent and will need to twist and the forearm will need to rotate. This could cause carpal tunnel syndrome and epicondylitis.
The work should be lowered. Two handles perpendicular to the wood surface should be attached to the saw handle to allow the saw to be operated with two hands with straight wrists and to turn the saw by extending one arm or the other. The saw should be mounted on a suspension device so that its weight is borne at the end of a cut. The automatic shutoff when the switch is released is a good safety feature; however, constant pressure with the forefinger could produce ischemia in the finger pad. The switch should be replaced by a pressure switch in one of the handles so that when the worker releases the tool it stops.

CHAPTER 17

1. First, the employee must have completed training that informs that person about the hazards of HBV and the efficacy, safety, benefits, and manner of administration of the vaccine. There is then a standard waiver the employee signs, a copy of which is in Appendix A of the regulations
2. This workplace should have the following:
   - A supply of protective clothing including gloves, face masks, and gowns.
   - Lab workers should have eye protection (needed in a laboratory whether body fluids are involved or not).
   - Facilities for storage, for laundering, or for disposal of contaminated protective clothing. The containers should be labeled as biohazard.
   - A mechanical device such as a brush and pan to pick up broken glass.
   - A labeled leak-proof and puncture-proof container to hold contaminated sharps.
   - A labeled leak-proof and puncture-proof container to hold used glass for decontamination.
   - A mechanical device for pipetting. Workers should never pipet by mouth suction.
   - Disinfectant to clean up any spills.
   - A sink and soap for hand washing.
   - All employees must have been trained in protection from biohazards and a plan should have been made covering normal operating practices and the handling of any accident such as a spill that would spread contamination.

CHAPTER 18

1. A. Part 56; B. Part 57; C. Part 71; D. Part 70
2. A. 30 CFR 56.5001. The limits are the ACGIH TLV of 1973.
   B. 30 CFR 56.5005. The standards are those set by the American National Standards Institute: ANSI 288.2-1969. The standard can be obtained by contacting ANSI at 1430 Broadway, New York, NY 10018.
3. A. 30 CFR 56.5050; 30 CFR 57.5050; 30 CFR 70.500; 30 CFR 71. They are all the same.
   B. They are the same.
   C. The standards are presented by the American National Standards Institute: ANSI-S1.4-1971.
4. A. 30 CFR 71.100
   B. 2.0 mg/m^3
   C. 30 CFR 71.101; silica
   D. 5%
   E. 10 mg/m^3/10 = 1 mg/m^3
   F. At 4% and 5%, 2 mg/m^3 is permitted. At 6%, 10 mg/m^3/6 = 1.7 mg/m^3 is permitted.
   0.04×2 mg/m^3 = 0.08 mg/m^3
   0.05×2 mg/m^3 = 0.10 mg/m^3
   0.06×1.7 mg/m^3 = 0.10 mg/m^3
   G. From 30 CFR 208: Bimonthly.

H. The operator must identify the mine and worker station in violation, indicate the measures to be taken to achieve compliance, and indicate the time, place, and manner these measures will be employed. In 30 CFR 71.301 we see that the District Manager, the manager of the Mine Safety and Health District in which the mine is located, must approve the plan, and may take air samples to ensure that the measures bring the mine into compliance.

5. A. Test methods and equipment are outlined in ANSI N13.8-1973, which may be obtained either at the Metal and Nonmetal Mine Safety and Health Subdistrict office or from the American National Standards Institute, 1430 Broadway, New York, NY, 10018.

B. "If concentrations of radon daughters in excess of 0.1 WL are found in an exhaust air sample . . . ."

C. The maximum permissible concentration (30 CFR 57.5039) is 1.0 WL and the maximum annual exposure (30 CFR 57.5038) is 4 WLM.

6. A. From 29 CFR 1910.1029(b) Definitions: "Coke oven emissions" means the benzene-soluble fraction of total particulate matter present during the destructive distillation or carbonization of coal for the production of coke.

B. From 29 CFR 1910.1029(b): "Green plush" means coke which when removed from the oven results in emissions due to the presence of unvolatilized coal. The hazard would therefore be a release of coke oven emissions.

C. From 29 CFR 1910.1029, Appendix A (II): Exposure to coke oven emissions is a cause of lung cancer and kidney cancer in humans. Although there has not been an excessive number of skin cancer cases in humans, repeated skin contact with coke oven emissions should be avoided.

D. From 29 CFR 1910.1029(c) or 29 CFR 1910.1029, Appendix A (I)(C): 150 mg/m^3 of air determined as an average over an 8-hr period.

E. From 29 CFR 1910.1029, Appendix B (I)(A): Samples collected should be full-shift (at least 7-hr) samples. Sampling should be done using a personal sampling pump with pulsation damper at a flow rate of 2 l/min. Samples should be collected on 0.8-µm pore size silver membrane filters (37 mm diameter) preceded by Gelman glass fiber type A-E filters encased in three-piece plastic (polystyrene) field monitor cassettes. The cassette face cap should be on and the plug removed. The rotometer should be checked every hour to ensure that proper flow rates are maintained.

A minimum of three full-shift samples should be collected for each job classification of each battery, at least one from each shift. If disparate results are obtained for a particular job classification, sampling should be repeated. It is advisable to sample each shift on more than one day to account for environmental variables (wind, precipitation, etc.) that may affect sampling. Differences in exposure among different work shifts may indicate a need to improve work practices on a particular shift. Sampling results for each job classification should not be averaged. Multiple samples from the same shift on each battery may be used to calculate an average exposure for a particular job classification.

F. From 29 CFR 1910.1029(b): "Beehive oven" means a coke oven in which the products of carbonization other than coke are not recovered, but are released into ambient air.

G. From 29 CFR 1910.1029(f)(1)(iii)(a): The employer shall institute engineering and work practice controls on all beehive ovens at the earliest possible time to reduce and maintain employee exposures at or below the permissible exposure limit, ...

From 29 CFR 1910.1029(f)(1)(iii)(b): If, after implementing all engineering and work practice controls required by paragraph (f)(1)(iii)(*b*) of this section, employee exposures still exceed the permissible exposure limit, the employer shall research, develop, and implement any other engineering and work practice controls necessary . . . . .

Wherever the engineering and work practice controls which can be instituted are not sufficient to reduce employee exposure to at or below the permissible exposure limit, the employer shall nonetheless use them to reduce exposures to the lowest level achievable by these controls, and shall supplement them by the use of respiratory protection which complies with paragraph (g) of this section.

7. A. 29 CFR 1910.251–.257

B. From 29 CFR.252(c)(2)(i): Mechanical ventilation shall be provided when welding or cutting is done on metals not covered in paragraphs (c)(5) through (c)(12) of this section

. . . . (A) in a space of less than 10,000 cubic feet (284 m^3) per welder, (B) in a room having a ceiling of less than 16 ft (5 m), (C) in confined spaces or where the welding space contains partitions, balconies, or other structural barriers to the extent that they significantly obstruct cross-ventilation. (ii) Such ventilation shall be at the minimum rate of 2000 cubic feet (57 m^3) per minute per welder . . . . .

C. From 29 CFR.252(c)(9)(i), General: Welding or cutting indoors or in confined spaces involving cadmium-bearing or cadmium-coated base metals shall be done using local exhaust ventilation or airline respirators unless atmospheric tests under the most adverse conditions have established that the workers' exposure is within the acceptable concentrations defined by §1910.1000 of this part.

D. From 29 CFR.252(c)(4), Ventilation in confined spaces, (i) Air replacement: All welding and cutting operations carried on in confined spaces shall be adequately ventilated to prevent the accumulation of toxic materials or possible oxygen deficiency.

8. From 29 CFR 1910.218: Guards to prevent physical injury during operation or changing of dies.

CHAPTER 19

1. Cans are added to molten primary aluminum. Moisture in cans may cause explosive steam production. Heat stress, noise exposure, and metal and metal oxide fumes may be produced. If chlorine is bubbled through the melt, exposure to the gas through leaks in the system or as it exits the melt are possible. If fluxes or alloying metals are added to the melt, dusts may reach the workers. Dross from the furnace contains an unpredictable mixture of compounds, some of which could be volatile.
2. During its purification; during the smelting of gold, lead, copper, cadmium, and zinc. When alloying with lead; in preparation of arsenic pesticides
3. Chromium plating, stainless steel production, welding and cutting, painting with chromate pigments, printing with chromate dyes, tanning leather, applying wood preservatives, cement production and handling, magnesium casting, chromate/dichromate production
4. Three metals may be present in quantity: nickel, copper and cobalt. Any dry residues in the vessel could produce airborne dust rich in metal salts. The leaching is done with hot acid, so there are acid residues. Workers should use respirators and should wear protective clothing.
5. A. From 29 CFR 1910.1025(c): PEL is 50 μg/m^3 TWA.

   B. From 29 CFR 1910.1025(d)(9) . . . which has an accuracy (to a confidence level of 95%) of not less than 20% for airborne concentrations of lead equal to or greater than 30μg/m^3.

   C. From 29 CFR 1910.1025(j)(1)(i): . . . all workers exposed above action level for more than 30 days per year. From (j)(2)(i)(A): At least every 6 months for each employee covered under paragraph From (j)(1)(i). (j)(2)(i)(B): At least every 2 months for each employee whose last blood sampling and analysis indicated a blood level at or above 40 μg/100 ml. From (j)(2)(i)(C): At least monthly during the removal period of each employee removed from exposure to lead due to an elevated blood level.

   D. From 29 CFR 1910.1025(j)(2)(iii): . . . have an accuracy (to a confidence level of 95%) within plus or minus 15% or 6 μg/m^3, whichever is larger.

   E. From 29 CFR 1910.1025(j)(3)(i) . . . to each employee covered under (j)(1)(i) of this section (A) At least annually.
6. As, Cr, Ni, Be, Cd, and Pb
7. 29 CFR 1910.1027, Appendix A. The EPA classes cadmium as B1, a probable human carcinogen. The International Agency for Research on Cancer classes cadmium as 2A, a probable human carcinogen. ACGIH has classed cadmium as a carcinogen.
8. 29 CFR 1910.1025, Appendix C
   29 CFR 1910.1025, Table II.

CHAPTER 20

1. Some biological hazard may exist due to residues of milk in the cartons having been subject to bacterial degradation. Transferring dyes into the mix may involve contact with potentially toxic chemicals. Proper guards must be placed around the chopping, grinding, and molding machinery. Heat stress is possible from the washing and drying operations and from the injection molding machines. Noise would be a problem in compression and chipping, blowing the chips dry, grinding, and from the injection molding machinery. Skin exposure to hot detergent solutions should be avoided. Final trimming of the fishing boxes may generate particulate.
2. A. Nylon (polyamide) strands can hydrogen bond to one another, whereas only very weak attractions can form between polyethylene strands.
   B. Slow cooling allows a greater proportion of the plastic to be in the form of crystallites.
   C. Cross-linking sharply increases rigidity.
   D. Plasticizers reduce crystallite formation, lowering rigidity.
3. *Polymerization adhesive.* Advantages: No solvent to evaporate. Disadvantages: Possible worker contact with reactive and toxic prepolymerization chemicals. Polymers *in solution.* Advantages: Far fewer (possibly none) of the reactive or toxic compounds used in the polymerization technique. Disadvantages: Evaporated solvent may require removal by ventilation.
4. A. 29 CFR 1910.1043
   B. From 29 CFR 1910.1043(b): "Lint free respirable cotton dust" means particles of cotton dust of approximately 15 microns or less aerodynamic equivalent diameter.
   C. From 29 CFR 1910.1043(c)(1): . . . measured by a vertical elutriator or a method of equivalent accuracy and precision. From 29 CFR 1910.1043(b): "Vertical elutriator cotton dust sampler" means a dust sampler which has a particle size cutoff at approximately 15 microns when operating at the flow rate of 7.4 ± 0.2 l/min.
   D. From 29 CFR 1910.1043(c)(1): The PEL for yarn manufacturing is 200 $\mu g/m^3$ of lint free respirable dust mean concentration averaged over an 8-hr day; for slashing and weaving to 750 $\mu g/m^3$ of lint free respirable dust mean concentration averaged over an 8-hr day; in other operations to 500 $\mu g/m^3$ of lint free respirable dust mean concentration averaged over an 8-hr day.
   E. From 29 CFR 1910.1043(d)(3)(i): The employer shall repeat the measurements required by paragraph (d)(2) at least every 6 months.
5. A. From 29 CFR 1910.1017(c)(1): PEL . . . 1 ppm averaged over any 8-hr period; rom 29 CFR 1910.1017(c)(2) . . . 5 ppm averaged over any period not exceeding 15 minutes; from 29 CFR 1910.1017(b)(1): Action level . . . 0.5 ppm averaged over an 8-hr work day.
   B. From 29 CFR 1910.1017(a)(2): Applies to the polyvinyl chloride not yet processed into product. From 29 CFR 1910.1045(a)(2)(i): Excludes polymers including acrylonitrile as a monomer from coverage.

APPENDIX A
SIMPLE STATISTICAL CALCULATIONS

STANDARD DEVIATION

The most common means of expressing the confidence limits on a set of experimental values is called the standard deviation. This is expressed by following the average value for the data plus or minus a number. If you had a large sample of data with a random error in the determination, you would expect 68% of the values to fall in a range from a value of the average minus the standard deviation to a value of the average plus the standard deviation. To obtain the standard deviation, follow these steps:

1. Average all of the experimental values.
2. Find the difference between the average value and each experimental value.
3. Square each of these differences.
4. Total the squares of the differences.
5. Divide the total of the squares of the differences by 1 less than the total number of data items.
6. Take the square root of the number obtained in step 5. This is the standard deviation.

95% CONFIDENCE LIMIT

95% confidence limits are often used in toxicology to express the likelihood that the averages obtained will fall inside a certain range. At the 95% limit, we are stating that 95% of a large number of averages taken of sets of data on a particular measurement will fall inside the stated range. Notice that we are not talking about the individual measurements, but of the averages of sets of data, a much more dependable set of numbers, falling within a stated range. 95% confidence limits are obtained as follows:

1. Take the standard deviation of the data as outlined above.
2. Divide the standard deviation by the square root of the number of items of data.
3. Multiply this value by the appropriate number (called the Student t value) from Table A.1. This is the desired confidence limit.

Table A.1. Student t Values.

Number of Items of Data	t Value
2	12.70
3	4.30
4	3.18
5	2.78
6	2.57
7	2.45
8	2.36
9	2.31
10	2.26
11	2.23
16	2.13
21	2.09
31	2.04
∞	1.96

APPENDIX B
**** MATERIAL SAFETY DATA SHEET ****

Benzene
02610

**** SECTION 1 — CHEMICAL PRODUCT AND COMPANY IDENTIFICATION ****

MSDS Name:
Benzene
Catalog Numbers:
B243 4, B243-4, B2434, B245 4, B245 500, B245-4, B245-500, B2454, B245500, B245J4, B411 1, B411 4, B411-1, B411-4, B4111, B4114, B414-1
Synonyms:
Benzol, coal naphtha, cyclohexatriene, phenyl hydride, pyrobenzol.
Company Identification:
Fisher Scientific
1 Reagent Lane
Fairlawn, NJ 07410
For information, call:
201-796-7100
Emergency Number:
201-796-7100
For CHEMTREC assistance, call:
800-424-9300
For International CHEMTREC assistance, call:
703-527-3887

**** SECTION 2 — COMPOSITION, INFORMATION ON INGREDIENTS ****

CAS#	Chemical Name	%	EINECS#
71-43-2	Benzene	>99%	200-753-7

Hazard Symbols: T F

**** SECTION 3 — HAZARDS IDENTIFICATION ****

EMERGENCY OVERVIEW

Appearance: Colorless. Flash Point: 12°F.
Danger! Extremely flammable liquid. Harmful if inhaled. May be harmful if absorbed through the skin. Aspiration hazard. Poison! May cause central nervous system effects. May cause eye and skin irritation. May cause respiratory and digestive tract irritation. May cause reproductive and fetal effects. Cancer hazard. May cause blood abnormalities. Harmful or fatal if swallowed.
Target Organs: Blood, central nervous system, bone marrow, immune system.

Potential Health Effects

Eye:
Produces irritation, characterized by a burning sensation, redness, tearing, inflammation, and possible corneal injury.
Skin:
Causes skin irritation. May be absorbed through the skin in harmful amounts.
Direct contact with the liquid may cause redness and vesiculation.

Ingestion:

Aspiration hazard. May cause central nervous system depression characterized by excitement, followed by headache, dizziness, drowsiness, and nausea. Advanced stages may cause collapse, unconsciousness, coma and possible death due to respiratory failure. May cause effects similar to those for inhalation exposure. Aspiration of material into the lungs may cause chemical pneumonitis, which may be fatal.

Inhalation:

May cause respiratory tract irritation. May cause adverse central nervous system effects including headache, convulsions, and possible death. May cause drowsiness, unconsciousness, and central nervous system depression. Exposure may lead to an irreversible injury to the bone marrow.

Exposure to high concentrations may be fatal. Brief exposure is irritating to the eyes and respiratory tract; continued exposure may cause euphoria, nausea, a staggering gait and coma. Exposure to lower concentrations produces vertigo, drowsiness, headache, and nausea. Typical symptoms of benzene exposure include light-headedness, headache, loss of apetite, and abdominal discomfort. With more severe intoxication, there may be weakness, blurring of vision, and breathing difficulties. The mucous membranes and skin may appear pale, and a hemorrhagic tendency may result in petechiae, easy bruising, nosebleeds, bleeding from the gums.

Chronic:

Possible cancer hazard based on tests with laboratory animals. Prolonged or repeated exposure may cause adverse reproductive effects. May cause bone marrow abnormalities with damage to blood forming tissues.

Chronic exposure to benzene has been associated with an increased incidence of leukemia and multiple myelomas. Benzene exposure has also been shown to have immunodepressive effects. Prolonged skin exposure may lead to the development of a dry, scaly dermatitis or secondary infections.

**** SECTION 4 — FIRST AID MEASURES ****

Eyes:

Flush eyes with plenty of water for at least 15 minutes, occasionally lifting the upper and lower lids. Get medical aid immediately.

Skin:

Get medical aid immediately. Immediately flush skin with plenty of soap and water for at least 15 minutes while removing contaminated clothing and shoes.

Ingestion:

Do NOT induce vomiting. If victim is conscious and alert, give 2-4 cupfuls of milk or water. Possible aspiration hazard. Get medical aid immediately.

Inhalation:

Get medical aid immediately. Remove from exposure to fresh air immediately. If not breathing, give artificial respiration. If breathing is difficult, give oxygen.

Notes to Physician:

Treat symptomatically and supportively.

**** SECTION 5 — FIRE FIGHTING MEASURES ****

General Information:

As in any fire, wear a self-contained breathing apparatus in pressure-demand, MSHA/NIOSH (approved or equivalent), and full protective gear. Water runoff can cause environmental damage. Dike and collect water used to fight fire. Vapors can travel to a source of ignition and flash back. Extremely flammable. Material will readily ignite at room temperature. Use water spray to keep fire-exposed containers cool.

Extinguishing Media:

Use water spray to cool fire-exposed containers. In case of fire use water spray, dry chemical, carbon dioxide, or chemical foam.

Autoignition Temperature:
928°F (497.78°C)
Flash Point:
12°F (–11.11°C)
NFPA Rating:
health-2; flammability-3; reactivity-0
Explosion Limits:
Lower: 1.3 Upper: 7.1

**** SECTION 6 — ACCIDENTAL RELEASE MEASURES ****

General Information:
Use proper personal protective equipment as indicated in Section 8.
Spills/Leaks:
Use water spray to dilute spill to a non-flammable mixture. Avoid runoff into storm sewers and ditches which lead to waterways. Wear a self contained breathing apparatus and appropriate Personal protection. (See Exposure Controls, Personal Protection section). Use water spray to disperse the gas/vapor. Remove all sources of ignition. Absorb spill using an absorbent, non–combustible material such as earth, sand, or vermiculite.

**** SECTION 7 — HANDLING and STORAGE ****

Handling:
Wash thoroughly after handling. Remove contaminated clothing and wash before reuse. Ground and bond containers when transferring material. Use spark-proof tools and explosion proof equipment. Do not get in eyes, on skin, or on clothing. Empty containers retain product residue, (liquid and/or vapor), and can be dangerous. Do not pressurize, cut, weld, braze, solder, drill, grind, or expose such containers to heat, sparks or open flames. Do not ingest or inhale. Use only in a chemical fume hood.
Storage:
Keep away from heat, sparks, and flame. Keep away from sources of ignition. Store in a tightly closed container. Store in a cool, dry, well-ventilated area away from incompatible substances.

**** SECTION 8 — EXPOSURE CONTROLS, PERSONAL PROTECTION ****

Engineering Controls:
Use only under a chemical fume hood.
Exposure Limits

Chemical Name	ACGIH	NIOSH	OSHA
Benzene	(10) ppm TWA; (32) mg/m3 TWA	0.1 ppm TWA; 1 ppm STEL	10 ppm TWA; 1 ppm TWA; 5 ppm STEL

OSHA Vacated PELs:
Benzene:
10 ppm TWA (unless specified in 1910.1028); 50 ppm STEL (10 min) (unless specified in 1910.1028); C 25 ppm (unless specified in 1910.1028)
Personal Protective Equipment
Eyes:
Wear appropriate protective eyeglasses or chemical safety goggles as described by OSHA's eye and face protection regulations in 29 CFR 1910.133.
Skin:
Wear impervious gloves.

Clothing:

Wear appropriate protective clothing to prevent skin exposure.

Respirators:

Follow the OSHA respirator regulations found in 29CFR 1910.134. Always use a NIOSH-approved respirator when necessary.

**** SECTION 9 — PHYSICAL AND CHEMICAL PROPERTIES ****

Physical State: Liquid
Appearance: Colorless.
Odor: Sweet, aromatic.
pH: Not available.
Vapor Pressure: 100 mm Hg
Vapor Density: 2.7 (Air=1)
Evaporation Rate: 2.8 (Ether=1)
Viscosity: 0.647mPa at 20C
Boiling Point: 176∞F
Freezing/Melting Point: 42°F
Decomposition Temperature: Not available.
Solubility: 0.18g/100g water at 25C.
Specific Gravity/Density: 0.88
Molecular Formula: C6H6
Molecular Weight: 78.042

**** SECTION 10 — STABILITY AND REACTIVITY ****

Chemical Stability:

Stable under normal temperatures and pressures.

Conditions to Avoid:

Incompatible materials, ignition sources, excess heat.

Incompatibilities with Other Materials:

Benzene is incompatible with arsenic pentafluoride + potassium methoxide, diborane, hydrogen + raney nickel, interhalogens, oxidants, uranium hexafluoride, bromine pentafluoride, chlorine, chlorine trifluoride, chromic anhydride, nitryl perchlorate, oxygen, ozone, perchlorates, perchloryl fluoride + aluminum chloride, permanganates + sulfuric acid, potassium peroxide and silver perchlorate.

Hazardous Decomposition Products:

Irritating and toxic fumes and gases.

Hazardous Polymerization:

Has not been reported.

**** SECTION 11 — TOXICOLOGICAL INFORMATION ****

RTECS#:

CAS# 71-43-2: CY1400000

LD50/LC50:

CAS# 71-43-2:
Inhalation, mouse: LC50 =9980 ppm
Inhalation, rat: LC50 =10000 ppm/7H
Oral, mouse: LD50 = 4700 mg/kg
Oral, rat: LD50 = 930 mg/kg
Skin, rabbit: LD50 = >9400 mg/kg.

Carcinogenicity:

Benzene -

ACGIH: (A2)-suspected human carcinogen

California: carcinogen
NTP: Known carcinogen
OSHA: Select carcinogen
IARC: Group 1 carcinogen

Epidemiology:

IARC has concluded that epidemiological studies have established the relationship between benzene exposure and the dev elopment of acute myelogenous leukemia, and that there is sufficient evidence that benzene is carcinogenic to hu mans. Animal studies have demonstrated fetoxicity (growth retardation) and teratogenicity (exencephaly, angulated ribs, dilated brain ventricles).

Teratogenicity:

Experimental teratogen. Animal studies have demonstrated fetoxicity (growth retardation) and teratogenicity (exencephaly, angulated ribs, dilated brain ventricles).

Reproductive Effects:

Experimental reproductive effects have been reported.

Neurotoxicity:

No information available.

Mutagenicity:

Chromosomal aberrations have been noted in animal tests.

Other Studies:

Please refer to RTECS CY1400000 for additional data.

**** SECTION 12 — ECOLOGICAL INFORMATION ****

Ecotoxicity:

Minnow (distilled water) lethal, 5 ppm/6H.
Sunfish (tap water) TLM=20 ppm/24H.
Striped bass TLm96=100-10 ppm.

Environmental Fate:

No information reported.

Physical/Chemical:

No information available.

Other:

None.

**** SECTION 13 — DISPOSAL CONSIDERATIONS ****

Dispose of in a manner consistent with federal, state, and local regulations.

RCRA D-Series Maximum Concentration of Contaminants:

waste number D018; regulatory level = 0.5 mg/L.

RCRA D-Series Chronic Toxicity Reference Levels:

chronic toxicity reference level = 0.005 mg/L.

RCRA F-Series: Not listed.

RCRA P-Series: Not listed.

RCRA U-Series:

waste number U019 (Ignitable waste; Toxic waste)

RCRA Substances Banned from Land Disposal

This material is banned from land disposal according to RCRA.

**** SECTION 14 — TRANSPORT INFORMATION ****

US DOT

Shipping Name: RQ, BENZENE
Hazard Class: 3
UN Number: UN1114
Packing Group: II

IMO

No information available.

IATA

No information available.

RID/ADR

No information available.

Canadian TDG

Shipping Name: BENZENE

Hazard Class: 3(9.2)

UN Number: UN1114

Other Information: FLASHPOINT -11 C

**** SECTION 15 — REGULATORY INFORMATION ****

US Federal

TSCA

CAS# 71-43-2 is listed on the TSCA inventory.

Health & Safety Reporting List

None of the chemicals are on the Health & Safety Reporting List.

Chemical Test Rules

Section 12b

None of the chemicals are listed under TSCA Section 12b.

TSCA Significant New Use Rule

None of the chemicals in this material have a SNUR under TSCA.

SARA

Section 302 (RQ)

final RQ = 10 pounds (4.54 kg)

Section 302 (TPQ)

None of the chemicals in this product have a TPQ.

SARA Codes

CAS # 71-43-2: acute, chronic, flammable.

Section 313

This material contains Benzene (CAS# 71-43-2, >99%),which is subject to the reporting requirements of Section 313 of SARA Title III and 40 CFR Part 373.

Clean Air Act:

CAS# 71-43-2 is listed as a hazardous air pollutant (HAP).

This material does not contain any Class 1 Ozone depletors.

This material does not contain any Class 2 Ozone depletors.

Clean Water Act:

CAS# 71-43-2 is listed as a Hazardous Substance under the CWA.

CAS# 71-43-2 is listed as a Priority Pollutant under the Clean Water Act.

CAS# 71-43-2 is listed as a Toxic Pollutant under the Clean Water Act.

OSHA:

None of the chemicals in this product are considered highly hazardous by OSHA.

US State

Benzene can be found on the following state right to know lists:

California, New Jersey, Florida, Pennsylvania, Minnesota, Massachusetts.

The following statement(s) is(are) made in order to comply with the California Safe Drinking Water Act:

WARNING: This product contains Benzene, a chemical known to the state of California to cause cancer.

California No Significant Risk Level:

CAS# 71-43-2: no significant risk level = 7 µg/day

European/International Regulations

European Labeling in Accordance with EC Directives

Hazard Symbols: irritating and toxic fumes and gases irritating and toxic fumes and gases

Risk Phrases:
Safety Phrases:
S 45 In case of accident of if you feel unwell, seek medical advice immediately (show the label where possible).
S 53 Avoid exposure – obtain special instructions before use.

Canada
CAS# 71-43-2 is listed on Canada's DSL/NDSL List.
CAS# 71-43-2 is listed on Canada's Ingredient Disclosure List.
CAS# 71-43-2: OEL-AUSTRALIA:TWA 5 ppm (16 mg/m^3);Carcinogen.

Exposure Limits
OEL-BELGIUM:TWA 10 ppm (32 mg/m^3);Carcinogen JAN9.
OEL-CZECHOSLOVAKIA:TWA 10 mg/m^3;STEL 20 mg/m^3.
OEL-DENMARK:TWA 5 ppm (16 mg/m^3);Skin;Carcinogen.
OEL-FINLAND:TWA 5 ppm (15 mg/m^3);STEL 10 ppm (30 mg/m^3);Skin; CAR.
OEL-FRANCE:TWA 5 ppm (16 mg/m^3);Carcinogen.
OEL-GERMANY;Skin;Carcinogen.
OEL-HUNGARY:STEL 5 mg/m^3;Skin;Carcinogen.
OEL-INDIA:TWA 10 ppm (30 mg/m^3);Carcinogen.
OEL-JAPAN:TWA 10 ppm (32 mg/m^3);STEL 25 ppm (80 mg/m3); CAR.
OEL-THE NETHERLANDS:TWA 10 ppm (30 mg/m^3);Skin.
OEL-THE PHILIPPINES:TWA 25 ppm (80 mg/m^3);Skin.
OEL-POLAND:TWA 30 mg/m^3;Skin.
OEL-RUSSIA:TWA 10 ppm (5 mg/m^3);STEL 25 ppm (15 mg/m^3);Skin; CAR.
OEL-SWEDEN:TWA 1 ppm (3 mg/m^3);STEL 5 ppm (16 mg/m^3);Skin; CAR.
OEL-SWITZERLAND:TWA 5 ppm (16 mg/m^3);Skin;Carcinogen.
OEL-THAILAND:TWA 10 ppm (30 mg/m^3);STEL 25 ppm (7 mg/m^3).
OEL-TURKEY:TWA 20 ppm (64 mg/m^3);Skin.
OEL-UNITED KINGDOM:TWA 10 ppm (30 mg/m^3).
OEL IN BULGARIA, COLOMBIA, JORDAN, KOREA check ACGIH TLV.
OEL IN NEW ZEALAND, SINGAPORE, VIETNAM check ACGI TLV

**** SECTION 16 — ADDITIONAL INFORMATION ****

Additional Information:
No additional information available.

MSDS Creation Date: 1/05/1995 **Revision #4 Date:** 10/08/1996

The information above is believed to be accurate and represents the best information currently available to us. However, we make no warranty of merchantability or any other warranty, express or implied, with respect to such information, and we assume no liability resulting from its use. Users should make their own investigations to determine the suitability of the information for their particular purposes. In no way shall Fisher be liable for any claims, losses, or damages of any third party or for lost profits or any special, indirect, incidental, consequential or exemplary damages, howsoever arising, even if Fisher has been advised of the possibility of such damages.

APPENDIX C
SAFETY PHONE NUMBERS

OSHA — Health Standards	1-202-523-7075
NIOSH	1-800-356-4674
National Safety Council	1-312-527-4800
National Response Center (to Report Releases)	1-800-424-8802
Chemical Emergency Preparedness Hotline (CERCLA)	1-800-535-0202
Department of Transportation Hotline	1-202-366-4488
Chemical Transportation Emergency Center	1-800-424-9300
EPA (RCRA) Hazardous Waste Hotline (Emergency Response)	1-800-424-9346
Chemical Manufacturers Association (Chemical Referral Center)	1-800-262-8200
Substance Identification	1-800-848-6538

APPENDIX D
USEFUL ADDRESSES

American Board of Industrial Hygiene
4600 W. Saginaw St.
Suite 101
Lansing, MI 48917-2737

American Conference of Government Industrial Hygienists
6500 Glenway Ave.
Bldg. D-5
Cincinnati, OH 45211

American Industrial Hygiene Association[1]
2700 Prosperity Ave.
Suite 250
Fairfax, VA 22031

American Public Health Association
1015 18th St., NW
Washington, DC 20036

American Society of Safety Engineers
1800 E. Oakton St.
Des Plaines, IL 60018

National Safety Council
1121 Spring Lake Dr.
Itasca, IL 60143

Health Physics Society
1340 Old Chain Bridge Rd.
McLean, VA 22101

[1] American Industrial Hygiene Foundation is also at this address.

APPENDIX E
FIRE DIAMOND

The fire diamond is a compact and standard symbol used to list the hazards of a chemical in compact fashion for container labels. The diamond has four zones: blue (A, left), red (B, top), yellow (C, right), and white (D, bottom). Numbers are used in zones A, B, and C to roughly quantify the degree of hazard, and abbreviations or symbols are placed in zone D to indicate specific hazards.

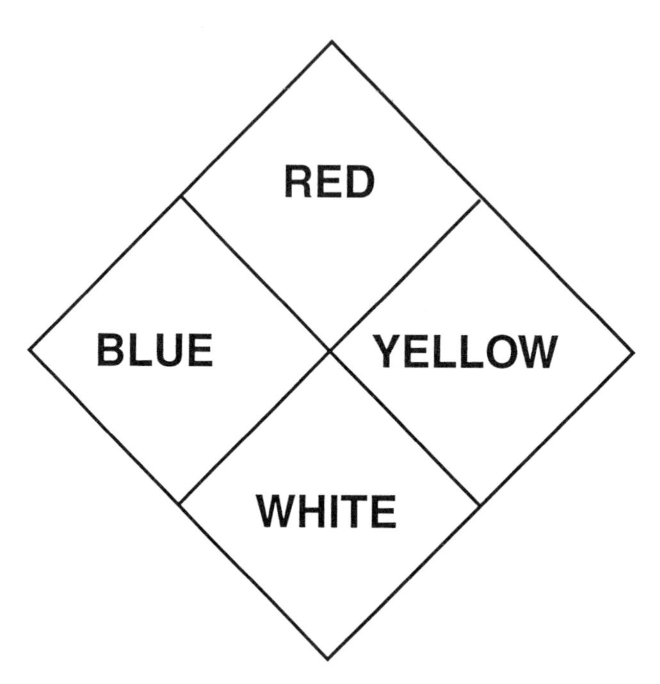

Zone A — Health

0 = very low
1 = slightly hazardous
2 = hazardous
3 = extreme danger
4 = deadly

Zone B — Flammability

0 = not flammable
1 = ignites if heated
2 = ignites if heated moderately
3 = ignites, ambient conditions
4 = burns readily, ambient conditions

Zone C — Reactivity

0 = stable, no reaction with water
1 = unstable if heated
2 = violent chemical reaction
3 = detonation on shock and heat
4 = detonates

Zone D — Specific Hazard

OXY oxidizer
ACID acid
ALKALI alkali
COR corrosive
W no water exposure; radiation hazard

APPENDIX F
CONFINED SPACE ENTRY PERMIT

Confined Space Entry Permit

Date & Time Issued: ____________ Date and Time Expires: ____________

Job site/Space I.D.: ____________ Job Supervisor ____________

Equipment to be worked on: ____________ Work to be performed: ____________

Stand-by personnel ____________ ____________ ____________

1. Atmospheric Checks: Time ______
   Oxygen ______ %
   Explosive ______ % L.F.L.
   Toxic ______ PPM

2. Tester's signature ____________

3. Source isolation (No Entry): N/A Yes No
   Pumps or lines blinded, ( ) ( ) ( )
   disconnected, or blocked ( ) ( ) ( )

4. Ventilation Modification: N/A Yes No
   Mechanical ( ) ( ) ( )
   Natural Ventilation only ( ) ( ) ( )

5. Atmospheric check after isolation and Ventilation:
   Oxygen ______ % > 19.5 %
   Explosive ______ % L.F.L. < 10 %
   Toxic ______ PPM < 10 PPM H_2S
   Time ______
   Testers signature ____________

6. Communication procedures: ____________

7. Rescue procedures: ____________

8. Entry, standby, and back up persons: Yes No
   Successfully completed required training?
   Is it current? ( ) ( )

9. Equipment: N/A Yes No
   Direct reading gas monitor - tested ( ) ( ) ( )
   Safety harnesses and lifelines for entry and standby persons ( ) ( ) ( )
   Hoisting equipment ( ) ( ) ( )
   Powered communications ( ) ( ) ( )
   SCBA's for entry and standby persons ( ) ( ) ( )
   Protective Clothing ( ) ( ) ( )
   All electric equipment listed Class I, Division I, Group D and Non-sparking tools ( ) ( ) ( )

10. Periodic atmospheric tests:
   Oxygen ___% Time___ Oxygen ___% Time___
   Oxygen ___% Time___ Oxygen ___% Time___
   Explosive ___% Time___ Explosive ___% Time___
   Explosive ___% Time___ Explosive ___% Time___
   Toxic ___% Time___ Toxic ___% Time___
   Toxic ___% Time___ Toxic ___% Time___

We have reviewed the work authorized by this permit and the information contained here-in. Written instructions and safety procedures have been received and are understood. Entry cannot be approved if any squares are marked in the "No" column. This permit is not valid unless all appropriate items are completed.

Permit Prepared By: (Supervisor) ____________ ____________

Approved By: (Unit Supervisor) ____________ ____________

Reviewed By (Cs Operations Personnel): ____________ ____________

(printed name) (signature)

This permit to be kept at job site. Return job site copy to Safety Office following job completion.

Copies: White Original (Safety Office) Yellow (Unit Supervisor) Hard(Job site)

Appendix D - 2

ENTRY PERMIT

PERMIT VALID FOR 8 HOURS ONLY. ALL PERMIT COPIES REMAIN AT SITE UNTIL JOB COMPLETED

DATE: - - SITE LOCATION/DESCRIPTION ____________

PURPOSE OF ENTRY ____________

SUPERVISOR(S) in charge of crews Type of Crew Phone #

____

COMMUNICATION PROCEDURES ____________

RESCUE PROCEDURES (PHONE NUMBERS AT BOTTOM) ____________

*** BOLD DENOTES MINIMUM REQUIREMENTS TO BE COMPLETED AND REVIEWED PRIOR TO ENTRY***

REQUIREMENTS COMPLETED	DATE	TIME	REQUIREMENTS COMPLETED	DATE	TIME
Lock Out/De-energize/Try-out	____	____	**Full Body Harness w/"D" ring**	____	____
Line(s) Broken-Capped-Blank	____	____	**Emergency Escape Retrieval Eq**	____	____
Purge-Flush and Vent	____	____	**Lifelines**	____	____
Ventilation	____	____	Fire Extinguishers	____	____
Secure Area (Post and Flag)	____	____	Lighting (Explosive Proof)	____	____
Breathing Apparatus	____	____	Protective Clothing	____	____
Resuscitator - Inhalator	____	____	Respirator(s) (Air Purifying)	____	____
Standby Safety Personnel	____	____	Burning and Welding Permit	____	____

Note: Items that do not apply enter N/A in the blank.

** RECORD CONTINUOUS MONITORING RESULTS EVERY 2 HOURS **

CONTINUOUS MONITORING** TEST(S) TO BE TAKEN	Permissible Entry Level									
PERCENT OF OXYGEN	**19.5% to 23.5%**	___	___	___	___	___	___	___	___	___
LOWER FLAMMABLE LIMIT	**Under 10%**	___	___	___	___	___	___	___	___	___
CARBON MONOXIDE	**+35 PPM**	___	___	___	___	___	___	___	___	___
Aromatic Hydrocarbon	+ 1 PPM * 5PPM	___	___	___	___	___	___	___	___	___
Hydrogen Cyanide	(Skin) * 4PPM	___	___	___	___	___	___	___	___	___
Hydrogen Sulfide	+10 PPM *15PPM	___	___	___	___	___	___	___	___	___
Sulfur Dioxide	+ 2 PPM * 5PPM	___	___	___	___	___	___	___	___	___
Ammonia	*35PPM	___	___	___	___	___	___	___	___	___

* Short-term exposure limit: Employee can work in the area up to 15 minutes.

\+ 8 hr. Time Weighted Avg.: Employee can work in area 8 hrs (longer with appropriate respiratory protection).

REMARKS: ____________

GAS TESTER NAME & CHECK # INSTRUMENT(S) USED MODEL &/OR TYPE SERIAL &/OR UNIT #

____

SAFETY STANDBY PERSON IS REQUIRED FOR ALL CONFINED SPACE WORK

SAFETY STANDBY PERSON(S) CHECK # CONFINED SPACE ENTRANT(S) CHECK # CONFINED SPACE ENTRANT(S) CHECK #

____

SUPERVISOR AUTHORIZATION - ALL CONDITIONS SATISFIED ____________ DEPARTMENT/PHONE ____________

AMBULANCE 2800 FIRE 2900 Safety 4901 Gas Coordinator 4529/5387

APPENDIX G
CONVERSION FACTORS

To convert from	To	Multiply by
calories	Btu	3.968×10^{-3}
centimeters	inches	0.3937
cubic centimeters	cubic inches	0.06102
cubic feet	liters	28.32
cubic inches	cubic meters	1.639×10^{-5}
cubic inches	liters	0.01639
cubic meters	cubic feet	35.31
feet	centimeters	30.482
gallons	cubic centimeters	3785
gallons	liters	3.785
grams	ounces	0.03527
grams	pounds	0.002205
inches	centimeters	2.540
kilograms	pounds	2.205
kilograms	tons (short)	1.102×10^{-3}
kilograms	metric tons	0.001000
liters	cubic centimeters	1000
liters	cubic feet	0.3532
liters	cubic inches	61.025
liters	gallons	0.2642
meters	inches	39.37
meters	feet	3.281
milligrams	ounces	3.527×10^{-5}
milligrams	pounds	2.205×10^{-6}
milliliters	cubic centimeters	1.000
milliliters	cubic inches	0.06103
milliliters	ounces (fluid)	0.03381
millimeters	inches	0.03937
ounces	grams	28.35
ounces (fluid)	cubic centimeters	29.57
ounces (fluid)	liters	0.02957
pounds	grams	453.6
pounds	kilograms	0.4536
quarts	liters	0.9463
tons (short)	kilograms	907.2
tons (short)	tons (metric)	0.9072

From *Lange's Handbook of Chemistry,* Handbook Publishers, Inc., Sandusky, OH.

Index

A

G

H

I

K

L

Q

R

S

T

U

V

W

X

Z